INDEX OF CACTUS ILLUSTRATIONS

By

John M. Evanich

INDEX OF CACTUS ILLUSTRATIONS

WITH

CROSS REFERENCES

First Edition

Published and Distributed by:

TORTILLA PRESS
8 Mulberry Lane
Bisbee, Arizona 85603
U.S.A.
(602) 432-7040
Library of Congress Reg. # TXU 338-656
ISBN 0-9622635-0-8
Printed in the USA

This compilation of information will help both the amateur and the knowledgeable cactophile find an illustration more easily.

Research completed in the many university and major city libraries over the past 10 years has shown that available references vary greatly.

Entries found in this index were taken from sources mentioned. Some species and varieties have no photograph references perhaps due to incorrect classification. Disputed nomenclature are included in this publication, using cross references, for possible future discussion concerning their legality.

I am deeply indebted to the following individuals for their help and support while compiling the "Index of Cactus Illustrations:"

Mr. Steven Brack, Mesa Gardens, Belen, New Mexico

Mr. Charles Glass, Editor, Cactus & Succulent Journal, Santa Barbara, California

Mr. Myron Kimnach, formerly of The Huntington, San Marino, California

Bruce MacBride, Ph.D., Department of the Interior, Office of Scientific Authority, U.S. Fish & Wildlife service, Washington, D.C.

Ed Maddox, D.V.M., Decatur, Texas

Mr. & Mrs. Joseph Mikulas, Albuquerque, New Mexico

Larry W. Mitich, Ph.D., F.C.S.S., Department of Botany, University of California, Davis, California

Donald J. Pinkava, Ph.D., Department of Botany & Microbiology, Arizona State University, Tempe, Arizona

Gordon Rowley, President of the British Cactus & Succulent Society

This book is dedicated to all of you and to my wife, Jean, who was so understanding and such an invaluable assistant throughout the years it has taken to complete this monumental project.

JME

July, 1988
New Port Richey, Florida

KEY **TITLE OF REFERENCE** **AUTHOR (S)**

A.....ARIZONA CACTI (Univ. of Ariz. ....Stockwell & Bull. # 1, Vol. 4, # 3)....1933...Breazeale

B.....ARIZONA'S CACTUSES..............1950...Marshall

C.....CACTI AND OTHER SUCCULENTS...Vol. 1.... 1955...E. & B. Lamb

D.....CACTI AND OTHER SUCCULENTS...Vol. 2.... 1959...E. & B. Lamb

E.....CACTI AND OTHER SUCCULENTS...Vol. 3.... 1963...E. & B. Lamb

F.....CACTI AND OTHER SUCCULENTS...Vol. 4.... 1966...E. & B. Lamb

G.....CACTI AND OTHER SUCCULENTS...Vol. 5.... 1978...E. & B. Lamb

H.....POCKET ENCYCL. OF CACTI IN COLOR..1969...E. & B. Lamb

I.....CACTI...........................1979...Barthlott

J*....CACTI AND SUCCULENTS............1968...Subik & Kublicka

K.....CACTI AND THEIR CULTIVATION.....1971...Martin, Chapman, & Auger

L.....CACTI...........................1937...Borg

M*....CACTI (A GOLDEN GUIDE)..........1974...Venning

N.....HENRIETTA'S NURSERY.............1980...(Catalogue USA)

O.....CACTI AND SUCCULENTS and how to grow them.....Haselton

P.....CACTI OF THE SOUTHWEST..........1963...Earle

Q.....CACTI OF ARIZONA................1969...Benson

R.....CACTI AND OTHER SUCCULENTS......1977...Kramer & Worth

S.....CACTI AND SUCCULENTS......... (Time-Life Book)...........1978...Perl

T.....CACTI AND SUCCULENTS............1963...Haage

U.....CACTI AND SUCCULENTS OF EL PASO........Champie

V.....CACTI, SUCCULENTS & TROPICAL PLANTS.... ...1975...Hayden

W.....CACTI AND SUCCULENTS............1975...Mulligan

X.....CACTI AND SUCCULENTS FROM.......1977...Gick MOTHER NATURE (Future Crafts today)

Y.....CACTI OF THE SOUTHWEST.................Weniger

Z.....CACTI AND SUCCULENTS FOR THE AMATEUR....................1977...Glass & Foster

AA....CACTI AND SUCCULENTS FOR .... MODERN LIVING..............1976...Rice

BB....CACTI AND SUCCULENTS............1978...(Sunset Books)

CC....CACTUS COLLECTOR'S HANDBOOK.....1975...Lawson

DD....CACTUS LEXICON..................1976...Backeberg

EE....CACTUS IDENTIFIER...............1977...Bechtel

FF....CALIFORNIA CACTUS...............1935...Baxter

GG....COLOURFUL CACTI OF THE DESERT...1974...E. & B. Lamb

HH....COMPLETE HANDBOOK OF CACTI & SUCCULENTS.................1977...Innes

II....FLORA OF TEXAS .... (Cactaceae Volume II)......1969...Benson

JJ....GUIA BOTONICA DE CACTACEAS Y
OTRAS SUCULENTAS DEL VALLE
DE TEHUACAN................1973...Meyran Garcia
KK....GYMNOCALYCIUMS...................1978...Putnam
LL....HOW TO GROW THE CACTI...........1963...Dawson
MM....HANDBOOK ON SUCCULENT PLANTS....1963...(Brooklyn Bot. Garden)
NN....KAKTEEN.........................1963...Cullman
OO....PLANTS OF BAJA CALIFORNIA.......1975...Coyle & Roberts
PP....POPULAR EXOTIC CACTI IN COLOUR..1975...E. & B. Lamb
QQ....SCHONE KAKTEEN UND ANDERE
SUKKULENTEN................1978...Rauh
RR....THE CACTACEAE...(4 Volumes 1963 Ed.)...Britton & Rose
SS....THE BOOJUM AND ITS HOME.........1974...Humphrey
TT....THE COLOR NATURE LIBRARY........1979...Pavey
UU....THE ILLUSTRATED ENCYCLOPEDIA OF
SUCCULENTS.................1978...Rowley
VV....THE MAMMILLARIA HANDBOOK........1945...Craig
WW*...THE PUNCTURED THUMB.............1977...Ashley
XX....WORLD OF CACTUS & SUCCULENTS....1977...Quiros & Young
(Ortho Books)
YY....CALIFORNIA EPI CENTER.................(Catalogues)
ZZ....CACTI AND OTHER SUCCULENTS FOR
YOUR HOME..................1976...Meadow
Aa....MAMMILLARIA SOCIETY JOURNAL...........(England)
Ab....ALTMAN SPECIALTY PLANTS...............(Catalogues)
Ac....MERITORIOUS MAMMILLARIAS........1980...Maddams
Ad....THE EXOTIC COLLECTION JOURNAL
- 1984..(England)
Ae....DEHERDT....(Catalogue)...1981 - 1983..(Belgium)
Af....CACTUS AND SUCCULENT SOCIETY OF ..... (U.S.A.)
AMERICA Journal.....1929 - 1988...
Ag....ABBEY GARDENS Catalogue..1981 - 1988...(U.S.A.)
Ah....GRIGSBY Catalogue........1979 - 1986...(U.S.A.)
Ai....ABBEY BROOK CACTUS NURSERY............(England)
(Catalogue).........1978 - 1979...
Aj....KAKTEEN UND ANDERE SUKKULENTEN..(Swiss, Austrian,
& German Cact. & Succ. Soc. Journ.)...1929 - 1983
Ak....CACTUS OF THE MONTH CLUB.1981 - 1982...(U.S.A.)
(Journal - Discontinued)
Al....CACTUS GUIDE....................1956...Cutak
Am....CANADIAN SUCCULENT INFORMATION
EXCHANGE Journal....1976 - 1986...(Canada)
An....CACTUS OF THE SOUTHWEST...(Revised).1980..Earle
Ao....UNIVERSITY OF ARIZONA Bull. #5
(Vol. XI #1)...............1940...Benson
Ap....OUR NATIVE CACTI................1931...Higgins, E. B.
Aq....MAMMILLARIA (A Collector's Guide) 1981.Pilbeam
Ar....COLOUR ENCYCLOPEDIA OF CACTI....1979...Hirao
As....CACTI OF ZION NATIONAL PARK.....1969...Trapp
At....CACTI DATA PLANTS...(Catalogue)...1982 - 1983..(U.S.A.)
Au....NATIONAL CACTUS & SUCCULENT

SOCIETY Journal.....1975 - 1982...(England)

Av....THE ILLUSTRATED ENCYCLOPEDIA OF
CACTI & OTHER SUCCULENTS...1981...Riha & Subik

Aw ...CACTACEAS Y SUCULENTAS MEXICANAS
(Journal...........1962 - 1988...(Mexico)

Ax....THE CACTI OF THE UNITED STATES
AND CANADA................1982...Benson

Ay....KAKTEEN AN IHREN STANDORTEN.....1979...Rauh

Az....LAS CACTACEAS DE MEXICO.........1978...Bravo-Hollis

Ba....SUCCULENTA...(Journal)...1974 - 1982...Holland

Bb....RAINBOW GARDENS (Catalogue) 1983 - 1987...(U.S.A.)

Bc....SOME PREHISPANIC USES OF CACTI
AMONG THE INDIANS..........1982...Sanchez-Mejorada

Bd....CACTACEAS Y SUCULENTAS DE LA
BARRANCA DE METZTITLAN.....1978...Sanchez-Mejorada

Be....CACTI................................Shurly

Bf....CYCADIA (Succ. & Cycads)...1984...(Catalogues - U.S.A.)

Bg....SULCOREBUTIA AND WEINGARTIA.....1985...Pilbeam

Bh....WHERE HAVE ALL THE WILD FLOWERS
GONE.......................1983...Mohlenbrock

Bi....BRITISH CACTUS AND SUCCULENT
JOURNAL............1983 - 1988...(England)

Bj....CACTUS AND ITS HOME....................Shreve

Bk....GESAMTBESCHREIBUNG DER KAKTEEN.........Schumann

Bl....MORPHOLOGY OF CACTI...................Buxbaum

Bm....CALENDARS (Kakteen und andere
Sukkulenten)......................(W. Germany)

Bn....THE FLOWERING CACTUS..................Carlson

Bo....DIE CACTACEAE.....(6 Volumes)..........Backeberg

Bp....THE GENUS REBUTIA..............1981...Fearn & Pearcy
(Abbey Brook Cactus Nursery

Bq....THE GENUS COCHEMIEA............1945...Lindsay
(Abbey Garden Press).......

Br....THE NIGHT BLOOMING CEREUS AND ITS
ALLIES.....................1945...Cutak

Bs....THE GENUS LOPHOCEREUS.(CSSA
Journal Vol. 35 #6)........1963...Lindsay

Bt....CACTI AND SUCCULENTS............1983...Andersohn

Bu....TROPICA.........................1978...Graf

Bv....EXOTICA.........................1970...Graf

Bw....DISCOCACTUS.....................1976...Buining

Bx....TEPHROCACTUS (The subgenus).....1973...Leighton-Boyce & Iliff

By....ECOLOGY OF THE SAGUARO (I & II)........Steenbergh & Lowe

Bz....TEXAS CACTI...........................Schultz & Runyon

Ca....THE NATIVE CACTI OF CALIFORNIA..1969...Benson

Cb....CACTI OF CALIFORNIA............1966...Dawson

Cc....PLANTAS Y FLORES DE MEXICO............O'Gorman

Cd....WILDFLOWERS OF THE UNITED STATES
(Vol. 4 - Part 1 - page 96).......Rickett

Ce....WILDFLOWERS OF THE UNITED STATES ......Rickett
(Vol. 1 - The Southeastern States)
Cf....CACTUS................................Armer
Cg....SUCCULENTS IN CULTIVATION.......1960...Higgins, Vera
Ch....FLORA OF THE GALAPAGOS ISLANDS.........Wiggins & Porter
Ci....FLORA OF TROPICAL FLORIDA.............Long & Lakela
Cj....GROWING CACTI AND SUCCULENTS...........Bartrum
Ck....HOW TO MAKE CACTI FLOWER..............E. Lamb
Cl....NEW ENCYCLOPEDIA OF GARDENING..........Everett
Cm....AMATEUR GARDENING..(Picture Book #11)..
(Cactus and Succulents)..........Huxley
Cn....CACTACEOUS PLANTS (Their culture
and history).....................Castle
Co....THE BOOK OF CACTI AND OTHER
SUCCULENTS.................1958...Chidamian
Cp....CACTI AS DECORATIVE PLANTS......1974...Kramer
Cq....CACTI AND SUCCULENTS (Indoors
and Outdoors)..............1971...Van Ness
Cr....CACTI.................................Kupper-Roshardt
Cs....DECORATIVE CACTI......................Subik & Kaplicka
Ct....BRADLEYA...(Yearbook of the British
Cactus & Succ. Soc.) 1983-1987...Hunt & Taylor
Cu....THE CACTUS AND SUCCULENT JOURNAL
OF GREAT BRITAIN..................
Cv....LOBIVIA......(Vol. 1, 2, 3).....1975...Rausch
Cw....SAGE OF THE DESERT....................Bonker &
Thornber
Cx....LAS CACTACEAS DE MEXICO.........1937...Bravo, H.
Cy....FLOWERING CACTI.......................Rose, H.
Cz....EXOTIC PLANT MANUAL.............1978...Graf
Da....SUCCULENTES....(Journal)...1977-1981...(France)
Db....CACTUS.........(Journal)...1979-1981...(Netherlands)
Dc....CACTUS.........(Journal)...1979-1980...(Belgium)
Dd....CACTUS AND SUCCULENT JOURNAL..........
1980-1982..(New South Wales)
De....CACTUS AND SUCCULENT JOURNAL..........
1980-1982...(New Zealand)
Df....CACTI OF TEXAS AND NEIGHBORING
STATES.....................1984...Weniger
Dg....DESERT PLANTS...(Univ. of Ariz.).......Boyce-Thompson
Quarterly Publication...... Arboretum
Dh....CALANDRINIA..(I & II) Succulent Public.
of South Australia....1980-1982...(Australia)
Di....HANDBOOK DER CACTEENKUNDE.............Forster
Dj....KAKTEENJAGD...........................Backeberg
Dk....REBUTIA...............................Marsden &
Jackson
Dl....CACTUS.........................1935...Van Laren
Dm....MAMMILLARIA...........................Marsden
Dn....CACTI FOR THE AMATEUR...........1958...Haselton
Do....CACTI OF SAN DIEGO COUNTY.............Lindsay
Dp....CACTI..........................1973...Leese

Dq....CACTACEAE.......................1941...Marshall & Bock
Dr....CACTACEAS.......................1984...Spegazzini
Ds....KAKTEEN UND ANDERER SUKKULENTEN........Rother
Dt....SUKKUL. JAHRB. SCHWEIZER KAKTEEN.(References taken from
(Gesellsch)...... Gray's Herbarium Card Index)
Du....KAKTEEN & ANDERE SUKK. .... "
Dv....MOLLERS DEUT. - GART. ZEIT. .... "

Dw....MONATS. DEUTSCH. KAKT............
(Gesellsch.)..........................."
Dx....ALLAN HANCOCK FOUND. ....
(Occas. Papers)......................"
Dy....ZEITSCHR. SUKK. ...."
Dz....BEITR. SUKK. ...."
Ea....KAKTEENKUNDE.........................."
Eb....ALAN. INST. BIOL. MEX. ...."
Ec....BOL. DIREC. ESTUD. BIOL. ...."
Ed....MONATS. KAKTEENKUNDE...................."
Ee....SUKKULENTENKUNDE........................"
Ef....BOL. ACAD. CIENC. FISIC. NAT.
(Venezuela)............................"
Eg....SCHWEIZER GART. ...."
Eh....DESERT PLANT LIFE..(Publication
by Boyce-Thompson Arboretum).........."
Ei....NOTIZBLATT.............................."
Ej....AMER. MIDL. NAT. ...."
Ek....KAKTEEN ....(Berger)................"
El....SPINE (Journal) Cactus & Succ.
Soc. of Australia...................."
Em....FEDDE. REP. ...."
En....NEW YORK BOT. GARD. (Journal)............."
Eo....MO. BOT. GARDEN REP. ...."
Ep....TORR. BOT. CLUB BULL. ...."
Eq....ANAL. MUS. NAC. BUENES AIRES.............."
Er....ANAL. SOC. CIENT. (Argentina)............."
Es....MUS. HIST. NAT. Bull. (Paris)............."
Et....SO. CALIF. ACAD. SCI. Bull. ...."
Eu....WASH. ACAD. SCI. (Journal) ...."
Ev....REV. CHIL. HIST. NAT. ...."
Ew....REVIST. AGRIC. COCHABAMBA................"
Ex....MISS. BOT. GARD. ANN. REPT. ...."
Ey....PROC. BIOL. SOC. WASH. ...."
Ez....SOUTHEAST FLORIDA MANUAL.................."
Fa....PROC. AM. ACAD. ARTS. & SCI. ....
New Mexico.........................."
Fb....U.S. NAT. HERB. ....1896.........."
Fc....LILLOA......................................"
Fd....ANAL. MUS. NAC. CHILE BOT. ...."
Fe....BOL. PRO. - CULT. REG. (Fam. Cact.)........"
Ff....INTERESTING NEWER MAMMILLARIAS...1973...Maddams
Fg....SMITHSONIAN INST. ANN. REPT. 1908 (Page 525)

Fh....ANAL. MUS. NAC. (Montevideo, Uruguay)....................References taken from Gray's Herbarium Card Index
Fi....CACT. HORT. DYCK. (Mexico)................"
Fj....NEVE KAKTEEN................1931.........."
Fk....MADRONO.....................1943.........."
Fl....BOIS. DICT. ....1898.........."
Fm....MITT. DEUTSCH. DENDR. GES. ..1925.........."
Fn....VEG. & FL. SONORAN DESERT (Shreve & Wiggins)................."

Fo....BLAT. KAKTEEN FORSCH......................"
Fp....JAHRB. DEUT. KAKT. (Gesel.) 1936.........."
Fq....KAKTUSY (Pazout, Valnicek, & Subik)................1960........."
Fr....KAKTUS - ABC................1935 (Backeberg & Knuth)................."
Fs....DIE KAKTEEN, LIEF...........1963 (Krainz)............................"
Ft....CURTIS BOT. MAG. ....1873........."
Fu....BRASIL & SEINE SAULENKAKT. ..1933........."
Fv....FORSTER'S HANDBOOK, ... CACTEENKUNDE...........1886........."
Fw....BOLL. R. ORT. GIARD. COL. ... (Palermo)........................."
Fx....CEIBA..............................."
Fy....KAKTEENFREUND........................"
Fz....ARCH. JARD. BOT. RIO DE JANEIRO..........."
Ga....ANAIS. PRIM. REUN. SUL. BOT. 1940........."
Gb....FUAUX HERB. Bull. ...."
Gc....KAKTUS ABC.............................."
Gd....BOL. SOC. ARGENTINA BOT. ...."
Ge....WERD. BRASILIEN & SEINE SAULENKAKTEEN....."
Gf....CACTACEAE...................1920........."
Gg....GESAMTB. KAKT. ...."
Gh....MART. FL. BRAS. ...."
Gi....BOL. SOC. VENEZOLANA, CIENC. NAT. ...."
Gj....KAKTUSAR................................"
Gk....ZIVOT V- PRIRODE (Kakt. a Succ.).........."
Gl....ARQUIV. JARD. BOT. RIO DE JANEIRO........."
Gm*...MELOCACTUS UND ECHINOCACTUS Reprint....1827...Link und Otto
Gn*...THE GENUS ECHINOCEREUS..........1985...Taylor
Go....COLORADO CACTI..................1940...Boissevain and Davidson
Gp....NOTOCACTUS......................1975...Mace
Gq....PEYOTE - The Divine Cactus......1980...Anderson
Gr....GUIDE TO CACTI AND SUCCULENTS (Edited by Schuler)........1985...Pizzettii

Gs....CARDENAS in CACTUS, ... (Paris, France)............(From Index Kewensis)

Gt....CACTUS (Paris, France)...Ritter........"
Gu....FLORA OF CUBA...(Vol. 3)........1953...Leon and Alain
Gv....HANDBOOK - KAKTEENK. Backeberg..(From Index Kewensis)
Gw....CARDENAS in SUCCULENTA.................."
Gx....DOLZ in KAKTEENKUNDE.....1942............"
Gy....BLATTER KAKTEENFROSCH, Backeberg........."
Gz....CARDENAS in REV. AGRIC., Cochabamba......"
Ha....POCKET ENCYCLOPEDIA OF CACTI
IN COLOUR........Revised - 1981...E. & B. Lamb
Hb....AN INTRODUCTION TO CACTI........1984...Schuster
Hc....DIE GATTUNG SULCOREBUTIA........1976...Brinkman
Hd....CACTUS.........(Luxembourg)...........DeLuetz
He....THE CHILIANS.........................
Hf....TEXAS HIGHWAYS (Magazine).............U.S.A.
Hg....KAKT. ORCH. RUNDSCHAU.................
Hh....ASHINGTONIA...(Vol. 1,2,&3) 1973-1980..(England)
(Journals).......................
Hi....KAKTEEN IN SUDAMERIKA..(4 Volumes),
1979...Ritter
Hj....KAKTUS...............................(Denmark)
Hk....CACTUS...............................Hoofstadt
Hl....KAKTUSY....(Journals)......1979-1983...(Czechoslovakia)
Hm....AZTEKIA....(Journals)......1978-1979...
and Yearbook..............1982...(Czechoslovakia)
Hn....10,000 MILES - HUNTING CACTUS IN
TEXAS.....................1984...Maddox
Ho....ROD ECHINOFOSSULOCACTUS...............
(Lawrence)................1984...Pechanek
Hp....DIE GATTUNG ASTROPHYTUM.........1979...Sadovsky & Schutz

Hq....KAKTEEN STERNE - Die Astrophytum 1957..Haage-Sadovsky
Hr....LOVEC KAKTUSU...................1983...Crkal, K
Hs....KAKTUSY SUKULENTY Journals 1980-1983...(Czechoslovakia)
Ht....MELOCACTUS NO BRASIL............1982...Rissini
Hu....200 KAKTEEN IN FARBE............1983...Bravenboer
Hv....CACTI (The Encyclopedia of).....1986...Cullman, Gotz, & Groner
Hw....WHITESTONE GARDENS Ltd. ....1981 - 1988.....
(Catalogue).......................(England)
Hx....KAKTEEN von A bis Z.............1983...Haage
Hy....THE CACTACEAE.....(Japan)..............Yoshio Ito
Hz....THE ILLUSTRATED STUDIES OF
GENUS OPUNTIA, III & IV,
21st & 22nd Ann. Rpt. of
Mo. Bot. Gar. ....1910-1912...Griffiths
Ia....CACTI AND SUCCULENTS...................Green
Ib....STRANGERS IN THE FRANKLINS......1973...Champie
Ic....SMITHSONIAN INST. ANN. RPT. ....1908...Safford
Id....CACTUS CULTURE..................1932...Schulz

Ie....NEW MEXICO AGRIC. EXPER. STATION
(Bull. # 78, Cacti in N. M.) 1911...Wooton

```
If....DAS KAKTEENBUCH.................1928...Kupper
Ig....CACTI...........................1960...Kupper
Ih....IN THE WORLD OF CACTUS -........1977...
          (Academy of Science U.S.S.R.).....Udalova &
          Translation.........................Vyugina
Ij....THE CACTUS PRIMER...............1986...Gibson & Nobel
Ik....KAKTEEN.........................1985...Moser
Il....BLV - HANDBUCH DER KAKTEEN......1982...Hecht
Im....LES CACTEES CULTIVEES...........1946...Guillaumin
In....COLLECTOR'S CACTI...............1960...Bloom
Io....A L'HISTOIRE DU REGNE VEGETAL..........DeCandolle
Ip....BEITRAGE ZUR SUKKULENTEN KUNDE
          UND PFLEGE - (Lfg-2).......1940...Neumann
Iq....KAKTEEN.........................1929...Berger
Ir....DIE SCHONHEIT UNSERER KAKTEEN..........Maasz
Is....MONOGRAPHIA GENERIS MELOCACTI...
          (Vol. XVIII Suppl.)........1838...Miquel
It....BRASILIEN UND SEINE SAULENKAKTEEN..1933...Werdermann
Iu....DEN NYE KAKTUSBOG...............1930...Knuthenborg
Iv....CACTUS - FRANCAISE DES AMATEURS
          DE CACTEES ET PLANTES GRASSES.....(France)
Iw....A COLOR PHOTO ALBUM OF CACTI AND
          SUCCULENTS (Vol. 1)...............(Japan)
          (Genshoku, Shaboten, Teniku, Shokubutsu, Daizukan)
Ix....FLOWERING CACTUS................1984...Rayzer
Iy....DESERT IN BLOOM.................1986...Eppele
Iz....CACTI FOR THE CONNOISSEUR.......1987...Pilbeam
Ja....HOUSE PLANTS, CACTI and SUCCULENTS..1972...Huxley
Jb....DIE GATTUNG MAMMILLARIA, nach dem heutigen
        Stand meines Wissens...........1987.....Reppenhagen
Jc....LOBIVIA 85.......................1985.....Rausch
Jd....DER KOSMOS KAKTEENFUHRER.........1983.....Haustein
Je....PERESKIA (Cactaceae).............1986.....Leuenberger
Jf....DIE GATTUNG PARODIA..............1987.....Weskamp
```

A COLOR PHOTO ALBUM OF CACTI & SUCCULENTS Genshoku, ...Iw
Shaboten
Teniku,
Shokubutsu,
Daizukan
A L'HISTOIRE DU REGNE VEGETAL............De Candolle.....Io
AN INTRODUCTION TO CACTI...........1984...Schuster........Hb
AMATEUR GARDENING (Cactus & Succulents)
Picture Book # 11.....................Huxley..........Cm
ARIZONA'S CACTUSES.................1950...Marshall........B
BEITRAGE ZUR SUKKULENTEN KUNDE UND
PFLEGE (Lfg-2)................1940...Neumann.........Ip
BLV - HANDBUCH DER KAKTEEN.........1982...Hecht...........Il
BRADLEYA (Yearbook of the British
Cactus & Succulent Society)..........Hunt & Taylor...Ct
BRASILIEN UND SEINE SAULENKAKTEEN..1933...Werdermann......It
CACTACEAE..........................1941...Marshall & Bock.Dq
CACTACEAS..........................1984...Spegazzini......Dr
CACTACEAS Y SUCULENTAS DE LA BARRANCA
DE METZTITLAN.................1978..Sanchez Mejorada.Bd
CACTACEOUS PLANTS (Their culture &
history)............................Castle..........Cn
CACTI..............................1979...Barthlott.......I
CACTI..............................1937...Borg............L
CACTI (The Encyclopedia of)........1986...Cullman, Gotz,
Groner.......Hv
CACTI..............................1960...Kupper..........Ig
CACTI.....................................Kupper-Roshardt.Cr
CACTI..............................1973...Leese...........Dp
CACTI.....................................Shurly..........Be
CACTI (A Golden Guide).............1974...Venning.........M
CACTI AND OTHER SUCCULENTS.........1977...Kramer & Worth..R
CACTI AND OTHER SUCCULENTS (Vol. 1) 1955..E. & B. Lamb....C
CACTI AND OTHER SUCCULENTS (Vol. 2) 1959..E. & B. Lamb....D
CACTI AND OTHER SUCCULENTS (Vol. 3) 1963..E. & B. Lamb....E
CACTI AND OTHER SUCCULENTS (Vol. 4) 1966..E. & B. Lamb....F
CACTI AND OTHER SUCCULENTS (Vol. 5) 1978..E. & B. Lamb....G
CACTI AND OTHER SUCCULENTS FOR YOUR HOME..Meadow..........ZZ
CACTI AND SUCCULENTS...............1983..Andersohn.......Bt
CACTI AND SUCCULENTS......................Green...........Ia
CACTI AND SUCCULENTS...............1963..Haage...........T
CACTI AND SUCCULENTS...............1975..Mulligan........W
CACTI AND SUCCULENTS...............1977..Pavey...........TT
CACTI AND SUCCULENTS...............1968..Subik & Kaplicka J
CACTI AND SUCCULENTS...............1978..(Sunset Books)..BB
CACTI AND SUCCULENTS AND HOW TO GROW THEM.Haselton.........O
CACTI AND SUCCULENTS FOR MODERN LIVING..1976...Rice.......AA
CACTI AND SUCCULENTS FOR THE AMATEUR.1977..Glass & Foster..Z

CACTI AND SUCCULENTS FROM MOTHER
NATURE.......................1977...Gick............X

CACTI AND SUCCULENTS OF EL PASO...........Champie.........U
CACTI AND SUCCULENTS (Indoors and
Outdoors)...........................Van Ness........Cq
CACTI AND SUCCULENTS (Time-Life Books) 1978....Perl.......S
CACTI AND THEIR CULTIVATION........1971...Martin, Chapman
& Auger.......K
CACTI AS DECORATIVE PLANTS................Kramer..........Cp
CACTI FOR THE AMATEUR.....................Haselton........Dn
CACTI FOR THE CONNOISSEUR..........1987...Pilbeam.........Iz
CACTI OF ARIZONA...................1969...Benson..........Q
CACTI OF CALIFORNIA................1966...Dawson..........Cb
CACTI OF TEXAS AND NEIGHBORING STATES.....Weniger.........Df
CACTI OF THE SOUTHWEST.............1963...Earle...........P
CACTI OF THE SOUTHWEST (Revised)...1980...Earle...........An
CACTI OF THE SOUTHWEST....................Weniger.........Y
CACTI OF SAN DIEGO COUNTY.................Lindsay.........Do
CACTI OF ZION NATIONAL PARK........1969...Trapp...........As
CACTI, SUCCULENTS, AND TROPICAL
PLANTS.......................1975...Hayden..........V
CACTUS....................................Armer...........Cf
CACTUS....................................De Leutz........Hd
CACTUS....................................Hoofstadt.......Hk
CACTUS....................................Van Laren.......Dl
CACTUS AND ITS HOME ....Shreve..........Bj
CACTUS COLLECTOR'S HANDBOOK........1975...Lawson..........CC
CACTUS CULTURE.....................1932...Schulz..........Id
CACTUS GUIDE.......................1956...Cutak...........Al
CACTUS IDENTIFIER..................1977...Bechtel.........EE
CACTUS LEXICON.....................1976...Backeberg.......DD
CALANDRINIA (I & II)........1980 & 1982...(Australia).....Dh
CALIFORNIA CACTUS..................1935...Baxter..........FF
COLLECTORS' CACTI..................1960...Bloom...........In
COLORADO CACTI..........1940...Boissevain & Davidson......Go
COLOURFUL CACTI OF THE DESERT......1974...E. & B. Lamb....GG
COLOUR ENCYCLOPEDIA OF CACTI (Japan)1979..Hirao...........Ar
COMPLETE HANDBOOK OF CACTI & SUCC. 1977...Innes...........HH
DAS KAKTEENBUCH....................1928...Kupper..........If
DECORATIVE CACTI.........................Subik & Kaplicka Cs
DEN NYE KAKTUSBOG..................1930...Knuthenborg.....Iu
DER KOSMOS KAKTEENFUHRER...........1983...Haustein........Jd
DESERT IN BLOOM....................1986...Eppele..........Iy
DIE CACTACEAE (6 Volumes).................Backeberg.......Bo
DIE GATTUNG ASTROPHYTUM............1979...Sadovsky-Schutz.Hp
DIE GATTUNG MAMMILLARIA............1987...Reppenhagen.....Jb
DIE GATTUNG PARODIA................1987...Weskamp.........Jf
DIE GATTUNG SULCOREBUTIA...........1976...Brinkman........Hc
DIE SCHONHEIT UNSERER KAKTEEN.............Maasz...........Ir
DISCOCACTUS........................1976...Buining.........Bw
ECOLOGY OF THE SAGUARO (I & II)...Steenbergh & Lowe.......By
EXOTICA..................................Graf............Bv
EXOTIC PLANT MANUAL................1978...Graf............Cz
FLORA OF CUBA (Vol. 3)............1953...Leon & Alain....Gu

FLORA OF TEXAS - CACTACEAE (Vol 2) 1969...Benson..........II
FLORA OF THE GALAPAGOS ISLANDS...........Wiggins & Porter.Ch
FLORA OF TROPICAL FLORIDA...............Long & Lakela....Ci
FLOWERING CACTI..........................Rose, H, ...Cy
FLOWERING CACTUS..................1984...Rayzer..........Ix
GESAMTBESCHREIBUNG DER KAKTEEN...........Shumann.........Bk
GROWING CACTI AND SUCCULENTS.............Bartrum.........Cj
GUIA BOTONICA DE CACTACEAS Y OTRAS 1973...
SUCULENTAS DEL VALLE DE TEHUACAN.....Meyran Garcia...JJ
GUIDE TO CACTI AND SUCCULENTS......1985...Pizzettii.......Gr
GYMNOCALYCIUMS....................1978...Putnam..........KK
HANDBOOK DER CACTEENKUNDE................Forster.........Di
HANDBOOK ON SUCCULENT PLANTS.......1963...(Brooklyn Bot.
Garden)......MM
HENRIETTA'S NURSERY CATALOG........1980...(U.S.A.).........N
HOUSE PLANTS, CACTI AND SUCCULENTS...1972...Huxley........Ja
HOW TO KNOW THE CACTI.............1963...Dawson..........LL
HOW TO MAKE CACTI FLOWER.................E. Lamb.........Ck
INTERESTING NEWER MAMMILLARIAS.....1973...Maddams.........Ff
IN THE WORLD OF CACTUS (Academy of
Science - U.S.S.R.)...1977...Udalova & Vyugina.......Ih
KAKTEEN...........................1929...Berger..........Ig
KAKTEEN...........................1963...Cullman.........NN
KAKTEEN...........................1985...Moser...........Ik
KAKTEEN AN IHREN STANDORTEN........1979...Rauh............Ay
KAKTEEN IN SUDAMERIKA (4 Volumes)..1979...Ritter..........Hi
KAKTEENJAGD..............................Backeberg.......Dj
KAKTEEN STERNE (Die Astrophytum)...1957...Haage-Sadovsky..Hq
KAKTEEN UND ANDERER SUKKULENTEN..........Rother..........Ds
KAKTEEN VON A bis Z...............1983...Haage...........Hx
KAKTUS...................................(Denmark).......Hj
KAKT. ORCH. RUNDSCHAU.....................................Hg
LAS CACTACEAS DE MEXICO...........1937...H. Bravo........Cx
LAS CACTACEAS DE MEXICO...........1978...Bravo-Hollis....Az
LES CACTEES CULTIVEES.............1946...Guillaumin......Im
LOBIVIA....(3 Volumes)............1975...Rausch..........Cv
LOBIVIA 85........................1985...Rausch..........Jc
LOVEC KAKTUSU.....................1983...K. Crkal........Hr
MAMMILLARIA..............................Marsden.........Dm
MAMMILLARIA (A Collector's Guide)..1981...Pilbeam.........Aq
MELOCACTUS NO BRASIL..............1982...Rizzini.........Ht
MELOCACTUS UND ECHINOCACTUS........1827...Link und Otto...Gm
MERITORIOUS MAMMILLARIAS...........1980...Maddams.........Ac
MONOGRAPHIA GENERIS MELOCACTI......1838...Miquel..........Is
(Vol. XVIII Suppl.)...........
MORPHOLOGY OF CACTI......................Buxbaum.........Bl
NEW ENCYCLOPEDIA OF GARDENING............Everett.........Cl
NOTOCACTUS........................1975...Mace............Gp
OUR NATIVE CACTI..................1931...Higgins, E. B...Ap
PERESKIA (Cactaceae)..............1986...Leuenberger.....Je
PEYOTE - The Divine Cactus.........1980...Anderson.........Gq
PLANTES Y FLORES DE MEXICO...............O'Gorman.........Cc

PLANTS OF BAJA CALIFORNIA..........1975...Coyle & Roberts..OO
POCKET ENCYCLOPEDIA OF CACTI IN
COLOR........................1969...E. & B. Lamb.....H
POCKET ENCYCLOPEDIA OF CACTI IN COLOUR
Revised 1981...E. & B. Lamb.....Ha
POPULAR EXOTIC CACTI IN COLOUR.....1975...E. & B. Lamb.....PP
THE PUNCTURED THUMB................1977...Ashley...........WW
REBUTIA...............................Marsden & Jackson..Dk
ROD ECHINOFOSSULOCACTUS (Lawr.)....1984...Pechanek.........Ho
SAGE OF THE DESERT.....................Bonker & Thornber..Cw
SCHONE KAKTEEN UND ANDERE SUKKULENTEN..1978...Rauh.........QQ
SOME PREHISPANIC USES OF CACTI AMONG
THE INDIANS..................1982...Sanchez Mejorada.Bc
STRANGERS IN THE FRANKLINS.........1973...Champie..........Ib
SUCCULENTS IN CULTIVATION.................Higgins, V. .....Cg
SULCOREBUTIA AND WEINGARTIA........1985...Pilbeam..........Bg
10,000 MILES - HUNTING CACTUS IN
TEXAS........................1984...Maddox...........Hn
TEPHROCACTUS (The Subgenus)........1973...Leighton-Boyce &
Iliff..........Bx
TEXAS CACTI...............................Schultz & Runyon Bz
THE BOOK OF CACTI AND OTHER
SUCCULENTS...................1958...Chidamian........Co
THE BOOJUM AND ITS HOME............1974...Humphrey.........SS
THE CACTACEAE...(4 Vol.)..........1920...Britton & Rose...RR
THE CACTACEAE...(Japan)...................Yoshio Ito.......Hy
THE CACTI OF THE UNITED STATES AND
CANADA.......................1982...Benson...........Ax
THE CACTUS PRIMER..................1986...Gibson & Nobel...Ij
THE CHILIANS...He
THE COLOR NATURE LIBRARY...........1979...Pavey............TT
THE FLOWERING CACTUS......................Carlson..........Bn
THE GENUS COCHEMIEA
(Abbey Garden Press)..........1945...Lindsay..........Bq
THE GENUS ECHINOCEREUS.............1985...Taylor...........Gn
THE GENUS LOPHOCEREUS (CSSA
Journal Vol. XXXV No. 6)......1963...Lindsay..........Bs
THE GENUS REBUTIA..................1981...Fearn & Pearcy...Bp
THE ILLUSTRATED ENCYCLOPEDIA OF
CACTI AND OTHER SUCCULENTS....1981...Riha & Subik.....Av
THE ILLUSTRATED ENCYCLOPEDIA OF
SUCCULENTS...................1978...Rowley...........UU
THE ILLUSTRATED STUDIES OF GENUS
OPUNTIA - (21st & 22nd Ann. Rpt. of
Mo. Bot. Gard.)........1910 & 1912...Griffiths........Hz
THE MAMMILLARIA HANDBOOK...........1945...Craig............VV
THE NATIVE CACTI OF CALIFORNIA.....1969...Benson...........Ca
THE NIGHT BLOOMING CEREUS AND ITS ALLIES..1945..Cutak......Br
TROPICA...................................Graf.............Bu
200 KAKTEEN IN FARBE...............1983...Bravenboer.......Hu
WHERE HAVE ALL THE WILD FLOWERS
GONE.........................1983...Mohlenbrock......Bh

WILDFLOWERS OF THE UNITED STATES
(Vol. I -The Southeastern States)....Rickett..........Ce
WILDFLOWERS OF THE UNITED STATES
(Vol. 4 Part 1 - Page 96)............Rickett..........Cd
WORLD OF CACTUS AND SUCCULENTS
(Ortho Books)...................1977...Quiros & Young...XX

KEY TO REFERENCE INDEX BY AUTHOR

AUTHOR (S)	TITLE OF REFERENCE		KEY
ANDERSOHN	Cacti and Succulents	1983	Bt
ANDERSON	Peyote - The Divine Cactus	1980	Gq
ARMER	Cactus		Cf
ASHLEY	The Punctured Thumb	1977	WW
BACKEBERG	Cactus Lexicon	1976	DD
BACKEBERG	Die Cactaceae..(6 Volumes)		Bo
BACKEBERG	Kakteenjagd		Dj
BARTHLOTT	Cacti	1979	I
BARTRUM	Growing Cacti & Succulents		Cf
BAXTER	California Cactus	1935	FF
BECHTEL	Cactus Identifier	1977	EE
BENSON	Cacti of Arizona	1969	Q
BENSON	Cacti of the United States & Canada	1982	Ax
BENSON	Flora of Texas (Cactaceae Vol. II)	1969	II
BENSON	Native Cacti of California	1969	Ca
BENSON	Univ. of Ariz. Bull. #5 (Vol. XI #1)	1940	Ao
BERGER	Kakteen	1929	Iq
BLOOM	Collector's Cacti	1960	In
BOISSEVAIN & DAVIDSON	Colorado Cacti	1940	Go
BONKER & THORNBER	Sage of the Desert		Cw
BORG	Cacti	1937	L
BRAVENBOER	200 Kakteen in Farbe	1983	Hu
BRAVO, H.	Las Cactaceas de Mexico	1937	Cx
BRAVO-HOLLIS	Las Cactaceas de Mexico	1978	Az
BRINKMAN	Die Gattung Sulcorebutia	1976	Hc
BRITTON & ROSE	The Cactaceae..(4 Vol.)	1920	RR
(FROM BROOKLYN BOT. GARDEN)	Handbook on Succ. Plants	1963	MM
BUINING	Discocactus	1976	Bw
BUXBAUM	Morphology of Cacti		Bl
CARLSON	The Flowering Cactus		Bn
CASTLE	Cactaceous Plants		Cn
CHAMPIE	Cacti and Succulents of El Paso		U
CHAMPIE	Strangers in the Franklins	1973	Ib
CHIDAMIAN	The Book of Cacti and other Succulents	1958	Co
COYLE & ROBERTS	Plants of Baja California	1975	OO
CRAIG	The Mammillaria Handbook	1945	VV
CRKAL, K.	Lovec Kaktusu	1983	Hr
CULLMAN	Kakteen	1963	NN
CULLMAN, GOTZ, & GRONER	The Encyclopedia of Cacti	1986	Hv
CUTAK	Cactus Guide	1956	Al
CUTAK	The Night Blooming Cereus and its Allies	1945	Br

DAWSON................How to Know the Cacti.........1963...LL
DAWSON................Cacti of California.................Cb
DE CANDOLLE...........A L'Histoire du Regne Vegetal........Io
EARLE.................Cacti of the Southwest........1963...P
EARLE.................Cacti of the Southwest........1980...An
EPPELE................Desert in Bloom...............1986...Iy
EVERETT...............New Encyclopedia of Gardening.......Cl
FEARN & PEARCY........The Genus Rebutia.............1981...Bp
FORSTER...............Handbook der Cacteenkunde............Di
GENSHOKU, SHABOTEN, A Colorful Album of Cacti and
TENIKU, SHOBUTSU, Succulents......(Japan)
DAIZUKAN...Iw
GICK..................Cacti and Succulents from
Mother Nature.................1977....X
GIBSON & NOBEL........The Cactus Primer.............1986...Ij
GLASS & FOSTER........Cactus and Succulents for the
Amateur.......................1977....Z
GRAF.................. .Exotica.............................Bv
GRAF..................Exotic Plant Manual...........1978...Cz
GRAF..................Tropica.............................Bu
GREEN.................Cacti and Succulents.................Ia
GRIFFITHS.............The Illustrated Studies of the
Genus Opuntia (Ann. Rpt. Mo.
Bot. Gard. ....1910-1912...Hz
GUILLAUMIN............Les Cactees Cultivees.........1946...Im
HAAGE.................Cacti and Succulents..........1963....T
HAAGE.................Kakteen von A bis Z...........1983...Hx
HAAGE-SADOVSKY........Kakteen - Sterne Die
Astrophytum...................1957...Hq
HASELTON..............Cacti and Succulents and
How to Grow Them.....................O
HASELTON..............Cacti for the Amateur.........1958...Dn
HAUSTEIN..............Der Kosmos Kakteenfuhrer......1983...Jd
HAYDEN................Cacti, Succulents and
Tropical Plants...............1975...V
HECHT.................BLV - Handbuch der Kakteen....1982...Il
HIGGINS, E. B. ....Our Native Cacti..............1931...Ap
HIGGINS, V. ....Succulents in Cultivation.....1960...Cg
HIRAO.................Colour Encyclopedia of Cacti..1979...Ar
HOOFSTADT.............Cactus...............................Hk
HUMPHREY..............The Boojum and its Home.......1974...SS
HUNT & TAYLOR.........Bradleya..(Yearbook of the
Brit. Cactus & Succ. Soc.)..........Ct
HUXLEY................Amateur Gardening (Picture
Book #11 Cact. & Succ.).............Cm
HUXLEY................House Plants (Cac. & Succ.)...1972...Ja
INNES.................Complete Handbook of Cacti and
Succulents....................1977...HH
ITO, JOSHIO...........The Cactaceae....(Japan).............Hy
KRAMER................Cacti as Decorative Plants....1974...Cp
KRAMER & WORTH........Cacti and other Succulents....1977....R
KNUTHENBORG...........Den Nye Kaktusbog.............1930...Iu

KUPPER...............Das Kakteenbuch................1928...If
KUPPER...............Cacti..........................1960...Ig
KUPPER-ROSHARDT......Cacti ....Cr
LAMB, E. ....How to Make Cacti Flower..............Ck
LAMB, E. & B. ....Cacti and other Succulents (1).1955...C
LAMB, E. & B. ....Cacti and other Succulents (2).1959...D
LAMB, E. & B. ....Cacti and other Succulents (3).1963...E
LAMB, E. & B. ....Cacti and other Succulents (4).1966...F
LAMB, E. & B. ....Cacti and other Succulents (5).1978...G
LAMB, E. & B. ....Pocket Encyclopedia of Cacti
& other Succulents in Color..1969...H
LAMB, E. & B. ....Colourful Cacti of the Desert..1974...GG
LAMB, E. & B. ....Pocket Encyclopedia of Cacti
and other Succulents in Colour
(Revised)...................1981...Ha
LAMB, E. & B. ......Popular Exotic Cacti in Colour..1975...PP
LAWSON...............Cactus Collector's Handbook....1975...CC
LEESE................Cacti..........................1973...Dp
LEIGHTON-BOYCE &
ILIFF.............Tephrocactus (The Subgenus)....1973...Bx
LEON & ALAIN.........Flora of Cuba..(Vol. 3)........1953...Gu
LEUENBERGER..........Pereskia (Cactaceae)...........1986...Je
LINDSAY..............Cacti of San Diego County.............Do
LINDSAY..............The Genus Cochemiea............1945...Bq
LINDSAY..............The Genus Lophocereus (CSSA
Journal Vol. XXXV #6)........1963...Bs
LINK UND OTTO........Melocactus und Echinocactus....1827...Gm
LONG & LAKELA........Flora of Tropical Florida.............Ci
MAASZ................Die Schonheit unserer Kakteen.........Ir
MACE.................Notocactus.....................1975...Gp
MADDAMS..............Interesting Newer Mammillarias.1973...Ff
MADDAMS..............Meritorious Mammillarias.......1980...Ac
MADDOX...............10,000 Miles Hunting Cactus
in Texas.....................1984...Hn
MARSDEN..............Mammillaria...........................Dm
MARSDEN & JACKSON....Rebutia...............................Dk
MARSHALL.............Arizona's Cactuses.............1950...B
MARSHALL & BOCK......Cactaceae......................1941...Dq
MARTIN, CHAPMAN &
AUGER.............Cacti and their Cultivation....1971...K
MEADOW...............Cacti and other Succulents for
Your Home...................1976...ZZ
MEYRAN GARCIA........Guia Botonica De Cactaceas Y
Otras Suculentas Del Valle
De Tehuacan.................1973...JJ
MIQUEL...............Monographia Generis Melocacti
(Vol. XVIII Suppl.)..........1838...Is
MOHLENBROCK..........Where have all the Wild
Flowers Gone................1983...Bh
MOSER................Kakteen........................1985...Ik
MULLIGAN.............Cacti and Succulents...........1975...W

NEUMANN..............Beitrage zur Sukkulenten kunde
und pflege (Lfg-2)...........1940...Ip
O'GORMAN.............Plantes Y Flores de Mexico...........Cc
PAVEY................Cacti and Succulents..........
(Color Nature Library).......1979...TT
PECHANEK.............Rod Echinofossulocactus (Lawr.)1984...Ho

PERL.................Cacti and Succulents
(Time-Life Books)............1978...S
PILBEAM..............Cacti for the Connoisseur......1987...Iz
PILBEAM..............Mammillaria (A Collector's
Guide)........................1981...Aq
PILBEAM..............Sulcorebutia & Weingartia......1986...Bg
PIZZETTII............Guide to Cacti & Succulents....1985...Gr
PUTNAM...............Gymnocalyciums..(Handbook #5)..1978...KK
QUIROS & YOUNG.......The World of Cacti & Succulents
(Ortho Books)................1977...XX
RAUH.................Kakteen an Ihren Standorten....1979...Ay
RAUH.................Schone Kakteen und andere
Sukkulenten..................1978...QQ
RAUSCH...............Lobivia (3 Volumes)............1975...Cv
RAUSCH...............Lobivia 85.....................1985...Jc
RAYZER...............Flowering Cactus...............1984...Ix
REPPENHAGEN..........Die Gattung Mammillaria........1987...Jb
RICE.................Cacti and Succulents for
Modern Living................1976...AA
RICKETT..............Wildflowers of the United States
(Vol. 1, Southeastern States).......Ce
RICKETT..............Wildflowers of the United States
(Vol. 4, Part 1, Page 96............Cd
RIHA & SUBIK.........The Illustrated Encyclopedia of
Cacti & Other Succulents.....1981...Av
RITTER...............Kakteen in Sudamerika.(4 Vol.).1979...Hi
RIZZINI..............Melocactus no Brasil...........1982...Ht
ROSE, H. ....Flowering Cacti.....................Cy
ROTHER...............Kakteen und anderer Sukkulenten.......Ds
ROWLEY...............The Illustrated Encyclopedia of
Succulents...................1978...UU
SADOVSKY-SCHUTZ......Die Gattung Astrophytum........1979...Hp
SAFFORD..............Smithsonian Inst. Ann. Rpt. ...1908...Ic
SANCHEZ MEJORADA.....Cactaceas Y Suculentas De La
Barranca De Metztitlan.......1978...Bd
SANCHEZ MEJORADA.....Some Prehispanic uses of Cacti
Among the Indians of Mexico..1982...Bc
SCHULZ...............Cactus Culture.................1932...Id
SCHULZ & RUNYON......Texas Cacti..........................Bz
SCHUMANN.............Gesamtbeschreibung der Kakteen........Bk
SCHUSTER.............An Introduction to Cacti.......1984...Hb
SHREVE...............Cactus and its Home..................Bj
SHURLY...............Cacti................................Be
SPEGAZZINI...........Cactaceas......................1984...Dr

STOCKWELL & BREAZEALE ....Arizona Cacti (Univ. of Ariz.) Bull. # 1, Vol. 4, # 3.......1933...A

STEENBERGH & LOWE....Ecology of the Saguaro (I & II).......By

SUBIK & KAPLICKA.....Cacti and Succulents...........1968...J

SUBIK & KAPLICKA.....Decorative Cacti......................Cs

(SUNSET BOOKS).......Cacti and Succulents...........1978...BB

TAYLOR...............The Genus Echinocereus.........1985...Gn

TRAPP................Cacti of Zion Nat. Park........1969...As

UDALOVA & VYUGINA....In the World of Cactus (Academy of Science U.S.S.R.)......... (Translation)................1977...Ih

VAN LAREN............Cactus.........................1935...Dl

VAN NESS.............Cacti and Succulents (Indoors & Outdoors).........1971...Cq

VENNING..............Cacti (A Golden Guide).........1974...M

WENIGER..............Cacti of the Southwest................Y

WENIGER..............Cacti of Texas and Neighboring States.......................1984...Df

WERDERMANN..........Brasilien und Seine Saulenkakteen................1933...It

WESKAMP..............Die Gattung Parodia............1987...Jf

WIGGINS & PORTER.....Flora of the Galapagos Islands........Ch

WOOTON...............New Mexico Agric. Exper. Sta. (Bull. # 78).................1911...Ie

REFERENCES FROM THE GRAY HERBARIUM CARD INDEX

REFERENCE		KEY
ALAN. INST. BIOL. MEX.		Eb
ALLAN HANC. FOUND. (Occas. Papers)		Dx
AMER. MIDL. NAT.		Ej
ANAIS. PRIM. REUN. SUL-AM. BOT.	1940	Ga
ANAL. MUS. NAC. MONTEVIDEO, URUGUAY		Fh
ANAL. MUS. NAC. CHILE BOT.		Fd
ANAL. MUS. NAC. BUENES AIRES		Eq
ANAL. SOC. CIENT. ARGENTINA		Er
ARCH. JARD. BOT. RIO DE JANIERO	1918	Fz
ARQUIV. JARD. BOT. RIO DE JANIERO		Gl
BACKEB. & KNUTH. KAKTUS - ABC	1935	Fr
BEITR. SUKK.		Dz
BLATT. KAKTEEN FORSCH		Fo
BOIS. DICT.	1898	Fl
BOL. ACAD. CIENC. FISIC. MAT. Y NAT. (Venezuela)		Ef
BOL. DIREC. ESTUD. BIOL.		Ec
BOL. PRO-CULT. REG. (Fam. Cact.)		Fe
BOL. SOC. ARGENTINA BOT.		Gd
BOL. SOC. VENEZOLANA, CIENC. NAT.		Gi
BOLL. R. ORT. BOT. GIARD. COL. (Palermo)		Fw
BRASIL & SEINE SAULENKAKTEEN	1933	Fu
BULL. SO. CALIF. ACAD. SCI		Et
CACTACEAE.....(Backeberg)	1920	Gf
CACT. HORT. DYCK. MEXICO		Fi
CEIBA		Fx
CURTIS BOT. MAG.	1873	Ft
DESERT PLANT LIFE ..(Boyce-Thompson Arboretum)		Eh
FEDDE. REP. SPEC. NOV.		Em
FORSTER'S HANDBOOK, CACTEENKUNDE	1886	Fv
FUAUX HERB. BULL.		Gb
GESAMTB. KAKT.		Gg
JAHRB. DEUT. KAKT. - Gesel.	1936	Fp
KAKTEEN ....(Berger)		Ek
KAKTEEN & ANDERE SUKK.		Du
KAKTEEN FREUND.		Fy
KAKTEENKUNDE		Ea
KAKTUS - ABC		Gc
KAKTUSAR		Gj
KAKTUSY - PAZOUT, VALNICEK, & SUBIK	1960	Fq
KRAINZ. DIE KAKTEEN, LIEF	1963	Fs

REFERENCES FROM INDEX KEWENSIS

REFERENCE			KEY
BLATTER KAKTEENFORSCH	(Backeberg)		Gy
CACTUS, PARIS, FRANCE	(Ritter)		Gt
CACTUS, PARIS, FRANCE	(Cardenas)		Gs
KAKTEENKUNDE,	(Dolz)	1942	Gx
KAKTEENKUNDE HANDBOOK	(Backeberg)		Gv
REV. AGRIC. COCHABAMBA	(Cardenas)		Gz
SUCCULENTA	(Cardenas)		Gw

THE FOLLOWING JOURNALS WERE USED AS REFERENCES:

JOURNAL	KEY
ASHINGTONIA The World of Succulents & Bulbs (England).1973-1980...	Hh
AZTEKIA Journal.............(Czechoslovakia)..1978-1979 & AZTEKIA YEARBOOK............................1982...	Hm
BRITISH CACTUS & SUCCULENT Journal..(England).1983-1988...	Bi
CACTACEAS Y SUCULENTAS MEXICANAS Journal...... (Mexico)..1962-1988...	Aw
CACTUS Journal....................(Belgium)..1979-1980...	Dc
CACTUS Journal................(Netherlands)..1979-1981...	Db
CACTUS AND SUCCULENT Journal...New South Wales (Australia)...1980-1982...	Dd
CACTUS AND SUCCULENT Journal...(New Zealand)..1980-1982...	De
CACTUS AND SUCCULENT Journal of Great Britain.............	Cu
CACTUS AND SUCCULENT Journal...(U.S.A.).......1929-1988...	Af
CACTUS OF THE MONTH CLUB....(Discontinued).... (U.S.A.).......1981-1982...	Ak
CANADIAN SUCCULENT INFORMATION EXCHANGE (C.S.I.E.) (Canada).......1976-1986...	Am
EXOTIC COLLECTION Journal......(England).........- 1984...	Ad
KAKTEEN UND ANDERE SUKKULENTEN Journal........1929-1983... Swiss, Austrian & German Cact. & Succ, Soc.	Aj
KAKTUS..........(Journal)....(Denmark)...................	Hj
KAKTUSY Journal.............(Czechoslovakia) 1979-1983...	Hl
KAKTUSY SUKULENTY Journal....(Czechoslovakia) 1980-1983...	Hs
MAMMILLARIA SOCIETY Journal....(England)......1980-1988...	Aa
NATIONAL CACTUS & SUCCULENT SOCIETY Journal (England)......1975-1982...	Au
SUCCULENTA Journal.............(Holland)......1974-1982...	Ba
SUCCULENTES Journal............(France).......1977-1981...	Da

BULLETINS, CALENDARS, MAGAZINES, PUBLICATIONS, & REPORTS

REFERENCE	KEY
ARIZONA CACTI (Univ. of Ariz. No. 1, Vol. 4, #3...1933...	A
CACTI IN NEW MEXICO (New Mexico Agric. Exp. Sta. Bull. # 78...............(Wooton)...........1911...	Ie
CALANDRINIA (I & II), Succulent Publication of South Australia Inc. ....1980-1982...	Dh
CALENDARS...(Kakteen und andere Sukkulenten) (Austria, Switzerland, W. Germany)............	Bm
DESERT PLANTS Magazine (Univ. of Ariz.)...... Boyce-Thompson Arboretum...............1979-1988...	Dg
SMITHSONIAN INST. Ann. Rpt. Page 525 ....1908...	Fg
TEXAS HIGHWAYS Magazine................................	Hf
THE ILLUSTRATED STUDIES OF THE GENUS OPUNTIA Mo. Bot. Gard. ....1911-1912...	Hz
THE NIGHT BLOOMING CEREUS & ITS ALLIES Mo. Bot. Gard. Bull. #33 ....(Cutak)....1945...	Br
THE UNIV. OF ARIZ. Bull. #5, Vol. XI, #1 (Benson) 1940...	Ao

NURSERY CATALOGUES USED AS REFERENCES

REFERENCE			KEY
ABBEY BROOK CACTUS NURSERY	(England)	1978-1979	Ai
ABBEY GARDENS NURSERY	(U.S.A.)	1981-1988	Ag
ALTMAN SPECIALTY PLANTS	(U.S.A.)		Ab
CACTUS DATA PLANTS	(U.S.A.)	1982-1983	At
CALIFORNIA EPI CENTER	(U.S.A.)		YY
CYCADIA (Succulents, Cycads)	(U.S.A.)	1984	Bf
DEHERDT SEEDLIST	(Belgium)	1981-1983	Ae
GRIGSBY CACTUS GARDENS	(U.S.A.)	1979-1986	Ah
HENRIETTA'S NURSERY	(U.S.A.)	1980	N
RAINBOW GARDENS	(U.S.A.)	1983-1987	Bb
WHITESTONE GARDENS Ltd.	(England)	1981-1988	Hw

CONTENTS

MEMORANDA

* ----Signifies drawings, paintings or sketches.
Hybrids or cultivars are not included to any degree, as the purpose of this compilation is limited to the recognized genera.

ACANTHOCALYCIUM

andreaeanum (See Neochilenia andreaeana, Neoporteria andreaeana, Pyrrhocactus andreaeanus), [Hh March 1975],
aurantiacum (See Echinopsis aurantiaca, Lobivia thionantha -v- aurantiaca), [Af 42-3], [Aj 19-5-1968], [Hh March 1974]
brevispinum (See A. thionanthum -v- brevispinum, Echinopsis brevispina, Lob. thionantha -v- brevispina)..[Bi Vol. 4, #1, 1986], [Iz],
catamarcense (See Lobivia thionantha -v- catamarcensis) [Hi Vol. 2],
chionanthum (See A. thionanthum -v- chionanthum, Lobivia thionantha -v- chionantha), [Aj 26-11-1975], [Aj 29-4-1978],
copiapoides (See A. thionanthum -v- copiapoides),
ferrarii (See Lobivia thionantha -var.)...[Ba 5-1976],
formosum (See Echinopsis formosus, Lobivia formosa, Soehrensia formosa),
glaucum (See A. thionanthum -v- glaucum, Lobivia thionantha -v- glauca)...[G], [XX], [Ar], [Hb], [Hi-2], [Hx*], [Ae-1981], [Hl 1-1974],
griseum (Invalid name - See Lobivia thionantha, A. thionanthum),.....[DD], [Hx], [Il],
klimpelianum (See A. spiniflorum -v- klimpelianum, Echinopsis klimpeliana, Lobivia klimpeliana, Lobivia spiniflora -v- klimpeliana), [Ar], [Bo-3], [Bu], [Gr], [Hb], [Hx*], [Jd], [Ba 7-1979], [Hw-1981],
oreopepon (See Echinopsis oreopepon, Lobivia oreopepon, Soehrensia oreopepon).....[E],
peitscherianum (See Echinopsis peitscherianum, Lobivia spiniflora -v- peitscheriana), [E], [EE], [Bo-3], [Hb], [Iz],
" -v- violaceum (See A. violaceum, Echinopsis violacea),
spiniflorum (See Echinopsis spiniflora, Lobivia spiniflora), [Bo-3], [Hb], [Hx*], [Hl 2-1972],
" -v- klimpelianum (See A. klimpelianum), [Hh Jan. 1975],
thionanthum (See L. thionantha), [DD], [Dq], [Gr], [Af 51-3], [Hh Jan. 1975]
" -v- aurantiacum (See A. aurantiacum),

ACANTHOCALYCIUM.....(Cont'd),

thionanthum.....(cont'd),
" -v- brevispinum (See A. brevispinum, Echinopsis brevispina).....[Hh March 1975],
" -v- chionanthum (See A. chionanthum), [Aj 4-1978],
" -v- copiapoides (See A. copiapoides),
" -v- glaucum (See A. glaucum), [Hh Jan. 1975],
" -v- griseum (See A. griseum), [Hh March 1975],
" -v- peitscherianum (See A. peitscherianum),
" -v- thionanthum.....[Hh Jan. 1975],
" -v- variiflorum (See A. variiflorum), [Hh Jan. 1975],
" -f. violaceum (See A. violaceum),
variiflorum (Invalid name - See A. thionanthum -v- variiflorum, Lobivia thionantha -v- variiflora),.....[DD],
violaceum (See A. peitscherianum -v- violaceum, Echinopsis violacea, Lobivia spiniflora -v- violacea), [QQ], [Ar], [Bo-3], [Bv-1], [Hb], [Hi-2], [Hv], [Hx], [Il], [Iz], [Jd], [Af 16-5], [Aj 26-9-1975], [Ba 7-1980],
" -v- P-204.....[NN],

ACANTHOCEREUS

acutangulus (See A. pentagonus, Selenicereus acutangulus),
albicaulis (See Hylocereus ocamponis), [RR], [Bo-4],
baxaniensis (See A. congatus, A. pentagonus, Cereus baxaniensis), [Az], [Bo-4], [Aw 2-1983],
brasiliensis (See Pseudoacanthocereus brasiliensis), [RR],
chiapensis.....[Aw 17 (4):117 1972 Description only]
columbianus.....[Ay], [Jd], [Af 15-6], [Af 55-1], [Ba 9-1979],
congatus (See A. pentagonus, Cereus pentagonus), [Hy],
floridanus (See A. pentagonus),
griseus (Invalid name).....[DD],
guatamalensis (See Hylocereus guatamalensis),
hondurensis (See A. pentagonus -v- hondurensis, Selenicereus hondurensis),
horribarbis (See Cereus horribarbis),
horridus.....[Ay], [Hy*], [RR],
maculatus (See Peniocereus maculatus).....[DD],

ACANTHOCEREUS.....(Cont'd),

occidentalis (See Peniocereus occidentalis),
[RR], [Az], [Bo-4], [Cx], [Af 14-4],
pentagonus (See A. baxaniensis, Cereus pentagonus),
[G], [Y], [LL*], [RR], [Bv-1], [Bz], [Ci],
[Df], [Dn], [Af 5-2], [Af 14-8], [Af 30-6],
[Aj 29-2-1978],
" -v- hondurensis (See Selenicereus hondurensis),
[Bv-1],
pitahaya (See Cereus pitahaja),
princeps (See A. pentagonus),
sicariguensis.....[Gi 11-75-1947],
subinermis.....[RR], [Az], [Bo-4], [Cx],
tetragonus (See A. pentagonus, Cereus pentagonus),
[Hx*],
" -v- micranthus.....
thalassinus (See Cereus jamacaru),
" -v- quadrangularis (See A. thalassinus),

ACANTHOLOBIVIA

ayacuchensis.....[Hy],
euanthema (See Lobivia euanthema, Mediolobivia
euanthema, Rebutia euanthema),
[Bp], [Hy],
incuiensis (See Lobivia tegeleriana -v- incuiensis),
[DD], [Ay], [Bo-3], [Aj 9-10-1958],
" -v- applanata.....
tegeleriana (See Echinopsis tegeleriana, Lobivia
tegeleriana),
[Bo-3], [Cv-1], [Hx*], [Hy], [Aj 16-10-1965],
" -v- eckardiana (See Lobivia tegeleriana),
[Cv-1],
" -v- medingiana (See Lobivia tegeleriana),
[Cv-1],
" -v- plominiana (See Lobivia tegeleriana),
[Cv-1],

ACANTHORHIPSALIS

brevispinum (See Lepismium brevispinum, Rhipsalis
brevispina),
[Hi-4], [Af 56-3],
conjugens (See A. monacantha, Rhipsalis monacantha),
crenata (See Hatiora crenata, Lepismium crenatum,
Rhipsalis crenata),
[RR],
incahuasina (See A. monacantha, Rhipsalis incahuasina,
Rhipsalis monacantha).....[Af 26-1954],
[Af 26-1954],

ACANTHORHIPSALIS.....(Cont'd),

micrantha (See Lymanbensonia micrantha, Rhipsalis micrantha),
[RR], [Dq*], [Hx*],
monacantha (See Lepismium monacanthum, Rhipsalis monacantha),
[K], [DD], [RR], [Ay], [Hx*], [Jd],
" -v- samaipatana.....[Hv],
paranganiensis (See Lepismium paranganiensis, Rhipsalis paranganiensis),
[Bo-2], [Hx*],
samaipatana (See A. monacantha -v- samaipatana, Rhipsalis monacantha -v- samaipatana),
" -v- incahuasina (See A. incahuasina),

AKERSIA

roseiflora (See Borzicactus roseiflorus),
[DD], [NN], [Bo-6], [Hx*], [Hy],
[Aj 30-5-1979], [Dg 7-1-1985],

ANCISTROCACTUS

brevihamatus (See Echinocactus brevihamatus),
[Y], [LL*], [RR], [Ar], [Bo-5], [Df],
[Hy], [Af 5-2],
crassihamatus (See Ferocactus crassihamatus, Glandulicactus crassihamatus, Hamatocactus crassihamatus, Sclerocactus crassihamatus, Thelocactus crassihamatus),
[Af 43-1], [Af 59-3], [Aj 23-11-1972],
megarhizus (See A. scheerii -f. megarhizus),
[RR], [Ar], [Bo-5], [Hy], [Iu], [Hl 6-1967],
scheerii (See Echinocactus scheerii, Thelocactus scheerii),
[E], [K], [M*], [Y], [DD], [GG], [LL], [NN],
[RR], [Ar], [Av], [Ax], [Ay], [Df], [Hy], [Il],
[Jd], [Af 5-2], [Af 59-3], [Aw 1-1966],
" -f. megarhizus (See A. megarhizus),
[Hv],
tobuschii (See Echinocactus tobuschii),
[M*], [Y], [Df], [Af 50-5], [Af 53-2],
[Hl 3-1965], [Hl 6-1983], [Bi 3-1987],
uncinatus (See Echinocactus uncinatus, Ferocactus uncinatus, Glandulicactus uncinatus, Hamatocactus uncinatus),
[U], [Hv],

(cont'd)

ANCISTROCACTUS.....(Cont'd),

uncinatus
" -v- wrightii (See Echinocactus uncinatus -v- wrightii, Ferocactus uncinatus -v- wrightii, Glandulicactus uncinatus -v- wrightii, Hamatocactus uncinatus -v- wrightii)...[G], [U], [Y], [Ax], [Df], [Ib], [Af 52-5], [Af-57-3], [Af 58-4], [Af 60-1-1988],

ANHALONIUM

engelmannii (See Ariocarpus fissuratus), [Di*],
prismaticum (See Ariocarpus prismaticus, Ariocarpus retusus),
" -v- minor (See Ariocarpus prismaticus, Ariocarpus retusus), [Gk 29-3-12-1925],

ANISOCEREUS

foetidus (See Pterocereus foetidus).....[Az],
gaumeri (See Pachycereus gaumeri, Pterocereus gaumeri) [Az], [Bo-4], [Hx*],
lepidanthus (See Escontria lepidanthus), [Bo-4], [Hx*], [Af 60-4],

APOROCACTUS

conzattii.....[M*], [DD], [RR], [Av], [Az], [Bo-2], [Bv-1], [Cx], [Dq], [Hx*], [Af 18-10], [Af 34-1], [Aw 3-1986], [Ba 10-1981], [Bm], [Hl 4-1982]
flagelliformis.....[C], [I], [L], [X], [LL*], [PP], [TT], [Ar], [Av], [Ay], [Be*], [Cc], [Cj], [Cm], [Cr], [Cx], [Gr], [Hw-1989], [Hx], [Hy], [Il], [Iu], [Iz], [Ja], [Jd], [Af 2-12], [Aw 3-1986], [Bb-1985], [Cu 8-4-1946], [Bi 5-4-1987],
" f. cristata.....[Af 57-1],
flagriformis.....[L], [Az], [Bd], [Bo-2], [Hx*], [Bb-1985],
knebelii.....[Bo-2], [Aj 16-12-1965],
leptophis.....[RR], [Az], [Hy], [YY-1980],
mallisonii (Hybrid - A. flagelliformis X Heliocereus speciosus), [C], [L], [Bt], [Hx],
martianus....[L], [M], [RR], [Az], [Dq], [Iu], [Jd], [Aj 16-12-1965], [Cu 3-1962],

APOROCACTUS.....(Cont'd),

Hybrid - (Aporocactus X smithii).....[Hu],

APOROHELIOCEREUS

'hasel'.....(X Helioporus).....[Bb-1986],
X Aporoheliocereus smithii.....[Bi 5-4-1987],

AREQUIPA
----

aurantiaca (See Borzicactus aurantiacus, Borzicactus calvescens, Matucana aurantiaca, Matucana calvescens, Submatucana aurantiacus),
[Ea 1939 - 77],
australis.....[Hi-3],
clavata (See Neoporteria clavata),

erectocylindrica (See A. rettigii, A. rettigii -v- erectocylindrica, Borzicactus hempelianus -v- rettigii),
[Ay], [Bo-2],
hempeliana (See Borzicactus hempelianus, Echinopsis hempelianus),
[Bo-2], [Hi-3], [Hx*], [Hy],
" -v- rettigii (See A. erectocylindrica, A. rettigii),
" -v- spinosissimus (See Borzicactus hempelianus -v- spinosissimus),
" -v- weingartiana (See A. weingartiana),
leucotricha (See A. neoleucotricha, Borzicactus leucotrichus),
[Ay], [Bv-1], [Cz], [Dq], [Hx],
" -v- erectocylindrica (See A. rettigii),
mirabilis (Borzicactus mirabilis),
[Hx*],
myriacantha (See Borzicactus myriacanthus, Matucana myriacantha, Submatucana myriacantha),
neoleucotricha (See A. leucotricha),
[Hy],
neoweingartiana (See A. weingartiana),
[Hy],
rettigii (See A. erectocylindrica, Borzicactus leucotrichus -v- rettigii, Oreocereus rettigii),
[DD], [Ay], [Bo-2], [Hi-4], [Jd], [Au 36-4], [Aw 4-1964], [Hl 4-1982], [Hl 6-1971],
" -v- borealis.....[Hi-4],
" -v- erectocylindrica (See A. erectocylindrica, A. rettigii),
[Gr],

AREQUIPA.....(Cont'd),

soehrensii (Invalid name - See A. rettigii, Borzicactus hempelianus -v- rettigii, Borzicactus leucotrichus -v- rettigii),
spinosissima (See Borzicactus hempelianus -v- spinosissimus, Borzicactus leucotrichus -v- spinosissimus),
[Bo-6], [Hi-4], [Aj 23-3-1972], [Hl 6-1971],
variicolor (See Matucana variicolor, Oreocereus variicolor, Submatucana variicolor).....[Bo-2],
weingartiana (See A. hempeliana -v- weingartiana, Borzicactus hempelianus -v- weingartianus, Borzicactus leucotrichus -v- weingartianus),
[Bl], [Bo-2], [Hv], [Hx*], [Il],
[Aj 11-6-1960],
" -v- carminanthema (See A. hempeliana -v- weingartiana, Borzicactus leucotrichus -v- weingartianus)
[Bo-2], [Gj 7-61-1936],
Spec. .....[Bo-2], [Bo-6],

ARIOCARPUS

agavoides (See Neogomesia agavoides),
[I], [K], [Z], [Ar], [Ax], [I], [Iz], [Jd],
[Aj 6-1981], [Aw 1-1989], [Bi 1-1-1983],
aselliformis (See Pelecyphora aselliformis),
disciformis (See Strombocactus disciformis),
[Iv 4-1949],
elongatus (See A. trigonus -v- elongatus)...[Af 9-5],
fissuratus (See Roseocactus fissuratus),
[D], [I], [T], [Y], [Z], [AA], [LL], [NN],
[TT], [Ar], [Ax], [Ay], [Bz], [Cz], [Df],
[Hn], [Hv], [Hx], [Ib], [Il], [Iz], [Jd],
[Af 16-12], [Af 50-4], [Af 52-5], [Af 56-3],
[Af 59-4], [Aj 28-7-1977], [Am 1-1980],
[Bi 1-1-1983], [Cu 4-1964], [Af 60-1-1988],
[Hw-1982],
" -v- lloydii (See A. lloydii),
[Ar], [Ay], [Iz], [Af 56-4],
[Au 37-4], [Aw 2-1965],
[Bi 1-1-1983], [Hl 1-1977],
" X retusus.....[Af 53-2],
" -f. cristata.....[Cu 2-4-1934],
" -f. monstrosus.....[Af 58-1],
furfuraceus (See A. retusus -v- furfuraceus),
[D], [Ar], [Av], [Gr], [Hx*], [Hy],
[Ad Nov. 1978], [Ae-1981], [Aj 6-1977],
" -v- rostratus.....[Bo-5], [Fg], [Hy],
[Bi 1-1-1983],
" -f. cristata.....[Gk 15-29 4:14 1925],

ARIOCARPUS.....(Cont'd),

intermedius (See A. fissuratus -v- lloydii, A. lloydii,
Roseocactus intermedius),
kotschoubeyanus (See Roseocactus kotschoubeyanus),
[C], [Z], [EE], [PP], [Ar], [Av], [Ay],
[Gr], [Hr], [Hv], [Il], [Iz], [Af 50-4], [Af 56-4], [Bi 1-1], [Bi 2-3], [Cu 4-1964], [Hl 3-1976], [Hw-1981],
" -v- albiflorus.....[Ar], [Iz],
" -v- macdowellii (See Roseocactus kotschoubeyanus -v- macdowellii),
[Ar], [Hy], [Cu 1-1977],
lloydii (See A. fissuratus -v- lloydii, Roseocactus fissuratus -v- lloydii, Roseocactus lloydii),
[F], [Ar], [Hr], [Au 37-4], [Hl 4-1981],
" -f. ....[Ar],
macdowellii (See A. kotschoubeyanus -v- macdowellii),
prismaticus (See Anhalonium prismaticum, A. retusus),
[Gk 29 3-12 Fig. 2, 1925],
retusus.....[D], [L], [V], [X], [Z], [EE], [UU], [Ar],
[Ay], [Bo-5], [Cm], [Cx], [Dq], [Hx], [Il],
[Iz], [Jd], [Ad Nov. 1978], [Af 59-4], [Aj 1-1982], [Aj 31-2-1980], [Aw 4-1969], [Bi 1-1], [Bm -], [Fg], [Hl 3-1970], [Hw-1986],
" -v- furfuraceus (See A. furfuraceus),
[Iz], [Aj 5-1982], [Hl 5-1976],
" -v- scapharostrus (See A. scapharostrus),
" -f. cristata.....[Ar],
rostratus (See A. furfuraceus -v- rostratus),
scapharostrus.....[G], [Ar], [Ay], [Bo-5], [Hv], [Hx*],
[Hy], [Iz], [Ae-1982], [Af 56-4],
[Aj 28-6-1977], [Aj 30-8-1979],
[Aw 3-1987], [Bi 1-1], [Hl 4-1974],
[Hl 2-1978],
trigonus.....[D], [K], [L], [R], [DD], [QQ], [Ar], [Ay],
[Az], [Bu], [Cr], [Gr], [Hx*], [Hy], [Il],
[Iz], [Jd], [Ae-1981], [Af 56-1],
[Aj 32-6-1981], [Aj 1-1982], [Aw 3-1987],
[Ba 10-1977], [Bi 1-1],
" -v- elongatus (See A. elongatus),
[Ay], [Bo-5], [NN], [Hl 5-1978]
" -v- minor.....[Ar], [Aj 27-11-1976],
[Hl 1-1976], [Hl 5-1978],
williamsii (See Lophophora williamsii),
(Hybrid).....[Bi 3-1],

ARMATOCEREUS

arboreus (See A. churinensis, A. ghiesbreghtii, A. matucanensis, A. riomajensis, Cereus ghiesbreghtii).....[Bo-2], [Aj 8-11-1957],

ARMATOCEREUS.....(Cont'd),

arduus.....
armatus.....(See A. procerus, A. procerus -v- armatus),
[Hy],
balsasensis (See A. rauhii -v- balsasensis),
[Hi-4],
cantaensis (See Haageocereus peculiaris, Loxanthocereus cantaensis, Loxanthocereus peculiaris),
cartwrightianus (See A. mocupensis),
[DD], [Ay]. [Bo-2], [Hy], [Jd], [Af 34 -3], [Aj 9-9-1958], [Hl 5-1982],
" -v- longispinus.....[Bo-2],
churinensis (See A. arboreus, A. ghiesbreghtii, A. matucanensis, A. riomajensis, Cereus ghiesbreghtii),
[Bo-2], [Aj 8-11-1957],
confusus (See A. procerus),
ghiesbreghtii (See A. arboreus, A. churinensis, A. matucanensis, A. riomajensis, Cereus ghiesbreghtii),
[Hi-4], [Aj 23-1-1972],
" -v- oligogonus (See A. oligogonus),
[Hi-4],
godingianus (See Lemaireocereus godingianus),
[Bo-2], [Hl 2-1969]
humilis (See Lemaireocereus humilis),
laetus (See Cereus laetus, Lemaireocereus laetus),
[RR], [Bo-2], [Hx*], [Aj 29-6-1978],
[Hl 5-1982], [N-1980],
maranionensis (See A. rauhii),
[Bo-2],
mataranus.....[Hy], [Hi-4],
" -v- ancashensis.....[Hi-4],
matucanensis (See A. arboreus, A. churinensis, A. ghiesbreghtii, A. riomajensis, Cereus ghiesbreghtii),
[DD], [Ay], [Bo-2], [Hx*], [Iv 7-1956],
microspermus (See Browningia microspermus, Gymnocereus microspermus),
mocupensis (See A. cartwrightianus),
nazcaensis (See A. procerus),
neolateus (See A. lateus),
[Hy],
oligogonus (See A. ghiesbreghtii -v- oligogonus),
[Bo-2],
parviflorus (See Cleistocactus parviflorus),
procerus (See A. armatus),
[Ay], [Bo-2], [Hi-4], [Aj 8-6-1957],
" -v- armatus (See A. armatus, A. procerus),
" -v- piscensis.....

ARMATOCEREUS.....(Cont'd),

rauhii (See A. maranionensis),
[Ay], [Bo-2], [Hi-4], [Jd], [Aj 8-7-1957],
" -v- balsasensis (See A. balsasensis),
riomajensis (See A. arboreus, A. churinensis,
A. ghiesbreghtii, A. matucanensis, Cereus
ghiesbreghtii).....[Bo-2], [Aj 8-10-1957],
rupicola.....[Hi-4],

ARROJADOA

albiflora.....[Hi-1], [Aj 29-10-1978],
auriazurea (See Micranthocereus auriazureus),
aureispina.....[DD-460], [Gr], [Hx*], [Aj 23-4-1972],
[Aj 29-10-1978], [Aj 34-3-1983], [Ct-1988],
" -v- anguinea.....[Af 60-4],
" -v- aureispina.....[Af 60-4],
" -v- guanambensis.....[Aj 31-11-1980],
canudosensis (See Cephalocereus canudosensis, A. rhodantha ssp. canudosensis)..[Aj 29-10-1978],
densiflora (See Micranthocereus densiflora),
dinae (See Cephalocereus dinae)...[DD-461], [Hx*], [Aj 24-5-1973], [Aj 32-11-1981], [Hs III-4-1982],
eriocaulis (See Cephalocereus eriocaulis).....[DD-462],
[Hx*], [Aj 24-11-1973], [Aj 32-2-1981],
" -v- albicoronata.....[Aj 33-11-1982],
" -v- eriocaulis.....[Aj 33-11-1982],
grandiflora.....[Bt],
horridispina.....
horstiana (See Cephalocereus horstiana),
[Aj 32-8-1981], [Ct-1988],
multiflora (See Cephalocereus multiflorus)
[Hi-1], [Aj 32-2-1981], [Aj 33-11-1982],
[Hl 4-1983],
oreophila (See albiflora),
penicillata (See Cephalocereus penicillatus),
[G], [I], [DD], [NN], RR], [Ar], [Ay],
[Bo-4], [Bt], [Hv], [Hx*], [Hy], [Il],
[Bm -], [Ad April, 1979],
" -v- decumbens.....[Hi-1], [Gl 9:164 1950],
" -v- spinosior.....[Ba 1-1980],
" -v- (H-443).....[Ba 10-1979],
polyantha (See Cephalocereus polyanthus, Micranthocereus polyanthus),
reflexa.....
rhodantha (See Cephalocereus rhodanthus),
[G], [RR], [Av], [Ay], [Bl], [Bo-4], [Cr],
[Dq], [Gr], [Hi-1], [Hv], [Hx*], [Hy], [Jd],
[Af 50-2], [Af 58-1], [Aj 31-11-1980],
[Hl 6-1979],
" ssp. canudosensis (See A. canudosensis)

ARROJADOA.....(Cont'd),

rhodantha.....(cont'd),
" ssp. reflexa (reflectus).....
" " rhodantha -v- occibahiensis.....
" " " -v- theunisseniana (See A. theunissensiana)
" -v- minensis.....[Hi-1],
" -v- robusta.....[Aj 29-10-1978],
segredensis.....[Aj 29-10-1978],
theunissensiana (See A. rhodantha -v- theunisseniana),
.....[Hx*],
Spec. .....[Ad -],
Spec. JMC - 27.....[Aj 33-6-1982],
Spec. HU-442.....[Aj 10-1978],

ARTHROCEREUS

akersii (See Haageocereus akersii, Pygmaeocereus akersii, Pygmaeocereus bylesianus, Pygmaeocereus rowleyanus),
aureispinus (See Haageocereus aureispinus, Haageocereus multangularis),
bylesianus (See Pygmaeocereus akersii, Pygmaeocereus bylesianus, Pygmaeocereus rowleyanus),
[Iz], [Jd],
campos-portoi.....[Bo-4], [Bu], [Hi-1], [Hv], [Hx*],
[Af 20-2], [Aj 34-1-1983], [Flower Ba 2-1979],
densiaculeatus (See Pygmaeocereus densiaculeatus),
glaziovii (See Leocereus glaziovi),
mello-barretoi (See Leocereus mello-barretoi),
[Bo-4], [Gl 9:157 Tab. 4 Fig. 2,1950],

microsphaericus (See Cereus microsphaericus, Epiphyllanthus microhaericus),
[Bo-4], [Hx*], [Hy*], [Jd], [Af 20-1],

mirabilis (See Echinopsis mirabilis, Setiechinopsis mirabilis),
[DD], [Dq], [Iz], [XX],
" -v- gracilior.....[DD],
odorus.....[Hi-1],
rondonianus.....[DD], [NN], [Bo-4], [Hi-1], [Hv], [Hx],
[Jd], [Af 23-4], [Af 41-5], [Af 56-5],
[Aj 32-9-1981],

rowleyanus (See Haageocereus akersii, Pygmaeocereus akersii, Pygmaeocereus bylesianus, Pygmaeocereus rowleyanus),

spinosissimus (See Eriocereus spinosissimus),

ASTROPHYTUM

asterias (See Echinocactus asterias),
[E], [X], [Y], [Z], [AA], [Ar], [Av], [Ay], [Bo-5], [Cr], [Cx], [Cz], [Gr], [Hx], [Ib], [Il], [Iz], [Ja], [Jd], [Ae-1981], [Af 16-10], [Af 60-6], [Ag-1985], [Aj 27-3-1976], [Aj 28-7-1977], [Aj 32-1-1981], [Aj 33-1-1982], [Ba 2-1978], [Dg 2-3-1980], [Hh Jan. 1975], [Bi 5-4-1987], [Bi 6-2-1988],
" -v- asterias.....[Hp],
" -v- multipunetatum.....[Hp], [Hl 1-1971],
" -v- nudicarpa.....[Hp],
" -v- nudum.....[Ar], [Hy],
" -v- " -f. variegata.....[Ar], [Aj 26-8-1975],
" -v- pubesente.....[Hy],
" -v- seminudum.....[Hp], [Hl 1-1971],
" -v- striatipetalum.....[Hy],
" -f. cristata.....[Ar],
" -f. variegata.....[Af 59-6],
" hybrid - X capricorne.....[Ar],
" hybrid - X myriostigma.....[Iz],
aureum (See A. senile -v- aureum),
capricorne.....[C], [L], [DD], [NN], [PP], [Ar], [Av], [Ay], [Be*], [Bo-4], [Bt], [Cr], [Cx], [Cz], [Dq], [Gr], [Hv], [Hx], [Hy], [Il], [Iz], [Jd], [Af 1-8], [Af 16-10], [Af 51-6], [Af 56-1], [Au 37-2],
" -v- aureum.....[CC], [Af 56-4], [N-1980],
" -v- crassispinum.....[Hx*], [Hy], [Aj 30-12-1979], [Hs III-3-1982],
" -v- " -f. monstrosus.....[Aj 5-1979],
" -v- major (See Echinocactus capricorne -v- major),
[Ar], [Bv-1], [CC], [Hp], [Hy], [Ae-1981], [Af 16-10], [Hl 4-1971],
" -v- minor (minus).....[AA], [CC], [Ar], [Hp], [Hq], [Hx*], [Hy], [Ae-1981], [Aj 26-6-1975], [Aw 3-1968], [Hl 4-1971],
" -v- niveum (See A. niveum),
[L], [Bo-5], [Hx*], [Hy], [Ae-1982], [Aw 3-1968],
" -v- senile (See A. senile),
[E], [H], [Ay], [Iu], [Z], [Af 16-10],
" -f. monstrosus.....[Aj 30-5-1979], [Hl 3-1965],
coahuilense (See A. myriostigma -v- coahuilensis),
[Ar], [Bo-5], [Cr], [Hq*], [Hr], [Hx], [Hy], [Ae-1981], [Aj 26-9-1975], [Bi 5-4],
" -f. cristata.....[Aj 27-7-1976],
" -f. monstrosus.....[Aj 30-5-1979],

ASTROPHYTUM.....(Cont'd),

coahuilense.....(cont'd),
" -f. (spiral).....[Aj 26-9-1975],
columnare (See A. myriostigma -v- columnaris),
[Hp], [Ba 10-1981],
crassispinoides.....
crassispinum (See A. capricorne -v- crassispinum),
[Ar], [Hp], [Hq], [Aj 8-9-1957],
glabrescens (See A. ornatum -v- glabrescens),
jaumavense (See A. myriostigma -v- jaumavense),
[Hp],
myriostigma.....[C], [I], [L], [X], [Z], [AA], [BB],
[CC], [UU], [Ar], [Ay], [Be*], [Bo-6],
[Bt], [Cj], [Cr], [Cx], [Cz], [Dq],
[Gr], [Hb], [Hu], [Hv], [Ib], [Il],
[Iz], [Jd], [Af 6-7], [Af 53-3], [Af 56-
1], [Af 56-6], [Af 57-6], [Af 59-2],
[Aj 11-7-1960], [Aj 32-7-1981], [Au 37-
2],
" -v- caespitosum.....[Hy],
" -v- circularis rotundula.....[Hp],
" -v- circumcostata strongy locostata.....[Hp],
" -v- coahuilensis (See A. coahuilense),
[F], [Af 10-12], [Af 16-10],
[Hl 2-1971],
" -v- columnaris (See A. columnare),
[H], [HH], [Ar], [Bo-6], [Hy], [Af 6-
7], [Af 16-10], [Hl 2-1971],
" -v- " -f. .....[Ar],
" -v- " -f. (espiralis).....[Af 16-10],
" -v- glabrum (invalid name).....[Hp],
" -v- jaumavense (See A. jaumavense),
[Hq], [Hl 1-1978],
" -f. nuda.....[Hq],
" -v- nudum....[E], [CC], [Ar], [Av], [Ay],
[Bv-1], [Hb], [Hu], [Aj 10-2-1959],
[Ba 1-1980], [Db 5-1979],
" -v- " -f. octogonum.....[Hy],
" -v- " -f. quadricostatum.....[Ar],
" -v- " -f. tricostatum.....[Ar],
" -v- potosinum.....[Ar], [Av], [Bo-5], [Hq],
[Iu], [Hl 2-1971],
" -v- " -f. nudum.....[Iz],
" -v- " -f. tulense (See A. myriostigma -v-
tulense),
[Bo-5], [Aj 30-5-1979],
" -v- " -f. turbinatum.....[Hl 2-1971],
" -v- quadricostatum (See A. myriostigma -v-
nudum -f. quadricostatum),
[Ar], [Ay], [Bo-5], [Bv-1],
(" Cont'd) [CC], [Cm], [E], [Hy],

ASTROPHYTUM.....(Cont'd),

myriostigma.....(cont'd),
" -v- quadricostatum .....(cont'd),
[Af 16-10], [Aj 29-4-1978],
[Hl 2-1971],
" -v- rotunda (rotundata).....[Hp],
" -v- senile (See A. capricorne -v- senile,
A. senile),
[Ar],
" -v- " -f. aureum.....[Ar],
" -v- strongylogonum.....[Ar], [Bo-5], [Hp],
[Ae-1981], [Aj 23-5-1972]
[Hl 2-1971], [Hw-1981],
" -v- " -f. nudum.....[Ar], [Bo-5],
[Aj 26-7-1975],
" -v- " -f. rotunda.....[Aj 26-7-1975],
" -v- " -f. cristata.....[Hx*], [Aj 2-1976],
" -v- tamaulipense.....
" -v- tricostatum (See A. myriostigma -v- nudum
-f. tricostatum),
[Ar], [Aj 29-4-1978],
" -v- tulense (See A. myriostigma -v- potosinum
-f. tulense),
[Ar], [Hp], [Hq], [Hx*], [Fy 1:58 -
1932], [Hl 2-1967], [Hl 5-1982],
" -f. vera.....[Af 16-10],
" -f. (de San Antonio).....[Ae-1982],
" -f. cristata.....[Hy], [Af 1-12], [Au 37-2],
[Iv 3-1950],
" -f. monstrosus.....[Ah-1985],
" X asterias.....[Gr],
niveum (See A. capricorne -v- niveum, A. ornatum -v-
niveum),
[Cr], [Hq*], [Aj 8-9-1957], [Aj 32-6-1981],
[Au 37-2], [Bi 2-1-1984], [Hl 4-1971],
ornatum.....[H], [L], [AA], [NN], [UU], [Av], [Ay],
[Be*], [Bo-5], [Bt], [Cj], [Cx], [Cz], [Gr],
[Hv], [Hx], [Hy], [Il], [Iz], [Ad Sept.
1978], [Af 16-10], [Aj 30-4-1979], [Aj 31-8-
1980], [Au 37-2], [Aw 4-1962], [Bi 2-1-1984]
[Bi 2-3-1984], [Bm-1988], [Da 2-1979],
[Hw-1982],
" -v- aureispinum.....[Bv-1],
" -v- glabrescens.....[Ar], [Hp], [Hy], [Ae-1981],
[Af 16-10], [Aj 11-3-1960],
" -v- mirabelli....[Ar], [Bv-1], [Hb], [Hp], [Hy],
[Af 16-10],
" -v- niveum (See A. niveum, A. capricorne -v-
niveum).....[Hp], [Hl 3-1971],
" -v- virens.....[Hp], [Aj 31-2-1980], [Aj 32-12-
1981], [Hl - 1971],

ASTROPHYTUM.....(Cont'd),

ornatum.....(cont'd),
" -f. cristata.....[Au 37-2],
" -f. monstrosus.....[Aj 30-5-1979],
" -f. (spiral).....[Af 16-10],
prismaticum (See A. myriostigma),
quadricostata (See A. myriostigma -v- quadricostatum, A. myriostigma -v- nudum -f. quadricostatum),
senile (See A. myriostigma -v- senile, A. capricorne -v- senile),
[Bo-5], [Hb], [Hq], [Hx*], [Ba 6-1981], [Hl 5-1971], [Bi 5-4-1987],
" -v- aureum (See A. myriostigma -v- senile -f. aureum),
[Hp], [Hu], [Hx], [Am 7-1978], [Hl 5-1971],
" -f. flavispina.....[Hq*],
" -f. .....[Hl 5-1971],
tulense (See A. myriostigma -v- tulense),
" -v- nudum.....[Hp],
(Hybrids).....[Gr], [Aj 30-10-1979],

AUSTROCACTUS

bertinii (See Notocactus bertinii),
[DD], [RR], [Bo-3], [Iu], [Af 59-1],
coxii (See A. dusenii, Cereus dusenii),
[Bo-3],
dusenii (See A. coxii, Cereus dusenii),
[F],
gracilis.....[Bo-3], [Hy], [Dz 1942:58 Fig.], [Jd],
hibernus.....[Hi-3], [Hx*], [Af 57-2], [Ee 7/8:34 Fig.-1963],
intertextus (See A. patagonicus, Notocactus patagonius),
[Bo-3], [Hv], [Hx*], [Il], [Iz], [Ae-1981], [Aj 28-2-1977],
philippii (See Erdisia philippii),
[Hi-3],
spiniflorus (See Erdisia spiniflorus),
[Hi -3],

AUSTROCEPHALOCEREUS

albicephalus (See Coleocephalocereus goebelianus, Micranthocereus albicephalus, Micranthocereus aureispinus, Micranthocereus monteazulensis).....[X], [Bt], [Hx*], [Af 58-1], [Aj 24-4-1973], [Aj 28-10-1977], [Aj 34-3-1983], [Ba 3-1980], [Ct-1988],

AUSTROCEPHALOCEREUS.....(Cont'd),

dolichospermaticus (See Micranthocereus dolichospermaticus),
[G], [Ct-1988], [Hl 3-1977],
dybowskii (See Cephalocereus dybowskii, Espostoa dybowskii, Espostoopsis dybowskii),
[G], [Ar], [Ay], [Bo-4], [Bt], [DD-463], [Gr], [Hx], [Hy], [Il], [Jd], [Af 52-1], [Af 56-6], [Aj 22-12-1971], [Aj 32-10-1981],
estevesii (See Coleocephalocereus estevesii, Micranthocereus estevesii),
[Hx*],
fluminensis (See Cephalocereus fluminensis, Coleocephalocereus fluminensis),
lehmannianus (See A. purpureus (Guerke) Buxb. 1970, Cephalocereus lehmannianus, Micranthocereus purpureus (Guerke) Ritt. 1968),
[Bo-4], [Hh Sept. 1975],
purpureus (Guerke) Backbg. (See Cephalocereus purpureus (Br. & R.) 1920 & sensu Werd. 1933, Cereus goebelianus (VPL), Coleocephalocereus goebelianus (Buin.).).....[DD-464], [HH], [Hx*], [Jd], [Ba 8-1981], [Ct-1988],
" -f. cristata.....[Ba 6-1974],
purpureus (Guerke) Buxb. 1970 (See A. lehmannianus (Werd.) Backbg. 1960, Cephalocereus lehmannianus (Werd.) 1933, Cephalocereus purpureus (Guerke) 1908, Micranthocereus lehmannianus (Werd.) Ritt. 1968, Micranthocereus purpureus (Guerke) Ritt. 1968),
salvadorensis (See Pilosocereus salvadorensis, Pseudopilocereus salvadorensis),

AUSTROCYLINDROPUNTIA

albiflora (See A. salmiana, Opuntia salmiana),
chuquisacana (See A. vestita -v- chuquisacana, Opuntia chuquisacana),
clavarioides (See Cylindropuntia clavarioides, Opuntia clavarioides),
[K], [RR], [Ay], [Bo-1], [Bt], [Hx*], [Aj 28-8-1977], [Aj 34-1-1983], [Ba 7-1976],
" -v- minima.....[Hx*],
" -v- ruiz-lealii.....[DD], [Bo-1], [Hx*],
" -f. cristata.....[L], [Ay],
" -f. monstrosus.....[Ay],
colubrina (See Opuntia colubrina),
[Bo-1],
cornephora (See A. cylindrica, Cylindropuntia cylindrica),
[Hy],

AUSTROCYLINDROPUNTIA.....(Cont'd),

cylindrica (See Cylindropuntia cylindrica, Opuntia cylindrica),
[Bo-1], [Hx*], [Da 1-1981],
" -f. cristata (See Cylindropuntia cylindrica -f. cristata),
[T],
" -f. monstrosus.....[Iv 5-1956],
exaltata (See Opuntia exaltata),
[Ay], [Bo-1], [Hi-4],
" -f. cristata (See Opuntia exaltata -f. cristata)
" -f. monstrosus.....[Bo-1], [Hy],
floccosa (See Opuntia floccosa, Tephrocactus floccosus)
[Hi-4],
haematacantha (See A. verschaffeltii),
[Bo-6],
humahuacana (See A. shaferi, A. vestita -v- shaferi, Cylindropuntia humahuacana, Opuntia shaferi),
[Bo-6],
inarmata (Invalid name - See Cylindropuntia inarmata, Opuntia inarmata),
[DD], [Bo-6],
intermedia.....[Bo-1],
ipatiana (See A. salmiana, Opuntia salmiana),
[Bo-1],
jilesii (See A. miquelii -v- jilesii),
jujuyense (See A. shaferi, Opuntia shaferi),
[Aw 1-1965],
lagopus (See Opuntia lagopus, Tephrocactus floccosus -v- lagopus, Tephrocactus rauhii),
[Hi-4],
" -f. rauhii (See Opuntia rauhii, Tephrocactus rauhii),
lauliacoana (See Opuntia lauliacoana),
[Hi-4],
machacana.....[Hi-4],
malyana (See Opuntia malyana, Tephrocactus malyanus),
miquelii (See Cylindropuntia miquelii, Miqueliopuntia miquelii, Opuntia miquelii),
[Bo-1], [Hx*], [Hy],
" -v- geissei (See Opuntia miquelii -v- geissei)
" -v- heteromorpha (See Opuntia miquelii -v- heteromorpha),
" -v- jilesii (See A. miquelii, Miqueliopuntia miquelii, Opuntia miquelii),
[Bo-1],
molesta (See Cylindropuntia molesta, Opuntia molesta),
[Ay],
pachypus (See Cylindropuntia pachypus, Opuntia pachypus),
[Bo-1], [Hi-4],

AUSTROCYLINDROPUNTIA.....(Cont'd),

pachypus.....(cont'd),
" f. cristata....[Ay], [Bo-1],
salmiana (See Opuntia salmiana),
[C], [T], [QQ], [RR], [Bl], [Bo-1], [Hx*],
" -v- albiflora.....[Bo-1],
" -f. atrovariegata.....
" -v- ipatiana (See A. ipatiana),
" -v- spegazzinii (See Opuntia salmiana -v- spegazzinii, Opuntia spegazzinii),
[Bo-1],
shaferi (See A. vestita -v- shaferi, Opuntia shaferi),
[Af 59-1],
spegazzinii (See Opuntia spegazzinii),
steiniana (See Opuntia steiniana),
[Bo-1],
subulata (See Cylindropuntia subulata, Opuntia subulata),
[C], [K], [DD], [EE], [Ar], [Ay], [Bo-1], [Bt], [Hy], [Aj 31-3-1980], [Da 1-1981],
" -f. cristata (See Opuntia subulata -f.- cristata),
" -f. monstrosus.....[Bo-1], [Ij], [Iv 5-1956],
tephrocactoides.....[Bo-1],
teres (See Cylindropuntia teres, Opuntia teres),
[Bo-1], [Bo-6], [Hi-2],
verschaffeltii (See Cylindropuntia verschaffeltii, Opuntia verschaffeltii),
[Bo-1], [Hi-2], [Hx], [Hy], [Aj 33-6-1982]
" -v- digitalis (See A. verschaffeltii, Opuntia verschaffeltii -v- digitalis),
" -v- hypsophila (See Opuntia verschaffeltii -v- hypsophila),
[Bo-1],
" -v- longispina (See A. verschaffeltii, Opuntia verschaffeltii -v- longispina),
[Bo-1],
vestita (See Cylindropuntia vestita, Opuntia vestita),
[RR], [Bo-6], [Hx*], [Hy], [Flwr. Aj 29-6-1978]
" -v- chuquisacana (See A. chuquisacana),
[Bo-1], [Bo-6],
" -v- intermedia (Invalid name).....[Bo-6],
" -v- major.....[Bo-1], [Bo-6],
" -v- shaferi (See A. shaferi, Opuntia shaferi),
[Hi-2],
" -f. cristata (See Opuntia vestita -f. cristata)
[Bo-1],
weingartiana (See Opuntia weingartiana),
[Bo-1], [Hy],
Spec. nov. .....[Bi 1-1986],
Spec. .....[Aj 3-1980], [Ba 5-1976],

AYLOSTERA

albiareolata (See Rebutia albiareolata),
albiflora (See Rebutia albiflora),
[DD], [NN], [Ar], [Ay], [Hx*], [Hy],
[Af 53-4], [Ba 9-1977],
albipilosa (See Rebutia albipilosa),
[Av], [Hb],
albopectinata (See Rebutia albopectinata),
[Aj 30-1-1979],
buiningiana (Rausch) (See Rebutia buiningiana),
[DD-465],
cajasensis (See A. heliosa -v- cajasensis, Rebutia cajasensis),
camargoensis (See Rebutia camargoensis),
[Hx*], [Ba 3-1976],
carminea (See Rebutia minuscula -v- violaciflora),
christinae (See Rebutia christinae),
cintiensis (See Rebutia cintiensis),
cyanensis.....
deminuta (See Rebutia deminuta),
[Ar], [Bo-3], [Bu], [Bv-1], [Hx*], [Hy],
" -v- pseudominuscula (See A. pseudominuscula, Rebutia deminuta -v- pseudominuscula).....[EE], [Hy],
densipectinata (See Rebutia albopectinata, Rebutia densipectinata),
" -v- donaldiana (See Rebutia donaldiana),
diersiana (See Rebutia diersiana),
durispina (See Rebutia durispina),
[Ba 12-1977],
espinosae (See Rebutia espinosae, Rebutia narvaecensis, A. narvaecense),
fabrisii (See Rebutia fabrisii),
" -v- flaviflora.....
" -v- nana.....
fiebrigii (See Rebutia fiebrigii).....[EE], [Ar],
[Bo-3], [Bu], [Hx*], [Hy], [Aj 7-6-1956],
" -v- densiseta (See Rebutia fiebrigii -v- densiseta).....[EE], [NN],
" -v- spinosior (See Rebutia fiebrigii),
flavistylus (See Rebutia flavistyla),
froehlichiana (See Rebutia froehlichiana),
fulviseta (See Rebutia fulviseta).....[Hb], [Hx*],
fusca (See Rebutia fusca),
graciliflora (See Rebutia graciliflora, Rebutia xanthocarpa -v-graciliflora),
heliosa (See Rebutia heliosa),
[DD-466], [Ar], [Bt], [Hx], [Aj 30-1-1979],
[Ba 7-1978],
" -v- cajasensis (See Rebutia cajasensis, Rebutia heliosa -v- cajasensis),

AYLOSTERA.....(Cont'd),

heliosa (cont'd),
" -v- condorensis (See Rebutia heliosa -v- condorensis),
" -v- heliosa (See Rebutia heliosa -v- heliosa),
" -f. cristata.....[Bt],
hoffmannii (See Rebutia hoffmannii),
[Aj 28-5-1977],
horstii.....
huasiensis (See Rebutia huasiensis),
[Aj 28-2-1977],
ithyacantha (See Mediolobivia ithyacantha, Rebutia ithyacantha),
jujuyana (See Rebutia jujuyana).....[DD-467],
kieslingii (See Rebutia kieslingii),
[Hx*], [Aj 28-8-1977],
krugerii (See Rebutia kruegeri, Sulcorebutia krugerii),
[Au 13-257-1957],
kupperiana (See Rebutia kupperiana),
[T], [NN], [Ar], [Cr], [Hb], [Hx], [Hy], [Jd],
leucanthema (See Rebutia leucanthema),
[Flower - Bt],
" -v- cocciniflora.....[Ba 8-1977],
mamillosa (See Rebutia mamillosa),
maxima.....[Hb],
muscula (See Rebutia muscula),
[Ar], [Hb], [Hx], [Ae-1981],
narvaecense (See Rebutia narvaecensis),
[Hx*],
odontopetala (See Rebutia friedrichiana, Rebutia odontopetala),
padcayensis (See Rebutia padcayensis),
[Hx*],
patericalyx (See Rebutia patericalyx),
paucicostata (See Rebutia paucicostata),
pseudodeminuta (See Rebutia pseudodeminuta),
[NN], [QQ], [Av], [Ay], [Bo-3], [Bt], [Bu], [Hb], [Hx*], [Hy],
" -f. albiseta (See Rebutia pseudodeminuta -f.- albiseta),
[Bo-3],
" -f. grandiflora (See Rebutia paeudodeminuta -f. grandiflora),
[Bo-3], [Hl 2-1972],
" -f. rubrifilamentosa (See Rebutia pseudodemin- uta -f.- rubrifilamentosa),
" -f. schneideriana (See Rebutia pseudodeminuta -f.- schneideriana),
" -f. schumanniana (See Rebutia pseudodeminuta -f. schumanniana),

AYLOSTERA.....(Cont'd),

pseudominuscula (See Rebutia deminuta -f. pseudominuscula, Rebutia pseudominuscula), [Bo-3], [Dr], [Hx], [Hy],
pulchella (See Rebutia pulchella),
pulvinosa (See Rebutia pulvinosa), [EE], [Hx*],
robustispina (See Rebutia robustispina), [Ba 8-1977],
" -v- minor (See Rebutia robustispina -v- minor), [Ba 8-1977],
rubiginosa (See Rebutia rubiginosa), [Hx*],
sanguinea (See Rebutia sanguinea),
" -v- minor (See Rebutia sanguinea -v- minor),
schatzliana (See Rebutia schatzliana),
spegazziniana (See Rebutia spegazziniana), [T], [Ar], [Bo-3], [Hx], [Hy], [Ia],
" -v- atroviridis (See A. walteriana, Rebutia spegazziniana -v- atroviridis),
spinosissima (See Rebutia spinosissima), [Ar], [Av], [Bo-3], [Jd],
" -v- cardenas (See Rebutia spinosissima -v- cardenas),
" -v- torquata (See Rebutia spinosissima -v- torquata, Rebutia torquata),
steinmannii (See Mediolobivia steinmannii, Rebutia brachyantha, Rebutia rauschii, Rebutia steinmannii).....[Hy],
supthutiana (See Rebutia supthutiana), [Aj 27-6-1976],
tarijensis (See Rebutia tarijensis),
tarvitensis (tarvitaensis) (See Rebutia tarvitaensis),
tuberosa (See Rebutia tuberosa),
vallegrandensis (See Rebutia vallegrandensis),
walteriana (waltheriana) (See A. spegazziniana -v- atroviridis), [Hy],
zavaletae (See Sulcorebutia zavaletae), [Aj 17-9-1966],
zecheri (See Rebutia zecheri), [Ba 2-1977],
Spec. L-405.....[Aj 31-3-1980],

AZTEKIUM

ritteri.....[D], [I], [X], [Z], [NN], [PP], [Ar], [Av], [Ay], [Bt], [Dq], [Gr], [Hv], [Hx], [Hy], [Il], [Iz], [Jd], [Ae-1981], [Af 5-3], [Aj 30-2-1979], [Aj 33-6-1982], [Aw 3-1987], [Ba 8-1978], [Bm-1989],

AZUREOCEREUS

ayacuchensis (See A. hertlingianus, Bolivicereus ayacuchensis, Browningia hertlingianus)
cillisquamus (See A. viridis, Browningia cillisquama)
columnaris (See Browningia columnaris)
deflexispinus (See Trichocereus knuthianus)
[Bo-2],
hertlingianus (See A. nobilis, Browningia hertlingianus, Browningia nobilis)
[EE], [NN], [QQ], [Av], [Ay], [Bt], [Bo-2], [Hu], [Hy], [Jd], [Af 51-5], [Db 4-1979],
" -v- nobilis (See A. nobilis),
[Iv 12-1956],
" -f. monstrosus.....
imperator (See A. hertlingianus),
nobilis (See A. hertlingianus, Browningia hertlingianus Browningia nobilis),
[QQ], [Ax], [Ay], [Af 21-4], [Hl 2-1971],
viridis (See Browningia viridis),
[DD], [Bo-2], [Bi Vol. 4, #6],

BACKEBERGIA

chrysomalla (See Backebergia militaris, Cephalocereus chrysomallus, Pilosocereus chrysomallus)
militaris (See Cephalocereus chrysomallus, Cephalocereus militaris, Mitrocereus militaris, Pilosocereus chrysomallus)
[DD], [HH], [Ar], [Ay], [Az], [Bo-4], [Bt], [Hv], [Hx*], [Flower - Jd], [Ad 4-1980], [Hh Nov., 1974], [Hl 2-1977], [Hs IV-1-1983],

BARTSCHELLA

insularis (See Mammillaria insularis),
[VV], [Aq], [Ar], [Af 51-6],
schumannii (See Mammillaria schumannii)...[L], [DD], [RR], [VV], [Ar], [Ay], [Bo-5], [Bt], [Bv-1], [Hu], [Hx], [Jd], [Af 26-3], [Af 50-4],
" -v- globosa.....

BERGEROCACTUS

emoryi (See Cereus emoryi),
[F], [M*], [DD], [FF], [GG], [OO], [Ar], [Ax], [Ay], [Bo-4], [Cb], [Do], [Dp], [Dq], [Hy], [Iz], [Jd], [Af 2-12], [Af 3-10], [Af 14-7]

BINGHAMIA

acanthura (See Haageocereus acanthurus, Loxanthocereus acanthurus),
acrantha (See Cereus acranthus, Haageocereus limensis), [RR],
aurivillus (See Borzicactus aurivillus, Borzicactus icosagonus, Cereus aurivillus, Cereus icosagonus, Seticereus aurivillus, Seticereus icosagonus).....[Af 4-12],
australis (See Haageocereus australis).....[Dq],
chosicensis (See Haagecereus chosicensis Haageocereus multangularis),
climaxantha (See Neobinghamia climaxantha, Peruvocereus albicephalus),
decumbens (See Borzicactus decumbens, Haageocereus decumbens, Haageocereus mamillatus, Haageocereus mamillatus -v- brevior, Haageocereus oconia-camanensis),
humboldtii (See Borzicactus humboldtii, Seticereus humboldtii, Seticereus neohumboldtii),
humifusa (See Haageocereus humifusus, Haageocereus versicolor -v- humifusus),
icosagona (See B. aurivillus, Borzicactus aurivillus, Borzicactus icosagonus, Cereus aurivillus, Seticereus aurivillus, Seticereus icosagonus)
laredensis (See Haageocereus laredensis, Haageocereus pacalaensis -v- laredensis),
melanostele (See B. chosicensis, B. multangularis, Haageocereus chosicensis, Haageocereus multangularis),
multangularis (See Haageocereus chosicensis, Haageocereus multangularis).....[RR]
olowinskianus (See Haageocereus clavispinus, Haageocereus lachayensis, Haageocereus limensis, Haageocereus olowinskianus),
pacalaensis (See Haageocereus pacalaensis)
plagiostoma (See Borzicactus plagiostoma, Cereus plagiostoma),
platinospina (See Haageocereus australis -v- albiflorus, Haageocereus platinospinus),
pseudomelanostele (See Cereus pseudomelanostele, Haageocereus multangularis -v- pseudomelanostele, Haageocereus pseudomelanostele, Haageocereus setosus),.....[Af 4-12],
versicolor (See Haageocereus versicolor),

BLOSSFELDIA

atroviridis.....[Bt], [Hi-2], [Il], [Au36-4]
" -v- intermedia.....

BLOSSFELDIA.....(Cont'd),

campaniflora (See B. liliputana -v- campaniflora),
[DD], [Bo-3],
cyathiformis (See B. campaniflora),
fechseri (Invalid name) (See B. liliputana -v- fechseri).....[G], [PP], [Bo-6],
flocculosa.....(See B. liliputana -v- formosa),
liliputana.....[E], [I], [DD], [NN], [QQ], [Ar], [Av], [Ay], [Bo-3], [Cz], [Hi-2], [Hv], [Hx], [Hy], [Iz], [Jd],
" -v- caineana.....
" -v- campaniflora (See B. campaniflora),
" -v- fechseri (See B. fechseri),
" -v- formosa.....[Hi-2],
minima.....[NN], [Hi-2], [Au 36-4], [Hl 4-1972],
pedicellata.....[Hi-2], [Hx*], [Aj 17-1-1966], [Aj 22-6-1971], [Au 36-4],
subterranea.....

BOLIVICEREUS

ayacuchensis (See Browningia hertlingiana),
brevicaulis.....[Hi-2],
croceus.....[Hi-2],
margaritanus (See Borzicactus margaritanus),
piscensis (See Corryocactus aureispinus, Corryocactus erectus, Erdisia aureispinus, Erdisia erecta),
rufus (See Borzicactus rufus),
[Hi-2],
samaipatanus (See Borzicactus samaipatanus, Cleistocactus samaipatanus),
[I], [DD], [NN], [Ar], [Ay], [Bo-2], [Bt], [Hi-2], [Hu], [Hx*], [Hy], [Jd], [Af 23-3], [Af 50-2], [Ba 1-1976],
" -v- divi-miseratus.....[Bo-2], [Aj 13-8-1962],
" -v- multiflorus.....[Af 23-3], [Aj 27-5-1976], [Flower - Ba 12-1978],
serpens (See Borzicactus serpens, Cleistocactus serpens).....[DD], [NN],
soukupi (See Borzicactus soukupi),
tenuiserpens (See Borzicactella tenuiserpens, Cleistocactus tenuiserpens),
[Bo-6],

BONIFAZIA

quelzalteca (See Disocactus quezaltecus).....[Jd],

BORZICACTELLA

serpens (See Bolivicereus serpens, Borzicactus serpens, Cleistocactus serpens),

BORZICACTELLA.....(Cont'd),

tenuiserpens (See Bolivicereus tenuiserpens, Borzicactus tenuiserpens, Cleistocactus tenuiserpens),
[Hi-4],

BORZICACTUS

acanthurus (See Borzicactus eriotrichus, Loxanthocereus acanthurus, Loxanthocereus gracilispinus),
[RR], [Bo-2], [Dq], [Jd], [Cu 3-2-1934],
aequatorialis (See B. sepium),
[DD], [Bo-2],
aurantiacus (See Arequipa aurantiaca, Borzicactus calvescens, Matucana aurantiaca, Matucana calvescens, Submatucana aurantiacus, Submatucana calvescens).....[F], [Hy], [Af 29-2],
" -v- calvescens (See B. aurantiacus, Matucana calvescens),
" -v- megalanthus (See Matucana megalantha),
aureiflorus (See Matucana aureiflora, Submatucana aureiflorus),
aureispinus (See Cleistocactus aureispina, Hildewinteria aureispina, Loxanthocereus aureispinus, Winterocereus aureispina),
[G], [DD], [EE], [HH], [UU], [Iz], [Jd],
[Af 52-3], [Af 58-2], [Ag-1984], [Am 7-1983],
[Dg 7-1],
" -f. cristata (See Hildewinteria aureispina -f.-cristata).....[Af 54-1]
aurivillus (See Binghamia aurivillus, Borzicactus icos-agonus, Cereus aurivillus, Seticereus aurivillus, Seticereus icosagonus),
[L], [Hi-4], [Hv], [Hx*], [Af 17-7],
australis (See Arequipa australis),
bicolor (See Haageocereus bicolor, Loxanthocereus bicolor),
blancii (See Borzicactus haynei -v- perplexus),
" -v- nigriarmata (See B. haynei -v- perplexus),
brevicaulis (See Bolivicereus brevicaulis),
brevispinus (See Loxanthocereus brevispinus, Loxanthocereus peculiaris),
cajamarcensis.....[Hi-4], [Hy],
callianthus (See Matucana calliantha),
" -v- prolifera (See Matucana calliantha -v-prolifera),
calocephalus (See Matucana calocephala, Matucana comacephala),
calvescens (See B. aurantiacus -v- calvescens, Matucana calvescens, Submatucana calvescens),
[Af 29-4],

BORZICACTUS.....(Cont'd),

calviflorus (See Clistanthocereus calviflorus)..[Hi-4],
camanaensis (See B. gracilis -v- camanaensis, Loxanthocereus camanaensis),
canetensis (See Loxanthocereus canetensis),
celendinensis (See Borzicactus intertextus -v- celendinensis, Matucana celendinensis, Submatucana celendinensis),
celsianus (See Oreocereus celsianus),
cereoides (See B. haynei),
chlorocarpus (See Browningia chlorocarpus, Gymnanthocereus chlorocarpus, Seticereus chlorocarpus),
comacephalus (See B. haynei -v- perplexus, Matucana comacephala),
convergens (See Haageocereus convergens, Loxanthocereus convergens),
crassiserpens (B. serpens X B. icosagonus) (See Cleistocactus crassiserpens, Loxanthocereus crassiserpens).....[QQ], [Ay],
criniferus (See Matucana crinifera),
croceus (See Bolivicereus croceus),
cullmannianus (See Loxanthocereus cullmannianus), [NN], [Hv],
currundayensis (See B. aurantiacus -v- calvescens, Matucana currundayensis, Submatucana currundayensis),
cutakii.....[DD],
decumbens (See Haageocereus decumbens, Haageocereus mamillatus, Haageocereus mamillatus -v- brevior, Haageocereus oconia-camanensis), [RR], [Hy], [Cu 8-1-1939]
deserticola (See Haageocereus deserticola, Loxanthocereus deserticola),
doelzianus (See Morawetzia doelzianus),
elongatus (See B. haynei -v- atrispinus, B. hystrix --v- atrispinus, Matucana elongata, Matucana haynei -v- elongata, Matucana multicolor),
erectispinus (See Loxanthocereus erectispinus),
eremiticus (See Loxanthocereus eremiticus),
erigens (See Loxanthocereus erigens),
eriotrichus (See Borzicactus acanthurus, Cereus eriotrichus, Loxanthocereus acanthurus, Loxanthocereus eriotrichus).....[Hy],
eulalianus (See Loxanthocereus eulalianus),
faustianus (See Haageocereus faustianus, Haageocereus paradoxus, Loxanthocereus faustianus, Loxanthocereus keller-badensis),
ferrugineus (See Loxanthocereus ferrugineus, Loxanthocereus variabilis).....[Il],
fieldianus (See Clistanthocereus fieldianus), [RR], [Ay], [Hi-4], [Ij],

fieldianus.....(cont'd),
" -v- tesselatus (See B. tesselatus).....[Hi-4],
formosus (See Matucana formosa).....[Af 51-3],
" -v- minor (See Matucana formosa -v- minor), [Af 51-3],
" -v- paucicostatus (See B. paucicostatus, Matucana paucicostata),
" -v- tuberculatus (See B. tuberculatus, Matucana tuberculata),
fossulatus (See Oreocereus celsianus -v- fossulatus, Oreocereus fossulatus),
fruticosus (See Matucana fruticosa),
gracilis (See Maritimocereus gracilis, Loxanthocereus aticensis, Loxanthocereus gracilis, Loxanthocereus nanus, Loxanthocereus splendens), [NN], [Aj 10-9-1959],
" -v- aticensis (See B. gracilis, Loxanthocereus gracilis, Loxanthocereus aticensis),
" -v- camanaensis (See Loxanthocereus camanaensis),
hastiferus (See Matucana hastifera),
haynei (See Matucana cereoides, Matucana haynei)..[Iz],
" -v- atrispinus (See Matucana haynei -v- atrispina, Matucana haynei -v- hystrix, Matucana hystrix),
" -v- breviflorus (See Matucana breviflora),
" -v- elongata (See Matucana elongata),
" -v- erectipetalus (See B. haynei, Matucana haynei -v- erectipetala),
" -v- perplexus (See Matucana herzogiana -v- perplexa, Matucana megalantha, Matucana yanganucensis -v- suberecta),
hempelianus (See Arequipa hempeliana, Arequipa weingartiana, Echinopsis hempeliana),
" -v- rettigii (See Arequipa erectocylindrica, Arequipa rettigii),
" -v- spinosissimus (See Arequipa spinosissima),
" -v- weingartiana (See Arequipa hempeliana, Arequipa weingartiana, Arequipa weingartiana -v- carminanthema),
hendricksenianus (See Oreocereus celsianus -v- henricksenianus, Oreocereus leucotrichus, Oreocereus hendricksenianus),
" -v- densilanatus (See Oreocereus hendricksenianus -v- densilanatus), [Af 56-5],
hoffmannii (See Loxanthocereus hoffmannii),
huagalensis (See Submatucana huagalensis), [Hh Jan. 1974],

BORZICACTUS.....(Cont'd),

humboldtii (See Seticereus humboldtii, Seticereus neohumboldtii),
[PP], [Bv-1], [Cz], [Gr], [Il],
hystrix (See Matucana hystrix, Matucana multicolor -v- hystrix),
" -v- brunnescens (See Loxanthocereus hystrix -v- brunnescens),
icosagonus (See Borzicactus aurivillus, Cereus aurivillus, Cereus icosagonus, Seticereus aurivillus, Seticereus icosagonus),
[I], [X], [Ay], [Bv-1], [Bt], [Il], [Iz], [Jd], [Af 4-12],
" -v- oehmeanus.....[Bt],
intertextus (See Matucana intertextus, Submatucana intertexta),
" -v- celendinensis (See Matucana celendinensis), [Hh July 1974],
jajoianus (See Loxanthocereus jajoianus),
keller-badensis (See Loxanthocereus faustianus, Loxanthocereus keller-badensis),
knizei.....
krahnii (See Matucana krahnii, Submatucana krahnii), [Af 51-2],
leucotrichus (See Arequipa leucotrichus, Arequipa neoleucotricha),
" -v- hempelianus (See Arequipa hempelianus, B. hempelianus),
" -v- rettigii (See Arequipa rettigii),
" -v- spinosissima (See Arequipa spinosissima),
" -v- weingartiana (See Arequipa weingartiana),
luteispina (See Cleistocactus luteispinus),
madisoniorum (See Eomatucana madisoniorum, Matucana madisoniorum, Submatucana madisoniorum),
[Z], [Gr], [Iz], [Af 35-6], [Af 50-1], [Af 51-3], [Af 56-5],
" -v- madisoniorum (See Matucana madisoniorum -v- madisoniorum, Submatucana madisoniorum -v- madisoniorum),
[Aj 4-1977],
" -v- pujupatii (See Submatucana paucicostata -v- pujupatii),
[Af 50-1], [Hh May, 1976],
margaritanus (See Bolivicereus margaritanus),
megalanthus (See Matucana herzogiana -v- perplexa, Matucana megalantha, Matucana yanganucensis),
mirabilis (See Arequipa mirabilis, Matucana mirabilis),
montanus (See Loxanthocereus montanus),

BORZICACTUS.....(Cont'd),

morleyanus (See B. sepium -v- morleyanus),
[RR], [Bo-2], [Jd], [Af 21-2], [Hl 2-1969],
multicolor (See Matucana elongata, Matucana multicolor)
multifloccosus (See Loxanthocereu multifloccosus, Loxanthocereus neglectus),
myriacanthus (See Matucana myriacantha),
[Hh Nov. 1974],
neglectus (See Loxanthocereus gracilispinus, Loxanthocereus neglectus),
" -v- chimbotensis (See Haageocereus pacaranensis -v- tenuispinus),
" -v- neglectus.....
neoroezlii (See Borzicactus roezlii, B. sepium, Cleistocactus roezlii, Seticereus roezlii),
[Hi-4],
oreodoxus (See Eomatucana oreodoxa, Matucana oreodoxa),
[Hl 5-1975],
otuscensis (See Loxanthocereus otuscensis),
pacaranensis (See Haageocereus pacaranensis, Loxanthocereus pacaranensis),
pachycladus (See Loxanthocereus pachycladus),
pallarensis (See B. weberbaueri -v- pallarensis, Matucana pallarensis),
parvitesselatus (See Loxanthocereus parvitesselatus),
paucicostatus (See Submatucana paucicostata),
[Iz], [Af 51-3],
" -f. robustispinus.....
peculiaris (See Loxanthocereus cantaensis, Loxanthocereus peculiaris),
pisacensis (See Bolivicereus pisacensis, Corryocactus erectus, Erdisia aureispina, Erdisia erecta)
piscoensis (See Loxanthocereus piscoensis, Oreocereus piscoensis),
plagiostomus (See Cereus plagiostoma),
[Hi-4],
platinospinus (See Haageocereus australis -v- albiflorus, Haageocereus platinospinus),
pseudothelogonus (invalid name).....[DD],
puquiensis (See Loxanthocereus puquiensis),
purpureoalbus (See Matucana purpureo-alba),
purpureus (See Borzicactus plagiostoma),
[Hi-4],
ritteri (See Matucana ritteri, Submatucana ritteri),
roezlii (See B. neoroezlii, B. sepium -v- roezlii, Seticereus roezlii),
[QQ], [Ay], [Hi-4], [Hv], [Jd],
roseiflorus (See Akersia roseiflorus),
[NN], [Il], [Bm-1986], [Dg 7-1],
rufus (See Bolivicereus rufus),

samaipatanus (See Bolivicereus samaipatanus,, Cleistocactus samaipatanus),
[I], [EE], [NN], [UU], [Ay], [Gr], [Hv],
[Il], [Jd], [Af 50-2],
" -v- multiflorus.....[F], [Aj 32-2-1981],
" -f. cristata.....
samnensis (See Clistanthocereus samnensis),
[Gr], [Hv], [Hi-4],
" -v- fuscatus (See B. variabilis),
sepium (See B. neoroezlii, B. ventimigliae),
[D], [H], [RR], [Bo-2], [Gr], [Hx*], [Jd],
[Af 32-5], [Hl 2-1969],
" -v- giganteus.....[Hl 2-1969],
" -v- morleyanus (See B. morleyanus).....[Jd],
" -v- roezlii (See B. roezlii, Seticereus roezlii),
sericatus (See Morawetzia sericata, Oreocereus sericatus),
serpens (See Bolivicereus serpens, Cleistocactus serpens),
[Af 32-3],
sextonianus (See Maritimocereus sextonianus, Loxanthocereus sextonianus),
[Bv-1], [Hv], [Hh March 1975],
soukupi (See Bolivicereus soukupi),
strausii (See Cereus strausii, Cleistocactus strausii)
[Dl], [Af 1-9], [Af 2-4],
sulcifer (See Loxanthocereus sulcifer),
[Af 32-3],
supertextus (See B. variabilis, Matucana supertexta, Matucana variabilis),
swartzii (See Pilosocereus swartzii, Cephalocereus swartzii),
" -f. cristata (See Cephalocereus swartzii -f. cristata),
[Af 2-2],
tacnaensis (See Oreocereus tacnaensis, Oreocereus variicolor -v- tacnaensis),
tenuiserpens (See Bolivicereus tenuiserpens, Borzicactella tenuiserpens, Cleistocactus tenuiserpens),
tesselatus (See B. fieldianus -v- tesselatus, Clistanthocereus tesselatus),
[Bo-2],
tominensis (See Cleistocactus tominensis),
trollii (See Oreocereus trollii),
trujilloensis (See Loxanthocereus trujilloensis),
tuberculatus (See Matucana tuberculata),
[Af 51-2],
tuberculosus (See Matucana tuberculosa),

BORZICACTUS.....(Cont'd),

variabilis (See Matucana supertexta, Matucana variabilis),
" -v- fuscata.....
ventimigliae (See B. sepium),
[Bo-2],
weberbaueri (See Matucana weberbaueri),
[Hh Nov. 1974],
" -v- flammeus (See Matucana weberbaueri -v- flammea),
[Hh Nov. 1974], [Hl 3-1978],
" -v- myriacanthus (See B. myriacanthus, Matucacana myriacantha),
" -v- pallarensis (See Matucana pallarensis),
[Hh Nov. 1974],
websterianus (See B. sepium),
[F], [Bo-2], [Cu 1-1960], [Fo 11-gen. 46, Sp. 2- Fig. 1937],
" -v- rufispinus (See B. sepium),
[Bo-2],
winteri (See Matucana yanganucensis),
xylorhizus (See Loxanthocereus xylorhizus),
yauyosensis (See Loxanthocereus yauyosensis),
Spec. Fr-173.....[NN],
Spec. .....[DD], [NN],

BRACHYCALYCIUM

tilcarense (See Gymnocalycium tilcarense),
[DD], [Ar], [Ay], [Bo-3], [Il], [Bi 6-3],
" -v- gigantea.....

BRACHYCEREUS

nesioticus (See Cereus nesioticus),
[I], [DD], [Ay], [Bo-2], [Ch], [Hx*], [Jd], [Aj 26-8-1975],
thouarsii (See Cereus thouarsii, Jasminocereus thouarsii),

BRASILICACTUS

elachisanthus (See Notocactus elachisanthus),
[Hx*]
graessneri (See Eriocactus graessneri, Notocactus graessneri),
[T], [NN], [QQ], [Ak], [Ar], [Hy], [Jd],
" -v- albisetus.....[NN], [Bo-3], [Hx],
" -v- albispinus.....
" -v- aureiflorus.....
" -v- flaviflorus.....

BRASILICACTUS.....(Cont'd),

graessneri.....(cont'd),
" -v- stellatus (See B. haselbergii -v- stellatus),
" -f. cristata.....
haselbergii (See Eriocactus haselbergii, Notocactus haselbergii).....[I], [T], [DD], [NN], [QQ], [Ar], [Av], [Ay], [Bo-3], [Bt], [Hi-1], [Hu], [Hx], [Hy], [Jd], [Hl 4-1972],
" -v- stellatus (See Notocactus haselbergii -v- stellatus),
" -v- .....[Hy],
" -f. monstrosus.....[Hy],

BRASILICEREUS

breviflorus (See B. phaeacanthus -v- breviflorus, Cereus phaeacanthus -v- breviflorus...[Hi-1],
markgrafii (See Cereus markgrafii).........[DD], [Ay], [Bo-4], [Ba 7-1974], [Jd],
parvisetus.....
phaeacanthus (See Cephalocereus phaeacanthus, Brasilicereus phaeacantha, Monvillea phaeacantha), [Hi-1], [Hx*], [Hy],
" -v- breviflorus (See B. breviflorus, Cereus phaeacanthus -v- breviflorus),

BRASILIOPUNTIA

bahiensis (See Opuntia bahiensis), [Bo-1], [Ba 4-1981],
brasiliensis (See Opuntia brasiliensis), [DD], [Bo-1], [Hx*], [Iu], [Da 1-1981],
" -v- minor.....
neoargentina.....[Bo-1],
schulzii (See Opuntia schulzii), [Bo-1],

BRASILIPARODIA

alacriportana (See Notocactus alacriportana, Parodia alacriportana),
brevihamata (See Parodia brevihamata), [Bo-3], [Hi-1],
" -f. conjungens (See Notocactus brevihamatus f.-conjungens),
" -v- molicula.....
" -v- molispinus (See Notocactus brevihamatus -v- molispinus),
buenekeri (See Parodia buenekeri, Notocactus buenekeri, Notocactus linkii -v- buenekeri),

BRASILIPARODIA.....(Cont'd),

buenekeri.....(cont'd),
" -v- conjungens (See B. brevihamata -v- conjungens),
[Hi-1],
" -v- intermedia (See Notocactus buenekeri -v- intermedia),
catarinensis (See Notocactus catarinensis),
[Hi-1],
chrysocoma (See Notocactus chrysocoma, Parodia chrysocoma),
[Hl 3-1982],
rechensis (See Notocactus rechensis),
[Hi-1],

BRITTONIA

davisii (See Hamatocactus hamatacanthus -v- davisii),
[Hy], [Af 2-8], [Cu 2-4-1934],

BROWNINGIA

albiceps.....
altissimus (See Gymnanthocereus altissimus, Gymnocereus altissimus),
[Gr], [Hv], [Fs 31-32, Gen. C, iv/1 Nov. 1965]
amstutziae (See Gymnanthocereus amstutziae, Gymnocereus amstutziae),
[Ay], [Jd],
caineana (See Castellanosia caineana),
candelaris.....[I], [DD], [QQ], [RR], [Ay], [Bl],
[Bo-2], [Hi-3], [Hi-4], [Jd], [Af 56-6],
[Dc 1-1979], [Hl 5-1982],
" -v- chilensis.....
" -v- nazca.....
chlorocarpus (See Gymnanthocereus chlorocarpus),
cillisquama (See Azureocereus cillisquama),
columnaris (See Azureocereus columnaris),
hertlingianus (See Azureocereus hertlingianus, Azureocereus nobilis,, B. nobilis),
[Av], [Ay], [Bt], [Gr], [Hi-4], [Hv], [Il],
[Iz], [Jd],
icaensis.....[Hi-4],
macracantha (See Gymnanthocereus macranthus),
microsperma (See Gymnocereus microspermus, Seticereus microspermus),
[Ay], [Jd],
nobilis (See Azureocereus hertlingianus, Azureocereus nobilis, B. hertlingianus),
[Ay],

BROWNINGIA.....(Cont'd),

pilleifer (pilleifera) (See Gymnanthocereus pilleiferus).....[Ay],
riosaniensis (See Rauhocereus riosaniensis), [Gr],
" -v- jaenensis (See Rauhocereus riosaniensis -v- jaenensis),
viridis (See Azureocereus viridis),

BUININGIA

aurea (See Coleocephalocereus aureus), [DD-468], [Bt], [Jd], [Ab-1985], [Au 26 -4-1971], [Ba 11-1976],
brevicylindrica (See Coleocephalocereus brevicylindricus), [G], [DD-469], [Av], [Ay], [Hv], [Hx*], [Il], [Jd], [Ad Nov., 1978], [Aj 27-4-1976], [Au 26-4-1971], [Au 37-1], [Hl 1-1980],
" -v- elongata (See Coleocephalocereus brevicylindricus -v- elongata), [Z], [Aj 24-4-1973], [Ba 1-1981],
" -v- longispina (See Coleocephalocereus brevicylindricus -v- longispina),
purpurea (See Buiningia roseiflora), [G], [DD-470], [Ay], [Hx*], [Jd], [Ad Nov., 1978], [Aj 24-6-1973], [Aj 30-11-1979],
roseiflora (See B. purpurea).....[Bt],

CALYMMANTHIUM

fertile.....[Hi - 4], [Ba Aug. 1966: 117],
substerile.....[DD], [NN], [Ay], [Bo-6], [Bu], [Hi-4], [Hx*], [Il], [Jd], [Hh 2-4-1976],

CARDENASIA

herzogiana (See Neocardenasia herzogiana, Neoraimondia herzogiana).....[Hl 2-1971],

CARNEGIEA

gigantea (See Cereus giganteus), [D], [I], [DD], [GG], [LL], [RR], [TT], [UU], [An], [Ar], [Ay], [Az], [By], [Co], [Cx], [Dq], [Gr], [Hy], [Il], [Iz], [Jd], [Flower - Aw 4-1983], [Bi 3-3], [Bi 6-2-1988],
" -f. cristata.....[GG], [Dp], [Gr], [Hx], [Hy], [Iz], [Af 26-2], [Bi 3-3], [Bi 6-3],
polylopha (See Neobuxbaumia polylopha)

CASTELLANOSIA

caineana (See Browningia caineana),
[DD], [Ay], [Bo-4], [Gr], [Hi-2], [Hx*], [Hy], [Ik], [Jd], [Af 23-3], [Aj 10-1-1959], [Aj 24-4-1973].
Spec. .....[Ik],

CEPHALOCEREUS

albicephalus (See Austrocephalocereus albicephalus, Coleocephalocereus goebelianus, Micranthocereus albicephalus),
albiflorus (See Arroyadoa albiflora),
albispinus (See Pilosocereus albispinus),
[L],
alensis (See Pilocereus alensis, Pilosocereus alensis)
[Az], [Af 34-4], [Cu 7-3-1939],
apicicephalium (See Neodawsonia apicicephalium),
[Az], [Jd], [Af 21-4], [Aj 15-3-1964],
arenicola (See Pilosocereus arenicola),
arrabidae (See Pilocereus arrabidae, Pilosocereus arrabidae, Pilosocereus exerens),
[C], [RR], [Af 2-10],
atroflavispinus (See Pilosocereus atroflavispinus),
atroviridis (See Subpilocereus atroviridis),
aureispinus (See Coleocephalocereus aureispinus),
aureus (See Buiningia aurea, Coleocephalocereus aureus)
auri-azureus (See Micranthocereus auri-azureus),
aurilanatus (See Pilosocereus aurilanatus, Pseudopilocereus aurilanatus),
azureus (See Cereus azureus, C. cyaneus, Pseudopilocereus azureus),
backebergii (See Cereus backebergii, Pilocereus backebergii, Pilosocereus backebergii),
bahamensis (See Pilosocereus bahamensis),
[RR],
barbadensis (See Pilosocereus barbadensis),
[RR],
blossfeldiorum (See Thrixanthocereus blossfeldiorum),
[Ea 1937:4 Fig.],
bradyi (See Pilosocereus bradyi, Pseudopilocereus bradyi),
bradypus (See Cephalocereus senilis),
[Hy],
" -f. cristata.....[Hy],
brasiliensis (See Pilosocereus brasiliensis, Pilosocereus sublanatus),
[RR],
braunii (See Coleocephalocereus braunii),
brevicylindricus (See Buiningia brevicylindrica, Coleocephalocereus brevicylindricus),

CEPHALOCEREUS.....(Cont'd),

brevicylindricus.....(cont'd),
" -v- elongatus (See Buiningia brevicylindrica -v- elongata, Coleocephalocereus brevicylindrica -v- elongatus),
" -v- longispinus (See Buiningia brevicylindrica -v- longispinus -v- longispinus),
brooksianus (See Pilocereus brooksianus, Pilosocereus brooksianus),
[RR]
" -f. cristata.....[Af 10-1],
canudosensis (See Arrojadoa canudosensis),
catalinii (catalanii) (See Pilosocereus catalanii),
catingicola (See Pilocereus catingicola),
[RR],
chrysacanthus (See Pilosocereus chrysacanthus),
[H], [L], [S], [RR], [Az], [Bv-1], [Cx], [Af 21-3], [Aw 2-1964],
" -f. cristata.....[Bv-1], [Ij],
chrysomallus (See Backebergia militaris, Mitrocereus militaris, Pilosocereus chrysomallus),
chrysostele (See Pilocereus chrysostele, Pilosocereus chrysostele, Pseudopilocereus chrysostele),
claroviridis (See Pilosocereus claroviridis),
coerulescens (See Pilosocereus aurisetus, Pilosocereus coerulescens, Pilosocereus glaucescens, Piptanthocereus aethiops),
collinsii (See Pilocereus collinsii, Pilosocereus collinsii),
[RR], [Az], [Bv-1], [Aw 2-1983],
columbianus (See Pilocereus columbianus, Pilosocereus columbianus),
columna-trajani (Karw.) (See Cephalocereus hoppenstedtii, Haseltonia columna-trajani (Karw.)),
cometes (See Pilosocereus cometes),
[CC], [RR], [Az], [Bv-1], [Cx],
cuyabensis (See Pilosocereus cuyabensis),
cyaneus (See Cereus azureus, Pseudopilocereus azureus)
decumbens (See Coleocephalocereus decumbens),
deeringii (See Cereus robinii -v- deeringii, Pilosocereus deeringii),
[CC], [RR], [Bj]
densiareolatus (See Pilosocereus densiareolatus),
diersianus (See Pseudopilocereus diersianus),
dinae (See Arrojadoa dinae),
dolichospermaticus (See Austrocephalocereus dolichospermaticus),

dybowskii (Austrocephalocereus dybowskii, Espostoopsis dybowski),
[RR], [Bv-1], [Hx*], [Af 56-6], [Cu 3-1967],
eriocaulis (See Arrojadoa eriocaulis),
" -v- albicoronata (See Arrojadoa eriocaulis -v- albicoronata),
estevesii (See Austrocephalocereus estevesii, Coleocephalocereus estevesii),
euphorbioides (See Neobuxbaumia euphorbioides, Pilosocereus euphorbioides, Rooksbya euphorbioides),
[L], [RR], [Bv-1], [Af 3-4], [Af 12-7],
exerens (See Cephalocereus arrabidae, Pilocereus arrabidae, Pilosocereus arrabidae),
flaviflorus (See Micranthocereus flaviflorus),
flavipulvinatus (See Pilosocereus flavipulvinatus, Pseudopilocereus flavipulvinatus),
" -v- carolinensis (See Pilosocereus flavipulvinatus -v- carolinensis),
flavisetus (See Coleocephalocereus flavisetus),
floccosus (See Pilosocereus floccosus, Pseudocephalocereus floccosus),
fluminensis (See Coleocephalocereus fluminensis),
[RR], [Bv-1], [Bu], [Dq], [It], [Af 5-11], [Af 8-10], [Af 22-2],
" -v- braamhaarii (See Coleocephalocereus fluminensis -v- braamhaarii),
fulviceps (See Mitrocereus fulviceps),
fulvilanatus (See Pilosocereus fulvilanatus, Pseudopilocereus fulvilanatus),
fulvispinosus (See Cephalocereus royenii),
gaturianensis (See Pilosocereus gaturianensis),
gaumeri (See Pilosocereus gaumeri),
[Az],
gironensis (See Pilosocereus gironensis),
glaucescens (See Pilocereus glaucescens, Pilosocereus glaucescens, Pilosocereus coerulescens),
glaucochrous (See Pilocereus glaucochrous, Pilosocereus glaucochrous, Pseudopilocereus glaucochrous),
[Bv-1], [Af 39-1],
goebelianus (See Austrocephalocereus purpureus (Guerke) sensu Backeb., Coleocephalocereus goebelianus, Cephalocereus purpureus (Br. & R.))
gounellei (See Pilocereus gounellei, Pilosocereus gounellei),
[RR], [Cu 3-1967],
" -v- zehntneri (See Pilosocereus gounellei -v- zehntneri),

CEPHALOCEREUS.....(Cont'd),

grenadensis (See Cereus grenadensis, Subpilocereus grenadensis),
gruberi (See Pilosocereus gruberi),
guentheri (See Espostoa guentheri, Vatricania guentheri),
guerreronis (See Pilosocereus guerreronis), [Az],
haematocarpus (See Micranthocereus haematocarpus),
hapalacanthus (See Pilocereus hapalacanthus, Pilosocereus hapalacanthus, Pseudopilocereus hapalacanthus),
hermentianus (See Pilosocereus hermentianus),
hoppenstedtii (See Haseltonia columna-trajani (Karw.), [E], [JJ], [Av], [Ay], [Az], [Bt], [Do], [Dq], [Hx], [Iu], [Af 25-3], [Af 41-5], [Af 56-5], [Af 57-6], [Aw 1-1964], [Hl 6-1982],
horrispinus (See Cereus horrispinus, Pilosocereus horrispinus, Subpilocereus horrispinus),
horstianus (See Arrojadoa horstiana),
houlletii (See Pilocereus houlletii, Pilocereus leucocephalus, Pilosocereus houlletii, Pilosocereus leucocephalus),
johansenii.....[Hr],
keyensis (See Cephalocereus robinii -v- keyensis, Cereus robinii -v- keyensis), [RR], [Af 6-12],
lanuginosus (See Pilosocereus lanuginosus), [RR], [Af 57-3],
lehmannianus (See Austrocephalocereus lehmannianus, Austrocephalocereus purpureus (Guerke) Bux., Cephalocereus purpureus (Guerke) Bux., Micranthocereus purpureus (Guerke) Bux.), [Bt], [It], [Ge 118 - Fig. 1933],
leucocephalus (See Pilocereus leucocephalus, Pilocereus houlletii, Pilosocereus houlletii, Pilosocereus leucocephalus), [NN], [RR], [Cx], [Dl*], [Af-15-4], [Aj 22-2-1971], [Dg 2-4, 1980-1981]
leucostele (See Stephanocereus leucostele), [RR], [Bu], [Bv-1], [Iz],
luetzelburgii (See Coleocephalocereus luetzelburgii, Pilocereus luetzelburgii, Pilosocereus luetzelburgii).....[Hl 2-1970],
machrisii (See Pseudopilocereus machrisii), [Ba 4-1977],
macrocephalus (See Neobuxbaumia macrocephala, Mitrocereus ruficeps), [RR], [Cx], [Af 25-3], [Af 33-6],

CEPHALOCEREUS.....(Cont'd),

magnificus (See Pilosocereus magnificus, Pseudopilocereus magnificus),
maxonii (See Pilosocereus maxonii),
[RR], [Ds], [Gr], [N-1980], [Af 37-3],
medirisii (See Pilosocereus medirisii, Pseudopilocereus medirisii),
melanostele (See Espostoa haagei, Espostoa melanostele, Pseudoespostoa melanostele),
mezcalaensis (See Neobuxbaumia mezcalaensis, Pilocereus mezcalensis),
[Bv-1], [Bz], [Cx]
" -v- multiareolatus (See Neobuxbaumia mezcalensis -v- multiareolata),
militaris (See Backebergia militaris, Mitrocereus militaris, Pilosocereus chrysomallus),
[Iz],
millspaughii (See Pilosocereus millspaughii),
[RR], [Af 52-1],
monoclonus (See Pilosocereus monoclonus),
monteazulensis (See Micranthocereus monteazulensis),
moritzianus (See Pilosocereus moritzianus),
[RR], [Af 16-5], [Af 24-1 & 2],
" -v- backebergii (See Pilosocereus backebergii),
mortensenii (mortensis) (See Pilosocereus mortensenii),
mucosiflorus (See Pilosocereus mucosiflorus, Pseudopilocereus mucosiflorus),
multiareolatus (See Neobuxbaumia mezcalaensis -v- multiareolata, Neobuxbaumia multiareolata),
multicostatus (See Pilosocereus multicostatus),
multiflorus (See Arrojadoa multiflora),
nizandensis (See Neodawsonia nizandensis).....[Az],
nobilis (See Pilocereus nobilis, Pilosocereus nobilis),
[RR], [Bv-1],
" -f. cristata.....[Bl],
nudus (See Cephalocereus tetetzu -v- nuda)....[Af 21-4],
oligolepis (See Pilosocereus oligolepis, Pseudopilocereus oligolepis),
oreus (See Pilosocereus oreus),
ottonis (See Subpilocereus ottonis),
pachycladus (See Pilosocereus pachycladus, Pseudopilocereus pachycladus),
palmeri (See Pilocereus palmeri, Pilosocereus palmeri, Cereus victoriensis),
[C], [H], [PP], [XX], [Az], [Bv-1], [Gr], [N-1980], [Af 1-9], [Af 16-5]
" -v- palmeri (See Pilosocereus palmeri -v- palmeri).....[Az]
" -v- sartorianus (See Pilosocereus palmeri -v- sartorianus).....[Aw 3-1986], [Jd],

CEPHALOCEREUS.....(Cont'd),

palmeri.....(cont'd),
" -v- victoriensis (See Cereus victoriensis, C. palmeri -v- palmeri),
paraguayensis (See Pilosocereus paraguayensis),
parvus (See Pseudopilocereus parvus),
paulensis (See Coleocephalocereus paulensis),
penicillatus (See Arrojadoa penicillata),
[It], [Iz],
" -v- decumbens (See Arrojadoa penicillata -v- decumbens),
" -v- spinosior (See Arrojadoa penicillata -v- spinosior),
pentaedrophorus (See Pilocereus pentaedrophorus, Pilosocereus pentaedrophorus),
[R], [RR],
pernambucoensis (See Pilosocereus pernambucoensis),
" -v- caesius (See Pilosocereus pernambucoensis -v- caesius),
" -v- montealtoi (See Pilosocereus pernambucoensis -v- montealtoi),
phaeacanthus (See Brasilicereus phaeacanthus),
[CC], [RR],
piauhyensis (See Pilocereus piauhyensis, Pilosocereus piauhyensis),
[RR], [Cu 3-1967],
pluricostatus (See Coleocephalocereus pluricostatus),
polyanthus (See Micranthocereus polyantha),
[It],
polygonus (See Pilocereus polygonus, Pilosocereus polygonus),
[RR], [Af 24-2], [Ct 2-1984*],
polylophus (See Neobuxbaumia polylophus, Pilosocereus polylophus),
[Az], [Bt], [Bv-1], [Iz], [N-1980],
purpureus, sensu Werd.1933 (See Austrocephalocereus purpureus (Guerke) Backg., Coleocephalocereus goebelianus (Buin.), Cereus goebelianus VPL - 1923-1924),
[RR], [It],
purpureus (Guerke) 1908 (See Austrocephalocereus lehmannianus, Cephalocereus lehmannianus, Micranthocereus purpureus (Guerke) 1968),
purpusii (See Pilosocereus purpusii),
[Az], [Bc], [Cx], [Af 15-4], [Af 21-2],
[Aw 4-1979],
quadricentralis (See Pilosocereus quadricentralis),
[Az], [Af 27-1],
quadricostatus (See Pilosocereus quadricostatus),
remolinensis (See Cereus remolinensis, Subpilocereus remolinensis),

CEPHALOCEREUS.....(Cont'd),

repandus (See Cereus repandus, Subpilocereus repandus),
rhodanthus (See Arrojadoa rhodantha),
[It], [Kakt. & Sukk. Pflanz, Taf. 44, 1933]
" -ssp. reflexus (See Arrojadoa rhodantha -ssp.- reflexus),
" -v- occibahiensis (See Arrojadoa rhodantha -v- occibahiensis),
robinii (See Cephalocereus keyensis, Cereus robinii, Pilosocereus robinii),
[RR],
" -v- keyensis (See Cephalocereus keyensis, Cereus robinii -v- keyensis, Pilosocereus robinii),
[RR],
robustus (See Pilosocereus robustus),
[RR],
" -v- ulei (See Cephalocereus ulei, Pilocereus ulei, Pilosocereus ulei),
[Bv-1],
rosae (See Pilosocereus rosae, Pseudopilocereus rosae)
royenii (See Pilosocereus royenii),
[LL], [RR], [Bv-1], [N-1980], [Af 4-8]
" -f. cristata.....[Af 4-8],
ruficeps (See C. macrocephala, Mitrocereus ruficeps, Neobuxbaumia macrocephala),
rupicolus (See Pilosocereus rupicola, Pseudopilocereus rupicola),
ruschianus (See Pseudopilocereus ruschianus),
russelianus (See Pilocereus russelianus, Pilosocereus russelianus, Subpilocereus russelianus),
[RR], [Bv-1], [Cz],
" -v- micranthus.....
salvadorensis (See Pilocereus salvadorensis, Pilosocereus salvadorensis),
sartorianus (See C. palmeri -v- sartorianus, Pilocereus sartorianus, Pilosocereus sartorianus),
[Cx], [Aj 15-5-1964], [Aj 16-5-1965],
saudadensis (See Pilosocereus saudadensis),
scoparius (See Neobuxbaumia scoparia),
[Az],
senilis.....[D], [I], [L], [T], [X], [Z], [DD], [NN], [RR], [UU], [Ar], [Ay], [Az], [Bo-4], [Bt], [Cj], [Cr], [Cx], Cz], [Gr], [Hb], [Hv], [Il], [Iz], [Ja], [Jd], [Ad Aug., 1978], [Af 1-5], [Af 6-10], [Af 32-6], [Af 56-6], [Ag-1988], [Aj 31-2-1980], [Am 8-1979], [Aw 4-1970], [Bi 2-3], [De 4-1981], [Hw-1986],
" -f. cristata.....[NN], [Ai], [Bv-1], [Af 16-12], [Af 25-2], [Aj 8-12-1957],

CEPHALOCEREUS.....(Cont'd),

sergipensis (See Pilosocereus sergipensis),
smithianus (See Monvillea smithiana),
[CC], [RR], [Bv-1],
splendidus (See Pilosocereus splendidus),
sublanatus (See C. brasiliensis, Pilosocereus brasiliensis, Pilosocereus sublanatus),
supthutianus (See Pilosocereus supthutianus)
superbus (See Cereus azureus, Pseudopilocereus azureus, Pilosocereus superbus),
" -v- gacapaensis (See Pilosocereus superbus -v- gacapaensis),
" -v- lanosior (See Pilosocereus superbus -v- lanosior),
" -v- regius (See Pilosocereus superbus -v- regius),
superfloccosus (See Pseudopilocereus superfloccosus),
swartzii (See Pilosocereus swartzii),
[RR], [Bj],
" -f. cristata.....[Af 2-2],
tehuacanus (See Pilosocereus tehuacanus),
tetetzo (See Neobuxbaumia tetetzo),
[Cx],
" -v- nuda (See C. nudus),
[Az],
theunissensianus (See Arrojadoa theunissensiana),
tillianus (See Pilosocereus tillianus),
totolapensis (See Neodawsonia totolapensis),
[Az],
tuberculatus (See Pilosocereus tuberculatus),
tuberculosus (See Pilosocereus tuberculosus),
tweedyanus (See Pilosocereus tweedyanus),
[RR], [Af 34-3],
uilianus (See Micranthocereus uilianus),
ulei (GUERKE) (See C. robustus -v- ulei, Espostoa ulei, Facheiroa ulei, Pilosocereus robustus -v- ulei),
[Bv-1],
urbanianus (See Pilosocereus urbanianus),
vilaboensis (See Pseudopilocereus vilaboensis),
violaciflorus (See Micranthocereus violaciflorus),
werdermannianus (See Pilosocereus werdermannianus, Pseudopilocereus werdermannianus),
" -v- densilanatus (See Pilosocereus werdermanianus -v- densilanatus),
" -v- diamantinensis (See Pilosocereus werdermannianus -v- diamantinensis),
zehntneri (See Pilosocereus gounellei -v- zehntneri, Pilosocereus zehntneri),
[RR - Flower], [Gf:2-35 Fig. 41, 1920],

CEPHALOCLEISTOCACTUS

chrysocephalus (See Cleistocactus chrysocephalus), [Ay], [Bo-6], [Hi-2], [Jd], [Da 4-1979]
pallidus (See Cleistocactus pallidus).....[DD],
ritteri (See Cleistocactus ritteri), [G], [DD], [NN], [Hx], [Il], [Af 53-1],
schattatianus (invalid name).....[DD], [Gr],
Spec. .....[Hu]

CEPHALOPHORUS
chrysacanthus.....[Ba 46, 107 July 1967],
palmeri.....[Ba 46, 107 July 1967],
" -v- sartorianus.....

CEREUS

acanthurus (See Borzicactus acanthurus, Borzicactus eriotrichus, Cereus eriotrichus, Cleistocactus anthurus, Loxanthocereus acanthurus, Loxanthocereus eriotrichus),
acida (See Eulychnia acida),
acranthus (See Binghamia acrantha, Haageocereus limensis),
adelmarii (See Monvillea adelmarii),
adscendens (See Harrisia adscendens, Eriocereus adscendens),
aethiops (See C. coerulescens, Piptanthocereus aethiops).....[RR], [Gr], [Hx], [Af 51-5],
" -v- landbeckii.....
" -v- melanacanthus.....[Bo-4],
alacriportanus (See C. paraguayensis, Piptanthocereus alacriportanus).....[RR],
" -v- bageanus (See Piptanthocereus bageanus),
alamosensis (See Rathbunia alamosensis, Stenocereus alamosensis),
alticostatus (See Monvillea alticostata),
anisitzii (See Monvillea spegazzinii),
apiciflorus (See Corryocactus apiciflorus, Erdisia apiciflora),
aragonii (See Lemaireocereus aragonii, Marshallocereus aragonii, Stenocereus aragonii)....[Aw 32-1],
arendtii (See Cereus tortuosus, Eriocereus arendtii, Eriocereus tortuosus),
areolatus (See Cleistocactus areolatus), [Cg 100 Fig. 20 - 1897],
" -v- herzogianus (See Cleistocactus areolatus -v- herzogianus),
arequipensis (See Neoraimondia arequipensis, Neoraimondia peruviana),
argentinensis (See Harrisia platygonus, Eriocereus platygonus).....[RR],[Bl],[Bo-4],

CEREUS.....(Cont'd)

assurgens (See Leptocereus assurgens), [Bk],
atacamensis (See Helianthocereus atacamensis, Trichocereus atacamensis),
aureus (See Cleistocactus aureus).....[Bv-1],
aurivillus (See Borzicactus aurivillus, Seticereus aurivillus, Seticereus icosagonus), [Af 4-12],
azureus.....[C], [Bk], [Hx*],
backebergii (See Pilocereus backebergii, Pilosocereus backebergii).....[Aj II-8-1930],
bageanus (See Piptanthocereus bageanus),
baumannii (See Cleistocactus baumannii).....[M*]
" -v- colubrinus (See Cleistocactus baumannii -v- colubrinus).....[Ds], [If],
" -v- flavispinus (See Cleistocactus baumannii -v- flavispinus),
baxaniensis (See Acanthocereus baxaniensis, Acanthocereus pentagonus, Cereus pentagonus),
beneckei (See Hertrichocereus beneckei, Lemaireocereus beneckei),
boeckmannii (See Selenicereus boeckmannii),
bolivianus (See Trichocereus macrogonus),
" -f. monstrosus.....[Af 2-7],
bonplandii (See Eriocereus bonplandii, Harrisia bonplandii).....[M*],
braunii (See Cephalocereus braunii, Coleocephalocereus braunii).....[Bo-4],
brevistylus (See Corryocactus brevistylus, Corryocactus krausii),
bridgesii (See Trichocereus bridgesii).....[Ds],
brookei (See Cleistocactus brookei),
cabralensis (See Piptanthocereus cabralensis),
caesius (See Cereus glaucus, Cereus pseudocaesius, Cereus pterophorus).....[Hx*],
" -f. monstrosus.....[Bo-4], [Bv-1],
calcaratus (See Hylocereus calcaratus),
calcirupicola (See Piptanthocereus calcirupicola),
" -v- cabralensis (See Piptanthocereus cabralensis)
" -v- cipoensis (See Piptanthocereus cipoensis),
" -v- pluricostatus.....
campinensis (See Pilocereus campinensis, Praeocereus campinensis),
" -v- piedadensis (See Monvillea campinensis -v- piedadensis, Monvillea piedadensis),
candelaris (See Browningia candelaris).....[Ds],
candicans (See Echinopsis candicans, Trichocereus candicans).....[If],

CEREUS.....(Cont'd)

cavendishii (See Monvillea cavendishii),
[M*],
celsianus (See Oreocereus celsianus),
" -v- lanuginosior (See Oreocereus celsianus -v- lanuginosior),
[Iq],
cephalomacrostibas (See Haageocereus ferox, Trichocereus cephalomacrostibas, Weberbauerocereus cephalomacrostibas),
chachapoyensis.....[Gv ii-859 1959],
chacoanus (See Monvillea chacoanus),
chalybaeus (See Piptanthocereus chalybaeus),
[Bk], [Bo-4], [Ia], [Ad July 1978],
[Hh Jan. 1975],
chende (See Heliabravoa chende, Lemaireocereus chende)
chichipe (See Lemaireocereus chichipe, Polaskia chichipe),
childsii (See Cereus hexagonus),
chilensis (See Trichocereus chilensis),
[Ds], [If], [Aj II-5-1930],
" -v- eburneus (See Eulychnia eburnea),
chlorocarpus.....
chosicensis (See Haageocereus chosicensis),
chrysacanthus.....[If],
cipoensis (See Piptanthocereus cipoensis),
coccineus (Heliocereus coccineus, Mediocactus coccineus)
cochabambensis (See Piptanthocereus forbesii -v- bolivianus).....[Af 42-1],
" -v- longicarpa.....[Bk], [Af 42-4],
cochal (See Myrtillocactus cochal),
coerulescens (See Cereus aethiops, Piptanthocereus aethiops),
" -v- melanacanthus (See Cereus aethiops -v- melanacanthus),
colosseus (See Piptanthocereus colosseus),
colubrinus (See Cleistocactus baumannii -v- colubrinus, Cleistocactus colubrinus),
" -v- flavispinus (See Cleistocactus baumannii -v- flavispinus),
columna-trajani (See Haseltonia columna-trajani)...[If]
comarapanus (See Piptanthocereus comarapanus),
[Gw 5-1956],
conoideus (See Echinocereus conoideus),
coquimbanus (See Trichocereus coquimbanus),
coryne (See Stetsonia coryne),
crassisepalus (See Piptanthocereus crassisepalus),
dayamii (See Piptanthocereus dayamii),
[RR], [Bv-1], [Af 2-10],

CEREUS.....(Cont'd),

deficiens (See Lemaireocereus deficiens, Rittero-
cereus deficiens),
" -f. monstrosus.....
deserticolus (See Trichocereus deserticola),
diguetii (See Neoevansia diguetii),
divaricatispinus (See Haageocereus divericatispinus),
donkelaarii (See Selenicereus donkelaarii),
dumortieri (See Isolatocereus dumortieri, Lemaireo-
cereus dumortieri),
dusenii (See Austrocactus coxii, Austrocactus dusenii)
eichlamii (See Lemaireocereus eichlamii, Ritterocereus
eichlamii, Stenocereus eichlamii),
elegantissimus (See Heliocereus elegantissimus),
" -v- helenae (See Heliocereus elegantissimus
-v- helenae),
" -v- stenopetalum (See Heliocereus elegantiss-
imus -v- stenopetalum),
emoryi (See Bergerocactus emoryi),
[M*], [Ax], [Ca],
erythrocephalus (See Denmoza erythrocephala),
[Iq],
euphorbioides (See Lemaireocereus euphorbioides,
Neobuxbaumia euphorbioides, Rooksbya
euphorbioides),
extensus (See Hylocereus extensus),
[Ft - 70:pl. 4066],
fascicularis (See Haageocereus fascicularis),
fernambucensis (See C. pernambucensis, C. obtusus,
Piptanthocereus obtusus),
[Bo-4], [Bv-1], [It],
flagelliformis (See Aporocactus flagelliformis),
[M*],
flavescens (See Cleistocactus flavescens),
fluminensis (See Cephalocereus fluminensis, Coleoceph-
alocereus fluminensis),
forbesii (See Piptanthocereus forbesii),
[Ds], [Hx], [Cu 1-1960],
" -v- bolivianus.....
formosus (See Echinopsis formosa, Lobivia formosa,
Soehrensia formosa),
fricii (See Cleistocactus fricii, Cleistocactus
strausii -v- fricii),
[Aj II-8-1930],
gatesii (See Lophocereus gatesii),
geometrizens (See Myrtillocactus geometrizens),
" -v- grandiareolatus.....
ghiesbreghtii (See Armatocereus ghiesbreghtii),
[Bk], [Bo-2],
giganteus (See Carnegiea gigantea),
[Ax],

CEREUS.....(Cont'd),

glaucus (See Echinopsis fulvilana, Trichocereus fulvilanus).....[Dn], [Af 10-1],
glaziovii (See Leocereus glaziovii),
goebelianus (See Austrocephalocereus purpureus (Guerke) Backg., Coleocephalocereus goebelianus),
goiasensis (See Piptanthocereus goiasensis),
gonacanthus (See Echinocereus triglochidiatus -v- gonacanthus, Echinocereus gonacanthus),
gonianthus (See Piptanthocereus gonianthus), [Bv-1], [Bi 4-1986],
gonzalezii (See Weberocereus tunilla),
gracilis (See Harrisia gracilis),
" -v- aboriginum (See Harrisia aboriginum, Harrisia gracilis -v- aboriginum), [Af 41:126 May, June 1969],
" -v- simpsonii (See Harrisia gracilis -v- simpsonii, Harrisia simpsonii), [Af 41:126 May, June 1969],
grandicostatus.....[Af 9-12],
grandiflorus (See Selenicereus grandiflorus, Selenicereus megalantha), [Ax], [Bk], [Af 1-7],
" -v- grandiflorus.....[Ax],
grandis (See Pachycereus grandis, Pachycereus weberi, Stenocereus weberi),
greggii (See Peniocereus greggii), [M*], [U], [An], [Ax], [Dq], [Ib], [Af 13-12],
" -v- greggii.....[Ax], [Df],
" -v- transmontanus.....[Ax], [Q],
grenadensis (See Subpilocereus grenadensis), [RR], [Af 57-3], [Ah-1986],
griseus (See Lemaireocereus griseus, Ritterocereus griseus, Stenocereus griseus),
guelichii (See Harrisia guelichii),
gummosus (See Machaerocereus gummosus, Stenocereus gummosus),
haematuricus.....[Bo-4],
hamatus (See Selenicereus hamatus),
hankeanus (See C. forbesii, Piptanthocereus forbesii), [Bo-4],
hertrichianus.....
hexagonus.....[M*], [RR], [Ay], [Bn], [Bo-4], [Bu], [Bv-2], [Cl], [Cz], [Hx], [M*], [Af 6-7], [Af 14-8], [Af 28-1], [Ba 9-1980], [Gu III-1953], [Hl 4-1981],
hildmannianus.....[RR], [Bu], [Hx*], [Gh 4-pt. 2:202 - pl. 41 Fig. 1, 1980],
hirschtianus (See Nyctocereus hirschtianus).....[Bk],
hollianus (See Lemaireocereus hollianus),
hondurensis (See Selenicereus hondurensis),

CEREUS.....(Cont'd),

hoppenstedtii (See Cephalocereus hoppenstedtii, Haseltonia columna-trajani),
[If],
horribarbis (See Acanthocereus horribarbis),
[Af 2-7], [Hl 6-1967],
horridus (See C. hexagonus),
[Bo-4],
" -v- alatosquamatus.....
horrispinus (See Pilosocereus horrispinus, Subpilocereus horrispinus),
[Aj II-8-1930],
houlletii (See Cephalocereus leucocephalus, Pilocereus houlletii, Pilocereus leucocephalus, Pilosocereus leucocephalus).....[If],
huascha (See Lobivia huascha, Trichocereus huascha - (Web.) Ritt.),
" -v- flaviflorus (See Trichocereus huascha (Web.) Ritt.),
" -v- rubriflorus (See Lobivia huascha -v- andalgalensis, Trichocereus andalgalensis),
huilunchu (See Piptanthocereus huilunchu).....[Bo-4],
humboldtii (See Borzicactus humboldtii, Seticereus humboldtii, Seticereus neohumboldtii),
humilis (See Armatocereus humilis, Lemaireocereus humilis),
huntingtonianus.....[Af 4-11], [Af 16-5], [Af 21-1 & 2],
" -f. monstrosus.....
hyalacanthus (See Cleistocactus hyalacanthus, Cleistocactus jujuyensis),
hystrix (See Lemaireocereus hystrix),
icosagonus (See Borzicactus aurivillus, Borzicactus icosagonus, Cereus aurivillus, Seticereus aurivillus, Seticereus icosagonus)....[Bk],
inermis (See Selenicereus anoplus, Selenicereus inermis),
insularis (Monvillea insularis).....[DD], [Bo-4],
iquiquensis (See Eulychnia iquiquensis),
jamacaru (See Piptanthocereus jamacaru),
[M*], [PP], [RR], [Ay], [Bk], [Bo-4], [Bv-1], [Cr], [Gr], [Hy], [If], [It], [Jd], [Af 1-7], [Cu 2-1967], [Hl 2-1970],
" -v- caesius (See C. caesius, Piptanthocereus jamacaru -v- caesius),
" -f. monstrosus.....[Be*], [Dp], [Hy], [Il],
johnstonii (See Peniocereus johnstonii),
jugatiflorus (See Cleistocactus jugatiflorus),
jusbertii (See Eriocereus jusbertii, Harrisia jusbertii).....[Bk],

CEREUS.....(Cont'd),

knuthianus (See Selenicereus knuthianus),
laboretianus (See Piptanthocereus laboretianus),
[Af 21, 1 & 2],
laetivirens (See Piptanthocereus jamacaru),
[Hy],
laetus (See Armatocereus laetus, Lemaireocereus laetus)
laevigatus (See Lemaireocereus longispinus, Rittero-cereus laevigatus, Stenocereus laevigatus),
lamprochlorus (See Echinopsis lamprochlora, Tricho-cereus lamprochlorus, Trichocereus neo-lamprochlorus).....[M*],
lamprospermus (See Cereus colosseus, Piptanthocereus colosseus, Piptanthocereus lamprospermus),
lanatus (See Espostoa lanata),
" subspec. sericatus, (See Espostoa sericata),
[Aj 5-1937],
" -f. cristata.....[If],
laniceps (See Cleistocactus laniceps),
lanosus (See Piptanthocereus lanosus),
lauterbachii (See Monvillea lauterbachii),
lemairei (See Hylocereus lemairei),
[M*],
leptacanthus (See Echinocereus leptacanthus, Echino-cereus pentalophus -v- leptacanthus),
leucostele (See Stephanocereus leucostele).....
limensis (See Cereus acranthus, Haageocereus clavi-spinus, Haageocereus lachayensis, Haageocereus limensis, Haageocereus olowinskianus),

lindenzweigianus (See Piptanthocereus lindenzweigianus)
litoralis (See Trichocereus litoralis),
[Aj 2-5-1951],
longispinus (See Eulychnia breviflora, Eulychnia longispinus),
macdonaldiae (See Selenicereus macdonaldiae),
macdougallii (See Peniocereus macdougallii),
macrostibas (See Cereus peruvianus, Neoraimondia arequipensis, Neoraimondia macrostibas),
maculatus (See Acanthocereus maculatus, Peniocereus maculatus),
margaritensis (See Subpilicereus margaritensis, Sub-pilocereus russelianus subspec. margarit-ensis),
marginatus (See Lemaireocereus marginatus, Marginato-cereus marginatus, Pachycereus marginatus, Stenocereus marginatus).....[Ax],
markgrafii (See Brasilicereus markgrafii)....[Ct-1988],
martianus (See Aporocactus martianus),
[M*], [If],
martinezii (See Lemaireocereus martinezii),

CEREUS.....(Cont'd),

martinii (See Eriocereus martinii, Harrisia martinii) [M*], [Aj II-6-1930],
melanostele (See Espostoa haagei, Espostoa melanostele, Pseudoespostoa melanostele),
micranthus.....
microspermus (See Browningia microsperma, Gymnocereus microspermus),
microsphaericus (See Arthrocereus microsphaericus, Epiphyllanthus microsphaericus),
milesimus.....[Aj 24-12-1973],
minensis (See Cipocereus minensis).....[It],
minutiflorus (See Hylocereus minutiflorus),
mixtecensis.....[If],
monacanthus (See Hylocereus monacanthus),
monoclonus (See Cephalocereus monoclonus),
monstrosus (Fairy Castle).....[Ah-1985],
montanus (See Lemaireocereus montanus, Stenocereus montanus),
moritzianus (See Cephalocereus moritzianus, Pilosocereus moritzianus, Pilosocereus backebergii),
" -v- curvispinus.....
" -v- robustus.....
multangularis (See Binghamia melanostele, Binghamia multangularis, Haageocereus chosicensis, Haageocereus multangularis),
myriophyllos (See Trichocereus strigosus),
nashii (See Harrisia nashii),
neonesioticus (See Piptanthocereus neonesioticus),
" -v- interior (See Piptanthocereus neonesioticus -v- interior),
neopitahaya (See C. pitahaya),
neotetragonus (Invalid name - See C. tetragonus),
nesioticus (See Brachycereus nesioticus), [Af 5-9],
nigripilis (See Trichocereus coquimbanus -v- nigripilis, Trichocereus nigripilis),
nitens (See Echinopsis lamprochlorus, Trichocereus candicans -v- nitens, Trichocereus lamprochlorus)
nycticallus (See Cereus pteranthus, Selenicereus pteranthus).....[Bk], [Cr], [If],
obtusangulus (See Epiphyllanthus obtusangulus).... [Bk],
obtusus (See C. fernambucensis, C. pernambucensis, Piptanthocereus obtusus), [RR], [Bo-4], Bv-1],
ocamponis (See Hylocereus ocamponis).....[M*],
octagonus (See Cereus hexagonus),
orcuttii (See Pachygerocereus orcuttii, Pachycereus orcuttii), [Af 34-3],

CEREUS.....Cont'd),

ovatus (See Opuntia aoracantha, Opuntia ovata,
Tephrocactus articulatus -v- ovatus,
Tephrocanthus aoracantha, Tephrocactus ovatus),
pachyrhizus (See Piptanthocereus pachyrhizus),
papillosus (See Wilcoxia papillosa, Echinocereus
papillosus),
paraguayensis (See C. alacriportanus)....[Af 2-10],
parviflorus (See Cleistocactus parviflorus),
pasacana (See Trichocereus pasacana),
" -f. cristata.....[Cu 4-2-1935],
paxtonianus (See C. cavendishii, Monvillea paxtoniana)
pectin-aboriginum (See Pachycereus pectin-aboriginum),
[Az], [By],
pensilis (See Echinocereus pensilis, Morangaya
pensilis).....[Aw 2-1977],
pentagonus (See Acanthocereus congatus, Acanthocereus
pentagonus, Cereus pentagonus),
[Ax], [Ci], [Df], [Hl 3-1972],
pepinianus (See Copiapoa pepiniana, Cereus chilensis),
perlucens (See Pilocereus perlucens, Pilosocereus
perlucens),
pernambucensis (See C. fernambucensis, C. obtusus,
Piptanthocereus obtusus),
[Bo-4], [Bv-1],
peruvianus (Werd) (See Cereus uruguayanus).....
peruvianus (Miller), (See Neoraimondia peruviana(L) Rit.
[M*], [AA], [RR], [Ax], [Ay], [Be*], [Cz],
Hv], [If], [Af 60-2], [Aj 29-4-1978],
" -v- glaucus.....[Bv-1],
" -v- ovicarpus.....[Af 11-1],
" -v- persicinus.....[Af 11-1],
" -v- piauhyensis (See Cephalocereus piauhyensis)
" -v- proferrens.....[Af 11-1],
" -v- reclinatus.....[Af 11-1],
" -v- tortuosus.....[Bo-4],
" -f. monstrosus.....[D], [H], [K], [X], [AA],
[EE], [WW], [Av], [Dp], [Bv-1],
[Cp], [Cz], [Hb], [Ij], [Il],
[Ja], [Jd], [N-1980], [Af 6-7],
[Af 11-1],
phaeacanthus (See Brasilicereus phaeacanthus),
" -v- breviflorus (See Brasilicereus phaeacanthus
-v- breviflorus, Brasilicereus
breviflorus),
piauhyensis (See Cephalocereus piauhyensis),
pitahaya (See Cereus neopitahaya).....[Hy],
" -v- bysiegeli.....
" -f. monstrosus.....[Be*], [Ir], [Af 1-7],
plagiostoma (See Binghamia plagiostoma, Borzicactus
plagiostoma),

CEREUS.....Cont'd),

platinospinus (See Haageocereus platinospinus)
platygonus (See Cereus argentinensis, Eriocereus platygonus, Harrisia platygona),
[Bk],
polyrhizus (See Hylocereus polyrhizus),
pomanensis (See Eriocereus pomanensis, Harrisia pomanensis),
portoricensis (See Eriocereus portoricensis, Harrisia portoricensis),
poselgeri (See C. tuberosus, Echinocereus tuberosus, Wilcoxia poselgeri, Wilcoxia tuberosa),
[M*], [Ax], [Df],
pringlei (See Pachycereus pringlei),
[Bk],
" -v- tortugaensis.....
pruinosus (See Lemaireocereus pruinosus),
pseudocaesius (See C. caesius, C. pterophorus),
pseudomelanostele (See Haageocereus multangularis -v- pseudomelanostele, Haageocereus pseudomelanostele),
[Af 4-12],
pseudothelegonus (See Borzicactus pseudothelegonus),
[Gv ii-1136 - 1959],
pteranthus (See Selenicereus pteranthus),
[Ax],
pterophorus (See C. caesius, C. jamacaru),
queretaroensis (See Lemaireocereus queretaroensis),
remolinensis (See Cephalocereus remolinensis, Pilosocereus remolinensis, Subpilocereus, remolinensis).....[Aj II-8-1930],
repandus (See Cephalocereus repandus, Subpilocereus repandus),
[M*], [RR], [Bj], [Dq], [Af 57-3],
" -v- weberi), Invalid name), (See Subpilocereus repandus -v- weberii),
rhodanthus (See Arrojadoa rhodantha, Cephalocereus rhodanthus),
ridlei (Invalid name).....
robinii (See Cephalocereus robinii, Cephalocereus keyensis, Pilosocereus keyensis, Pilosocereus robinii).....[Ax],
" -v- deeringii (See Cephalocereus deeringii, Pilosocereus deeringii),
[Ax],
" -v- keyensis (See Cephalocereus keyensis, Cephalocereus robinii -v- keyensis, Cephalocereus robinii),
" -v- robinii.....[Ax],
roezlii (See Borzicactus neoroezlii, Seticereus roezlii).....[Hi-4],

rosei (See Peniocereus rosei)
roseiflorus (See Cereus peruvianus (Linn.) Miller, Neoraimondia peruviana (L) Ritter, Neoraimondia roseiflora, Neoraimondia arequipensis -v- roseiflora).....[Dr],
saddianus (See Monvillea saddiana),
santiaguensis (See Trichocereus santiaguensis),
schickendantzii (See Trichocereus schickendantzii), [Ds],
schottii (See Lophocereus schottii), [Q], [An], [Ax], [By], [Af 13-12],
" -v- australis (See Lophocereus australis, Lophocereus schottii -v- australis)
seidelii (See C. azureus, Pseudopilocereus azureus),
sericatus (See Cereus lanatus -v- sericatus, Espostoa sericata),
sericifer (See Piptanthocereus sericifer),
serpens (See Borzicactella serpens, Borzicactus serpens, Bolivicereus serpens, Cleistocactus serpens),
serpentinus (See Nyctocereus serpentinus),
setaceus (See Mediocactus setaceus, Mediocactus coccineus),.....[M*],
silvestrii (See Chamaecereus silvestrii, Lobivia silvestrii, Echinopsis chamaecereus),
" -f. cristata (See Chamaecereus silvestrii -f. cristata)...[Ds], [Ir], [Cu 2-4-1934],
spachianus (See Trichocereus spachianus),
speciosus (See Heliocereus speciosus), [M*], [Bk], [If],
spegazzinii (See Monvillea spegazzinii), [M*],
" -v- hassleri (See Piptanthocereus lindenzwiegianus),
spinibarbis (See Trichocereus spinibarbis),
spinulosus (See Selenicereus spinulosus), [Ax],
squamulosa (See Zehntnerella squamulosa),
stellatus (See Lemaireocereus stellatus, Ritterocereus stellatus, Stenocereus stellatus).....[If],
stenogonus (See Piptanthocereus stenogonus), [RR], [Bo-4],
stenopterus (See Hylocereus stenopterus), [M*],
striatus (See Neoevansia striata, Neoevansia diguettii, Peniocereus diguettii, Wilcoxia striata), [Q], [An], [Ax],
strigosus (See Trichocereus strigosus),
subrepanda (See Cereus gracilis, Harrisia gracilis),

CEREUS.....(Cont'd),

tacuaralensis.....[Gs No. 80-81, 19 - 1964],
tamaulipensis (See Wilcoxia tamaulipensis),
tarijensis (See Helianthocereus tarijensis, Lobivia tarijensis, Trichocereus tarijensis),
tehuacanus (See Pilosocereus tehuacanus),
tephracanthus (See Eriocereus tephracanthus, Roseocereus tephracanthus),
[Bk],
terscheckii (See Trichocereus terscheckii),
[Cu 4-2-1935],
testudo (See Deamia testudo, Selenicereus testudo),
tetetzo (See Neobuxbaumia tetetzo),
[Dp], [If],
tetragonus (See Acanthocereus pentagonus, Acanthocereus tetragonus, C. pentagonus, C. pitahaja),
[Bv-1], [Ed 12:158],
thalassinus (See C. jamacaru),
thouarsii (See Jasminocereus thouarsii),
[Af 5-8], [Af 5-9],
thurberi (See Lemaireocereus thurberi, Marshallocereus thurberi, Stenocereus thurberi),
[Q], [An], [Ax], [By], [Af 14-1],
tominensis (See Cleistocactus tominensis),
tonduzii (See Werkleocereus tonduzii),
[M*],
tortuosus (See Eriocereus tortuosus, Harrisia tortuosa, Harrisia arendtii),
[Ik], [Aj 24-10-1973],
triangularis (See Hylocereus triangularis),
trigonodendron.....[NN], [RR], [Bo-4], [BOT. JAHRB. Engl. 40:pl. 10], [Af 50-3],
trigonus (See Hylocereus trigonus),
trollii (See Oreocereus trollii),
tuberosus (See Cereus poselgeri, Echinocereus poselgeri, Wilcoxia poselgeri, Wilcoxia tuberosa),.....
tunilla (See Weberocereus tunilla),
[M*],
tweedyanus (See Cephalocereus tweedyanus, Pilocereus tweedyanus, Pilosocereus tweedyanus),
undatus (See Hylocereus undatus),
[M*], [Ax],
uruguayanus (See Cereus peruvianus (Werd.), Piptanthocereus uruguayanus),.....[Gr],
" -f. cristata.....
" -f. monstrosus.....[Gr],
urbanianus (See Selenicereus urbanianus),
[M*],
vagans (See Selenicereus vagans),

CEREUS.....(Cont'd),

validus (See Piptanthocereus validus),
[M*], [CC], [RR], [Bv-1], [Bo-4], [Af 2-10],
[Bi 2-4], [Dg 3-3-1981],
vargasianus.....[Bo-4], [Gw 34 - 1951],
variabilis (See Cereus fernambucensis, Cereus obtusus,
Cereus pernambucensis, Cereus neopitahaya,
Cereus pitahaya, Piptanthocereus obtusus),
[D], [Bo-4], [Bv-1],
versicolor (See Haageocereus versicolor),
" -v- humifusus (See Haageocereus humifusus),
victoriensis (See Cephalocereus palmeri, Pilocereus
palmeri, Pilosocereus palmeri).....[Az],
viperinus (See Wilcoxia viperina),
weberbaueri (See Trichocereus fascicularis, Tricho-
cereus weberbaueri, Weberbauerocereus
weberbaueri, Weberbauerocereus fascicularis,
Weberbauerocereus seyboldianus, Haageocereus
weberbaueri),
weberi (See Lemaireocereus weberi, Pachycereus weberi,
Stenocereus weberi).....[RR], [Ax],
werklei (See Selenicereus werklei),
wittii (See Selenicereus wittii, Strophocactus wittii)
[M*],
xanthocarpus (See Piptanthocereus xanthocarpus),
[Bo-4], [Af 9-7], [Af 26-3], [Hl 6-1967],
" -f. monstrosus.....
yunckeri.....

CHAMAECEREUS

cv. - "Antoinette".....[Bi 4-1, 1986],
cv. - "Wotan".....[Hw-1981],
silvestrii (Invalid name - See Echinopsis chamaecereus,
Lobivia silvestrii).....[L], [M*], [R], [X],
[AA], [HH], [NN], [PP], [UU], [Ar], [Ay],
[Be*], [Bn], [Bt], [Cj], [Cm], [Cr], [Cz],
[Dq*], [Gr], [Hu], [Hx], [Il], [Ja], [Jd], [Ad
3-1978], [Ad 4-1978], [Ad 6-1978], [Af 16-8],
[Ba 8-1979 - Flower], [Hw-1982],
" -v- aurea.....[Ay], [Ba 7-1976],
" -v- crassicaulis.....[DD],
" -f. cristata (See Cereus sylvestrii -f.-
cristata).....[Bo-2], [Bv-1],
" -v- grandiflora (See Lobivia grandiflora -v-
lobivioides),

CHIAPASIA
nelsonii (See Disocactus nelsonii).....[F], [I], [DD],
[PP], [RR], [Ay], [Bo-2], [Bt], [Flwr. Dq],
[Hx*], [Jd], [Aw 4-1963],

CHILENIA

heteracantha (See Neoporteria heteracantha, Neoporteria subgibbosa -v- subgibbosa -f. heteracantha),
nigrihorrida (See Neoporteria nigrihorrida, Neoporteria subgibbosa -v- nigrihorrida),
" -v- nigrihorrida (See Neoporteria subgibbosa -v- nigrihorrida).
" -v- major (See Neoporteria nigrihorrida -v- major),
" -v- minor (See Neoporteria nigrihorrida -v- minor),
subcylindrica (See Neoporteria subcylindrica, Neoporteria subgibbosa -v- subgibbosa -f. subcylindrica)

CHILEOREBUTIA

aerocarpa (See Neochilenia aerocarpa, Neoporteria reichei -v- reichei -f. aerocarpa, Thelocephala aerocarpa).....[Hl 5-1970], [Iv 3-1960],
aricensis (See Pyrrhocactus aricensis),
carneoflora (See Neochilenia carneoflora),
duripulpa (See Thelocephala duripulpa),
esmeraldana (See Neochilenia esmeraldana), [Hl 5-1970],
fulva (See Neochilenia aerocarpa -v- fulva, Neochilenia fulva, Thelocephala fulva).....[Hl 5-1970],
imitans (See Neochilenia imitans, Neoporteria reichei -v- imitans),
glabrescens (See Neochilenia glabrescens, Thelocephala glabrescens), [Hl 5-1970], [Iv 3-1960],
krausii (See Neochilenia krausii, Thelocephala krausii) [Hl 5-1970], [Iv 3-1960],
" -v- albispina.....
lembckei (See Neochilenia lembckei, Neoporteria reichei -v- reichei -f. lembckei, Thelocephala lembckei)
malleolata (See Neochilenia malleolata, Neoporteria malleolata, Thelocephala malleolata), [Hl 5-1970],
" -v- solitaria (See Neochilenia malleolata -v- solitaria, Neoporteria reichei -v- malleolata -f. solitaria, Thelocephala malleolata -v- solitaria) [Hl 5-1970],
mitis (See Neochilenia mitis, Neoporteria mitis, Thelocephala mitis),
napina (See Neochilenia napina, Neoporteria napina, Thelocephala napina), [Hl 5-1970],
" -v- fracosa.....[Hl 5-1970],
" -v- lanigera (See Neochilenia napina -v- lanigera)

CHILEOREBUTIA.....(Cont'd),

napina.....(cont'd),
" -v- spinosior (See Neochilenia napina -v- spinosior),
[Hl 5-1970],
odierii (See Neochilenia odierii, Neoporteria odierii, Thelocephala odierii),
[Hl 5-1970], [Iv 3-1960],
reichei (See Neochilenia imitans, Neochilenia lembckei, Neochilenia neoreichei, Neoporteria reichei -v- reichei -f. lembckei, Reicheocactus neoreichei, Thelocephala lembckei),
saxifraga (See Neochilenia saxifraga, Pyrrhocactus saxifragus),
[Hl 5-1970],
solitaria (See C. malleolata -v- solitaria, Thelocephala malleolata -v- solitaria),
tenebrica (See Neochilenia tenebrica, Thelocephala tenebrica),

CINNABARINEA

acanthoplegma (See Lobivia acanthoplegma, Lobivia cinnabarina ssp. acanthoplegma, Pseudolobivia acanthoplegma),
" -v- leucosiphus (See Lobivia taratensis -v- leucosiphus),
boedeckeriana (See Lobivia boedeckeriana, Lobivia rossii -v- boedeckeriana),
[Hi-2],
cinnabarina (See Echinopsis cinnabarina, Lobivia cinnabarina),
graulichii (See Lobivia tiegeliana -v- cinnabarina),
neocinnabarina (See Lobivia cinnabarina ssp. acanthoplegma -v- oligotricha, Lobivia neocinnabarina, Lobivia oligotricha),
" -v- microthelis (See Lobivia microthelis),
oligotricha (See Lobivia cinnabarina ssp. acanthoplegma -v- oligotricha, Lobivia neocinnabarina, Lobivia oligotricha),
prestoana (See Lobivia cinnabarina ssp. prestoana, Lobivia prestoana),
pseudocinnabarina (See Lobivia cinnabarina subspec. acanthoplegma -v- oligotricha, Lobivia oligotricha),
[Hi-2],
" -v- microthelis (See Lobivia microthelis),
purpurea (See Sulcorebutia purpurea, Weingartia purpurea),
torotorensis (See Sulcorebutia torotorensis, Weingartia torotorensis),

CINNABARINEA.....(Cont'd),

walterspielii (See Lobivia cinnabarina -v- walter-
spiellii, Lobivia walterspielii),
[Hi-2],
" -v- sanguiniflora (See Lobivia cinnabarina
-v- walterspielii, Lobivia
sanguiniflora),
[Hi-2],
zudanensis (See Lobivia cinnabarina -v- zudanensis,
Lobivia zudanensis),

CIPOCEREUS

aurisetus (See Pilocereus aurisetus, Pilosocereus
aurisetus, Pilosocereus coerulescens),
minensis (See Cereus minensis, Pilosocereus minensis,
Pseudopilocereus minensis),
[Hi-1], [Aj - 1979 Braz. Expedition Report],
[Aj 31-4-1980], [Aj 33-9-1982],
pleurocarpus (See Coleocephalocereus pleurocarpus,
Pilosocereus pleurocarpus),
[Hi-1], [Af 58-1],
werdermannianus (See Pilosocereus werdermannianus,
Pseudopilocereus werdermannianus),

CLEISTOCACTUS

acanthurus (See Borzicactus acanthurus, Borzicactus
eriotrichus, Cereus acanthurus, Haageocereus
acanthurus, Loxanthocereus acanthurus),
albisetus (See Cleistocactus hyalacanthus, Cleisto-
cactus jujuyensis),
angosturensis.....[Bo-2], [Af 28-2],
anguinus (See Cleistocactus baumannii),
[Bo-2], [Hx], [Iu],
apurimacensis (See C. morawetzianus),
[Bo-2],
areolatus (See Cereus areolatus),
[Ay], [Bo-2], [Hx*], [Af 1-7],
" -v- herzogianus (See C. herzogianus, C. parvi-
florus -v- herzogianus),
aureispinus (See Borzicactus aureispinus, Hildewinteria
aureispina, Loxanthocereus aureispinus,
Winteria aureispina, Winterocereus aurei-
spinus),
[Hr], [Af 2-10],
aureus (See Erdisia meyenii),
ayopayanus (See Borzicactus strausii, C. strausii),
[Hi-2], [Af 28-2],
azerensis.....[DD], [Hi-2], [Af 33-3],

CLEISTOCACTUS.....(Cont'd),

baumannii (See Cereus baumannii),
[C], [L], [M*], [T], [RR], [Be*], [Bo-2],
[Bv-1], [Dq], [Hx], [Ik], [Af 6-11],
[Af 15-7], [Af 29-5], [Af 59-1], [Ag-1988],
[Bi 2-4],
" -v- colubrinus (See Cereus baumannii -v- colubrinus).....[Hl 6-1967],
" -v- flavispinus (See Cereus baumannii -v- flavispinus),
[Bv-1],
" -v- grossei (See C. grossei),
" -f. cristata.....[Af 15-7],
brevispinus.....[Hi-4],
brookei (See Cereus brookei, C. flavescens, C. wendlandiorum),
[DD], [NN], [Hv], [Af 24-5],
" -v- flavispinus.....[Hi-2],
bruneispinus (See C. baumannii),
[Bo-2],
buchtienii (See C. ressinianus, C. sucrensis, C. tupizensis -v- sucrensis),
[Ay], [Bo-2], [Bt], [Bv-1], [Hi-2], [Af 16-1],
" -v- flavispinus.....[Af 24-6],
" -f. cristata.....[Ay],
candelilla.....[H], [Av], [Ay], [Bu], [Hi-2], [Hv],
[Hy], [Il], [Jd], [Af 24-5], [Aj 33-3-1982], [Bi 3-2], [Hw-1989],
" -v-pojoensis (See C. pojoensis),
capadalensis.....
chacoanus (See C. baumannii),
[Hi-2], [Af 22-12], [Aj 6-1977],
" -v- santacruzensis.....[Hi-2],
choatensis (See Seticereus choatensis),
chrysacanthus (See Cephalocleistocactus chrysocephalus)
" -v- multiareolatus.....
clavicaulis (See C. crassicaulis)
[Gs No. 80, 81 -20, 1964],
colubrinus (See C. baumannii -v- colubrinus, Cereus baumannii -v- colubrinus, C. bruneispinus),
" -v- flavispinus (See C. baumannii -v- flavispinus),
" -v- grossei (See C. baumannii -v- grossei, C. grossei),
compactus.....[Bo-2],
crassicaulis (See C. clavicaulis),
" -v- paucispinus.....
crassiserpens (See Borzicactus crassiserpens, Loxanthocereus crassiserpens),
[Bo-2],
croceiflorus (See C. baumannii),

CLEISTOCACTUS.....(Cont'd),

dependens (See Seticleistocactus dependens),
[G], [Af 24-5],
fieldianus (See Borzicactus fieldianus, Clistanthocereus fieldianus),
flavescens (See C. brookei, C. wendlandiorum),
[T], [NN], [Hx],
" -v- wendlandiorum (See C. wendlandiorum),
flavispinus (See C. baumannii).....[T], [Bo-2], [Hx],
fossulatus (FR-100a).....
fossulatus (KK888).....
fricii (See Cereus fricii, C. strausii -v- fricii),
fusiflorus.....[Hw-1981], [Iv 12-1957],
glaucus.....[Hi-2],
" -v- plurispinus.....[Hi-2],
granjaensis.....
grossei (See C. baumannii -v- grossei),
[Av], [Bo-2], [Hx], [Hl 4-1971], [Hl 5-1980],
herzogianus (See Cereus areolatus -v- herzogianus, C. parviflorus -v- herzogianus)....[Hx*],
hildegardiae.....[Hi-2],
" -v- flavirufus.....
hildewinterae (See C. hildegardiae),
" -v- flavirufus (See C. hildegardiae -v- flavirufus),
horstii.....[AJ 33-10-1982],
humboldtii (See Binghamia humboldtii, Borzicactus humboldtii, C. neohumboldtii, Seticereus humboldtii, Seticereus neohumboldtii),
hyalacanthus (See C. jujuyensis),
[Hb], [Af 33-1], [Af 57-3],
" -f. cristata.....[Hy], [Af 57-3],
ianthinus.....[Bo-2], [Af 28-2],
icosagonus (See Seticereus icosagonus)
ipotanus (See C. crassicaulis),
jugatiflorus (Invalid name - See Cereus jugatiflorus),
jujuyensis (See C. hyalacanthus),
[H], [Ar], [Bo-2], [Hi-2], [Hx], [Af 51-5],
[Aj 22-12-1971], [Bi 2-4],
" -v- fulvus.....
laniceps (See Cereus laniceps)..[Hi-2], [Il], [Af 39-1],
" -v- plurispinus.....
luminosus.....
luribayensis.....[Af 28-2],
luteispinus (See Borzicactus luteispina, C. strausii -v- luteispinus),
[Hl 4-1971],
margaritanus (See C. chacoanus),
mendozae.....[Af 35-6],
micropetalus.....[Hi-2],
monvilleanus (See Cereus monvilleanus),

CLEISTOCACTUS.....(Cont'd),

morawetzianus (See C. villaazulensis),
[Ay], [Bl], [Bo-2], [Hi-4], [Ij],
[Iv 6-1957],
" -v- arboreus.....
" -v- aureiflorus.....
" -v- pycnacanthus (See Cereus pycnacanthus),
[Bo-2],
muyurinensis.....[Hi-2],
neohumboldtii (See Binghamia humboldtii, Borzicactus humboldtii, Seticereus humboldtii),
[Hy],
neoroezlii (See Borzicactus neoroezlii, Borzicactus roezlii, Seticereus roezlii),
nivosus (See C. strausii -v- fricii),
orthogonus.....[Iv 7-1959],
otuyensis (See C. buchtienii, Corryocactus otuyensis),
palhuayensis.....[Hi-2],
" -v- camachoensis.....[Hi-2],
pallidus (See Cephalocleistocactus pallidus),
paraguariensis.....
parapetiensis.....[Bo-6], [Hy], [Af 24-6], [Aj 22-12-1971],
parviflorus (See C. herzogianus),
[Ad Oct. 1978],
" -v- aiquilensis.....[Hi-2], [Aj 14-6-1963],
" -v- comarapanus.....
" -v- herzogianus (See C. herzogianus),
pilcomayoensis.....
piraymirensis (See Seticleistocactus piraymirensis),
[Af 33-3],
plagiostoma (See Borziczctus plagiostoma, Cereus plagiostoma),
pojoensis (See C. candelilla -v- pojoensis),
pseudostrausii.....
pungens.....[Hi-4],
" -v- roseiflora.....
pycnacanthus (See C. morawetzianus -v- pycnacanthus),
reae.....[Aj 21-3-1970], [Iv 12-1957],
ressinianus (See C. buchtienii, C. sucrensis, C. tupizensis -v- sucrensis),
[Af 28-2],
rhodacanthus (See Denmoza rhodacanthus),
[Hy],
ritteri (See Cephalocleistocactus ritteri)...[I], [Ay], [Hi-2], [Iz], [Aj 10-11-1959], [Hw-1989],
roezlii (See Borzicactus roezlii, Borzicactus neo-roezlii, Seticereus roezlii),
rojoi (See C. smaragdiflorus -f. rojoi),
[Af 28-2],

CLEISTOCACTUS.....(Cont'd),

samaipatanus (See Bolivicereus samaipatanus, Borzicactus samaipatanus),
[TT],
santacruzensis (See C. chacoanus -v- santacruzensis),
[DD], [Iz],
serpens (See Bolivicereus serpens, Borzicactella serpens, Borzicactus serpens, Cereus serpens),
[Bo-2],
schattatianus (See Cephalocleistocactus schattatianus),
smaragdiflorus.....[I], [RR], [Ay], [Bu], [Bv-1], [Cy], [Cz], [Hi-2], [Hv], [Jd], [Af 29-5], [Af 58-2], [Hs IV 3-1983],
" -v- gracilior.....
" -v- rojoi (See C. rojoi),
sotomayorensis.....[Hl 4-1977],
strausii (See Borzicactus strausii),
[C], [L], [AA], [HH], [NN], [Ar], [Ay], [Bo-2], [Cj], [Cp], [Cr], [Cz], [Gr], [Hb], [Ja], [Hi-2], [Hr], [Hx], [Hy], [Il], [Jd], [Af 12-12], [Af 13-4], [Af 29-5], [Af 50-2], [Af 59-1], [Af 59-2], [Db 6-1979], [Dh-I], [N-1980], [Hw-1982],
" -v- fricii (See Cereus fricii),
[NN], [Ah], [Bo-2], [Hr], [Hv], [Hy],
" -v- jujuyensis.....[Bu], [Bv-I], [Hy], [Af 29-5]
" -v- luteispinus (See C. luteispinus),
[Hr],
" -f. (hybrid).....[Aj 30-9-1979],
" -f. cristata.....[Hy], [Af 13-4],
sucrensis (See C. buchtienii, C. ressinianus, C. tupizensis -v- sucrensis),
[Af 24-5],
tarijensis.....[Hi-2], [Af 28-2],
tenuiserpens (See Bolivicereus tenuiserpens, Borzicactella tenuiserpens, Borzicactus tenuiserpens).....[Bo-2],
tominensis (See Borzicactus tominensis, Cereus tominensis),
[Bo-2], [Af 17-7],
tupizensis.....[F], [NN], [Bo-2], [Hi-2], [Hv], [Hx*], [Hy],
" -v- sucrensis (See C. buchtienii, C. ressinianus, C. sucrensis),
[Bo-6],
vallegrandensis.....[DD], [Af 33-3],
variispinus.....[Hi-2],
villaazulensis (See C. morawetzianus).....[Hi-4],
villamontesii.....[Hi-2], [Af 33-3],
" -v- longiflorior (invalid name).....

CLEISTOCACTUS.....(Cont'd),

viridialabastri.....[Af 35-6],
viridiflorus.....[DD], [Il],
vulpis-caudae.....[Bt], [Hi-2], [Hv], [Hx*], [Iz],
[Aj 13-3-1962],
wendlandiorum (See C. brookei, C. flavescens -f.- wendlandiorum),
[T], [NN], [PP], [QQ], [Av], [Ay], [Bo-2], [Bu], [Gr], [Hu], [Il], [Jd], [Aj 6-2-1955], [Dc 5-1979*],
winterae (See Hildewintera aureispina)
Spec. FR-106.....[Bo-6],
Spec. FR-108.....[Bo-6],
Spec. HU-373.....[Aj 27-6-1976],

CLISTANTHOCEREUS

calviflorus (See Borzicactus calviflorus),
fieldianus (See Borzicactus fieldianus),
[Ay], [Bo-2], [Hx*], [Hy],
hertlingianus (See Azureocereus hertlingianus, Azureocereus nobilis, Browningia hertlingianus, Browningia nobilis),
plagiostoma (See Borzicactus plagiostoma),
roezlii (See Borzicactus roezlii, Borzicactus neoroezlii, Seticereus roezlii),
samnensis (See Borzicactus samnensis),
tesselatus (See Borzicactus fieldianus -v- tesselatus, Borzicactus tesselatus)..[Flower - DD], [Bo-4],

COCHEMIEA

halei (See Mammillaria halei),
[RR], [Bo-5], [Bq], [Hx*], [Jd], [Aa 1-1984], [Af 2-4], [Af 16-2], [Cu 7-4-1939], [Hl 2-1974],
maritima (See Mammillaria maritima),
[Ar], [Ay], [Bq], [Hy], [Il], [Aa 1-1984], [Ae-1981], [Au 35-2], [Aw 1-1972],
pondii (See Mammillaria pondii),
[Hx*], [Pq], [Ae-1983], [Aw 2-1988], [Hm 3-1979], [Iv 12-1957],
poselgeri (See Mammillaria poselgeri),
[D], [I], [DD], [NN], [XX], [Ar], [Ay], [Bo-5], [Bq], [Bt], [Bv-1], [Cz], [Gr], [Hx], [Hy], [Iu], [Jd], [Aa 1-1984], [Au 35-2], [Aw 3-1979], [Db 6-1979],
roseana (See C. poselgeri, Mammillaria poselgeri),
setispina (See Mammillaria setispina),
[XX], [Ar], [Ay], [Bo-2], [Bq], [Bv-1], [Dq], [Gr], [Hx*], [Hy], [Jd], [Aa 1-1984], [Ae-1981], [Af 54-6], [Aw 1-1965], [Cu 6-2-1937],

COCHISEIA

robbinsorum (See Coryphantha robbinsorum, Escobaria robbinsorum),
[An], [Hx*], [Af 50-6],

COLEOCEPHALOCEREUS

aureispinus (See Pilosocereus aureispinus).....[Jd],
aureus (See Buiningia aurea),
[S], [DD-468], [Hi-4], [Ct-1988],
braunii (See Cephalocereus braunii).....[Ct-1988],
brevicylindricus (See Buiningia brevicylindrica),
[G], [DD-469], [Av], [Ay], [Bt],
[Au 37-1],
" -v- elongatus (See Buiningia brevicylindrica -v- elongata),
" -v- longispinus (See Buiningia brevicylindrica -v- longispina),
buxbaumianus.....[Ba 2-1974],
decumbens (See Cephalocereus decumbens),
[Hi-1], [Aj 160-161, 1968],
diersianus.....
elongatus (See Buiningia brevicylindrica -v- elongata),
estevesii (Austrocephalocereus estevesii, Micranthocereus estevesii),
[Aj 29-9-1978],
flavisetus (See Cephalocereus flavisetus),
fluminensis (See Austrocephalocereus fluminensis, Cephalocereus fluminensis).....[DD], [Bo-4],
[Hx*], [Hy], [Jd], [Aj 27-4-1976],
" -v- braamhaarii (See Cephalocereus fluminensis -v- braamhaarii)..[Aj 33-6-1982],
goebelianus (See Austrocephalocereus purpureus sensu Backbg., 1938, Cephalocereus purpureus sensu Br. & R. 1920, Cephalocereus purpureus sensu Werd. 1933, Also shown as Austrocephalocereus albicephalus in [Af 58-1]),
[Hi-1], [Hx*], [Il], [Aj 34-3-1983],
[Ba 12-1981],
luetzelburgii (See Cephalocereus luetzelburgii, Pilocereus luetzelburgii, Pilosocereus luetzelburgii),.....[Jd],
pachystele (See Coleocephalocereus goebelianus),
paulensis (See Cephalocereus paulensis),
[Hi-1], [Hx*], [Ba 12-1981],
pleurocarpus (See Cipocereus pleurocarpus),
pluricostatus (See Cephalocereus pluricostatus),
[NN], [Ay], [Jd],
purpureus (See Buiningia purpureus),
[G], [DD-470], [Ay], [Ct-1988],
roseiflora (See Buiningia roseiflora),

COLORADOA

mesae-verdae (See Echinocactus mesae-verdae, Pediocactus mesae-verdae, Sclerocactus mesae-verdae),
[M*], [P], [Y], [DD], [Ar], [Ay], [Bo-5], [Dq], [Go], [Hx*], [Hy], [Jd], [Ad -], [Aj 30-3-1979],
" -f. cristata.....[Go],
Spec. .....[Ar], [Hy],

CONSOLEA

acaulis (See Opuntia acaulis),
corallicola (See C. spinosissima, Opuntia spinosissima).....[Af 2-5],
falcata (See Opuntia falcata),
[Cz],
leucacantha (See Opuntia leucacantha),
macracantha.....[Gu III - 1953], [Hl 3-1972], [Hs IV 3-1983],
millspaughii (See Opuntia millspaughii),
moniliformis (See Opuntia moniliformis),
[Bo-1], [Af 56-3], [Aj 30-2-1979],
multispina (See Opuntia spinosissima, C. spinosissima),
nashii (See Opuntia nashii).....[Ay],
rubescens (See Opuntia rubescens),
[DD], [Ar], [Bl], [Bo-1], [Hx], [Hy], [Af 2-5], [Da 1-1981],
spinosissima (See C. corallicola, Opuntia spinosissima),
[TT], [Ax], [Bo-1], [Af 2-5],
urbaniana (See O. urbaniana),
Spec. .....[Af 56-3],

COPIAPOA

albatus (See C. cinerea -v- albata),
albispina (See C. cinerea -v- albispina),
alticostata.....[Ar], [Hi-3],
" -v- minima.....
applanata (Invalid name - See C. cinerascens -v- intermedia, C. intermedia),
[Ar], [Bo-3], [Hy],
atacamensis (See C. calderana),.....[Cu 2-1981],
aurata.....[Hl 3-1971],
aureispina.....[Ar],
barquitensis (See C. hypogaea -v- barquitensis),
[Ae-1981], [Hx],
boliviana (See C. echinoides, or C. atacamensis)..Hi-3]
borealis (See C. echinata -v- borealis, C. totoralensis),

COPIAPOA.....(Cont'd),

bridgesii.....[Ar], [Bo-3], [Hb], [Hi-3], [Hy],
[Cu 2-1981],
brunescens.....[Bo-3],
calderana (See C. lembckei),
[Hi-3], [Gt xiv No. 65, 197 - 1959],
[Iv 12-1959],
" -v- spinosior.....[Hi-3],
carrizalensis....[Ar], [Hi-3], [Af 53-6], [Aj 26-2-1975]
" -v- gigantea.....[Hi-3], [Af 50-5], [Af 53-4],
" -f. cristata.....[Af 53-6], [Aj 22-11-1971],
castanea (See C. serpentisulcata -v- castanea),
[Bo-6],
chaniaralensis.....[DD], [Bo-6], [Gr], [Hi-3], [Hx*],
[Hy],
cinerascens.....[Hi-3], [Hy], [Af 56-3], [Cu 2-1981],
" -v- intermedia (See C. applanata, C. intermedia)
[Hi-3],
cinerea.....[I], [T], [Z], [NN], [QQ], [RR], [Ar], [Av],
[Ay], [Bo-3], [Bt], [Dq], [Gr], [Hi-3], [Hv]
[Hx], [Hy], [Il], [Iz], [Jd], [Ad Sept.,
1980], [Ae-1981], [Af 50-5], [Af 59-2], [Ah-
1986], [Aj 8-10-1957], [Am 9-1981], [Au 37-
1], [Bm-1987], [Cu 2-1981], [Dh-2], [Hh Jan.
- 1975],
" -v- albispina (See C. albispina),
[Ar], [Bo-3], [Hi-3], [Hv],
[Aj 22-11-1971], [Cu 2-1981],
[Hl 6-1978],
" -v- columna-alba (See C. columna-alba),
[Ar], [Hb], [Ad Sept. 1980],
[Af 56-2], [Af 57-3],
[Cu 2-1981],
" -v- cremnophilla.....[Am 9-1981],
" -v- dealbata (See C. dealbata),
[Gr], [Hy], [Aj 24-8-1973],
[Am 9-1981], [Bi 4-#1, 1986],
[Cu 2-1981],
" -v- flavescens (See C. columna-alba),
" -v- gigantea (See C. gigantea),
" -v- haseltoniana (See C. haseltoniana, C.
gigantea -v- haseltoniana)
[Ij], [Am 9-1981], [Cu 2-1981],
" -v- uhligiana (See C. uhligiana),
" -v- viderispina.....
" -f. (spiral).....[Bi 3-3],
columna-alba (See C. cinerea -v- columna-alba),
[Av], [Bo-3], [Hi-3], [Ae-1981],
[Hl 4-1970], [Iv 12-1959],
" -v- nuda (See C. cinerea -v- columna-alba),
[Hi-3], [Aj 22-12-1971],

columna-alba.....(cont'd),
" -f. cristata.....[Aj 22-12-1971],
conglomerata (See C. ferox, C. solaris, Pilocopiapoa solaris),
[Aj 17:29-31 Feb. 1966],
coquimbana (See C. pseudocoquimbana, Echinocactus coquimbanus),
[RR], [Bo-3], [Dq], [Gr], [Hi-3], [Hx], [Iz], [Hl 4-1967],
" -v- armata (See C. coquimbana -v- wagenknechtii) [Hi-3],
" -v- coquimbana.....
" -v- domeykoensis (See C. coquimbana -v- domeykoensis),
" -v- vulgata.....
" -v- wagenknechtii (See C. coquimbana),
[DD], [Hb], [Cu 2-1981],
cuprea.....[Bo-3], [Hi-3], [Af 53-6], [Aj 22-11-1971]
cupreata (See Neoporteria tuberisulcata -v- cupreata),
[Ad-1981], [Bo-3], [Af 53-6],
dealbata (See C. cinerea -v- dealbata),
[Av], [Bo-3], [Hi-3], [Ae-1981], [Hl 4-1970],
desertorum.....[Bo-6], [Hi-3], [Af 50-5],
domeykoensis (See C. pseudocoquimbana -v- domeykoensis)
dura.....[Bo-6], [Hi-3], [Hy],
" -v- carrizalensis (See C. carrizalensis),
echinata.....[Bo-3], [Bo-6], [Hi-3], [Iv 4-1959],
" -v- borealis (See C. borealis, C. totoralensis) [Hi-3],
" -v- pulla.....
echinoides.....[Hx], [Hy], [Cu 2-1981],
" -v- salm-dyckiana.....[Hy],
echinus (See C. echinata),
eremophila.....[Hi-3],
" -f. cristata.....[Hi-3],
esmeraldana.....[Hi-3], [Dh-II],
fiedleriana (See pepiniana -v- fiedleriana),
[E], [Hi-3], [Hx], [Cu 2-1981], [Dh-II],
ferox (Invalid name - See C. conglomerata, C. solaris),
[Bo-3], [Aj 26-4-1975],
gigantea (See C. cinerea -v- gigantea),
[Ar], [Bo-3], [Hx*], [Ae-1981],
" -v- gigantea.....[Hi-3],
" -v- haseltoniana (See C. haseltoniana, C. cinerea -v- haseltoniana),
[Hi-3],
goldii (See C. aurata),
grandiflora.....[DD], [Ar], [Hi-3], [Hx*], [Hy],
[Cu 2-1981], [Dh-II], [Hl 6-1967],

haseltoniana (See C. cinerea -v- haseltoniana, C. gigantea -v- haseltoniana),
[F], [T], [Ar], [Bo-3], [Hx*], [Hy], [Ad-1983], [Hh March 1975], [Hl 4-1979],
hornilloensis (See C. rupestris),
[Hi-3, A-973],
humifusa (See C. humilis),
[Hy],
humilis (See C. humifusa),
[E], [H], [DD], [Av], [Bo-3], [Bo-6], [Hb], [Hi-3], [Hx], [Ba-9-1978], [Ct-1983], [Cu 2-1981],
hypogaea.....[G], [Z], [NN], [Ar], [Bo-6], [Hi-3], [Hv], [Hx], [Iz], [Af 52-4], [Ag-1981], [Aj 31-8-1980], [Cu 1-1979], [Dh-II], [Hl 3-1974],
" -v- barquitensis (See C. hypogaea)....
[Hi-3],
" -f. (grafted).....[Hi-3],
imbricata.....(See C. wagenknechtii -v- imbricata),
[Dh-II],
" -v- rubriflora (See C. rubriflora),
intermedia (See C. applanata, C. cinerascens -v- intermedia),
[Bo-6], [Aj 24-5-1973],
krainziana.....[NN], [Av], [Bo-3], [Bo-6], [Hi-3], [Hu], [Hx*], [Hy], [Iz], [Jd], [Af 56-2], [Ah-1980], [Aj 26-2-1975], [At-1982], [Cu 2-1981], [Dh-II],
" -v- scopulina (See C. scopulina),
[Af 53-2], [Af 56-4],
laui.....[Bf], [Af 54-4], [Aj 31-12-1980],
lembckei (Invalid name - See C. calderana),
[Ar], [Ay], [Bo-3], [Hx*], [Hy], [Am 9-1983], [Dh-II], [Hl 1-1967], [Hl 4-1970], [Hl 6-1971]
longispina.....[DD], [Hi-3], [Hx], [Dh-II],
longistaminea (See C. microsperma),
[Ar], [Hi-3], [Hx*], [Af 56-3],
magnifica.....(See C. marginata -v- magnifica),
magnimamma.....[Hy],
malletiana.....[Bo-3], [Hx*], [Hy], [Aj 6-1937],
marginata (See C. streptocaulon),
[DD], [RR], [Bo-3], [Bo-6], [Hi-3], [Hx*], [Hy], [Cu 2-1981],
" -v- magnifica (See C. magnifica),
megarhiza (See C. echinata -v- borealis),
[Hi-3], [Hy], [Aj 26-3-1975], [Cu 2-1981],
" -v- microrhiza.....
melanohystrix.....[Hi-3],
microsperma (See C. longistaminea),
mollicula.....[Hi-3], [Ad 3-1982], [Af 56-4],

COPIAPOA.....(Cont'd),

montana.....[J*], [EE], [Ar], [Bo-3], [Bo-6], [Hi-3], [Hy], [Il], [Iz], [Cu 2-1981],
neocoquimbana (See C. coquimbana), [Hy],
" -v- wagenknechtii (See C. coquimbana -v- wagenknechtii),
olivana.....[Hi-3], [Af 56-2],
paposoensis.....[Hi-3],
pendulina.....[F], [Bo-3], [Bo-6], [Hi-3], [Aj 22-12-1971],
pepiniana.....[Bo-3], [Hy],
" -v- fiedleriana (See C. fiedleriana), [Bo-3], [Hx], [Jd],
pseudocoquimbana (See C. coquimbana), [T], [Hy], [Hl 2-1967],
" -v- chaniaralensis (See C. chaniaralensis), [Hi-3],
" -v- domeykoensis (See C. domeykoensis), [Hi-3],
" -v- pseudocoquimbana.....[Hi-3],
" -v- vulgata (See C. coquimbana), [Hi-3],
rarissima.....[Hi-3],
rubriflora.....[Bo-3], [Hi-3], [Hy],
rupestris.....[G], [Ar], [Bo-3], [Hi-3], [Dh-II],
" -v- hornilloensis (See C. hornilloensis).....
" -v- rubriflora.....
scopulina (See C. krainziana -v- scopulina), [Bo-3], [Bo-6],
serenana.....[Aj 27-8-1976]
serpentisulcata.....[Bo-3], [Bo-6], [Hi-3], [Hx*], [Af 50-6], [Cu 2-1981],
" -v- castanea (See C. serpentisulcata)..[Hi-3],
solaris (See C. ferox, Pilocopiapoa solaris), [Ay], [Hi-3], [Jd], [Aj 2-12-1971], [Cu 2-1981], [Dh-II],
" -v- luteispina.....
streptocaulon (See C. marginata), [DD],
taltalensis.....[Bo-6], [Hi-3],
tenebrosa.....[Hi-3],
tenuissima.....[R], [X], [NN], [Ar], [Bo-6], [Hi-3], [Hv], [Hx], [Hy], [Iz], [Ae-1981], [Aj 31-8-1980], [Am 8-1979], [Am 9-1981], [Cu-2-1981], [Dh-II],
" -v- monstrosus.....
tocopillana.....[Hi-3], [Af 55-3], [Dh-II],
totoralensis (See C. echinata -v- borealis), [Bo-6], [Iv 3-1960],

COPIAPOA.....(Cont'd),

uhligiana (See C. cinerea -v- uhligiana),
[Da 1-1979],
vallenarensis.....[Ar], [Hi-3],
variispinata.....[Hi-3], [Aj 31-8-1980],
" -f. (fine-spined).....[Hi-3],
vulgata (See C. coquimbana -v- vulgata),
wagenknechtii (See C. coquimbana -v- wagenknechtii),
[DD],
" -v- imbricata (See C. imbricata),

Spec. .....[Bo-6],
Spec. de Chanaral.....[Hy],
Spec. KK-1385.....[Ar],
Spec. Tigrillo.....[Ar],
Natural Hybrid - C. cinerea X C. haseltoniana...[Bo-6],

CORRYOCACTUS

acervatus.....[Hi-4],
apiciflorus (See Erdisia apiciflora, Erdisia maxima),
[Jd],
apurimacensis (See C. ayacuchoensis),
aureus (See Erdisia meyenii),
[Hx*],
aureispinus (See C. erectus, Erdisia aureispina,
Erdisia erectus),
ayacuchoensis.....[Bo-2], [Hi-4],
" -v- fuscus.....
" -v- leucacanthus (See C. ayacuchoensis),
[Bo-2],
ayopayanus (See C. tarijensis),
[PP], [Bo-2], [Hy], [Il], [Af 39-1],
brachycladus.....
brachypetalus.....[DD], [RR], [Ay], [Bo-2], [Hi-4],
[Hy], [Jd], [Aj 23-1-1972],
brevispinus.....[Bo-2],
brevistylus (See C. krausii).....[RR], [Ay], [Hi-3],
[Hx*], [Hl 2-1971], [Hl 4-1982],
" -v- puquiensis (See C. pachycladus, C.
puquiensis),
chachapoyensis (Invalid name - See Cereus chachapoy-
ensis).....[Bo-2],
charazanensis (See C. melanotrichus),
[Bo-2], [Iv 12-1957],
chavinilloensis.....
corroanus.....(See Samaipaticereus corroanus),
cuajonesensis.....
erectus (See Erdisia aureispina, Erdisia erecta),
[Hi-4], [Iz],
gracilis.....[Hi-4],

CORRYOCACTUS.....(Cont'd),

heteracanthus (See C. ayacuchoensis),
huincoensis.....
incuiensis (See C. pachycladus),
inquisivensis (See Samaipaticereus inquisivensis, Yungasocereus inquisivensis, Yungasocereus microcarpus),
krausii (See C. brevistylus),
[Bo-2], [Hl 2-1967], [Hl 4-1967],
matucanensis.....[Hi-4],
maximus (See Erdisia apiciflorus, Erdisia maxima),
megarhizus.....
melaleucus.....
melanotrichus (See C. otuyensis),
[F], [Bo-2], [Dq], [Hi-2], [Hx*], [Hy], [Ij], [Jd], [Af 15-11],
" -v- caulescens (See C. melanotrichus),
[Bo-2]. [Ew -Rev. Agric. Cochabamba, Bol. No. 7:20 Fig. 1952],
odoratus.....[Hi-4, page 1285, Fig. A-1150],
otuyensis (See C. melanotrichus),
[Gs No. 78, 88, - 1963],
pachycladus (See C. brevistylus -v- puquiensis, C. puquiensis),
[Bo-2], [Hx*],
perezianus (See C. tarijensis),
[Ew -Rev. Agric. Cochabamba Bol. No. 7:22 Fig. 1952],
pilispinus.....
prostratus.....[Hi-4],
pulquinensis (See C. melanotrichus),
[Da],
puquiensis (See C. brevistylus -v- puquiensis, C. pachycladus),
[Ay], [Bo-2], [Hx*], [Af 51-5], [Aj 16-6-1965], [Aw 4-1964], [Hl 4-1966],
pyroporphyranthus.....[Hi-4],
quadrangularis (See Erdisia quadrangularis),
quivillanus.....
serpens.....[Hi-4],
solitarius.....
spiniflorus (See Austrocactus spiniflorus, Erdisia spiniflora),
squarrosus (See Erdisia squarrosa),
tarijensis.....[Bo-2], [Hi-2],
tenuiculus (See Erdisia fortalezensis, Erdisia tenuicula).....[Hi-4],

CORYNOPUNTIA

agglomerata (See Opuntia agglomerata).....[Bo-1],

CORYNOPUNTIA.....(Cont'd),

barkleyana (See Micropuntia barkleyana, O. barkleyana),
bulbispina (See Opuntia bulbispina),
[Ay], [Bo-1], [Hx*], [Aj 29-3-1978],
clavigera (See C. clavata, Opuntia clavata),
[Hy],
clavata (See C. clavigera, Opuntia clavata),
[An], [Bo-1], [Hx*],
dumetorum (See Opuntia dumetorum, Platyopuntia dumetorum),
gracilicylindrica (See C. pulchella, Opuntia pulchella),
grahamii (See Grusonia grahamii, Opuntia grahamii, Opuntia schottii -v- grahamii),
[Bo-1],
invicta (See Cylindropuntia invicta, Opuntia invicta),
[Ar], [Ay], [Bo-1], [Bt], [Hx*], [Hy],
moellerii (See Opuntia moellerii),
[Bo-1], [Hy], [Ba 1-1974],
parishii (See Opuntia parishii, Opuntia stanlyi -v- parishii),
[An], [Hy*],
planibulbispina (See Opuntia schottii -v- grahamii),
[Bo-6],
pulchella (See Micropuntia brachyrhopalica, Opuntia pulchella),
[An], [Bo-1],
reflexispina (See Opuntia reflexispina),
schottii (See Grusonia schottii, Opuntia schottii),
[Hy*],
" -v- grahamii (See C. grahamii, Opuntia grahamii, Opuntia schottii -v- grahamii),
stanlyi (See Cylindropuntia stanlyi, Opuntia stanlyi),
[An], [Bo-1],
" -v- kunzei (See Opuntia kunzei, Opuntia stanlyi, -v- kunzei).....[An],
" -v- parishii (See C. parishii, Opuntia parishii, Opuntia stanlyi -v- parishii),
[Hy*],
" -v- peeblesiana (See Opuntia stanlyi -v- peeblesiana),
[An],
" -v- wrightiana (See Grusonia wrightiana, Opuntia stanlyi -v- wrightiana, Opuntia wrightiana),
[Bo-1],
vilis (See Opuntia vilis),
[Bo-1],
" -v- bernhardinii (Invalid name)......[Bo-6],
(cultivated),

CORYPHANTHA

aggregata (See Echinocereus triglochidiatus -v- melanacanthus, C. vivipara -v- aggregata, C. vivipara -v- bisbeeana, Escobaria bisbeeana, Escobaria vivipara -v- bisbeeana), [A], [RR], [Af 3-9], [Af 6-6], [Cu 4-1970],
albicolumnaria (See C. orcuttii, Escobaria albicolumnaria, Escobaria orcuttii -v- albicolumnaria, Escobaria orcuttii, Mammillaria albicolumnaria),
alversonii (Invalid name - See C. vivipara -v- alversonii, Escobaria vivipara -v- alversonii), [P], [FF], [An], [Ar], [Bo-5], [Cv], [Ad May - 1979], [Af 5-10], [Af 55-2],
ancistrocantha (See C. clavata -v- ancistrocantha),
andreae.....[Ar], [Ay], [Bv-1], [Cx], [Gr], [Hu], [Hw-1981], [Hx*], [Hy], [Af 29-5], [Aj 24-1-1973], [Aj 28-1-1977], [Ba 1-1978], [Flower - Ba 3-1979],
arizonica (See C. vivipara -v- arizonica, Mammillaria arizonica, Mammillaria vivipara -v- arizonica) [P], [FF], [LL*], [RR], [An], [Ay], [Cb], [Df], [Af 5-10], [Hl 5-1968],
asperispina (See Neobesseya -v- asperispina), [Aj I-10-1929],
asterias.....[Bv-1], [Hx*], [Af 29-5], [Aw 3-1971],
bergeriana.....[Ar], [Bo-5], [Hy],
bernalensis.....[Af 56-4],
bonantii.....[V],
borwigii (See C. difficilis), [W], [Ar], [Bo-5], [Iz], [Af 56-4], [Aw 3-1980], [Bi 2-1987],
bumamma.....[NN], [RR], [Ar], [Bv-1], [Cx], [Hx], [Hy], [N-1980], [Ad May 1979], [Af 53-4], [Aw 2-1967], [Ba 3-1979], [Bm-1986], [Hl 5-1979],
calcarata (See C. sulcata, Mammillaria sulcata),
" -v- sulcata (See C. sulcata), [Aj IV-4-1932],
calipensis.....[Av], [Hv], [Iz], [Aj 23-3-1972], [Aw 4-1964], [Bi-2-1987], [Bm-1989], [Hl 2-1974],
calochlora.....
caupescens.....[NN],
ceratites (See Neolloydia connoidea -v- ceratites),
chaffeyi (See Escobaria chaffeyi),
chihuahuensis (See Escobaria chihuahuensis),
chlorantha (See C. vivipara -v- desertii, C. desertii, Escobaria chlorantha, Escobaria desertii), [RR], [Bo-5], [Af 4-1],
" -v- desertii (See C. desertii, C. vivipara -v- desertii).....[Ar], [Bo-5], [Hy],

CORYPHANTHA.....(Cont'd),

clava.....[H], [I], [L], [S], [Ar], [Bv-1], [Ck], [Cx], [Dq], [Hx*], [Hy], [Iu], [Af 1-10], [Cu 4-1970],
" -v- schlechtendalii.....
clavata (See C. rhaphidacantha, Neolloydia clavata), [EE], [QQ], [RR], [Ay], [Bo-5], [Gr], [Hy], [Jd],
" -v- ancistracantha.....
" -v- radicantissima.....[Bo-6], [Af 55-6],
columnaris (See C. vivipara -v- radiosa), [Bo-5], [Cz], [Af 6-2],
compacta.....[A], [RR], [Ay], [Bo-5], [Bv-1], [Cx], [Hy], [Af 29-5], [Aw 2-1978],
conimamma.....
connivens.....[Cx], [Aa 6-1982], [Iv 1-1948],
conoidea (See Echinocactus conoideus, Neolloydia conoidea),
cornifera (See C. pfeifferiana), [L], [RR], [XX], [Ar], [Bd], [Be*], [Bk], [Bo-5], [Bv-1], [Cx], [Hb], [Hx*], [Af 29-5], [Hl 3-1965],
" -v- echinus (See Mammillaria echinus), [Ax], [Df],
cornuta.....
cubensis (See Escobaria cubensis, Neobesseya cubensis, Neolloydia cubensis), [RR], [Gu III-1953],
cuencamensis.....[Af 52-4],
daimonoceras.....[Ar], [Ay], [Hy],
" -v- jaumavei (See C. jaumavei),
dasyacantha (See Escobaria dasyacantha, Escobesseya dasyacantha, Mammillaria dasyacantha), [Df], [Af 52-6],
" -v- dasyacantha.....[Ax],
" -v- varicolor (See Escobaria varicolor, Mammillaria varicolor), [Y], [Ax], [Bo-5], [Df],
delaetiana.....[Bo-5], [Hy],
delicata.....[Af 51-2],
densispina.....[Hl 5-1983],
desertii (See C. chlorantha, C. chlorantha -v- desertii, C. vivipara -v- desertii, Escobaria chlorantha, Escobaria desertii), [E], [P], [Q], [CC], [FF], [LL], [RR], [An], [Cb], [Cf], [Af 5-10],
difficilis (See C. borwigii), [Bo-5], [Hr], [Hx*], [Hy], [Af 56-4], [Aw 4-1978], [Cu 3-1962], [Cu 4-1970],
duncanii (See Escobaria duncanii, Escobesseya dun-duncanii, Mammillaria duncanii).....[Ax], [Df],

CORYPHANTHA.....(Cont'd),

durangensis.....[RR], [Ar], [Bo-5], [Hx*], [Hy],
[Af 57-5], [Aj 24-1-1973], [Aw 3-1971],
echinoidea.....[Aa 6-1982], [Aw 2-1979], [Cu 2-1972],
" -v- paradensis.....
" -f. cristata.....[Hy],
echinus (See C. cornifera -v- echinus, Mammillaria echinus),
[E], [Y], [DD], [GG], [LL*], [RR], [Ar], [Bo-5],
[Bv-1], [Df], [Hn], [Hy], [Af 6-7], [Hl 5-1983]
[Af 60-1-1988],
elephantidens.....[L], [AA], [RR], [Ar], [Ay], [Bd],
[Bo-5], [Bu], [Cm], [Cz], [Dq], [Hv], [Hx],
[Il], [Iu], [Iz], [Ja], [Jd], [Af 57-4],
[Ag-1988], [Ah-1986], [Aj 5-1981], [Aw 2-1969], [Aw 2-1973], [Bi 2-1987], [Bm-1987],
[Hl 5-1979], [Bi Vol. 4, #6],
" -v- barciae.....
" -v- elephantidens.....[Hl 5-1979],
" -v- roseiflora.....[Hy],
erecta.....[L], [RR], [Ar], [Ay], [Bd], [Bk], [Bo-5],
[Cm], [Cx], [Gr], [Hx], [Hy], [Ja], [Jd],
[Aw 3-1965], [Aw 4-1965],
" -v- albispina.....
exsudans.....[Bo-5], [Hx*],
fragrans (See C. vivipara -v- radiosa, Mammillaria fragrans),
[Y], [Df],
garessii.....[Af 52-2],
georgii....[Bo-5], [Hx*], [Hy], [Aw 4-1966], [Aw 3-1979], [Bi 2-1987],
gladiispina.....[Bo-5], [Hy], [Aw 4-1972],
gracilis.....[Hx*], [Aj 30-1-1979],
grahamii (See Mammillaria grahamii),
[Q], [An],
grandiflora (See Neolloydia grandiflora),
grandis.....[Ae-1981], [Af 50-3],
grata.....[Af 53-6],
greenwoodii.....[At], [Ae-1982], [Af 55-3], [Aj 23-7-1972], [Aj 24-8-1973], [Aw 2-1970],
[Aw 4-1987], [Hl 1-1972], [Hl 5-1979],
guerkeana.....[Bo-5], [Af 54-6],
henricksonii (See Escobaria henricksonii)....[Af 51-3],
hesteri (See Escobaria hesteri, Mammillaria hesteri),
[D], [H], [Y], [Ax], [Df], [Af 4-5], [Af 20-11],
[Af 55-6], [Hl 4-1974],
heyei.....[Hr],
impexicoma.....[Bo-5],
indensis.....[Aw 4-1977], [Aw 1980-1983 Index],
jalpanensis.....[Aw 2-1965],
jaumavei (See C. daimonoceras -v- jaumavei),

CORYPHANTHA.....(Cont'd),

kieferiana.....[Bo-5],
laredoi (See Escobaria laredoi, Escobaria rigida),
laui.....[Af 51-6],
lloydii (See Escobaria lloydii, Mammillaria lloydii),
longicornis.....[Bo-5], [Hb], [Af 57-5], [Aw 3-1981],
macdougalii (See Ortegocactus macdougalii),
macromeris (See Lepidocoryphantha macromeris, Mammillaria macromeris),
[E], [M*], [P], [U], [Y], [LL*], [An], [Ar], [Av], [Bv-1], [Bz], [Cz], [Df], [Iz], [N-1980] [Af 52-4], [Af 53-3], [Af 57-3], [Af 57-6], [Af 60-1-1988], [Bi 2-1987],
" -v- macromeris.....[Ax],
" -v- runyonii (See C. runyonii, Lepidocoryphantha runyonii, Mammillaria runyonii),
[M*], [Ar], [Ax], [Df], [Hv], [Hy],
" -v- viola.....
" -f. monstrosus.....
macrothele.....[Cu 3-1962],
maiz-tablasensis.....[NN], [Ar], [Iz], [Af 56-4], [Aw 2-1979],
maliterrarum.....[Af 56-2],
marstonii (See C. missouriensis -v- marstonii),
melleospina.....
michoacanensis.....[Hb],
minima (See C. nellieae, Escobaria minima, Escobaria nellieae, Mammillaria nellieae),
[M*], [Y], [Z], [Ax], [Df], [Af 55-6], [Ag-1985-1986], [Hw-1986],
missouriensis (See Escobaria missouriensis, Neobesseya missouriensis),
[An], [Af 53-3],
" -v- caespitosa (See Mammillaria similis, Neobesseya similis),
[Ax], [Df],
" -v- marstonii (See Escobaria missouriensis -v- marstonii, Neobesseya missouriensis -v- marstonii)....[An], [Ax],
" -v- missouriensis.....[Ax],
" -v- robustior (See Neobesseya similis -v- robustior),
" -v- similis (See C. missouriensis -v- caespitosa),
muhlbaueriana.....[Aj 2-1-1951],
muehlenpfordtii (See C. scheeri -v- scheeri),
[B], [CC], [Bo-5], [Bv-1], [Bz], [Af 39-2], [Am 1-1985],
" -v- robustispina (See C. scheeri -v- robustispina).....[B], [P], [An], [Ar],

CORYPHANTHA.....(Cont'd),

neglecta.....[Aw 1-1979],
nellieae (See C. minima, Escobaria nellieae, Mammillaria nellieae),
[F], [H], [Ar], [Af 55-6], [Am 9-1978],
neomexicana (See C. vivipara -v- neomexicana, Mammillaria vivipara -v- neomexicana),
[D], [U], [LL*], [RR], [Bv-1], [Bz], [Df],
[Af 6-6], [Af 29-5],
neoscheeri (See C. scheeri -v- scheeri),
[Bo-5], [Hx*],
neovivipara (See C. vivipara).....[Hy],
nickelsae (See C. sulcata -v- nickelsae),
[RR], [Ar], [Bo-5], [Hy], [Aw 1-1979],
[Aw 4-1980],
nivosa (See Mammillaria nivosa).....[Hy],
nuttallii (See C. missouriensis -v- missouriensis),
obscura.....[Bo-5], [Af 4-4], [Aj 2-2-1951],
octacantha.....[Bo-5], [Cx], [Hy],
odorata (See Cumarinia odorata, Neolloydia odorata),
oklahomensis (See C. radiosa, C. vivipara -v- radiosa),.....[Af 22-6],
orcuttii (See C. strobiliformis -v- orcuttii, Escobaria orcuttii).....[An],
" -v- koenigii.....
organensis (See C. sneedii -v- sneedii),
[Ax], [Hx*], [Af 56-2], [Hl 5-1974]
orthacantha (See C. muehlenpfordtii, C. neoscheeri, C. scheeri).....[Hy],
ottonis (ottonianus).....[Bo-5], [Hy],
pallida.....[C], [H], [Ar], [Hx*], [Hy], [Iu], [Jd],
[Af 1-9], [Aj 28-12-1977], [Aw 2-1964],
[Ba 4-1980],
palmeri.....[E], [CC], [NN], [Ar], [Bo-5], [Bu], [Gr],
[Hx*], [Jd], [Cu 3-1962], [Cu 4-1970],
[Hl 3-1979],
pectinata (See C. cornifera -v- echinus),
[GG], [LL*], [RR], [Bo-5],
pfeifferiana (See C. cornifera).....[Hy],
" -f. cristata.....[Hy],
pirtlei (See C. macromeris -v- runyonii),
poselgeriana.....[G], [W], [CC], [Ar], [Hv], [Hy], [Iu],
[Af 58-2], [Am 9-1978], [Aw 1-1977],
[Hw-1982],
" -v- saltillensis.....[Aw 1-1977],
" -v- valida (See C. valida),
[Bo-5], [Hx], [Ae-1981],
potosiana.....[Af 42-6], [Af 43:7 Jan.-Feb. - 1977],
[Aw 1-1976],
pottsii (See Mammillaria leona, Mammillaria pottsii),
[Df],

CORYPHANTHA.....(Cont'd),

pseudoechinus.....[Ar], [Bo-5], [Hy], [Af 54-3],
[Af 56-4], [Hl 3-1979],
pseudonickelsae.....[NN], [Ar], [Bo-5], [Hy],
pseudoradians.....[Eb - Anal. Inst. Biol. Mex. 25:528
Fig. 25],
pulleineana (See Neolloydia pulleineana),
[Z], [Hx*], [Af 42-4], [Aw 2-1968],
[Hl 2-1972],
pusilliflora.....
pycnacantha (See Mammillaria aureilanata),
[RR), [Bo-5], [Ck], [Hx], [Iz],
[Aa 6-1982], [Af 3-9], [Aj 15-3-1964],
[Cu 4-1970],
radians (See C. cornifera -v- echinus),
[E], [H], [RR], [Ar], [Ay], [Bo-5], [Bv-1],
[Cx], [Gr], [Hy], [Jd], [Af 29-5], [Aw 2-1967],
" -v- echina.....[Hy],
" -v- pectinata (See C. cornifera -v- echinus,
C. pectinata),
" -v- pectinoides.....
" -v- pseudoradians (See C. pseudoradians),
radiosa (See C. vivipara -v- radiosa, Mammillaria
vivipara -v- radiosa),
[Df], [Go], [Af 6-6], [Af 29-5],
ramillosa (See Mammillaria ramillosa),
[G], [M*], [Y], [Ax], [Df], [Bo-5],
[Af 14-12], [Af 60-1-1988],
raphidacantha.....[See C. clavata),
[Bk],
" -v- ancistracantha (See C. clavata -v-
ancistracantha),
recurvata.....[B], [D], [P], [Q], [CC], [RR], [An],
[Ar], [Ax], [Bh], [Bo-5], [Hy], [Af 52-1],
[Aw 2-1979],
" -f. cristata.....[Hy],
recurvispina.....[Aw 1-1976],
reduncispina.....[NN], [Av], [Bo-5], [Bt], [Gr],
retusa (See C. sulcifera),
[RR], [Ar], [Cx], [Hy], [Ag-1984-1985], [Aj 23-
7-1972],
" -v- melleospina.....
" -v- pallidispina (invalid name).....[Bo-5],
rhaphidacantha (raphidacantha) (See C. clavata),
robbinsorum (See Escobaria robbinsorum),
[An], [Af 50-6],
robertii (See Escobaria runyonii),
[M*], [Ax], [Df],
robustispina (See C. scheeri -v- robustispina),
[A], [D], [Ar], [Dq*],
roederiana.....[Bo-5], [Bv-1],

CORYPHANTHA.....(Cont'd),

runyonii (See C. macromeris -v- runyonii, Lepidocoryphantha runyonii, Mammillaria runyonii), [Y], [NN], [Ar], [Bz], [Df], [Hn],
salm-dyckiana (See Mammillaria deherdtiana), [CC], [RR], [Bo-5], [Jd],
scheeri (See C. muehlenpfordtii, C. neoscheeri, Mammillaria muehlenpfordtii, Mammillaria scheeri).....[M*], [Y], [An], [Df], [Iz],
" -v- robustispina (See C. robustispina), [Q], [An], [Ax],
" -v- scheeri.....[Df], [Af 57-3],
" -v- uncinata.....
" -v- valida.....[Q], [U], [An], [Ax], [Af 58-2],
schwarziana.....[NN],
scolymoides (See Mammillaria scolymoides), [Y], [Bo-5], [Df], [Hu], [Hx*], [Y], [Hl 5-1983],
similis (See C. missouriensis -v- caespitosa, C. missouriensis -v- similis, Mammillaria similis, Neobesseya similis), [Df],
sneedii (See Escobaria sneedii, Mammillaria sneedii), [E], [M*], [U], [Y], [Ax], [Bh],
" -v- leei (See Escobaria leei, Mammillaria leei) [Bh], [Df], [Af 56-2],
" -v- sneedii (See C. organensis, Escobaria sneedii -v- sneedii, Mammillaria sneedii), [Ax], [Df], [Ib], [Af 57-2],
speciosa.....[F], [Bo-5], [Aj 2-2-1951],
strobiliformis (See C. tuberculosa, Escobaria strobiliformis, Escobaria tuberculosa, Mammillaria tuberculosa), [M*], [Q], [U], [An],
" -v- durispina (See Escobaria strobiliformis -v- durispina),
" -v- orcuttii (See C. orcuttii, Escobaria orcuttii), [An], [Ax],
" -v- strobiliformis (See C. tuberculosa, Escobaria strobiliformis -v- strobiliformis, Escobaria strobiliformis, Mammillaria tuberculosa), [Ax], [Df], [Ib],
stuetzlei (See Neolloydia conoidea),
sulcata (See C. calcarata, Mammillaria sulcata), [G], [M*], [Y], [LL], [RR], [Ax], [Bo-5], [Bz], [Df], [Aj 15-3-1964], [Db 10-1979],

CORYPHANTHA.....(Cont'd),

sulcata.....(cont'd),
" -v- nickelsiae.....
" -v- sulcata (See C. calcarata -v- sulcata),
[Ax],
sulcolanata.....[RR], [Bd], [Hx*], [Il], [Af 55-4],
[Aw 4-1970], [Cu 1-1962],
sulcifera (See C. retusa).....[Hy],
tripugionacantha.....[Aw 1-1988],
tuberculosa (See C. strobiliformis -v- strobiliformis,
Escobaria strobiliformis, Escobaria tuberculosa, Mammillaria tuberculosa),
[Df],
" -v- varicolor (See C. dasyacantha -v- varicolor, Escobaria dasyacantha -v- varicolor),
unicornis.....[Bo-5], [Aa 5-1980], [Aw 2-1978],
valida.....[Aw 1-1977], [Hl 5-1965],
vaupeliana.....[L], [Bo-5], [Hx*], [Hy],
villardii (See Escobaria villardii),
villarensis.....[Ar], [Bo-5],
" -v- morellii.....
vivipara (See Escobaria vivipara, Mammillaria vivipara)
[D], [M*], [UU], [Ar], [Ay], [Bo-5], [Bv-1],
[Bz], [Cd], [Go], [Hn], [Hv], [Ij], [Ja], [Jd]
[Aa 6-1982], [Am 3-1984], [Hh May 1974],
" -v- aggregata (See C. aggregata, C. vivipara -v- bisbeeana, Echinocereus triglochidiatus -v- melanacanthus),
[Bn],
" -v- alversonii (See C. alversonii, Escobaria vivipara -v- alversonii),
[An], [Ax], [Az], [Bo-5], [Ca]
" -v- arizonica (See C. arizonica, Mammillaria vivipara -v- arizonica),
[B], [F], [H], [P], [Q], [Y], [GG],
[PP], [An], [Ax], [Df], [Ae-1981],
[Af 52-2], [Af 52-4], [Hl 2-1973],
" -v- bisbeeana (See Escobaria bisbeeana, Escobaria vivipara -v- bisbeeana),
[Q], [An], Ax], [Af 52-1], [Au 37-2]
" " " -f. bisbeeana.....[Af 52-4],
" " " -f. sonorensis.....[Af 52-4],
" -v- borealis (See Mammillaria vivipara -v- borealis),
" -v- buoflama.....[Af 52-1],
" -v- desertii (See C. desertii, Escobaria desertii, Escobaria vivipara -v- desertii),
[An], [Ax], [Az], [Ca], [Af 52-4],
" -v- exaltissima (See C. vivipara -v- rosea),

CORYPHANTHA.....(Cont'd),

vivipara.....(cont'd),
" -v- kaibabensis (See C. vivipara -v- rosea, Escobaria vivipara -v- rosea) [An], [Af 51-6],
" -v- neomexicana (See C. neomexicana, Escobaria neomexicana, Escobaria vivipara -v- neomexicana, Mammillaria vivipara -v- neomexicana), [Bo-5], [Hb], [Af 52-4],
" -v- radiosa (See C. radiosa, Escobaria radiosa, Escobaria vivipara -v- radiosa, Mammillaria vivipara -v- radiosa), [U], [Y], [Ax], [Df], [Ib], [Af 52-3] [Af 52-4], [Aj 29-12-1978],
" -v- rosea (See C. vivipara -v- kaibabensis, Escobaria vivipara -v- rosea), [An], [Ax], [Az], [Ca], [Af 52-4],
" -v- vivipara (See Escobaria vivipara -v- vivipara, Mammillaria vivipara -v- vivipara), [P], [X], [Y], [LL], [Ax], [Df], [Af 52-4],
voghtherriana.....[Aj IV-2-1932],
werdermannii (See Mammillaria hahniana -v- werdermanniana, Mammillaria werdermanniana), [Ar], [Bo-5], [Hx*], [Hy], [Iz], [Jd], [Aj II-10-1930], [Aj 26-11-1975],
wissmannii (See C. missouriensis -v- robustior, Neobesseya wissmannii).....[Iq],
zilziana (See Escobaria zilziana, Neobesseya zilziana) [Aj II-11-1930),
Spec. .....[Ar], [Hv],

CRYPTOCEREUS

anthonyanus.....[E], [I], [DD], [Ay], [Bo-2], [Bt], [Bv-1], [Cz], [Hx*], [Hy], [Jd], [N-1980], [Aj 32-7-1981], [Aw 2-1974], [Bb-1985], [Hh July, 1973],
imitans (See Eccremocactus imitans, Werkleocereus imitans), [Bo-2], [Aj 33-8-1982],
rosei (See Eccremocactus rosei),

CULLMANNIA

viperina (See Wilcoxia viperina -v- viperina, Peniocereus viperinus), [Ay],

CUMARINIA

odorata (See Coryphantha odorata, Neolloydia odorata),
[Ar], [Hx], [Jd],

CUMULOPUNTIA

alboareolata (See Opuntia alboareolata, Tephrocactus alboareolata),
[Hi-4],
berteri (See Opuntia corotilla, Opuntia sphaerica, Tephrocactus berteri, Tephrocactus sphaericus),
[Hi-3],
boliviana (See Opuntia boliviana, Tephrocactus alboscoparius, Tephrocactus bolivianus),
[Hi-2],
crassicylindrica (See Opuntia crassicylindrica, Tephrocactus crassicylindrica),
echinacea (See Opuntia echinacea, Tephrocactus echinacea),
[Hi-3],
famatimensis.....[Hi-2],
frigida.....(See Opuntia frigida),
galerasensis (See Opuntia galerasensis),
hystrix (See Opuntia hystrix),
[Hi-3],
ignescens (See Opuntia ignescens, Tephrocactus ignescens),
[Hi-3],
ignota (See Opuntia ignota, Tephrocactus ignotus),
[Hi-4],
kuehnrichiana (See Opuntia kuehnrichiana, Tephrocactus kuehnrichianus),
multiareolata (See Opuntia multiareolata),
[Hi-4],
pampana (pampeana) (See Opuntia maculacantha, Opuntia pampeana, Opuntia sulphurea, Opuntia vulpina),
[Hi-2],
pentlandii (See Opuntia pentlandii, Tephrocactus pentlandii),
[Hi-2],
" -v- colchana (See Opuntia chichensis -v- colchana, Tephrocactus chichensis -v- colchanus),
[Hi-2],
" -v- dactylifera (See Opuntia dactylifera, Tephrocactus dactyliferus),
[Hi-2],
pyrrhacantha (See Opuntia pyrrhacantha, Tephrocactus pyrrhacanthus),

CUMULOPUNTIA.....(Cont'd),

rauppiana (See Opuntia rauppiana, Tephrocactus sphaericus -v- rauppianus),
rossiana (See Tephrocactus microcladus, Tephrocactus pentlandii -v- rossianus),
[Hi-2],
subterranea (See Opuntia subterranea, Tephrocactus subterraneus, Tephrocactus variiflorus),
[Hi-2],
ticnamarensis (See Opuntia ticnamarensis),
[Hi-3],
tortispina (See Tephrocactus tortispinus),
[Hi-3],
tubercularis.....
tumida.....[Hi-4],
unguispina (See Opuntia unguispina, Tephrocactus sphaericus -v- unguispinus),
" -v- major.....
" -v- unguispina.....
zehnderi (See Tephrocactus zehnderi),

CYLINDROPUNTIA

abyssi (See Opuntia abyssi, Opuntia whipplei -v- multigeniculata),
[Bo-1],
acanthocarpa (See Opuntia acanthocarpa),
[An], [Ba 10-1981],
" -v- coloradoensis (See Opuntia canthocarpa -v- coloradoensis),
[An],
" -v- ganderi (See Opuntia acanthocarpa -v- ganderi),
[An],
" -v- major (See Opuntia acanthocarpa -v- major)
[An],
" -v- ramosa (See Opuntia acanthocarpa -v- major)
[An], [Bo-1],
" -v- thornberi (See Opuntia acanthocarpa -v- thornberi),.....[An], [Bo-1],
alamosensis (See Opuntia alamosensis, Opuntia thurberi -v- alamosensis).....[Az],
alcahes (See Opuntia alcahes, O. prolifera),
[Bo-1], [Hy],
arborescens (See Opuntia arborescens, Opuntia imbricata),
arbuscula (See Opuntia arbuscula),
[An], [Az], [Hy], [Ba 10-1981],
" -v- congesta (See Opuntia congesta, Opuntia kleiniae -v- tetracantha),
[Az], [Bo-1],

CYLINDROPUNTIA.....(Cont'd),

bigelovii (bigelowii)...(See Opuntia bigelovii),
[An], [Ar], [Ay], [Bo-1], [Bt],
[Hx*], [Hy], [Aj 31-7-1980],
[Ba 10-1981], [Hs IV-2-1983],
brevispina (See Opuntia brevispina),
[Bo-1], [Bo-6],
brittonii (See Nopalea brittonii, Opuntia brittonii,
Opuntia leptocaulis -v- brittonii),
burrageana (See Opuntia burrageana),
[Bo-1],
californica (See Opuntia parryi -v- serpentina, Opuntia serpentina),
[Bo-1],
calmalliana (See Opuntia molesta),
caribaea (See Opuntia caribaea),
[Af 56-3],
cholla (See Opuntia cholla),
[Bo-1],
ciribe (See Opuntia ciribe),
[Bo-1], [Hy],
clavarioides (See Austrocylindropuntia clavarioides,
Opuntia clavarioides),
" -f. monstrosus (See Austrocylindropuntia
clavarioides -f. monstrosus),
" -v- ruiz-leallii.....[Fs 33- July, 1966],
clavellina (See Austrocylindropuntia molesta,
C. molesta, Opuntia molesta),
cylindrica (See Austrocylindropuntia cylindrica,
Opuntia cylindrica),
" -f. cristata.....[T],
davisii (See Opuntia davisii, Opuntia tunicata -v-
davisii),
[Bo-1], [Df],
densiaculeata (See Opuntia densiaculeata),
[Bo-1],
echinocarpa (See Opuntia echinocarpa),
[An], [As], [Bo-1],
" -v- echinocarpa (See Opuntia echinocarpa -v-
echinocarpa),
exaltata (See Opuntia exaltata),
" -f. monstrosus.....[Hy],
fulgida (See Opuntia fulgida -v- fulgida),
[DD], [An], [Ay], [Bo-1], [Hy], [Ba 10-1981],
[Hs IV-2-1983],
" -v- mamillata.....[An], [Bo-1], [Bo-6],
" -f. cristata.....[Hy],
" -f. monstrosus.....[An], [Bo-6],
hamiltonii (See Grusonia hamiltonii, Opuntia hamiltonii),.....
heteracantha.....

hualpaensis (See Opuntia hualpaensis, Opuntia whipplei -v- whipplei),
[Bo-1],
humahuacana (humahuacensis) (See Austrocylindropuntia humahuacana, Austrocylindropuntia shaferi, Austrocylindropuntia vestita -v- shaferi, Opuntia humahuacana),
[Af 51-5],
imbricata (See Opuntia imbricata),
[An], [Bo-1], [Hx*], [Hy], [Aj 29-3-1978], [Hs II-2-1981],
" -v- argentea (See Opuntia imbricata -v- argentea).....
" -v- imbricata.....
inarmata (See Austrocylindropuntia inarmata, Opuntia inarmata),
invicta (See Corynopuntia invicta, Opuntia invicta),
[E], [NN], [RR],
kelvinensis (= C. fulgida X C. spinosior),
kleiniae (See Opuntia kleiniae),
[An],
" -v- tetracantha (See C. arbuscula, Opuntia arbuscula, Opuntia tetracantha),
" -v- kleiniae.....
leptocaulis (See Opuntia leptocaulis),
" -v- badia.....
" -v- brevispina.....
" -v- glauca.....[Bo-6],
" -v- longispina (See C. mortolensis),
" -v- pluriseta.....
" -v- robustior.....
" -v- tenuispina.....[Bo-6],
" -v- vaginata.....[Bo-1],
lloydii (See Opuntia lloydii),
[Bo-1],
metuenda.....
miquelii (See Austrocylindropuntia miquelii, Miqueliopuntia miquelii, Opuntia miquelii),
[Aj 9-3-1958],
molesta (See Austrocylindropuntia molesta, Opuntia molesta),
[Bo-1],
mortolensis (See Opuntia leptocaulis -v- longispina, Opuntia mortolensis),
[Bo-1], [Bo-6],
multigeniculata (See Opuntia multigeniculata, Opuntia whipplei -v- multigeniculata),
[As],
" -v- multigeniculata.....

myriocaulis (See C. prolifera, Opuntia prolifera),
[Hy],
pachypus (See Austrocylindropuntia pachypus, Opuntia pachypus),
[Ay],
pallida (See Opuntia pallida),
parryi (See Opuntia parryi),
[Bo-1], [Hy],
" -v- parryi (See Opuntia parryi),
" -v- serpentina (See Opuntia serpentina),
prolifera (See Opuntia prolifera),
[Bo-1],
ramosissima (See Opuntia ramosissima),
[An], [Bo-1], [Iu],
recondita (See Opuntia recondita),
" -v- perrita (See Opuntia recondita -v- perrita)
[Bo-1],
rosarica (See Grusonia rosarica, Opuntia rosarica),
[Bo-1],
rosea (See Opuntia rosea),
[Bo-1], [Hx*], [Aj 28-11-1977], [Aj 33-6-1982],
" -v- atrorosea (Invalid name).....[DD],
schottii (See Opuntia schottii),
serpentina (See C. californica, Opuntia parryi -v- serpentina, Opuntia serpentina),
[Hy*],
spinosior (See Opuntia spinosior),
[An], [Bo-1], [Bt], [Df],
stanlyi (See Corynopuntia stanlyi, Opuntia stanlyi),
[Ba 10-1981],
" -v- kunzei (See Opuntia kunzei),
subulata (See Austrocylindropuntia subulata, Opuntia subulata),
[Ay],
teres (See Austrocylindropuntia teres, Opuntia teres),
tesajo (See Opuntia tesajo),
" -v- cineracea (See Opuntia cineracea),
[Az],
tetracantha (See C. kleiniae -v- tetracantha, Opuntia kleiniae -v- tetracantha, Opuntia tetracantha),
[Bo-1],
thurberi (See Opuntia thurberi),
tunicata (See Opuntia tunicata),
[NN], [Av], [Ay], [Bo-1], [Bt], [Df], [Hi-3],
[Hy], [Ba 5-1980], [Gu III-1953],
" -v- aricensis.....[Hi-3],
" -v- chilensis.....[Hi-3],
" -v- davisii (See Opuntia tunicata -v- davisii),
" -v- tunicata (See Opuntia tunicata -v- tunicata)

CYLINDROPUNTIA.....(Cont'd),

verschaffeltii (See Austrocylindropuntia verschaffeltii, Opuntia verschaffeltii),
[Aj 27-10-1976],
" -v- hypsophila (See Austrocylindropuntia verschaffeltii -v- hypsophila),
" -v- longispina (See Austrocylindropuntia verschaffeltii -v- longispina),
[Fs 33, July 1966],
versicolor (See Opuntia versicolor),
[DD], [An], [Be*], [Hx*],
vestita (See Austrocylindropuntia vestita, Opuntia vestita),
[NN], [Cr], [Ig],
viridiflora (See Opuntia viridiflora, Opuntia whipplei -v- viridiflora),
vivipara (See Cylindropuntia arbuscula, Opuntia arbuscula).....[Bo-1],
whipplei (See Opuntia whipplei),
[An], [Bo-1], [Df],
" -v- enodis (See Opuntia whipplei).....[Az],
" -v- multigeniculata (See Opuntia whipplei -v- multigeniculata),
[An], [Az],
wigginsii (See Opuntia wigginsii).....[An],

CYLINDROREBUTIA

einsteinii (See C. pseudoeinsteinii, Rebutia einsteinii)
[Hr],
karreri (See Rebutia einsteinii -f. karreri).....[Hr],
nicolai.....[Hr],
pseudoeinsteinii.....[Hr],
rubriviride (See R. einsteinii -f. rubriviridis),.....
[Hr],
schmiedchenianum (See Mediolobivia schmiedcheniana, Rebutia einsteinii -f. schmiedcheniana, Rebutia schmiedcheniana),
[NN],
spiralisepala (See Mediolobivia spiralisepala, Rebutia aureiflora),
[DD], [NN],

DEAMIA

diabolica.....[Ep Bull. Torr. Bot. Club. - 65:570 Fig. 5 1938],
testuda (See Selenicereus testuda),
[G], [I], [DD], [RR], [Ay], [Bu], [Cx], [Hv], [Hx], [Jd], [Af 31-5], [Af 52-3],

DELAETIA

woutersiana (Invalid name - See Neoporteria woutersiana, Pyrrhocactus neohankeanus -f. woutersianus), [DD], [Bo-6],

DENDROCEREUS

nudiflorus.....[M*], [DD], [RR], [Av], [Bl], [Bo-4], [Hx*], [Jd], [Af 10-4], [Af 27-2], [Bi 6-1], [Gu III-1953], [Hl 2-1972], [Hl 3-1972],
undulosus.....[Af 56-3], [Ct 2-1984],

DENMOZA

erythrocephala (See D. rhodacantha), [L], [PP], [Ar], [Av], [Bo-2], [Hu], [Il], [Jd], [Af 56-1], [Bi 2-4], [Hl 6-1971],
rhodacantha (See D. erythrocephala, Pilocereus erythrocephala), [F], [L], [NN], [Ar], [Bo-2], [Dq], [Hi-2], [Hv], [Hx], [Iu], [Af 6-7], [Aj 33-5-1982], [Hl 4-1980],
" -v- (Famatina-Gebirge).....
spec. .....[Bo-2],

DIGITOREBUTIA

albopectinata (See Rebutia pectinata), [Aj 23-9-1972],
atrovirens (See Rebutia pygmaea -v- atrovirens), [Hl 1-1976],
brachyantha (See Mediolobivia brachyantha, Rebutia brachyantha),
" -v- ritteri (See Rebutia ritteri),
brunescens (See Mediolobivia brunescens, Rebutia brunescens), [Aj 23-9-1972],
brunneoradicata (See Rebutia brunneoradicata),
canacruzensis (See Rebutia canacruzensis), [Aj 27-3-1976],
carmeniana (See Rebutia carmeniana), [Aj 29-5-1978],
christinae (See Rebutia christinae).....[Aj 26-7-1975],
cincinnata (See Rebutia cincinnata).....[Aj 27-1-1976],
colorea (See Rebutia colorea),
costata (See Mediolobivia costata, Rebutia costata),
" -v- eucalyptana (See Rebutia costata -v- eucalyptana),
" -v- pilifera (See Rebutia costata -v- pilifera)

DIGITOREBUTIA.....(Cont'd),

densipectinata.....
diersiana (See Rebutia diersiana),
" -v- atrovirens (See Rebutia diersiana -v- atrovirens),
" -v- minor (See Rebutia diersiana -v- minor),
digitiformis (See Mediolobivia digitiformis, Rebutia digitiformis, Rebutia pygmaea),
[Hl 4-1982],
donaldiana (See Rebutia steinmannii -v- cincinnata),
eos (See Mediolobivia eos, Rebutia eos),
euanthema (See Mediolobivia euanthema, Rebutia euanthema).....[Hy],
" -v- friciana (See Rebutia euanthema -f. fricii)
" -v- longispina (See Rebutia euanthema),
" -v- oculata (See Rebutia euanthema -f. oculata)
[NN],
eucalyptana (See Mediolobivia eucalyptana, Rebutia costata -f. eucalyptana, Rebutia eucalyptana)
fricii (See Rebutia euanthema -f. fricii),
[NN],
friedrichiana (See Rebutia friedrichiana),
[Ba 6-1976],
graciliflora (See Rebutia graciliflora),
gracilispina (See Rebutia gracilispina),
haagei (See Mediolobivia haagei, Mediolobivia pygmaea, Rebutia haagei, Rebutia pygmaea),
[Hr], [Hu],
" -v- atrovirens (See Rebutia pygmaea -v- atrovirens),
" -v- chamelon.....[Hr],
" -v- digitiformis (See Rebutia pygmaea),
" -v- orutensis (See Rebutia pygmaea),
" -v- pectinata (See Rebutia pygmaea),
" -v- pygmaea (See Rebutia pygmaea),
[Cu 2-1961],
" -v- salmonea (See Rebutia xanthocarpa -f. salmonea),
[Hr],
haefneriana (See Mediolobivia haefneriana, Rebutia pygmaea -f. haefneriana),
hirsutissima (See Mediolobivia hirsutissima, Rebutia hirsutissima),
iridescens (See Rebutia iridescens),
iscayachensis (See Rebutia iscayachensis),
[Ba 1-1977],
lanosiflora (See Rebutia lanosiflora),
minutissima (See Rebutia minutissima),
mixta (See Rebutia mixta),
mixticolor (See Rebutia mixticolor),
mudanensis (See Rebutia mudanensis).....[Aj 27-8-1976],

DIGITOREBUTIA.....(Cont'd),

nazarenoensis (See Rebutia nazarenoensis),
[Ba 8-1979],
nigricans (See Mediolobivia nigricans, Rebutia ritteri -v- nigricans),
oculata (See Rebutia euanthema -f. oculata, Rebutia oculata),
" -v- friciana (See Mediolobivia euanthema -f. fricii, Rebutia euanthema -f. fricii),
odontopetala (See Aylostera odontopetala, Rebutia friedrichiana, Rebutia odontopetala),
orurensis (See Mediolobivia orurensis, Rebutia pectinata -v- orurensis, Rebutia orurensis),
[Hu],
pallida (See Rebutia pallida),
[Ba 10-1977],
patericalyx (See Rebutia patericalyx),
pauciareolata (See Rebutia pauciareolata),
paucicostata (See Rebutia paucicostata),
pectinata (See Mediolobivia pectinata, Mediolobivia pygmaea, Rebutia pectinata, Rebutia pygmaea),
[Hu],
peterseimii (See Rebutia peterseimii, Rebutia ritteri -v- nigricans -f. peterseimii),
[Hr],
pilifera (See Rebutia costata -v- pilifera),
poecilantha (See Rebutia poecilantha),
potosina (See Rebutia potosina),
pseudopygmaea (See R. iscayachensis),
pygmaea (See Digitorebutia haagei, Mediolobivia haagei, Mediolobivia pygmaea, Rebutia haagei, Rebutia pygmaea),
[NN],
" -v- haefneriana (See Digitorebutia haefneriana, Mediolobivia haefneriana, Rebutia pygmaea -f. haefneriana),
raulii (See Rebutia raulii),
[Aj 31-6-1980],
rauschii (See Mediolobivia rauschii, Rebutia rauschii),
[Aj 28-4-1977],
ritteri (See Mediolobivia ritteri, Rebutia ritteri),
[NN],
" -v- nigricans (See Mediolobivia nigricans, Rebutia ritteri -v- nigricans),
rosalbiflora (See Rebutia rosalbiflora),
ruberrima.....[Hr],
rutiliflora (See Rebutia rutiliflora),
salpingantha (See Rebutia salpingantha),
steinmannii (See Rebutia steinmannii),
tarvitaensis (See Rebutia tarvitaensis),

DIGITOREBUTIA.....(Cont'd),

torquata (See Rebutia torquata),
tropaeoliptica (See Rebutia tropaeoliptica),
violascens (See Rebutia violascens),
yuquinensis (See Rebutia yuquinensis),
[Bt], [Aj 31-10-1980],
Spec. .....[Hu],
Spec. FR-770.....[Hu],

DISCOCACTUS

albispinus (See D. zehntneri),
[G], [Av], [Bw]. [Af 46-], [Af 50-3],
[Aj 28-8-1977], [Ba 11-1978],
alteolens (See D. placentiformis),
[DD], [RR], [Ar], [Bo-4], [Bw], [Dq*],
[Af 50-3], [Aj 33-9-1982], [Hl 2-1972],
araneispinus (See D. zehntneri),
[Ar], [Bw], [Gr], [Ab-1985], [Af 50-3],
[Af 58-5], [Ba 1-1982], [Hl 4-1983],
bahiensis (See D. placentiformis),
[RR], [Bw], [Af 51-3], [Aj 28-8-1977],
[Aj 30-8-1979],
besleri (See D. insignis -f. besleri),
[Hw],
boliviensis (Invalid name -See D. heptacanthus),
[G], [DD], [Bw], [Hx], [Hy], [Ik],
[Af 50-4], [Aj 30-2-1977], [Ba 4-1976],
[Hl 2-1972],
" -f. cristata.....[Bw],
boomianus (See D. zehntneri),
[Ar], [Bw], [Hv], [Hx*], [Af 50-4], [Af 51-3]
[Aj 28-8-1977], [Ba 50:26-29, Feb. 1971],
[Hl 2-1972],
cangaensis (See D. heptacanthus),
[Af 52-3],
carasolensis (See D. magnimammus),
[Hl 2-1972],
catingicola (See D. heptacanthus),
[Bw],
cephaliaciculosus (See D. heptacanthus),
[Bw], [Af 50-4], [Aj 26-5-1975],
cipolandensis.....[Hl 2-1972],
crystollophorus (See D. placentiformis)....[Ag-1988],
[Aj 32-11-1981], [Aj 33-9-1982],
diersianus (See D. heptacanthus),
[Af 51-4],
estevesii (See D. heptacanthus),
[Af 51-2], [Af 51-3], [Ba 2-1981], [Hl 4-1983]
ferricola (See D. heptacanthus).....[Bw], [Af 50-4],
[Aj 28-8-1977], [Ba 4-1976], [Hl 6-1980],

DISCOCACTUS.....(Cont'd),

flavispinus (See D. heptacanthus),
[Bw],
giganteus (See D. pulvicapitatus),
goianus (See D. heptacanthus),
[Aj 3-1980],
griseus (See D. heptacanthus),
[Bw], [Hl 4-1983],
grossoanus (See D. silicicola),
hartmannii.....[G], [I], [RR], [Ay], [Bw], [Hv], [Hx*],
[Hy], [Af 50-5], [Aj 28-8-1977],
heptacanthus (See D. griseus) [G], [RR], [Bo-4], [Bw], [Hx*],
[Hy], [Aj 15-5-1964], [Aj 28-8-1977],
[Cu-2-1981], [Hl 5-1978],
horstii.....[I], [Z], [Ar], [Av], [Ay], [Bw], [Hi-1],
[Hv], [Hx*], [Iz], [Jd], [Ab-1985], [Ae-
1983], [Af 50-5], [Af 55-2], [Ah-1986], [Aj
28-8-1977], [Am 8-1981], [Ba 11-1979], [Ba
8-1981], [Hl 5-1973], [Hl 1-1981],
insignis (See D. placentiformis),
[Bw], [Af 51-2], [Af 58-1], [Aj 28-9-1977],
[Aj 33-12-1982], [Aw 2-1981],
" -f. besleri (See D. besleri),
latispinus (See D. placentiformis),
[Bw], [Aj 31-2-1980], [Aj 32-11-1981],
lehmannii (See D. insignis, D. placentiformis),
lindaianus (See D. heptacanthus).....[Af 53-2],
linkii (See D. insignis, D. placentiformis),
magnimammus (See D. hartmannii),
[Bw], [Hv], [Af 50-5], [Af 53-1],
" -v- magnimammus.....[Bw],
" subspec. bonitoensis.....[Bw], [Ba 4-1981],
mamillosus (See D. hartmannii),
[Bw], [Hl 4-1983]
melanochlorus (See D. heptacanthus),
minimus.....[Af 50-6],
multicolorispinus (See D. placentiformis),
[Aj 32-3-1981],
mutuensis.....[Hl 6-1980],
nigrisaetosus (See D. heptacanthus),
[Bw], [Af 50-6],
pachythele (See D. hartmannii or D. heptacanthus),
[Bw], [Af 47-], [Aj 28-9-1977],
paranaensis (Invalid name - See D. heptacanthus),
[Bo-4],
patulifolius.....[Bw],
placentiformis (See D. subviridigriseus),
[RR], [Ar], [Bo-4], [Bw], [Hi-1], [Hx*],
[Iz], [Af 50-6], [Af 51-2], [Af 58-1],
[Aj 28-9-1977], [Aj 33-9-1982],
[Cu 2-1981],

DISCOCACTUS.....(Cont'd),

pugionacanthus (See D. placentiformis),
[Bw], [Af 50-6],
pulvinicapitatus.....[Ar], [Bw], [Af 50-6], [Aj 33-7-1982],
rapirhizus (See D. heptacanthus),
[Ar], [Bw], [Af 50-6], [Aj 28-9-1977],
[Hh Nov. 1975], [Hl 4-1982],
rarissimus.....[Af 50-6],
semicampaniflorus (See D. heptacanthus),
[Bw], [Af 47-],
silicicola (See D. heptacanthus),
[Ar], [Av], [Bw], [Hx*], [Jd], [Af 47-],
[Af 51-1], [Hl 5-1977], [Hl 5-1981],
silvaticus (See D. heptacanthus),
[Bw], [Af 51-1], [Ba 4-1981],
spinosior (See D. heptacanthus),
[Bw], [Af 51-1],
squamibaccatus (See D. heptacanthus),
[Bw], [Hl 4-1983],
subnudus.....[RR],
subterraneo-proliferans (See D. heptacanthus),
[Aj 31-9-1980],
subviridigriseus (See D. placentiformis),
[Bw], [Af 51-1], [Af 51-2],
[Hl 4-1983],
tricornis (See D. alteolens, D. placentiformis),
[DD], [Ar], [Bo-4], [Bt], [Hx*], [Af 51-1],
[Af 51-3], [Aj 28-9-1977], [Aj 30-12-1979],
" -v- ammophilus.....
" -v- deflexispinus.....[Hl 2-1972],
" -v- giganteus.....
woutersianus (See D. horstii),
[Ba 9-1980],
zehntneri (See D. araneispinus),
[Bw], [Hx*], [Iz], [Cu 2-1981], [Hl 2-1983],
[Hl 4-1983],
" -f. albispinus.....[Hl 4-1983],
" -v- boomianus.....[Sm 8-1981],

Spec. HB-4.....[Aj 32-7-1981],
Spec. HU-232.....[Aj 28-9-1977],
Spec. HU-325.....[Hl 2-1972],
Spec. Hu-437.....[Aj 30-8-1979],
Spec. HU-438.....[Aj 28-9-1977],
Spec. HU-440.....[Aj 28-9-1977],
Spec. HU-461.....[Af 51-2],
Spec. HU-485.....[Aj 32-7-1981],
Spec. W-27.....[Af 5103], [Aj 28-9-1977],
Spec. .....[Ik],

DISOCACTUS

acuminatus (See Pseudorhipsalis acuminata),
alatus (See Pseudorhipsalis alata),
amazonicus (See Wittia amazonica),
biformis (See Epiphyllum biformis, Phyllocactus biformis),
[RR], [Be*], [Bo-2], [Hx], [Hy],
eichlamii.....[DD], [RR], [Ax], [Bo-2], [Bu], [Dq*], [Hx], [Il], [Af 29-3],
himantocladus (See Pseudorhipsalis himantoclada),
[Aj 26-11-1975], [Aj 26-12-1975],
horichii (See Nopalxochia horichii).....[Af 51-4],
lankesteri.....[Af 51-4*],
macranthus (See Pseudorhipsalis macrantha, Rhipsalis macrantha).....[E], [I], [H], [Ay], [Az], [Aw 1-1986], [Hw-1982],
nelsonii (See Chiapasia nelsonii, Epiphyllum nelsonii, Phyllocactus chiapensis, Phyllocactus nelsonii)
[I], [HH], [Ay], [Az], [Iz], [Jd], [Ab-],
[Aw 4-1963], [Hh Sept., 1974],
" -v- hondurensis.....[Ad Jan. 1980], [Af 37-2],
panamensis (See Wittia panamensis, Wittia amazonica),
quezaltecus (See Bonifazia quelzalteca),.....[Ay], [Az], [Bo-6], [Aw 2-1975],
ramulosus (See Rhipsalis jamaicensis, Rhipsalis ramulosa),
[Ay], [Az], [Jd], [Af 33-1], [Aj 27-3-1976],
[Aw 1-1986],
" -v- angustissimus (See Rhipsalis angustissima),

DOLICHOTHELE

albescens (See Mammillaria albescens, Mammillaria camptotricha),
[VV], [Bo-5], [Hy],
balsasoides (See Mammillaria beneckei),
[VV], [Ae-1981],
baumii (See Mammillaria baumii),
[E], [NN], [Ar], [Ay], [Bo-5], [Gr], [Hx*], [Hy], [Aj 9-6-1958],
beneckei (Mammillaria beneckei)
[VV], [Bo-5], [Hu], [Hy],
camptotricha (See Mammillaria camptotricha, Mammillaria decipiens -v- camptotricha),
[E], [NN], [RR], [VV], [Ay], [Hx*], [Hy], [Il], [Jd],
" -v- albescens (See D. albescens),
[Hy],
" -v- longithele.....[Bo-5],
decipiens (See Mammillaria decipiens),
[C], [W], [RR], [Ar], [Hy],

DOLICHOTHELE.....(Cont'd),

heidiae (See Mammillaria heidiae),
[Aq],
longimamma (See Mammillaria longimamma),
[E], [I], [M*], [DD], [PP], [QQ], [RR],
[VV], [Ar], [Av], [Ay], [Be*], [Bo-5], [Bt],
[Bv-1], [Cm], [Cr], [Cx], [Cz], [Gr], [Hn],
[Hu], [Hy], [Il], [Jd], [N-1980], [Af 58-1],
[Aw 3-1979], [Aw 4-1962], [Flower - Ba 4-1978],
" -v- gigantothele.....
" -v- globosa (See Mammillaria longimamma),
[Bo-5],
melaleuca (See Mammillaria melaleuca),
[VV], [Cr], [Ig], [Iu], [Ae-1983],
nelsonii (See Mammillaria beneckei),
[Bo-5], [Hh 1-8-1974],
saffordii (See Mammillaria carretii),
sphaerica (See Mammillaria longimamma -v- sphaerica, Mammillaria sphaerica),
[E], [H], [L], [Y], [GG], [RR], [VV], [XX],
[Ar], [Bo-5], [Bv-1], [Bz], [Df], [Gr], [Hn],
[Hx*], [Hy], [Am 2-1980], [Dg 2-3-1980],
surculosa (See Mammillaria surculosa),
[D], [H], [NN], [VV], [Ar], [Ay], [Bo-5],
[Hu], [Hx*], [Hy], [Il], [Jd], [Ad Feb., 1981],
uberiformis (See Mammillaria longimamma -v- uberiformis, Mammillaria uberiformis),
[RR], [VV], [Al*], [Ar], [Gr],
zephyranthoides (See Mammillaria zephyranthoides),
[NN], [Bo-5], [Hx], [Hy],

ECCREMOCACTUS

bradei.....[DD], [HH], [RR], [Ay], [Bo-2], [Dq*], [Hx*],
[Hy], [Jd], [Af 28-6], [Aw 2-1985],
imitans (See Cryptocereus imitans, Werkleocereus imitans),
[Ij],
rosei (See Cryptocereus rosei),
[Ij], [Af 34-3],

ECHINOCACTUS

acanthodes (See Ferocactus acanthodes, Ferocactus cylindricus),
[Iq],
" -v- rostii (See Ferocactus rostii),
acuatus (See Wigginsia acuata).....[Gm*],
" -v- arechavaletai (See Wigginsia arechavaletai)

ECHINOCACTUS.....(Cont'd),

andreae (See Gymnocalycium andreae).....[Aj 2-10-1951],
araneifer (See Neoporteria odierii -v- araneifera)
asterias (See Astrophytum asterias),
[Ax], [Df],
bicolor (See Thelocactus bicolor),
[Fg],
" -v- flavidispinus (See Thelocactus bicolor -v- flavidispinus, Thelocactus flavidispinus),
[Df],
" -v- schottii (See Thelocactus bicolor -v- schottii),
[Df],
brevihamatus (See Ancistrocactus brevihamatus),
[Df],
caespitosus (See Parodia caespitosa),
capricorne (See Astrophytum capricorne),
" -v- major (See Astrophytum capricorne -v- major).....[Aj II-9-1930],
centeterius (See Pyrrhocactus centeterius),
[Aj 22-9-1971],
concinnus (See Notocactus concinnus, Parodia concinna)
conoideus (See Neolloydia conoidea),
[Y], [Df],
coquimbanus (See Copiapoa coquimbana),
[Aj 2-7-1951],
cornigerus (See Ferocactus latispinus)
[Fg], [Io*],
" -v- flavispinus (See Ferocactus latispinus -v- flavispinus),
cupreatus (See Copiapoa cupreata),
ebenacanthus (K. Schum.) (See Neoporteria fusca - (Br. & R)),
ebenacanthus (Monv.) (See Echinocactus fuscus (Muehl.)),
[Cu 4-1962],
echinoides (See Copiapoa echinoides)
ehrenbergii (See Thelocactus ehrenbergii, Thelocactus leucacanthus -v- ehrenbergii),
emoryi (See Ferocactus covillei, Ferocactus emoryi),
erectocentrus (Invalid name) (See Echinomastus erectocentrus, Gymnocactus beguinii, Neolloydia beguinii, Neolloydia erectocentra, Sclerocactus erectocentrus, Thelocactus erectocentrus).....[Hn],
" -v- pallidus (See Echinomastus pallidus),
[Y],
erinaceus (See Malacocarpus erinaceus),
[Iq],
famatimensis (See Lobivia famatimensis, Reicheocactus pseudoreicheanus).....[Dr],

ECHINOCACTUS.....(Cont'd),

floricornus.....[Af 4-3],
fobeanus (See Neochilenia fobeana, Neoporteria jussieui -v- fobeana),
froehlichianus (See Horridocactus froehlichianus),
fuscus (Muehlpf.) (See Echinocactus ebenacanthus - (Monv.)),
geissei (See Horridocactus geissei, Neoporteria geissei),
gladiatus (See Echinofossulocactus gladiatus), [Gm*],
grandis (See E. platyacantha -f. grandis), [JJ], [Ar], [Ay], [Bo-5], [Bu], [Bv-1], [Cx], [Ds], [Hy], [If], [Ad Aug. 1978], [Ba 2-1980], [Ba 4-1980], [Cu 1-1968], [Db 6-1979], [Hl 1-1978],
grusonii.....[C], [X], [AA], [BB], [DD], [TT], [Ar], [Ay], [Bt], [Cq], [Cr], [Cx], [Cz], [Gr], [Hv], [Hx], [Hy], [Il], [Iz], [Ja], [Jd], [Ad 8-1978], [Ag-1984], [Aj 29-10-1978], [Aj 30-12-1979], [Ba 2-1977], [Bi 6-2-1988], [Bm-1988], [Bm-1989], [N-1980],
" -v- albispinus (white spined form)...... [Ar], [Hy],
" -v- horridus.....[Hy],
" -v- intertextus.....[Hy],
" -v- subinermis.....[Hy],
" -f. variegated.....[Ar],
" -f. cristata.....[Ar], [Hy],
" -f. monstrosus.....[Ar],
" -f. (reduced spination).....[Af 58-4],
haematanthus (See Lobivia haematantha).....[Dr],
hamatacanthus (See Ferocactus hamatacanthus, Hamatocactus hamatacanthus),
" -v- insignis (See Hamatocactus hamatacanthus -v- insignis),
hankeanus (Foerst.) (See E. humilis (Ruempl.)),
herteri (See Notocactus herteri),
heteracanthus (See Stenocactus heteracanthus), [Iq],
heterochromus (See Thelocactus heterochromus)....[Ir],
horizonthalonius.....[F], [M*], [P], [U], [LL*], [NN], [PP], [An], [Ar], [Ay], [Bm], [Bo-4], [Bz], [Cx], [Df], [Hn], [Hv], [Hx], [Hy], [Il], [Iz], [Jd], [Ad 9-1979], [Ae-1982], [Af 52-3], [Af 57-3], [Af 58-4], [Af 60-1-1988], [Am 6-1983], [Au 37-4], [Aw 2-1962], [Aw 4-1966], [Db 7-1979],
" -v- curvispina (See E. horizonthalonius), [Y], [Df],
" -v- horizonthalonius.....[Ax],

ECHINOCACTUS.....(Cont'd),

horizonthalonius (cont'd),
" -v- moelleri (See E. horizonthalonius -v- nicholii),
[Y], [Df], [Hy],
" -v- nicholii....(See E. horizonthalonius -v- moelleri), [Q], [Ax], [Ay], [Bh],
" (8 photographs of spine differentiation).....
[Aj 27-9-1976],
humilis (Ruempl.) (See E. hankeanus (Foerst.)),
hybocentrus (See Uebelmannia gummifera),
[Aj 22-9-1971],
hyptiacanthus (See Gymnocalycium hyptiacanthus),
[Bk],
ingens.....[C], [I], [L], [Ar], [Ay], [Bd], [Bo-5],
[Bv-1], [Bt], [Cx], [Cz], [Gr], [Hv], [Hx],
[Hy], [Jd], [Ad Aug. 1978], [Aw 1-1962],
[Bi 3-3], [Bm-1987], [Da 2-1977], [Hl 1-1978]
" -v- grandis (See E. grandis),
[Bu],
" -v- palmeri (See E. palmeri),
intertextus (See Echinomastus intertextus, Neolloydia intertextus, Sclerocactus intertextus),
[Hy],
" -v- dasyacanthus (See Echinomastus intertextus -v- dasyacanthus, Neolloydia intertextus -v- dasyacanthus, Sclerocactus intertextus -v- dasyacanthus),
[Y], [Df],
" -v- intertextus (See Echinomastus intertextus -v- intertextus, Neolloydia intertextus -v- intertextus),
[Y], [Df],
intricatus (See Malacocarpus arechaveletai, Notocactus intricatus, Notocactus arechaveletai, Wigginsia arechaveletai),
[Gm*],
knuthianus (See Gymnocactus knuthianus, Neolloydia knuthianus, Thelocactus knuthianus),
[Aj II-7-1930],
kurtzianus (See Gymnocalycium kurtzianus).....[Cy],
longihamatus (See Ferocactus hamatacanthus, Hamatocactus hamatacanthus),
[Fg], [Ir],
" -v- gracilispinus (See E. hamatacanthus),
maassii (maasii) (See Parodia maassii),
[Aj 2-9-1951], [Cu 2-2-1933],
macrocanthus (See Malacocarpus macrocanthus),
malletianus (See Copiapoa malletiana),

ECHINOCACTUS.....(Cont'd),

mariposensis (See Echinomastus mariposensis, Neolloydia mariposensis, Sclerocactus mariposensis), [Y], [Df],
meonacanthus (See Melocactus meonacanthus), [Gm*],
mesae-verdae (See Coloradoa mesae-verdae, Pediocactus mesae-verdae, Sclerocactus mesae-verdae), [Df],
microspermus (See Hickenia microsperma, Parodia microsperma).....[Ir],
" -v- macrancistrus.....[Iq], [Ir], [Cu 4-1962],
" " " -f. cristata.....[Ds],
mondragona.....[CC],
multiformis.....[Cu 4-1962],
nidulans (See Thelocactus nidulans, Thelocactus rinconensis -v- nidulans).....[Ds],
nigricans (See Horridocactus nigricans, Neoporteria nigricans, Pyrrhocactus nigricans),
obvallatus (See Echinofossulocactus obvallatus), [Cu 2-1979],
odierii (See Chileorebutia odierii, Neochilenia odierii, Neoporteria odierii, Thelocephala odierii),
" -v- mebbesii (See Neoporteria mebbesii, Neochilenia mebbesii).....[Aj 10-10-1959],
orthacanthus (See Notocactus orthacanthus).....[Gm*],
palmeri (See E. ingens -v- palmeri).....[RR], [Ar], [Cx], [Fg], [Hy], [Jd], [Ad Aug., 1978],
parryi.....[Ar], [Bo-5], [Hy], [Af 37-5], [Aj 30-7-1979] [Aj 30-11-1979], [Hl 2-1977],
pepinianus (See Copiapoa pepiniana),
pilosus (See Ferocactus pilosus, Ferocactus stainesii),.....[If],
pfeifferi (See Ferocactus pfeifferi, Ferocactus glaucescens).....[Dn],
platyacanthus.....[Cx], [Dq], [Gm*], [Ij], [Jd], [Cu 3-1982],
" -f. grandis (See E. grandis).....[Il],
" -f. ingens (See E. ingens)...[Hl 6-1983],
" -v- platyacanthus.....
" -f. visnaga (See E. visnaga),
poliacanthus (See Malacocarpus polyacanthus, Melocactus poliacanthus, Malacocarpus erinaceus), [Gm*],
polycephalus.....[G], [M*], [P], [Q], [CC], [FF], [LL], [RR], [An], [Ar], [Av], [Ax], [Ay], [Bo-5], [Ca], [Cb], [Hn], [Hx*], [Hy], [Af 3-11], [Af 53-6], [Au 37-2], [Hl 1-1971],
" -v- polycephalus.....
" -v- xeranthemoides (See E. xeranthemoides), [P], [An], [Ax], [Af 39-2],

ECHINOCACTUS.....(Cont'd),

pottsii (See Thelocactus bicolor -v- pottsii, Thelocactus pottsii),
[Bk],
pringlei (See Ferocactus stainesii -v- pringlei),
[Fg],
quehlianum (See Gymnocalycium quehlianum),
[Ds], [Ir],
recurvus (See Ferocactus recurvus),
[Gm*],
" -v- spiralis (See Ferocactus latispinus -v- spiralis),
salmianus (See Melocactus salmianus),
[Gm*],
scheeri (See Ancistrocactus scheeri, Sclerocactus scheeri),
[Df],
schwebsianus (See Parodia schwebsianus),
[Aj 2-9-1951],
sellowii (See Melocactus sellowii, Malacocarpus sellowii),
[Gm*],
scopa (See Notocactus scopa),
" -v- candida (See Notocactus scopa -v- candida),
[Ds],
" -v- rubra (See Notocactus scopa -v- rubra),
" -v- " -f. cristata.....[Ds],
setispinus (See Ferocactus setispinus, Hamatocactus setispinus).....[Df],
sinuatus (See Ferocactus hamatacanthus -v- sinuatus, Hamatocactus hamatacanthus -v- sinuatus, Hamatocactus sinuatus),
[Df],
subuliferus.....[Gm*],
tenuispina (See Notocactus ottonis -v- tenuispinus),
[Gm*],
" -v- minor.....[Gm*],
tephracanthus (See Malacocarpus sellowii).....[Gm*],
texensis (See Homalocephala texensis)....[Ax], [Ay], [Df], [Gr], [Hw], [Iz], [Jd], [Af 57-3], [Af 60-1-1988], [Hw-1981], [Hw-1982],
tobuschii (See Ancistrocactus tobuschii).....[Df],
tuberculatus.....[Gm*],
uncinatus (See Ancistrocactus uncinatus, Ferocactus uncinatus, Glandulicactus uncinatus, Hamatocactus uncinatus),
" -v- wrightii (See Ancistrocactus uncinatus -v- wrightii, Ferocactus uncinatus -v- wrightii, Glandulicactus uncinatus -v- wrightii, Hamatocactus uncinatus -v- wrightii)....[Df],

ECHINOCACTUS.....(Cont'd),

valdezianus (See Gymnocactus valdezianus, Neolloydia valdeziana, Normanbokea valdeziana, Pelecyphora valdeziana, Thelocactus valdezianus, Turbinicarpus valdezianus),
[Aj 2-7-1951],
visnaga (See E. platyacantha -f. visnaga, Ferocactus visnaga),
[RR], [Bo-5], [Cn], [Aj 23-7-1972],
warnockii (See Echinomastus warnockii, Neolloydia warnockii, Sclerocactus warnockii),
[Df],
whipplei (See Sclerocactus whipplei),
" -v- heleii (See Sclerocactus heleii, Sclerocactus whipplei -v- heleii),
[Df],
" -v- intermedius (See Sclerocactus intermedius, Sclerocactus parviflorus -v- intermedius).....[Df],
wislizenii (See Ferocactus wislizenii),
[Df],
xeranthemoides (See E. polycephalus -v- xeranthemoides)
[RR], [An], [Hx*], [Hy],

ECHINOCEREUS

acifer (See E. polyacanthus -v- densus),
[Hx*], [Il], [Aw 2-1972],
" -v- durangensis (See E. durangensis, E. polyacanthus).....[Ip], [Bm-1987],
" -v- trichacanthus (See E. polyacanthus),
adustus.....[Z], [RR], [Iz], [Af 59-4],
" -v- schwarzii (See E. radians, E. schwarzii),
[Gn*], [Af 58-2],
albatus (Invalid name - See E. nivosus, E. longisetus -v- albatus),
[Av], [Bo-4], [Bo-6], [Hx], [Hy], [Aj 16-4-1965],
albiflorus (See E. leucanthus, Wilcoxia albiflora),
albispinus (See E. baileyi, E. baileyi -v- albispina, -v- brunispinus, -v- roseispinus, E. longispinus, E. oklahomensis, E. reichenbachii - -v- baileyi, E. reichenbachii -v- albispinus)
[D], [M*], [Y], [Bu], Bv-1], [Df], [Af 6-9], [Af 33-4],
amoenus (See E. pulchellus -v- amoenus),
[RR], [Bo-4], [Hx*], [Aj 23-9-1972], [Aj 29-10-1978], [Hl 6-1973],
" -v- albiflorus.....[Ag-1988],
angusticeps (See E. berlandieri -v- angusticeps, E. blanckii -v- angusticeps, E. papillosus -v- angusticeps),

ECHINOCEREUS.....(Cont'd),

arizonicus (See E. triglochidiatus -v- arizonicus),
armatus (See E. reichenbachii -v- armatus),
[Bo-6], [Bv-1], [Ae-1981],
baileyi (See E. albispinus, E. baileyi -v- albispinus, -v- brunispinus, -v- caespiticus, -v- roseiispinus, E. longispinus, E. oklahomensis, E. reichenbachii -v- albispinus, E. reichenbachii -v- baileyi),
[D], [M*], [T], [Y], [GG], [LL], [NN], [RR], [TT], [Bo-4], [Bv-1], [Df], [Gr], [Hn], [Hx], [Hy], [Ja], [Af 6-9], [Ah-1985],
" -v- albispinus (See E. albispinus, E. reichenbachii -v- albispinus),
[Y], [Bo-4], [Hy],
" -v- brunispinus (See E. reichenbachii -v- albispinus, E. brunispinus) spinus),
[Bo-4], [Hy],
" -v- caespiticus (Invalid name - See E. reichenbachii -v- albispinus),
[Bo-4], [Hy],
" -v- roseispinus (See E. reichenbachii -v- albispinus),
[Bo-4],
barthelowanus.....[Bo-4], [Gn*], [Gr], [Hx], [Af 2-4], [Aw 4-1967],
berlandieri (See E. blanckii, E. blanckii -v- berlandieri) (E. berlandieri sensu Weniger is E. pentalophus),
[Y], [CC], [Ax], [Bv-1], [Bz], [Df], [Hx*], [Hy], [Iz], [Jd], [Bi 6-1], [Hs IV-4-1983],
" -v- angusticeps (See E. blanckii -v- angusticeps, E. papillosus -v- angusticeps),
" -v- berlandieri.....[Ax],

blanckii (See E. berlandieri -v- berlandieri),
[I], [L], [M*], [V], [Y], [LL*], [Ar], [Bn], [Bu], [Bv-1], [Bz], [Cm], [Df], [Ds], [Hx*], [Hy], [Aj 28-9-1977], [Ba 3-1978],
" -v- angusticeps (See E. berlandieri -v- angusticeps, E. papillosus -v- angusticeps).....[M*],
" -v- berlandieri (See E. berlandieri -v- berlandieri, E. blanckii),
[T], [QQ], [Ay], [Bo-4], [Di*],
" -v- leonensis.....[Hx*], [Hy],
" -v- poselgerianus.....[Hx*],
bonkerae (See E. fasciculatus -v- bonkerae, E. fendleri -v- bonkerae)....[G], [P], [Q], [GG], [An],

ECHINOCEREUS.....(Cont'd),

boyce-thompsonii (See E. fasciculatus -v- boyce-thompsonii, E. fendleri -v- boyce-thompsonii).....[G], [P], [Q], [GG], [Cb],
brandegeei.....[F], [NN], [OO], [Bo-4], [Gn*], [Hv], [Hx*], [Ae-1982], [Aj 29-2-1978], [Aw 4-1967], [Bi 3-3], [Hh July, 1976],
" -f. cristata.....
bristolii (See E. pectinata -v- minor), [Dq], [Ae-1981], [Af 9-10], [Au Vol. 3 - 1978 Zone 3 Compendium],
" -v- pseudopectinatus.....[Gn*],
brunispinus.....[Ia],
caespitosus (See E. reichenbachii -v- reichenbachii), [T], [NN], [Bo-4], [Bo-6], [Go], [Hn], [Hx*], [Hy], [Ae-1982],
" -v- albiflora (See E. reichenbachii).....[Y],
" -v- aureiflora (See E. reichenbachii).....[Y],
" -v- caespitosus (See E. reichenbachii), [Y], [Df],
" -v- minor (See E. reichenbachii).....[Y], [Df],
" -v- perbellus (See E. reichenbachii -v- perbellus).....[Y], [Df],
" -v- purpureus (See E. reichenbachii), [Y], [Df],
" -v- tamaulipensis.....[Ay],
canyonensis.....
chihuahuaensis.....[Af 33-4],
chisoensis (See E. reichenbachii -v- chisoensis), [Y], [Al*], [Df],
" -v- fobeanus (See E. fobeanus).....[Gn*],
chloranthus (See E. viridiflorus -v- chloranthus), [M*], [U], [LL*], [NN], [An], [Bv-1], [Bz], [Cf], [Hv], [Hy], [Jd], [Af 14-10], [Am 9-1981], [Hl 4-1972], [Hl 2-1983],
" -v- chloranthus.....[Y], [Ax], [Df],
" -v- cylindricus (See E. viridiflorus -v- cylindricus).....[Af 58-4],
" -v- " -f. cristata.....[Af 58-4],
" -v- davisii (See E. viridiflorus -v- davisii, E. davisii),
" -v- neocapillus.....[Y], [Ax], [Df], [Af 41-1],
" -v- roseiflorus.....[Hy],
" -v- russanthus (See E. russanthus, E. chloranthus -v- chloranthus), [Gn], [Iz],
chlorophthalmus (See E. cinerascens), [RR], [Hx*], [Aj 19-8-1968],
cinerascens.....[Bd], [Bv-1], [Cx], [Gn*], [Hx*], [Hy], [Ij], [Aw 2-1969], [Ba 5-1980], [Hl 5-1979],
(cont'd)

ECHINOCEREUS.....(Cont'd),

cinerascens (cont'd),
" -v- ehrenbergii.....[Gn*],
cirrhiferus (See E. cinerascens),
coccineus (See E. triglochidiatus -v- melanacanthus),
[L], [T], [Ar], [Ay], [Df], [Go], [Hx], [Ie],
[Aj 22-8-1971], [Ba 9-1981],
" -v- coccineus (See E. triglochidiatus -v-
melanacanthus),
[Y], [Df],
" -v- conoideus (See E. conoideus, E. trigloch-
triglochidiatus -v- melanacanthus),
[Y], [Df],
" -v- gurneyi (See E. triglochidiatus -v-
gurneyi).....[Af 60-1-1988],
" -v- inermis (mutant) (See E. triglochidiatus
-v- inermis),
[An], [Ar], [Go], [Ae-1981], [Hl 1-
1973],
" -v- kunzei (See E. kunzei, E. triglochidiatus
-v- melanacanthus),
" -v- octacanthus (See E. octacanthus, E. tri-
glochidiatus -v- melanacanthus,
E. triglochidiatus -v- octa-
canthus),
[Go],
conglomeratus (See E. enneacanthus -v- conglomeratus),
[T], [RR], [Cx],
conoideus (See E. coccineus -v- conoideus, E. tri-
glochidiatus -v- melanacanthus),
[CC], [RR], [Bv-1], [Ie],
ctenoides (See E. dasyacanthus -v- ctenoides, E. pec-
tinatus -v- pectinatus),
[LL*], [RR], [Af 60-1-1988],
cucumis (cultivar) (See E. gentryi, E. scheeri -v-
gentryi),
[T], [Gn*],
dasyacanthus (See E. pectinatus -v- dasyacanthus,
E. pectinatus -v- neomexicanus),
[D], [H], [I], [P], [GG], [NN], [Ar], [Bn],
[Bo-4], [Bz], [Cd], [Cx], [Cz], [Df], [Hn],
[Il], [Ad Sept. 1980], [Af 13-3], [Af 52-5],
[Af 57-3], [Af 58-4], [Af 60-1-1988],
[Aj 33-10-1982], [Hl 4-1972],
" -v- ctenoides (See E. ctenoides, E. pectinatus
-v- pectinatus),
[Y],
" -v- dasyacanthus (See E. pectinatus -v-
dasyacanthus, E. pectinatus -v-
neomexicanus),
[Y], [Df],

dasyacanthus (cont'd),
" -v- hildmannii (See E. pectinatus -v- neomexcanus),
[M*], [U], [Df],
" -v- minor (See E. pectinatus -v- minor)...[H],
" -v- steereae (See E. pectinatus -v- pectinatus, E. steereae),
davisii (See E. chloranthus -v- davisii, E. viridiflorus -v- davisii),
[E], [Y], [DD], [Ay], [Af 2-11], [Ag-1988],
" -v- viridiflorus (See E. viridiflorus),
[Hl 3-1978],
decumbens (See E. engelmannii -v- decumbens, or E. triglochidiatus -v- melanacanthus),
delaetii.....[F], [L], [M*], [HH], [NN], [Ar], [Ay], [Bo-4], [Cx], [Dq], [Gr], [Hb], [Hx], [Hy], [Ia], [Iu], [Iz], [Ae-1982], [Af 4-6], [Af 58-3], [Au 37-4], [Aw 4-1986], [Cu 8-4-1946], [Hl 4-1972], [N-1980],
" -v- freudenbergeri (See E. freudenbergeri),
depressus.....[Bo-6],

dubius (See E. enneacanthus -v- dubius),
[M*], [U], [Y], [CC], [GG], [RR], [Bo-6], [Df], [Fg], [Hx*], [Hy], [Ae-1981],
durangensis (See E. acifer -v- durangensis, E. polyacanthus),
ehrenbergii (See E. cinerascens -v- ehrenbergii, E. pentalophus -v- ehrenbergii),
[Bd], [Bu], [Bv-1], [Cx], [Iu],
engelmannii.....[B], [F], [M*], [P], [Q], [R], [AA] [BB], [FF], [GG], [LL], [OO], [An], [Ar], [As], [Bn], [Bu], [Cb], [Do], [Ha], [Hx], [Hy], [Iz], [Ad March 1978], [Af 2-12], [Aj 10-8-1959], [Aw 4-1967], [Hh 1-9-1974],
" -v- acicularis.....[Q], [An], [Ax], [Ca], [Af 5-7],
" -v- armatus.....[Ax], [Ca],
" -v- chrysocentrus....[Ax],
" -v- decumbens (See E. triglochidiatus -v- melanacanthus),
" -v- engelmannii.....[An], [Ax],
" -v- howei.....[An],
" -v- munzii.....[Ax], [Ca],
" -v- nicholii.....[P], [GG], [An],
" -v- purpureus.....[Bh],
" -v- robustior (See E. engelmannii -v- engelmannii),
" -v- variegatus.....[An], [Ax],
" -f. cristata.....[Hy],

ECHINOCEREUS.....(Cont'd),

enneacanthus.....[D], [M*], [P], [LL*], [RR], [Ar], [Bz], [Cr], [Gr], [Ha], [Jd], [Af 5-2], [Af 58-4], [Am 1-1980],
" -v- brevispinus.....
" -v- carnosus (See E. enneacanthus -v- enneacanthus).....[Y], [Df],
" -v- conglomeratus (See E. conglomeratus),
" -v- dubius (See E. dubius), [U], [Ax], [Ib],
" -v- enneacanthus.....[Y], [Ax], [Df], [Af 60-1-1988],
" -v- stramineus (See E. stramineus), [M*], [U], [Ax], [Df], [Ib]

fasciculatus (See E. fendleri -v- fasciculatus), [Q], [An],
" -v- bonkerae (See E. bonkerae, E. fendleri -v- -v- bonkerae), [Q], [An], [Ax], [Af 52-1], [Au 37-2],
" -v- boyce-thompsonii (See E. boyce-thompsonii, E. fendleri -v- boyce-thompsonii), [Q], [An], [Ax], [Dg 7-1],
" -v- fasciculatus (See E. fendleri -v- fasciculatus), [Q], [Ax], [Au-37-1],

fendleri.....[B], [D], [M*], [P], [Q], [LL*], [RR], [Bo-4], [Bz], [Co], [Hx*], [Hy], [Ib], [Il], [Jd], [Af 16-9], [Af 52-2], [Ba 3-1979], [Hl 5-1968],
" -v- albiflorus.....[Bo-4],
" -v- bonkerae (See E. bonkerae, E. fasciculatus -v- bonkerae), [G], [GG], [An], [Bn], [Bo-4],
" -v- boyce-thompsonii (See E. boyce-thompsonii, E. fasciculatus -v- boyce-thompsonii), [GG], [LL], [An], [Bn], [Ad Aug. 1980],
" -v- fasciculatus (See E. fasciculatus), [Gn*],
" -v- fendleri.....[Q], [U], [Y], [Ax], [Df],
" -v- kuenzleri (See E. hempelii, E. kuenzleri),
" -v- ledingii (See E. ledingii),
" -v- pauperculus (See E. fendleri).....[An],
" -v- rectispinus (See E. rectispinus), [Q], [Y], [PP], [An], [Ax], [Df], [Af 58-2], [Au 37-1],
" -v- robustus (See E. robustus, E. fasciculatus -v- fasciculatus),

ECHINOCEREUS.....(Cont'd),

ferreirianus.....[Bv-1], [Bv-3], [Iz], [Jd], [Af 33-4], [Aj 5-1978], [Aj 9-1981], [Aw 4-1967], [Bi 5-1],
" -v- lindsayi (See E. lindsayi).....[Bi 6-1],
fitchii (See E. reichenbachii -v- fitchii), [F], [H], [M*], [T], [X], [Y], [EE], [NN], [Av], [Ay], [Bo-4], [Cr], [Df], [Gr], [Hx], [Hy], [Ia], [Il], [Ae-1982], [Am 3-1984], [Au Vol. 3-1978, Zone 3 Compendium], [Cu 1-1968],
floresii (See E. sciurus -v- floresii), [Bo-4], [Hy], [Af 49-5],
fobeanus (See E. chisoensis -v- fobeanus, E. reichenbachii -v- chisoensis), [Aj 33-7-1982],
freudenbergeri (See E. delaetii -v- freudenbergeri), [Aj 32-5-1981], [Ct-1988],
gentryi (See E. cucumis, E. scheeri -v- gentryi), [G], [T], [Ar], [Bo-4], [Hw-1981], [Hx],
glycimorphus (See E. cinerascens -v- cinerascens), [L] ,
gonacanthus (See E. triglochidiatus -v- gonacanthus), [P], [CC], [An], [Ie],
grandis.....[G], [RR], [Cx], [Iz], [Ae-1982], [Af 19-12] [Aj 24-4-1973], [Aw 4-1967],
hancockii (See E. maritimus -v- hancockii) [Desert Plant Life - 21:89 Fig. 2-4, 1949], [Bi 2-3],
hempelii (See E. fendleri -v- kuenzleri), [RR], [Ar], [Bo-4], [Il], [Ae-1981], [Aj 30-4-1979], [Aj 33-9-1982],
hidalgensis.....[Bo-6],
huitcholensis (See E. polyacanthus -v- densus, E. leeanus, E. polyacanthus -v- huitcholensis.....[Aj 30-1-1979 ?],
knippelianus.....[F], [T], [W], [RR], [UU], [Ar], [Bk], [Bo-4], [Gr], [Hv], [Hx], [Hy], [N-1980], [Ad April-1980], [Af 3-7], [Af 50-4], [Af 58-3], [Af 59-1], [Aj 28-7-1977], [Aw 2-1987], [Bi 1-4], [Bi-3-3],
" -v- kruegeri (See E. knippelianus -v- reyesii), [Ad -], [Ae-1983], [Af 50-2], [Bi 3-3], [Hw-1989],
" -v- reyesii (See E. knippelianus -v- kruegeri) [Ae-1982], [Af 52-6], [Hs III-4-1982],
" -v- pacificus (See E. pacificus, E. polyacanthus -v- pacificus),

ECHINOCEREUS.....(Cont'd),
----

kuenzleri (See E. fendleri -v- kuenzleri),
[Bh], [Df], [Hx*], [Aj 28-9-1977],
[Hh May-June, 1977],
kunzei (See E. coccineus -v- kunzei, E. kunzei, E. triglochidiatus -v- melanacanthus),
[Aj 24-3-1973],
laui.....[P], [Hx*], [Iz], [Ae-1981], [Af 50-4],
[Af 52-2], [Aj 29-4-1978], [Bi 6-1],
[Bi 5-4-1987], [Hm 2-1979],
ledingii (See E. fendleri -v- ledingii),
[M*], [P], [Q], [An], [Ax], [Dq], [Af 8-3],
leeanus (See E. polyacanthus -v- densus),
[RR], [Bk], [Bo-4], [Hx*],
" -v- multicostatus.....
leonensis (See E. pentalophus -v- leonensis),
[Af 50-4], [Aj 20-6-1969],
leptacanthus (See E. pentalophus -v- leptacanthus, E. procumbens)..[Ds], [Hy], [If], [Aj 26-7-1975],
leucanthus (See Wilcoxia albiflora, Wilcoxia leucanthus).....[Iz],
lindsayi (See E. ferreirianus -v- lindsayi),
[Af 54-5], [Aw 4-1975], [Hl 4-1977],
lloydii.....[M*], [Y], [RR], [Bh], [Df], [Af 57-3],
[Af 58-4],
longisetus.....[RR], [Bo-6], [Cx], [Gr], [Hy], [Jd],
[Aj 27-2-1976],
" -v- albatus (See E. albatus, E. nivosus),
[Ba 1-1980],
" -v- delaetii.....[Ct-1988],
" -v- longisetus.....[Ct-1988],
longispinus (See E. reichenbachii -v- albispinus),
[Bu], [Bv-1], [Hn],
luteus (See E. subinermis -v- luteus) [RR], [Ar], [Cl],
mamillatus (See E. brandegeei).....[RR],
mariae (See E. reichenbachii -v- albispinus)..[Af 52-6]
" -f. monstrosus.....[Af 52-6],
maritimus.....[F], [OO], [Ar], [Bo-4], [Dp], [Gr], [Hy],
[Aw 4-1967], [Aw 2-1988] [Ba 5-1981], [Ct-1988],
" -v- hancockii (See E. hancockii),
marksianus (Invalid name - See E. polyacanthus),
[Bo-4], [Hx*], [Hy],
matthesianus (Invalid name - See E. scheeri),
[DD], [Aj 18-7-1967],
matudae (See E. enneacanthus),
melanacanthus (See E. triglochidiatus -v- melanacanthus)
[P], [An],
melanocentrus (See E. reichenbachii -v- albertii),,[G],
[T], [Y], [GG], [Ar], [Bo-4], [Df], [Hx], [Hy],
[Jd], [N-1980], [Aj 10-2-1959], [Ba 11-1978],

merkeri (See E. enneacanthus),
[RR], [Hx*], [Hy], [Ad Sept. 1978],
mojavensis (See E. triglochidiatus -v- mojavensis),
[P], [FF], [An], [Bo-4], [Cb], [Hy], [Af 5-7],
morricalii (See E. viereckii -v- morricalii),
[Af 54-2], [Aw 1-1979], [Hl 4-1975],
[Hh March, 1976],
munzii (See E. engelmannii -v- munzii),
neomexicanus (Coulter) L. Benson, (See Echinocereus pectinatus -v- neomexicanus),
neomexicanus (Standley) Bull. Torrey Club 35:87, 1908, (See E. polyacanthus, E. polyacanthus -v- neomexicanus, E. rosei, E. triglochidiatus -v- neomexicanus).....[AA],
" -v- neoroemeri (See E. roemeri).....[Hy],
nivosus (See E. albatus, E. longisetus -v- albatus),
[Hx*], [Ae-1981], [Af 50-1], [Ah-1986], [Ct-1988]
noctiflorus (See E. gentryi),
ochoteranae (See E. subinermis -v- ochoteranae),
[Cx], [Af 56-1],
octacanthus (See E. coccineus -v- octacanthus, E. triglochidiatus -v- melanacanthus, E. triglochidiatus -v- octacanthus),
[Ar], [Bo-4], [Hn], [Hu], [Hx*], [Hy], [Iu],
oklahomensis (See E. albispinus, E. baileyi, E. baileyi -v-albispina, E. longispina, E. reichenbachii -v- albispinus),
[F], [Af 6-9], [Cu 5-4-1937],
ortegae.....[Hx*],
pacificus (See E. polyacanthus pacificus, E. triglochidiatus -v- pacificus),
[Ar], [Ad March 1978], [Aw 4-1967],
palmeri.....[Ar], [Bo-4], [Hb], [Hx*], [Hy], [Iz],
[Bi 6-1], [Hl 6-1973],
pamanesiorum.....[Aw 2-1981],
papillosus (See E. berlandieri -v- papillosus, E. blanckii -v- papillosus, Wilcoxia papillosa),
[H], [BB], [NN], [Ar], [Bo-4], [Bu], [Bv-1],
[Bz], [Df], [Gn*], [Gr], [Hx*], [Hy], [Af 24-4]
" -v- angusticeps (See E. blanckii -v- angusticeps)
[Y], [Df], [Gn*],
" -v- papillosus.....[Y], [Df], [Gn*],
" -v- rubescens (See E. pectinatus -v- neomexicanus)
[Q],
parkerii -v- parkerii.....[Ct-1988],
" -v- gonzalezii var. nov. .....[Ct-1988],
paucispinus (See E. gonacanthus, E. triglochidiatus - -v- gonacanthus).....[An], [Di*],
" -v- gonacanthus (See E. triglochidiatus -v- gonacanthus),

ECHINOCEREUS.....(Cont'd),

paucispinus (cont'd),
" -v- hexaedra (See E. triglochidiatus -v- hexaedrus, E. triglochidiatus -v- melanacanthus),
pectinatus.....[C], [H], [I], [M*], [P], [T], [U], [NN], [An], [Ar], [Av], [Ay], [Bt], [Cd], [Cx], [Gr], [Hn], [Hr], [Hu], [Hx*], [Hy], [Il], [Jd], [Ae-1982], [Af 53-2], [Af 57-5], [Aw 4-1986], [Bi 3-3], [Bm-1986], [Hf 4-1986],
" -v- caespitosus (See E. reichenbachii -v- reichenbachii).....[Bn],
" -v- castaneus (See E. reichenbachii -v- reichenbachii), [Bo-4], [Bo-6], [Hu], [Hx*], [Hy],
" -v- ctenoides (Invalid name) (See E. dasyacanthus -v- ctenoides, E. pectinatus -v- pectinatus), [Y], [Df],
" -v- dasyacanthus (See E. dasyacanthus), [Gn*], [Iz],
" -v- major.....[U],
" -v- minor (See E. bristolii, E. roetteri), [Y], [Ax],
" -v- neomexicanus (See E. dasyacanthus, E. pectinatus -v- dasyacanthus), [M*], [Q], [U], [An], [Ax], [Ib] [Af 52-1], [Am 9-1983], [Hl 6-1983],
" -v- pectinatus (See E. steereae).[Hv], [Ct-1988]
" -v- reichenbachii (See E. reichenbachii), [Gr],
" -v- rigidissimus (See E. rigidissimus), [B], [M*], [P], [R], [Y[, [DD], [LL*], [RR], [An], [Ar], [Ax], [Ay], [Bo-4], [Df], [Hu], [Hx], [Hy], [Af 52-1], [Af 56-6], [Da 2-1979],
" -v- " -f. cristata.....
" -v- rubescens (See E. pectinatus -v- neomexicanus),
" -v- rubrispinus (See E. rigidissimus -v- rubrispinus), [Gn*], [Ae-1983], [Af 54-3], [Aj 33-2-1982],
" -f.- rufispinus (See E. pectinatus -v- pectinatus).....[Bv-1], [Af 33-3], [Cu 1-1964],
" -v- sciurus (See E. sciurus).....[Iv 1-1959],
" -v- sonorensis.....[Hl 6-1973],

ECHINOCEREUS.....(Cont'd),

pectinatus.....(cont'd),
" -v- wenigeri.....[M*], [Y], [Df],
" -f. cristata.....[E],
pensilis (See Cereus pensilis, Morangaya pensilis), [Bv-1], [Af 6-10], [Af 15-1], [Af 33-4], [Aw 4-1967],
pentalophus (See E. procumbens), [C], [K], [M*], [S], [Y], [LL], [RR], [Ar], [Ax], [Ay], [Bo-4], [Bz], [Cm], [Df], [Gr], [Hv], [Hx], [Hy], [Il], [Iz], [Jd], [Ag-1984, 1988], [Au 37-4], [Ba 10-1980], [Dg 2-3-1980], [Hh July 1973],
" -v- ehrenbergii (See E. ehrenbergii), [Bo-4], [Aj 28-6-1977],
" -v- leonensis (See E. leonensis), [L] = E. dubius sensu Borg],
" -v- leptacanthus (See E. pentalophus), [Bv-1],
" -v- procumbens (See E. pentalophus), [Hu], [Hy], [Aj 27-9-1976], [Aj 28-6-1977],
" -f. white-flowered.....[Aj 9-1981],
" -f. cristata.....[Hy],
" -f. monstrosus.....[Bv-1],
perbellus (See E. caespitosus -v- perbellus, E. reichenbachii -v- perbellus), [M*], [CC], [RR], [Bv-1], [Bu], [Bz], [Cx], [Hx], [Ae-1983], [Af 6-11],
phoeniceus (See E. triglochidiatus -v- melanacanthus), [Hy],
-v- inermis (See E. coccineus -v- inermis, E. triglochidiatus -v- inermis, E. triglochidiatus -v- melanacanthus)..[Hy],
pleiogonus.....[Bo-6],
polyacanthus (See E. triglochidiatus -v- neomexicanus), [P], [An], [Ar], [Bt], [Cx], [Ds], [Aw 3-1966], [Cu 4-1962], [Hl 5-1977],
" -v- densus (See E. leeanus), [Gn*], [Ct-1988],
" -v- galtieri (See E. triglochidiatus -v- neomexicanus).....[F],
" -v- huitcholensis (See E. huitcholensis),
" -v- neomexicanus (See E. triglochidiatus -v- neomexicanus).....[Y], [Df],
" -v- pacificus (See E. pacificus, E. triglochidiatus -v- pacificus).....[Iz],
" -v- polyacanthus (See E. leeanus).....[Ct-1988]
" -v- rosei (See E. triglochidiatus -v- neomexicanus).....[Df],

ECHINOCEREUS.....(Cont'd),

poselgeri (See Cereus tuberosus, Wilcoxia poselgeri),
[Gn*], [Iz],
primolanatus.....[Bo-4], [Bo-6], [Hn*], [Hx*], [Hy],
[Af 50-3], [Bi 2-3],
" -f.- cristata.....[Ar], [Hy],
procumbens (See E. pentalophus),
[Di*], [Hb], [Flower - Ba 10-1980],
[Aj 8-10-1957],
" -f. cristata.....
pulchellus.....[F], [J*], [L], [RR], [Ar], [Bo-4], [Cs],
[Hx], [Hy], [Jd], [Ab-1985], [Af 42-3],
[Af 51-1], [Af 57-6], [Aw 3-1966],
[Aw 4-1969], [Bi 6-1], [Cu 1-2-1932],
" -v- amoenus (See E. amoenus),
[Bo-4], [Gn*], [Iz], [Aw 1-1985],
" -v- weinbergii (See E. weinbergii),
[Gn*], [Iz],
" -f. cristata.....
purpureus (See E. caespitosus -v- purpureus, E. reichenbachii -v- reichenbachii)....[D], [M*], [X],
[Bo-4], [Cz], [Hx*], [Hy], [Af 6-9], [Af 23-2],
[Am 4-1984], [Ba 9-1979], [Cu 5-4-1937],
" -f. cristata.....[Ag-1987],
radians (See E. adustus -v- schwarzii, E. schwarzii),
[Bo-4], [Hy], [Af 54-1], [Flower - Ba 11-1980],
[Bi 3-3],
rayonesensis.....[Ct-1988],
rectispinus (See E. fendleri -v- rectispinus),
[Q], [Y], [PP], [An],
reichenbachii (See E. caespitosus, E. pectinatus -v- caespitosus, E. pectinatus -v- reichenbachii),
[F], [H], [M*], [LL*], [RR], [UU], [Ar],
[Ax], [Bo-4], [Bo-6], [Cx], [Cz], [Hx*],
[Af 12-6], [Af 16-9], [Af 21-3], [Af 52-3],
[Af 52-6], [Ba 7-1981], [Bi 6-1],
" -v- albertii (See E. melanocentrus),
[Ax], [Bh], [Df],
" -v- albispinus (See E. albispinus, E. baileyi),.....[F], [Df], [Hv],
" -v- anigosanthus.....[Hy],
" -v- armatus (See E. armatus).....[Bi 5-4-1987],
[Bi 6-1],
" -v- baileyi (See E. baileyi, E. baileyi -v- albispinus, E. reichenbachii -v- albispinus).....[Iz],
" -v- caespiticus (See E. baileyi -v- caespiticus),
" -v- chisoensis (See E. chisoensis, E. fobeanus).....[Hv], [Af 52-2],

reichenbachii.....(cont'd),
" -v- fitchii (See E. fitchii),
[AA], [Df], [Hv],
" -v- perbellus (See E. perbellus),
[Ax],
" -v- reichenbachii (See E. caespitosus),
[Ax],
" -f. cristata.....[Hy],
rigidissimus (See E. pectinatus -v- rigidissimus),
[GG], [An], [Cr], [Ds], [Hx], [Ad Jan. - 1980], [Af 2-9],
" -v- rubispinus.....[Gn*],
robustus (See E. fendleri -v- robustus, E. fasciculatus -v- fasciculatus),
[Q], [An],
roemeri (See E. coccineus -v- octacanthus, E. octacanthus, E. triglochidiatus -v- melanacanthus, E. triglochidiatus -v- octacanthus),
[Q], [Go], [Aj 22-8-1971],
roetteri (See E. pectinatus -v- minor),
[M*], [Y], [CC], [Bo-4], [Bv-1], [Df], [Gr], [Hx*],
" -v- lloydii (See E. lloydii).....
rosei (See E. polyacanthus -v- rosei, E. triglochidiatus -v- neomexicanus),
[P], [U], [CC], [LL], [RR], [Bm-], [Hx], [Hy], [Af 32-5], [Bi 3-3], [Cu 3-1973],
rubescens (See E. dasyacanthus, E. papillosus -v- rubescens, E. pectinatus -v- neomexicanus),
[Q], [An],
rufispinus (See E. pectinatus -f. rufispinus),
[Bo-6],
russanthus (See E. chloranthus -v- chloranthus, E. chloranthus -v- russanthus),
[Y], [Df], [Il], [Af 52-5], [Af 60-1-1988, [Bi-4-3],
salm-dyckianus (See E. salmianus, E. scheeri),
[L], [NN], [RR], [Ay], [Be*], [Bt], [Cr], [Gr], [Hv], [Hx], [Hy], [Il], [Jd], [Af 14-10], [Bm-1987],
salmianus (See E. salm-dyckianus, E. scheeri),
[Hx*], [Aj 20-5-1969],
sanborgianus (See E. brandegeei),
[Aj 23-9-1972],
sarissophorus (See E. enneacanthus).....[R],
scheeri (See E. salmianus, E. salm-dyckianus),
[M*], [T], [NN], [Be*], [Bk], [Bo-6], [Bv-1], [Cm], [Hx], [Iz], [N-1980], [Af 15-4],
" -v- australis (See E. scheeri -v- koehresianus)
" -v- cucumis (See E. cucumis, E. gentryi),

ECHINOCEREUS.....(Cont'd),

scheeri.....(cont'd),
" -v- gentryi (See E. cucumis, E. gentryi), [Gn*], [Hw-1986],
" -v- koehresianus.....[Ct-1988],
" -V- scheeri.....[Ct-1988],
" -f. cristata.....[Hy],
schmollii (See Wilcoxia schmollii),
schwarzii (See E. radians, E. adustus -v- schwarzii), [Af 54-1],
sciurus.....[L], [NN], [RR], [Bo-4], [Bu], [Bv-1], [Gr], [Hx], [Iz], [Aj 26-8-1975], [Aw 4-1967], [Bi 6-1],
" -v- floresii (See E. floresii),
scopulorum (See E. pectinatus -v- minor), [RR], [Bo-4], [Hy], [Ad Feb. 1979], [Aw 4-1967], [Hl 6-1973],
spinibarbis (See E. cinerascens -v- ehrenbergii, E. pentalophus -v- ehrenbergii)..[Bo-4], [Hx*],
spinigemmatus.....
spinosissimus (See E. dasyacanthus, E. pectinatus -v- neomexicanus).....[Hy],
" -v- ctenoides (See E. ctenoides, E. pectinatus -v- pectinatus).....[Hy],
" -v- steereae (See E. pectinatus -v- pectinatus, E. steereae), [Hy],
" -f. cristata.....[Hy],
standleyi (See E. viridiflorus -v- viridiflorus, E. viridiflorus -v- standleyi), [RR],
steereae (See E. dasyacantha -v- steereae, E. pectinata -v- pectinata),
stoloniferus.....[T], [Bo-4], [Bv-1], [Dq], [Hx*], [Af 9-10], [Af 34-4], [Af 35-1],
" -v- stoloniferus.....[Gn*],
" -v- tayopensis (See E. tayopensis).....[Iz],
stramineus (See E. enneacanthus -v- stramineus), [F], [P], [Y], [GG], [LL*], [RR], [An], [Ar], [Bz], [Ce], [Cx], [Df], [Ds], [Ha], [Hx], [Hy], [Il], [Iz], [Jd], [N-1980], [Af 5-2], [Af 57-3], [Af 57-5], [Af 58-4], [Af 60-1-1988], [Ba 2-1979], [Ba 7-1980], [Bm-1988], [Cu 3-1973], [Hl 3-1979],
" -v- occidentalis.....[Ct-1988],
subinermis (See E. luteus), [G], [EE], [RR], [Ay], [Bn], [Bt], [Gr], [Hu], [Hx], [Hy], [Iz], [Ja], [Jd], [Ad May, 1978], [Af 16-9], [Bm-], [Hl 2-1970],
" -v- luteus (See E. luteus), [Bo-4], [Il],

ECHINOCEREUS.....(Cont'd),

subinermis.....(cont'd),
" -v- ochoteranae (See E. ochoteranae),
[Mexico Forestal 6:88-90, 1928],
[An. Inst. Biol. Mex. 1:177-180,
w/Fig. 1930],
subterraneus (Invalid name - See E. sciurus),
[Bo-4], [Hx], [Hy],
" -f. cristata.....
tayopensis (See E. stoloniferus -v- tayopensis),
[F], [Hb], [Af 35-1], [Aw 4-1976], [Aj 3-1980],
triglochidiatus.....[B], [E], [M*], [P], [R], [An],
[Ar], [Ax], [Bz], [Cd], [Df], [Go],
[Hn], [Hx*], [Hy], [Iz], [Ad March-1979], [Af 5-2], [Af 21-6], [Au - 37-2], [Ba 9-1981], [Dg 7-1],
[Hf 4-1987],
" -v- alcifer (See E. polyacanthus),
" -v- arizonicus.....[Q], [An], [Ax], [Bh],
" -v- coccineus (See E. coccineus, E. triglochidiatus -v- melanacanthus),
" -v- gonacanthus (See E. gonacanthus),
[Y], [An], [Ax], [Df],
[Hl 1-1973],
" -v- gurneyi.....[U], [Ax], [Ib], [Af 57-3],
[Af 58-4],
" -v- hexaedrus (See E. triglochidiatus -v- melanacanthus),
[Df],
" -v- inermis (mutant) (See E. triglochidiatus -v- melanacanthus),
[UU], [Ar], [Bh], [Af 53-2],
[Af 56-1],
" -v- melanacanthus (See E. coccineus, E. melanacanthus, E. roemeri)
[GG], [An], [As], [Ax], [Bn],
[Ca], [Hv], [Au 37-2], [Hh Vol. 3 #2, 1978],
" -v- mojavensis (See E. mojavensis),
[M*], [LL], [An], [As], [Ax],
[Ca], [Af 57-4],
" -v- neomexicanus (See E. neomexicanus),
[M*], [An], [Ax], [Df],
" -v- octacanthus (See E. octacanthus, E. coccineus -v- octacanthus, E. triglochidiatus -v- melanacanthus),
[Y], [GG], [Df],
" -v- pacificus (See E. pacificus, E. polyacanthus -v- pacificus),

ECHINOCEREUS.....(Cont'd),

triglochidiatus.....(cont'd),
" -v- paucispinus (See E. paucispinus),
[H], [Ax], [Bo-4],
" -v- polyacanthus (See E. neomexicanus, E. polyacanthus, E. triglochid-iatus -v- neomexicanus),
[B], [M*], [LL], [Df],
" -v- triglochidiatus.....[Ax], [Df],
tuberosus (See Cereus poselgeri, Cereus tuberosus, E. poselgeri, Wilcoxia poselgeri, Wilcoxia tuberosa),
tulensis (See E. enneacanthus -v- brevispinus, E. pentalophus),
[Af 33-5], [Aw 4-1973],
uehrii (See E. cinerascens),
uspenskii (See E. triglochidiatus),
[Bo-4], [Ds], [Hx],
viereckii.....[L], [Bo-4], [Hx], [Il], [Aj 7-1976],
" -v- morricalii (See E. morricalii)...[Ct-1988],
" -v- viereckii.....[Gn*], [Ct-1988],
viridiflorus.....[D], [L], [M*], [P], [GG], [HH], [LL*], [NN], [An], [Ay], [Bo-4], [Bz], [Cz], [Df], [Go], [Hv], [Hx*], [Hy], [Af 56-1], [Aj 33-11-1982], [Hl 2-1980], [Hw-1982],
" -v- chloranthus (See E. chloranthus -v- chlor-anthus, E. russanthus),
[Ar], [Bo-4], [Bo-6], [Gr],
" -v- correllii (See E. viridiflorus -v- standleyi)
" -v- cylindricus (See E. chloranthus -v- cylind-ricus),
[E], [U], [Y], [Ax], [Df], [Ib], [Af 57-3], [Am 4-1984], [Hl 3-1973],
" -v- davisii (See E. davisii),
[Bh], [Df], [Hu], [Iz], [Jd], [Af 55-6], [Af 56-3], [Ba 9- 1981], [Bi-1], [Bi 6-1], [Hl 1-1971],
" -v- intermedius (Invalid name - See E. chlor-anthus -v- cylindricus, E. vir-idiflorus -v- cylindricus),
[Bo-4],
" -v- montanus.....[Hx*],
" -v- standleyi (See E. standleyi, E. viridiflorus -v- viridiflorus),
[Y], [Df],
" -v- viridiflorus.....[Ax], [Df], [Iz],
websterianus.....[EE], [Bo-4], [Hx*], [Af 19-10], [Af 34-4], [Aj 31-4-1980], [Db - 26th Anniv. Edition],
webeserii.....[TT],

ECHINOCEREUS.....(Cont'd),

weinbergii (See E. pulchellus -v- weinbergii),
[RR], [Ah], [Ar], [Ay], [Bo-6], [Bm-],
[Hx*], [Ae-1981], [Af 51-5], [Aw 1-1978],
Spec. .....Lau 088.....[Bt], [Am 5-1980],
Spec. .....Lau 091.....[Ae-1981],

ECHINOFOSSULOCACTUS

acroacanthus (See Stenocactus acroacanthus),
[Bv-1], [Ho], [Af 29-3], [Hl 3-1969],
[Hl 1-1977],
albatus (See Stenocactus albatus),
[F], [Ar], [Bo-5], [Gr], [Hb], [Ho], [Hw-1981],
[Hx*], [Hy], [Ba 3-1974],
" -f. nigrispinus.....
" -f. cristata.....[Hy],
anfractuosus (See Stenocactus anfractuosus),
[Ar], [Bd], [Cx], [Ho], [Hx], [Aj 28-10-1977], [Aw 3-1969], [Aw 4-1973],
arrigens (See Stenocactus arrigens),
[RR], [Ho], [Hx*], [Il], [Af 2-11), [Aw 3-1969], [Hl 3-1969], [Hl 2-1977], [Hl 1-1983],
" -v- atropurpureus.....[Ho], [Hl 2-1977],
" -v- xiphacanthus.....
boedekerianus (See Stenocactus boedekerianus),
brachycentrus (See Stenocactus brachycentrus, S. oligacanthus).....[Ho],
bravoae.....[Ho],
bustamantei (See Stenocactus bustamantei, E. lexarzai, Stenocactus lexarzai),
[Ay], [Cx], [Ho], [Aw 2-1969],
caderoyi.....[Bo-5],
caespitosus (Backbg) (See E. caespitosus -v- gracilispinus (H. Bravo)),
[Bo-5], [Gr], [Aw 2-1970],
[Cu 2-1979],
" -v- gracilispinus (See E. caespitosus),
[Ho], [Aw 14:34 Apr,-June, 1969], [Hl 2-1977],
confusus (See Stenocactus confusus),
[RR], [Bo-5], [Ho], [Hl 4-1977],
coptonogonus (See Stenocactus coptonogonus),
[L], [Z], [NN], [Av], [Bv-1], [Cx], [Gr],
[Ho], [Hx*], [Hy], [Iu], [Jd], [Af 56-5],
[Aw 1-1969], [Hl 2-1977], [Hw-1986
" -v- major.....[Hl 5-1978],
crispatus (See Stenocactus crispatus),
[D], [H], [NN], [RR], [Ar], [Bo-5], [Bv-1],
[Cm], [Gr], [Ho], [Hx], [Hy], [Af 2-11],
[Af 29-3], [Af 52-6], [Aw 3-1969],

ECHINOFOSSULOCACTUS.....(Cont'd),

debilispinus (See Stenocactus debilispinus),
[Ho], [Hl 7-1965], [Hl 3-1969], [Hl 4-1977],
densispinus (See Stenocactus densispinus),
[Bo-5], [Ho], [Af 6-3], [Hl 7-1965],
[Hl 4-1977],
dichroacanthus (See Stenocactus dichroacanthus),
[Av], [Ho], [Af 55-4], [Aw 4-1966],
[Aw 1-1969], [Aw 1-1983], [Hl 3-1969],
[Hl 2-1978], [Hl 1-1983],
" -v- violaciflorus.....
erectocentrus (Invalid name),
(See Stenocactus erectocentrus),
[Z], [NN], [Ay], [Bo-5], [Ho],
esperanzaensis (See Stenocactus esperanzaensis),
flavispinus (See Stenocactus flavispinus).....[Ho],
flexispinus (See Stenocactus flexispinus),
[Ho], [Aw 14 Apr.-June, 1969], [Hl 3-1969],
flexuosus.....
fluctuosus.....
foersteri.....
geminispinus (See Stenocactus geminispinus),
[Bv-1],
gladiatus (See Stenocactus gladiatus),
[RR], [Bo-5], [Ho], [Hx*], [Hl 4-1977],
" -v- carneus (See Stenocactus gladiatus -v- carneus),
[Bo-5], [Ho], [Af 6-3], [Hl 3-1977],
grandicornis (See Stenocactus grandicornis),
[Hy],
guerraianus (See Stenocactus guerraianus),
[Ho], [Af 43-3],
hastatus (See Stenocactus hastatus),
[M*], [X], [Ar], [Ho], [Hy], [Il], [Ae-1981],
[Af 29-3], [Aj 9-2-1958], [Aj 11-5-1960],
heteracanthus (See Stenocactus heteracanthus),
[Bv-1], [Ho], [Hx*], [Af 29-3], [Aw 2-1975], [Hl 6-1977],
hookeri (See Stenocactus hookeri),
[Ho], [Hl 4-1977],
kellerianus (See Stenocactus kellerianus),
[Ar], [Hx*], [Hy],
lamellosus (See Stenocactus lamellosus],
[V], [Ar], [Bd], [Cx], [Gr], [Ho], [Hx],
[Il], [Jd], [Af 2-11], [Aw 1-1969], [Aw 2-1973], [Hl 4-1978],
lancifer (See Stenocactus lancifer),
[RR], [Bo-5], [Cr], [Ho], [Ig], [Aw 4-1979],
lexarzai (See Stenocactus lexarzai),
[Ar], [Cx], [Ho], [Aw 2-1972], [Hl 6-1977],

lloydii (See Stenocactus lloydii),
[M*], [RR], [Ar], [Ho], [Hx*], [Hy], [Ae-1981], [Aw 3-1969], [Aw 1-1983], [Cu 2-1979], [Hl 3-1973], [Hl 2-1982],
" -f. .....[Ho],
longispinus (See Stenocactus longispinus),
[Ho], [Hl 4-1973],
mocianus (H. Bravo).....
multiareolatus (H. Bravo).....
multicostatus (See Stenocactus multicostatus),
[F], [H], [Ay], [Bt], [Ck], [Cx], [Ho], [Hv], [Hx*], [Hy], [Iu], [Ae-1981], [Af 2-11], [Af 29-3], [Af 52-6], [Af 56--1], [Aw 1-1969], [Aw 4-1981], [Hl 2-1973],
" -v- coahuilensis.....[Af 2-11],
" -v- elegans.....
" -v- potosinus.....
nerispinus (See Stenocactus nerispinus),
[Ho], [Hl 3-1977],
obvallatus (See Echinocactus obvallatus, Stenocactus obvallatus),
[RR], [Bo-5], [Ho], [Hx], [Aw 2-1969], [Aw 2-1972], [Ba 10-1981], [Hl 6-1973], [Hl 4-1977],
" -v- arrigens (See E. arrigens, Stenocactus arrigens).....[Aw 2-1972],
ochoterenaus (See Stenocactus ochoterenaus),
[DD], [Ay], [Az], [Bo-5], [Cx], [Hb], [Ho], [Hx], [Hy], [Af 6-3], [Cu 2-1979], [Hl 3-1977],
" -v- parksianus (See E. parksianus),
" -v- sphacelatus (See E. sphacelatus),
parksianus (See Stenocactus parksianus),
[Bo-5], [Ho], [Af 6-3], [Hl 3-1977],
pentacanthus (See Stenocactus pentacanthus),
[D], [J*], [PP], [Bo-5], [Cs], [Ho], [Hx], [Hy], [Jd], [Ae-1981], [Af 2-11], [Aw 3-1965], [Aw 2-1969], [Db 2-1979], [Hl 1-1983],
" -v- david-boudetianus.....[Aw 14:36 Apr. -June, 1969],
phyllacanthus (Mart.) Lawr. (See Stenocactus phyllocanthus, E. tegelbergii, Stenocactus tegelbergii),
[T], [RR], [Ar], [Bo-5], [Cx], [Ho], [Hv], [Hx*], [Hy], [Af 52-6], [Aw 2-1972], [Cu 4-1978], [Hl 6-1977],
" -v- hookeri.....[Aw 14:19 Jan.-Mar.-1969],
" -v- tricuspidatus (See E. tricuspidatus),

ECHINOFOSSULOCACTUS.....(Cont'd),

rectospinus (See E. vaupelianus -v- rectospinus, Stenocactus vaupelianus -v- rectospinus)....[Ho],
rentacantha (See Stenocactus rentacantha), [Hl 2-1968],
rosasianus (See Stenocactus rosasianus), [Bo-5], [Ho], [Af 6-3], [Hl 3- 1977],
sphacelatus (See E. ochoterenaus -v- sphacelatus, Stenocactus sphacelatus), [Bo-5], [Ho], [Af 6-3], [Hl 3-1977],
sulphureus (See Stenocactus sulphureus), [Ho], [Hy], [Aw 2-1977],
tegelbergii (See E. phyllacanthus, Stenocactus tegelbergii),.....[X], [Af 52-6], [Hl 4-1972],
tellii (See Stenocactus tellii).....[Ar], [Hy],
tetraxyphus (See Stenocactus tetraxyphus), [Hb], [Ho], [Aw 4-1969],
" -v- longiflorus.....[Aw 14:84 Oct.-Dec.,1969]
tricuspidatus (See Stenocactus tricuspidatus), [QQ], [Jd], [Cu 1-1962], [Db 12-1979], [Hl 1-1972], [Hl 5-1978],
" -v- longispinus.....[Aw 14 Jan.-Mar.,1969],
vaupelianus (See Stenocactus vaupelianus), [T], [TT], [Ar], [Gr], [Ho], [Hx*], [Hy], [Il], [Af 52-6], [Aw 2-1976],
" -v- rectospinus (See E. rectospinus, Stenocactus rectospinus), [Aw 14:65 July-Sept. 1969],
violaciflorus (See Stenocactus violaciflorus), [M*], [NN], [RR], [Ar], [Cg], [Gr], [Ho], [Hx*], [Hy], [Af 2-11], [Aj 9-4-1958], [Aw 2-1969], [Aw 1-1981], [Hl 5-1977],
wippermannii (See Stenocactus wippermannii), [Gr], [Aw 2-1967], [Hl 5-1972],
xiphacanthus (See Stenocactus xiphacanthus), [EE], [Hy], [Aj 11-5-1960], [Ba 3-1974],
zacatecasensis (See Stenocactus zacatecasensis), [RR], [Bl], [Bv-1], [Cz], [Hb], [Ho], [Hx*], [Af 8-12], [Af 29-3],
" -v- brevispina.....[Af 2-11],
" -v- moranensis.....[Aw 14:67 July-Sept. 1969], [Hl 5-1977],
Spec. Lau-738.....[Ae-1981],
Spec. Lau-1088.....[Ar],
Spec. Lau-1092.....[Au 37-4],
Spec. Lau-1341.....[Ba 7-1981],
Spec. Lau-1377.....[Ba 7-1981],
Spec. Lau-1398.....[Bs 8-1981],
Spec. from Fresnillo.....[Ho],
Spec. .....[AA], [JJ], [NN], [Hu],

ECHINOMASTUS

acunensis (See E. erectocentrus -v- acunensis, Neolloydia erectocentra -v- acunensis),
[B], [P], [An], [Hl 1-1975],
centralis (See E. intertextus).....[Hy],
dasyacanthus (See Echinocactus intertextus -v- dasyacanthus, E. intertextus -v- dasyacanthus, Neolloydia intertextus -v- dasyacanthus),
[U], [DD], [RR], [Hy],
durangensis (See E. unguispinus -v- durangensis, Sclerocactus unguispinus -v- durangensis),
[Bo-5], [Hx], [Hy], [Aj 29-2-1978],
" -v- mapimiensis (See E. mapimiensis, Thelocactus mapimiensis),
erectocentrus (See Echinocactus erectocentrus, Gymnocactus beguinii, Neolloydia beguinii, Neolloydia erectocentra, Thelocactus erectocentrus).....[A], [B], [P], [NN], [Bn],
[Dq], [Hx*], [Hy], [Jd], [Af 21-2],
" -v- pallidus (See E. pallidus, E. erectcentrus -v- pallidus, Neolloydia erectocentrus -v- acunensis),
" -v- acunensis See E. acunensis, Neolloydia erectocentra -v- acunensis),
intertextus (See Echinocactus intertextus, Neolloydia intertexta, Sclerocactus intertextus),
[B], [D], [P], [Y], [LL*], [Ar], [Bo-5],
[Bz], [Df], [Il], [Jd], [Af 5-2], [Hl 1-1973],
" -v- dasyacanthus (See Echinocactus intertextus -v- dasyacanthus, E. dasyacanthus, Neolloydia intertexta -v- dasyacantha, Sclerocactus intertextus -v- dasyacanthus)
[Y], [Ay], [Df],
" -v- intertextus (Echinocactus intertextus -v- intertextus, Neolloydia intertexta -v- intertexta),
[Y], [RR], [An], [Df],
johnsonii (See Neolloydia johnsonii, Sclerocactus johnsonii),
[B], [GG], [Ar], [Ay], [Cb], [Hy], [Af 5-7],
[Aj 31-5-1980], [Cu 1-1970], [Hl 3-1979],
" -v- lutescens (See Neolloydia johnsonii -v- lutescens, Sclerocactus johnsonii -v- lutescens)....[P], [FF], [Bn],
[Hy], [Af 27-4], [Aj 31-5-1980],
" -v- multicolor.....
kakui (Invalid name - See E. pallidus, E. erectocentrus -v- pallidus, E. erectocentrus -v- acunensis, Neolloydia erectocentra -v- acunensis)..[DD], [Hx*],

ECHINOMASTUS.....(Cont'd),

krausei (See E. erectocentrus -v- erectocentrus, Neolloydia erectocentra -v- erectocentra), [Bo-5],
lauii.....[Ae-1983], Af 50-4], [Aj 33-5-1982], [Hm 1-1979],
macdowellii (See Neolloydia macdowellii, Thelocactus macdowellii, Thelocactus conothelos -v- macdowellii).....[H], [NN], [RR], [Ar], [Bo-5], [Bv-1], [Bu], [Cz], [Gr], [Hx*], [Hy], [Jd], [Hs II-2-1981],
mapimiensis (See E. durangensis -v- mapimiensis, Thelocactus mapimiensis), [Hx*], [Hy], [Hl 1-1975],
mariposensis (See Echinocactus mariposensis, Neolloydia mariposensis, Sclerocactus mariposensis, Thelocactus mariposensis), [M], [Y], [Bo-5], [Df], [Hy], [Af 52-6], [Af 56-3], [Af 60-1-1988], [Hl 1-1975],
" -v- .....[Y],
pallidus (See Echinocactus erectocentrus -v- pallidus, Neolloydia erectocentra -v- acunensis).[Y], [Bo-5],
parviflorus (See Sclerocactus parviflorus),
unguispinus (See Neolloydia unguispina, Sclerocactus unguispinus, Thelocactus unguispinus), [RR], [Ar], [Bo-5], [Ct-1986], [Hx*], [Hy], [Il], [Aw 1-1967], [Hl 5-1972], [Hl 5-1977],
" -v- crassihamatus .....[Ar], [Hy],
" -v- durangensis (See E. durangensis, Sclerocactus unguispina -v- durangensis),
" -v- laui.....[Av], [Af 51-3],
warnockii (See Echinocactus warnockii, Neolloydia warnockii, Sclerocactus warnockii), [Ar], [Av], [Ct-1986], [Df], [Af 52-5], [Af 60-1-1988], [Aj 32-10-1981], Hl 4-1979],
Spec. (von Salinas SLP).....[Aj 28-11-1977],

ECHINOPSIS

adolpho-friedrichii.....[Ik], [Au 37-2],
albiflora.....
albispina.....[Bo-2],
albispinosa.....[Bo-2],
" -v- fauxiana.....[Bo-2],
amblayensis (See Lobivia amblayensis, Lobivia haematantha ssp. kuehnrichii -v- amblayensis), [Hx*],
" -v- albispina (See Lobivia amblayensis -v- albispina),

ancistrophora (See Pseudolobivia ancistrophora), [H], [RR], [Av], [Bo-2], [Cv-3], [Hx*], [Jd], [Af 6-11], [Hi-2],
" -v- hamatacantha (See E. hamatacantha, Lobivia hamatacantha, Pseudolobivia hamatacantha),
" -v- kratochviliana (See E. kratochviliana, Lobivia kratochviliana, Pseudolobivia kratochviliana),
" -v- polyancistra (See E. polyancistra, Pseudolobivia polyancistra),
" subspec. cardenasiana (See E. cardenasiana, L. cardenasiana)...[Cv-3], [Iz],
" " " -v- rubriflora.....
" " pojoensis (See E. pojoensis, Lobivia pojoensis, Echinopsis rauschii) [Cv-3],
" " " -v- grandiflora (See Lobivia pojoensis -v- grandiflora),
" " " -v- megalocephala (See Echinopsis rauschii -v- megalocephala)
" " arachnacantha (See E. arachnacantha, Lobivia arachnacantha), [Cv-3],
" " " -v- arachnacantha.....[Iz],
" " " -v- densiseta (See Lobivia arachnacantha).....[Cv-3],
" " " -v- sulphurea (See Lobivia arachnacantha -v- sulphurea),
" " " -v- torrecillasensis (See E. torrecillasensis, Lobivia torrecillasensis, Pseudolobivia torrecillasensis, Lobivia arachnacantha -v- torrecillasensis), [Cv-3],
" " " -v- vallegrandensis (See E. vallegrandensis, Lobivia arachnacantha -v- vallegrandensis), [Cv-3], [Aj 32-11-1981],
andalgalensis (See Trichocereus andalgalensis),
antezanae (See Helianthocereus antezanae),
arachnacantha (See E. ancistrophora ssp. arachnacantha, Lobivia arachnacantha), [X], [EE], [NN], [Ar], [Ay], [Bo-2], [Cv-3], [Hv], [Aj 26-1-1975], [Hi-2],
" -v- densiseta (See E. ancistrophora ssp. arachnacantha -v- densiseta, Lobivia arachnacantha -v- densiseta),

ECHINOPSIS.....(Cont'd),
arachnacantha....(cont'd),
" -v- minor.....
" -v- sulphurea (See E. ancistrophora ssp. arachnacantha -v- sulphurea, Lobivia arachnacantha -v- sulphurea),
" -v- torrecillasensis (See E. ancistrophora ssp. arachnacantha -v- torrecillasensis, Lobivia arachnacantha -v- torrecillasensis, E. torrecillasensis, Pseudolobivia torrecillasensis, Lobivia torrecillasensis), [Hx*], [Hh March, 1976],
" -v- vallegrandensis (See E. ancistrophora ssp. arachnacantha -v- vallegrandensis, Lobivia arachnacantha -v- vallegrandensis, E. vallegrandensis), [Aj 26-1-1975], [Hw-1988],
" -v- FR-360a.....[Hi-2],
arebaloi.....[Af 28-2],
atacamensis (See Trichocereus atacamensis),
aurantiaca (See Acanthocalycium aurantiacum),
aurea (See Lobivia aurea, Pseudolobivia aurea, Salpingolobivia aurea).....[M*], [NN], [RR], [Av], [Cr], [Cv-3], [Hv], [Hx*], [Hw-1981], [Il], [Iu], [Jd], [Af 4-3], [Aj 16-11-1965],
" -v- elegans (See E. aurea),
" -v- fallax (See E. fallax, Lobivia calochrysea),
" -v- grandiflora (See E. aurea),
" -v- leucomalla (See Hymenorebutia leucomalla, Lobivia aurea -v-leucomalla, Lobivia famatimensis -v- leucomalla, Lobivia leucomalla).....[Aj 16-11-1965],
" -v- quinesensis (See E. quinesensis, Lobivia quinesensis, Pseudolobivia aurea -v- quinesensis).....[Aj 16-11-1965],
" -v- robustior.....
" -v- shaferi (See Lobivia aurea -v- shaferi, Lobivia shaferi).....[Aj 16-11-1965],
ayopayana (See Echinopsis megacarpa), [Hi-2], [Ba 47:-June 1968],
backebergii (See Lobivia backebergii, Pseudolobivia backebergii),
baldiana (See Pseudolobivia baldiana),
berlingii.....[Hy],
bertramiana (See Helianthocereus bertramianus, Lobivia formosa ssp. tarijensis -v- bertramianus)
boliviensis (See Lobivia boliviensis),
boyuibensis (See E. obrepanda, Pseudolobivia boyuibensis).....[Hi-2],
brasiliensis....(See E. multiplex, E. oxygona),
brevispina (See Acanthocalycium brevispinus),

ECHINOPSIS.....(Cont'd),
bridgesii (See Trichocereus bridgesii).....[L], [Bo-2], [Dq], [Hi-2], [Hx*], [Hy], [Af 12-8], [Hw-1989]
" -v- "arebaloi".....(See E. arebaloi),
" -v- "cochabambensis".....(See E. cochabambensis)
bruchii (See Lobivia bruchii, Soehrensia bruchii), [Hx*],
" -v- nivalis.....
caespitosa.....
cajasensis.....[Hi-2],
calliantha (See Trichocereus callianthus),
calliantholilacina (See E. obrepanda, Pseudolobivia frankii)..... [Hi-2],
callichroma (See E. obrepanda -v- purpurea, Pseudolobivia callichroma), [L], [M*], [T], [RR], [Ar], [Dq], [Hx*], [Hy] [Af 6-11], [Aj 16-3-1965], [Bi 3-1987],
" -v- albispina (Invalid name).....[Bo-6],
calochlora.....[L], [M*], [T], [RR], [Ar], [Dq], [Hx*], [Hy], [Af 6-11], [Bi 3-1987],
" -v- albispina.....(Invalid name).....[Bo-6],
calochrysea (See Lobivia aurea -v- fallax, Lobivia calochrysea),
calorubra (See E. obrepanda -v- calorubra), [Cv-1], [Gr], [Hx],
camarguensis (See Trichocereus camarguensis),
campylacantha (See E. leucantha, E. spegazziniana, E. xiphacantha)...[L], [Bk], [Bv-1], [Jd],
candicans (See Cereus candicans, Trichocereus candicans)
" -v- gladiatus (See Trichocereus candicans -v- gladiatus),
" -v- nitens (See Trich. candicans -v- nitens),
" -f. rubriflora (See Trichocereus candicans -v- rubriflorus),
" -v- tenuispinus (See Trichocereus candicans -v- tenuispinus),
cardenasiana (See E. ancistrophora ssp. cardenasiana, Lobivia cardenasiana).....[Hv], [Hx*],
carmineoflora (See E. obrepanda -v- calorubra, Lobivia carmineoflora, Pseudolobivia carmineoflora)
catamarcensis (See Lobivia huascha -v- crassicaulis, Lobivia andalgalensis (Br. & R.), Trichocereus catamarcensis),
cephalomacrostibas (See Haageocereus ferox, Trichocereus cephalomacrostibas, Weberbauerocereus cephalomacrostibas),
cerdana (See E. potosina, Lobivia ferox -v- potosina, Pseudolobivia potosina).....
chacoana (See E. rhodotricha -v- chacoana), [Ar], [Y], [Hx*], [Hy], [Ik], [Aj 9-1981], [Am 7-1979], [Hl 2-1967],

ECHINOPSIS.....(Cont'd),
chacoana.....(cont'd),
" -v- spinosior (See E. rhodotricha -v- spinosior).....[Ba Feb. 1965],
chalaensis (See Trichocereus chalaensis),
chamaecereus (See Chamaecereus silvestri),
chereauniana (See Lobivia cinnabarina),
chilensis (chiloensis) (See Trichocereus chilensis),
" -v- eburneus (See Trichocereus chilensis -v- eburneus, Trichocereus chilensis),
chrysantha (See Hymenorebutia chrysantha, Lobivia chrysantha),
chrysochete (See Lobivia chrysochete),
cinnabarina (See Cinnabarinea cinnabarina, Lobivia cinnabarina).....[Bk],
" -v- grandiflora.....
clavata (See Trichocereus clavatus),
cochabambensis.....[Bo-2],
colmari (See Lobivia corbula, Lobivia maximiliana, Lobivia pseudocariquinensis),
comarapana.....[Au 12:61, 1957],
coquimbana (See Trichocereus coquimbanus, Trichocereus serenanus),
cordobensis.....[Hy],
coronata (See E. obrepanda).....
cotacajensis.....[Af 42-4],
courantii (See Trichocereus courantii),
crassicaulis (See Helianthocereus crassicaulis, Lobivia crassicaulis, Lobivia huascha -v- crassicaulis),
cuzcoensis (See Trichocereus cuscoensis),
cylindracea (See Lob. aurea -v- shaferi, Lob.shaferi),
decaisneana.....[Di*], [Ds], [Af 17-4],
dehrenbergii (See E. paraguayensis),
" -v- blossfeldii.....[Bo-2],
densispina (See Lobivia densiapina, Lobivia haematantha -v- densispina, Hymenorebutia densispina, Lobivia famatimensis -v- densispina, Salpingo-lobivia densispina),
" -v- rebutioides.....
deserticola (See Trichocereus deserticolus),
elongata (See Lobivia elongata, Lobivia haematantha -v- elongata),
eremophila (See Trichocereus eremophilus),
escayachensis (See Helianthocereus escayachensis, Trichocereus escayachensis, Trichocereus werdermannianus),
eyriesii.....[L], [J], [Ar], [Cj], [Cm], [Cs], [Hi-1], [Hx], [Jd], [Am 4-1983],
" -v- duvalii.....
" -v- grandiflora (See E. oxygona -f. brevispina, E. werdermannii), [Bo-2], [Aj 32-11-1981],

eyriesii.....(cont'd),
" -v- inermis.....
" -v- muelleri.....
" -v- nigerrima.....
" -v- pudantii.....(See E. pudantii),
" -v- quehlii.....
" -v- roseum.....[Ds], [Hy], [Ir],
" -v- schelhasei.....
" -v- tettavii.....[Di*],
" -v- triumphans.....
" -v- wilkensii.....
" -f. cristata.....[Hy], [Hl 5-1967],
fabrisii (See E. volcanensis, Trichocereus fabrisii, Trichocereus volcanensis),
falcata (See Pseudolobivia falcata)
fallax (See E. aurea -v- fallax, Lobivia aurea -v- fallax, Lobivia calochrysea),
ferox (See Lobivia ferox, Pseudolobivia ferox, Soehrensia ferox),
[Hv],
" -v- longispina (See Lobivia aureolilacina, Lobivia claeysiana, Lobivia durispina, Lobivia hastifera, Echinopsis lecoriensis, Echinopsis nigra, Pseudolobivia longispina, Pseudolobivia wilkeae, Pseudolobivia wilkeae -v- carminantha),
" -v- potosina (See Lobivia ferox -v- potosina, E. potosina, Pseudolobivia potosina)
fiebrigii (See E. obrepanda, Pseudolobivia obrepanda -v- fiebrigii),
[RR], [Hy],
forbesii (See E. rhodotricha).....[Ds],
frankii (See E. obrepanda, Pseudolobivia frankii),
fricii (See Lobivia fricii, E. tiegeliana -v- fricii),
friedrichii (See Trichocereus shaferi),
fulva (See Lasiocereus fulvus),
fulvilana (See Trichocereus fulvilanus, Trichocereus glaucus).....[Hw-1981],
gemmata (See E. turbinata).....[Be*], [Ir],
gigantea (See Trichocereus gigantea)
glaucina (See Acanthocalycium glaucum),
glaucum (See Acanthocalycium glaucum)....
grandiflora (See E. rowleyi, Helianthocereus grandiflorus, Lobivia grandiflora, Lobivia huascha -v- grandiflora),
grandis (See Lobivia grandis, Lobivia formosa ssp. grandis, Soehrensia grandis).....[Hx*],
haematantha (See Lobivia haematantha)

ECHINOPSIS.....(Cont'd),

haku-jo (hybrid).....[R], [AA], [Ar], [N-1980], [Hl 1-1981],
" -f. cristata.....
hamatacantha (See E. ancistrophora -v- hamatacantha, Pseudolobivia hamatacantha), [Ay], [Bu], [Bv-1], [Gr], [Hx],
hamatispina (See E. ancistrophora -v- kratochviliana, E. kratochviliana, Lobivia kratochviliana, Pseudolobivia kratochviliana),
hammerschmidii....[Gr], [Hy], [Af 28-2], [Aj 18-10-1967]
hardeniana (See Lobivia pentlandii -v- hardeniana, Lobivia rossii, Lobivia pugionacantha -v- rossii),
hempeliana (See Arequipa hempeliana, Arequipa weingartiana, Arequipa weingartiana -v- carminanthema, Borzicactus hempelianus),
herbasii (See E. mamillosa, Pseudolobivia mamillosa) [Af 28-4],
hertrichiana (See Lobivia hertrichiana),
herzogiana (See Lobivia formosa ssp. tarijensis -v- bertramiana, Helianthocereus herzogianus, Helianthocereus orurensis, Trichocereus bertramianus, Trichocereus herzogianus),
hualfinensis (See Lobivia hacmatantha -v- hualfinensis, Lobivia hualfinensis),
huascha (See Helianthocereus huascha, Helianthocereus pecheretianus, Lobivia huascha, Lobivia hyalacantha, Trichocereus callianthus),
" -v- andalgalensis (See Lobivia andalgalensis, Helianthocereus andalgalensis, Lobivia huascha -v- crassicaulis, Trichocereus andalgalensis),
" -v- auricolor (See Helianthocereus huascha -v- auricolor, Lobivia huascha, Trichocereus andalgalensis -v- auricolor, Trichocereus auricolor, Trichocereus huascha),
" -v- macranthus (See Helianthocereus huascha -v- macranthus, Lobivia huascha, Trichocereus huascha),
" -v- rosiflorus (See Lobivia huascha),
" -v- rubriflora (See Helianthocereus huascha -v- rubriflorus, Lobivia huascha -v- andalgalensis (Web.) Rausch),
huottii.....[QQ], [Bk], [Hy], [Af 6-11], [Af 38-2],
" -v- longispina.....[Hi-2],
" -v- oropezana.....
hyalacantha (See Helianthocereus hyalacantha, Lobivia hyalacantha, Lobivia huascha, Helianthocereus huascha),

ECHINOPSIS.....(Cont'd),

hystrichoides (See E. mamillosa -v- hystrichoides),
[Hi-2],
ibicuatensis.....[Af 28-2],
imperialis (hybrid).....[Bo-2],
ingens (See Lobivia formosa, Soehrensia ingens),
intricatissima (See E. leucantha),
[If], [Af 6-7],
jajoiana (See Lobivia jajoiana),
johnsonii.....[Am 7-1984],
kermesina (See E. mamillosa -v- kermesina, Lobivia
kermesina, Pseudolobivia kermesina, Pseudo-
lobivia mamillosa -v- kermesina).....[Ay],
[Bv-1], [Cp], [Cv-2], [Gr], [Hy], [Il], [Jd],
" -v- cylindrica.....[Hy],
kladiwaiana (See E. ayopayana).....[Aj 23-9-1972],
klimpeliana (See Acanthocalycium klimpelianus),
klingeriana (See E. chacoana),
[Af 54-6], [Ba 1-1976], [Bi 4-1986],
knuthiana (See Trichocereus cuzcoensis -v- knuthianus,
Trichocereus knuthianus),
korethroides (See Soehrensia korethroides, Lobivia
formosa ssp.bruchii -v- nivalis),
krainziana.....[D],
kratochviliana (See E. ancistrophora -v- kratochvil-
iana, Lobivia kratochviliana, Pseudo-
lobivia kratochviliana),
[C], [NN], [Hx*], [Il], [Jd], [Af 6-11],
kuehnrichii (See Lobivia haematantha -v- kuehnrichii,
Lobivia kuehnrichii, Lobivia nigrispina),
lagenaeformis (See E. bridgesii, Trichocereus
bridgesii),
" -f. monstrosus.....
lamprochlora (See Cereus lamprochlorus, Trichocereus,
lamprochlorus, Trichocereus neolamprochlorus)

lateritia (See Lobivia lateritia, Lobivia camataquien-
sis, Lobivia imporana, Lobivia scopulina)..[Cv-2]
lecoriensis (See Lobivia ferox -v- longispina, Lobivia
longispina, Pseudolobivia wilkeae),

leucantha (See E. intricatissima),
[Ar], [Bo-2], [Ck], [Hi-2], [Hx*], [Iu],
[Af 6-7], [Af 51-4], [Bi 6-3],
" -v- brasiliensis.....
" -v- brevispina.....[Iu],
" -v- lobivioides.....
" -v- volliana.....[Bo-2],
leucomalla (See Echinopsis aurea -v- leucomalla, Lob-
ivia famatimensis -v- leucomalla, Hymeno-
rebutia leucomalla, Lobivia leucomalla),

ECHINOPSIS.....(Cont'd),

leucorhodantha (See E. ancistrophora, Pseudolobivia leucorhodantha, Pseudolobivia pelecyrhachis),
[Bv-1], [Iv 3-1947],
litoralis (See Trichocereus litoralis),
lobivioides (See E. ancistrophora, E. leucorhodantha, Pseudolobivia pelecyrhachis)
longispina (See Lobivia ferox -v- longispina, Lobivia longispina, Pseudolobivia longispina),
[H], [S], [Gr],
" -v- nigra (See Echinopsis nigra, Lobivia ferox -v- longispina),
luteiflora (See Lobivia aurea, Lobivia luteiflora, Pseudolobivia luteiflora),
macrogona (See Trichocereus macrogonus),
mamillosa (See E. herbasii, Pseudolobivia mamillosa),
[Bo-2], [Cv-2], [Hx*], [Hy], [Aj 27-11-1976],
" -v- flexilis.....
" -v- hystrichoides (See E. hystrichoides),
[Cv-2],
" -v- kermesina (See E. kermesina, Pseudolobivia mamillosa -v- kermesina, Pseudolobivia kermesina),
[Bt], [Cv-2], [Hx*], [Iz],
[Aj 22-1-1971],
" -v- mamillosa.....[Hi-2],
" -v- orozasana (See E. orozasana, Pseudolobivia orozasana, Echinopsis mamillosa, E. herbasii),
[Hi-2],
" -v- ritteri (See E. ritteri, E. mamillosa, E. herbasii),
[Gr], [Hi-2], [Ba Feb. 1965:24],
" -v- tamboensis (See E. mamillosa, E. tamboensis),
[Hi-2], [Af 42:184-187 July-Aug. 1970],
manguinii (See Trichocereus manguinii),
mataranensis (See E. obrepanda, E. fiebrigii, Pseudolobivia obrepanda),
maximiliana (See Lobivia corbula, Lobivia maximiliana, Lobivia pseudocariquinensis),
megacarpa (See Lobivia megacarpa. E. ayopayana).
[Hy],
melanopotamica.....[Bo-2], [Af 56-1], [Bi 2-4], [Bi 4-1986],
meyeri (See E. rhodotricha, E. forbesii),
[RR], [Bo-2],
minuana.....[Bo-2],
mirabilis (See Setiechinopsis mirabilis).....[RR],

ECHINOPSIS.....(Cont'd),

mistiensis (See Lobivia mistiensis, Lobivia pampana),
molesta.....
muhriae.....(See Lobivia muhriae),
multiplex (See E. oxygona),
[C], [M*], [R], [LL], [RR], [Bv-1], [Gr],
[Hy], [Af 5-5], [Am 7-1981], [Ba 7-1979],
[Dc 4-1979],
" -v- floribunda.....
" -v- picta.....
" -f. cristata.....[Di*], [Gr], [Hy], [N-1980],
narvaecensis (narvaezensis) (See Helianthocereus narvaecensis, Lobivia huascha ssp. narvaezensis, Trichocereus narvaezensis),
nealeana (See Hymenorebutia nealeana, Lobivia nealeana, Lobivia pseudocachensis, Lobivia cachensis, Lobivia saltensis -v- nealeana).....[Cv-2],
" -v- grandiflora (See Lobivia nealeana -v- grandiflora),
" -v- purpureiflora.....
nigra (See E. ferox -v- longispina, Lobivia ferox -v- longispina, Lobivia longispina, Lobivia longispina -v- nigra, Pseudolobivia longispina -v- nigra),
nigripilis (See Trichocereus coquimbanus -v- nigripilis, Trichocereus nigripilis),
obrepanda (See Pseudolobivia obrepanda),
[L], [M*], [RR], [Cv-1], [Hi-2], [Hx], [Hy],
[Jd], [Hl 3-1973],
" -v- aguilari (See Lobivia aguilari),
[Cv-1],
" -v- calliantholilacina (See E. obrepanda),
" -v- calorubra (See E. calorubra, Pseudolobivia carmineoflora),
[Cv-1], [Iz],
" -v- mizquensis (See Lobivia mizquensis),
[Cv-1],
" -v- purpurea (See E. callichroma, E. toralapana, Hymenorebulobivia purpurea, Lobivia purpurea, Pseudolobivia purpurea),
[Cv-1], [Jd],
omasuyana (See Lobivia omasuyana)
oreopepon (See Acanthocalycium oreopepon, Lobivia oreopepon, Soehrensia oreopepon),
orozasana (See E. mamillosa -v- orozasana, Pseudolobivia orozasana).....
orurensis (See Helianthocereus orurensis, Lobivia formosa ssp. tarijensis -v- bertramiana, Trichocereus orurensis, Trichocereus tarijensis -v- orurensis),

ECHINOPSIS.....(Cont'd),

orurensis.....(cont'd),
" -v- albiflora (See Helianthocereus orurensis, Trichocereus orurensis -v- albiflora),
oxygona (See E. multiplex),
[M*], [RR], [Hx*], [Aj 8-3-1957], [Hl 3-1983],
" -f. brevispina (See E. eyriesii -v- grandiflora, E. paraguayensis),
[Hi-1],
" -f. oxygona.....[Hi-1],
pachanoi (See Trichocereus pachanoi),
" -v- peruviana (See Trichocereus peruvianus),
pamparuizii.....[Af 42-1],
paraguayensis (See E. oxygona -f. brevispina, E. eyriesii -v- grandiflora),
[Hi-1],
pasacana (See Helianthocereus pasacana, Trichocereus pasacana),
pecheretiana (See Helianthocereus pecheretiana, Lobivia huascha),
" -v- viridior (See Helianthocereus pecheretianus -v- viridior),
peclardiana (See Lobivia peclardiana),
peitscherianum (See Acanthocalycium peitscherianum),
pelecyrhachis (See E. ancistrophora, Pseudolobivia pelecyrhachis),
" -v- lobivioides (See E. lobivioides, Pseudolobivia pelecyrhachis -v- lobivioides),
pentlandii (See Lobivia pentlandii, Lobivia boliviensis),
" -v- tricolor (See Lobivia maximiliana, Lobivia corbula),
pereziensis.....
peruviana (See E. pachanoi -f. peruviana, Trichocereus peruvianus),
pojoensis (See E. ancistrophora ssp. pojoensis, Lobivia pojoensis),
[Hx*], [Iv 7-1959],
polyancistra (See E. ancistrophora -v- polyancistra, Pseudolobivia polyancistra),
[I], [Cy], [Hi-2], [Hv], [Hx*],
[Am 7-1984], [Am 6-1986],
potosina (See E. ferox -v- potosina, Lobivia ferox -v- potosina, Lobivia potosina, Pseudolobivia potosina),

pseudocachensis (See Lobivia pseudocachensis, Lobivia nealeana, Hymenorebutia nealeana, E. nealeana).....[Hx*],

pseudocandicans (See Trichocereus pseudocandicans, Helianthocereus pseudocandicans),
pseudomamillosa (See E. obrepanda).. [Iv 7-1959],
pudantii (See E. eyriesii -v- pudantii),
puquiensis (See Trichocereus puquiensis),
purpureopilosa (See Trichocereus purpureopilosa),
pusilla (See Lobivia pusilla),
quadratiumbaccata (See Trichocereus quadratiumbaccata),
quiabayensis (See Lobivia quiabayensis),
quinesensis (See E. aurea -v- quinesensis, Lobivia aurea -v- quinesensis),
randallii (See Helianthocereus randallii, Lobivia formosa ssp. tarijensis, Trichocereus poco)
rauschii (See E. ancistrophora ssp. pojoensis, Lobivia pojoensis, E. pojoensis).....[Cv-3], [Hx*],
" -v- grandiflora (See E. ancistrophora ssp. pojoensis -v- grandiflora, Lobivia pojoensis -v- grandiflora),
" -v- megalocephala (See E. ancistrophora ssp. pojoensis -v- megalocephala),
rebutioides (See Hymenorebutia rebutioides, Lobivia chlorogona, Lobivia haematantha ssp. densispina -v- rebutioides, Lobivia rebutioides), [Hx*],
" -v- chlorogona (See Lobivia chlorogona, Lobivia haematantha ssp. densispina, -v- rebutioides, Lobivia rebutioides -v- chlorogona),
" -v- citriniflora (See Lobivia rebutioides -v- citriniflora, Lobivia haematantha ssp. densispina -v- rebutioides),
" -v- kraussiana (See Lobivia rebutioides -v- kraussiana, Lobivia haematantha ssp. densispina -v- rebutioides),
" -v- sublimiflora (See Lobivia sublimiflora, Lobivia haematantha ssp. densispina -v- sublimiflora),
rhodotricha (See E. forbesii), [K], [RR], [Cp], [Hx*], [Af 2-10],
" -v- argenteniensis.....
" -v- brevispina.....
" -v- chacoana (See E. chacoana),
" -v- droegana.....
" -v- robusta.....
" -v- rohlandii.....
" -v- spinosior (See E. chacoana -v- spinosior), [Hi-1], [Hl 1-1979],
riograndensis.....
riomizquensis (See Trichocereus riomizquensis),

ECHINOPSIS.....(Cont'd),

ritteri (See E. mamillosa -v- ritteri).....[Cv-II],
riviere-de-caraltii (See E. obrepanda).....[Hx*],
rivierei (See Leucostele rivierei, Trichocereus rivierei),
robinsoniana.....[Em - Fedde. Rep. Spec. Nov. Sonder-Beiheft C Lief. 20 tab. 80, 1934],
rohlandi (See E. rhodotricha -v- rohlandi),
rojasii (See E. obrepanda, Pseudolobivia rojasii),
" -v- albiflora (See E. obrepanda), [Hy],
roseo-lilacina (See E. obrepanda).....[Iv 12-1957],
rossii (See Lobivia rossii, L. pugionacantha -v- rossii),
rowleyi (rowleyana) (See Lobivia huascha -v- grandiflora, Lobivia grandiflora, Trichocereus grandiflorus)...[UU],
" -f. alba.....
rubinghiana (See Trichocereus rubinghiana),
rupicola (See Lasiocereus rupicola),
salmiana (See E. bridgesii, Pseudolobivia bridgesii, Trichocereus bridgesii),
salmonia.....[Bo-2],
salpingophora (See E. spegazziniana)......[Hr], [Hy],
saltensis (See Lobivia saltensis),
santaensis (See Trichocereus santaensis),
santiaguensis (See Trichocereus santiaguensis),
scheeri.....
schickendantzii (See Trichocereus schickendantzii),
schieliana (See Lobivia backebergii ssp. schieliana, Lobivia schieliana),
schoenii (See Trichocereus schoenii),
schreiteri (See Lobivia schreiteri),
schwantesii (See E. paraguayensis),
scoparia (See Hymenorebutia scoparia, Lobivia haematantha ssp. densispina, Lobivia scoparia),
scopulicola (See Trichocereus scopulicola),
shaferi.....(See Echinopsis friedrichii, Trichocereus friedrichii, Trichocereus shaferi),
silvatica.....[Hi-2],
silvestrii.....[M*], [DD], [RR], [Ar], [Bo-2], [Cv-3], [Hi-2],
skottsbergii (See Trichocereus skottsbergii),
" -v- breviata (See Trichocereus skottsbergii -v- breviatus),
smrziana (See Trichocereus smrzianus),
spachiana (See Trichocereus spachianus),
spachianoides (See Trichocereus spachianoides),
spegazziniana (See E. campylacantha, E. leucantha), [Hy],

ECHINOPSIS.....(Cont'd),
spiniflora (See Acanthocalycium spiniflorum), [O], [Af 4-3],
stilowiana (See Lobivia stilowiana),
stollenwerkiana (See Lobivia hardeniana, Lobivia rossii, Lobivia stollenwerkiana),
strigosa (See Trichocereus strigosus),
" -v- flaviflora (See Trichocereus strigosus -v- flaviflorus),
subdenudata (See E. pamparuizi)...[Hi-2], [Ag-1986], [Hw-1986],
sucrensis.....
tacaquirensis (See Trichocereus tacaquirensis),
tacnaensis (See Trichocereus tacnaensis),
tamboensis (See E. mamillosa -v- tamboensis, E. herbasii, E. ritteri).....[Hy],
tapecuana (See E. obrepanda)..... [Hi-2],
" -v- tropica.....
taquimbalensis (See Trichocereus taquimbalensis),
" -v- wilkeae (See Trichocereus taquimbalensis -v- wilkeae),
taratensis (See Trichocereus taratensis),
tarijensis (See Helianthocereus tarijensis),
" -v- orurensis (See Lobivia formosa ssp.tarijensis -v- bertramiana, Lobivia tarijensis -v- orurensis),
" -v- poco (See Lobivia formosa ssp. tarijensis, Lobivia tarijensis -v- poco),
tarmaensis (See Trichocereus tarmaensis),
tegeleriana (See Acantholobivia tegeleriana, Lobivia tegeleriana),
tenuispina (See Trichocereus tenuispinus),
terscheckii (See Trichocereus terscheckii),
" -v- montana (See Trichocereus terscheckii -v- montanus),
terscheckioides (See Trichocereus terscheckioides),
thelegona (See Trichocereus thelegonus),
thelegonoides (See Trichocereus thelegonus),
thionantha (See Acanthocalycium thionanthum),
tiegeliana (See Lobivia tiegeliana),
" -v- ruberrima (See Lobivia ruberrima),
toralapana (See E. obrepanda -v- purpurea, Pseudolobivia toralapana),
torataensis (See Trichocereus torataensis),
torrecillasensis (See E. ancistrophora ssp. arachnacantha -v- torrecillasensis, Lobivia arachnacantha -v- torrecillasensis, Lobivia torrecillasensis, Pseudolobivia torrecillasensis)...[F], [Bo-2], [Cv-3], [Hx*], [Af 28-4], [Aj 10-7-1959],
totorensis (See Helianthocereus herzogianus -v- totorensis, Trichocereus herzogianus -v- totorensis, Trichocereus totorensis),

ECHINOPSIS.....(Cont'd),

totorillana (See Trichocereus totorillanus, Trichocereus tarijensis -v- totorillanus),
trichosa (See Trichocereus trichosus),
tricolor (See Lobivia maximiliana),
tubiflora...[M*], [RR], [Bv-1], [Dq], [Hx*], [Hy], [Il], [Jd], [Aj 8-3-1957], [Flower - Aj 30-9-1979],
" -v- rosea.....[T],
" -f. cristata.....[If],
tucamanensis.....[Hy],
tulhuayacensis (See Trichocereus tulhuayacensis),
tunariensis (See Trichocereus tunariensis),
turbinata.....[RR], [Hy], [Af 24-5], [Af 55-6],
" -v- falcata (See E. falcata, Pseudolobivia falcata),
" -v- gemmata (See E. gemmata, E. turbinata),
" -v- tacuarembense.....
" -v- tephracantha.....
uyupampensis (See Trichocereus glaucus, Trichocereus uyupamensis),
valida (See Trichocereus validus).....[Hy],
vallegrandensis (See E. ancistrophora ssp. arachnacantha -v- vallegrandensis), [H], [NN], [RR], [Av], [Bo-2], [Cv-3],
variiflorum (See Acanthocalycium variiflorum),
vasquezii (See Trichocereus vasquezii),
vatteri (See Lobivia vatteri, Trichocereus vatteri),
violacea (See Acanthocalycium peitscherianum -v- violacea, Acanthocalycium violaceum).....[L],
volcanensis (vulcanensis) (See E. fabrisii, Trichocereus fabrisii, Trichocereus volcanensis),
volliana (See Trichocereus vollianus),
" -v- rubrispinus (See Trichocereus vollianus -v- rubrispinus, Trichocereus spachianus),
walteri (See Soehrensia walteri, Lobivia walteri),
werdermannii (See E. backebergii, E. oxygona -f. brevispina, E. eyriesii -v- grandiflora), [Av], [Hl 1-1972],
wilkeae (See Lobivia ferox -v- longispina, Pseudolobivia wilkeae),
" -v- carminantha (See Lobivia ferox -v- longispina, Pseudolobivia wilkeae -v- carminantha),
winteriana (See Lobivia winteriana),
wrightiana (See Lobivia wrightiana),
xiphacantha (See E. campylacantha).....[Hy],
" -f. cristata.....[Hy],
yungasensis.....
zuccarinii (zuccariniana) (See E. tubiflora),
Spec. -f. cristata.....[Ik],
Hybrid - (Echinopsis X Lobivia).....[Iz]), [Gr],,

ECHINOREBUTIA

caespitosa.....
grandiflora (See R. pseudodeminuta -f. grandiflora)...
pectinata (See R. pygmaea).....
pseudodeminuta (See R. pseudodeminuta).....
robustiflora.....

ENCEPHALOCARPUS

strobiliformis (See Pelecyphora strobiliformis),
[F], [L], [Z], [DD], [HH], [NN], [PP],
[Ar], [Bt], [Bo-5], [Cr], [Gr], [Hv],
[Hx], [Il], [Jd], [Ae-1983], [Af 1-7],
[Ag-1987], [Ag-1988], [Aj 28-10-1977],
[At-1982-1983], [Bi 2-3], [Bm-1986],
[Hm 3-1979],
" -f. cristata.....[Ar], [Hy], [Hl 3-1968],

EOMATUCANA

madisoniorum (See Borzicactus madisoniorum, Matucana madisoniorum, Submatucana madisoniorum),
oreodoxa (See Borzicactus oreodoxus, Matucana oreodosa)
[Ar], [Hi-4], [Aj 17-2-1966], [Aj 26-5-1975],
[Bm-1986],

EOPUNTIA

douglasii.....[Bo-1],

EPIPHYLLANTHUS

candidus (See Zygocactus candidus).....[Bo-2],
microsphaericus (See Arthrocereus microsphaericus, Epiphyllanthus obtusangulus, Schlumbergera obtusangula),
[Iu],
obovatus (See Schlumbergera opuntioides, Zygocactus opuntioides),
[HH], [RR], [Bo-2], [Bu], [Dq*], [Hv], [Jd],
obtusangulus (See Cereus obtusangulus, Epiphyllanthus microsphaericus, Schlumbergera obtusangula)
[DD], [Bo-2], [Hx*], [Aj IV-2-1932],

EPIPHYLLOPSIS

gaertneri (See Rhipsalidopsis gaertneri, Schlumbergera gaertneri),
[DD], [Bo-2], [Cr], [Hb], [Hx*], [Jd],
" -v- serrata.....
" -v- tiburtii.....

EPIPHYLLUM

ackermannii (hybrid) (See Nopalxochia ackermannii, Phyllocactus ackermannii),
[I], [L], [M*], [LL], [Cx], [Ja], [Bb-1984],
" -f. candida.....[Af 19-4],
alba.....[HH],
altensteinii (See Zygocactus truncatus).....[Id],
anguliger (See E. darrahii, Phyllocactus darrahii),
[LL], [Av], [Az], [Bo-2], [Hx*], [Jd], [Af 16-8], [Af 20-11], [Af 32-1*], [Aj 31-6-1980],
anthonyanum (See Cryptocereus anthonyanus),
biformis (See Disocactus biformis, Phyllocactus biformis),
[Be*],
bridgesii (See Schlumbergera bridgesii),
cartagense (See Phyllocactus cartagensis),
[Ad -], [Aj 9-2-1958],
caudatum (See E. pumilum).....[Az], [32-1*],
caulorhizum (See E. crenatum),
chrysacardium (See Marniera chrysocardium),
[E], [DD], [QQ], [Ay], [Az], [Bu], [Bv-1], [Hx*], [Jd], [Ad -], [Bb-1986], [Aj 32-6-1981], [Aw 1-1985], [Hh Nov. 1973],
costaricense (See E. thomasianum -v- costaricense),
crenatum.....[R], [RR], [Bv-1], [Af 32-1], [Af 57-3], [Aj 9-11-1958],
" -v- chichicastandango.....
" -v- crenatum.....[Az],
" -v- kimnachii (See Marniera macroptera -v- -v- kimnachii),
[Az], [Eb 35:77-80, 1965],
darrahii (See Epiphyllum anguliger, Phyllocactus darrahii).....[L], [M*], [RR], [Bl], [Bo-2], [Dq*], [Af 16-8], [Af 32-1*],
gaertneri (See Schlumbergera gaertneri, Phyllocactus gaertneri).....[Ds],
gigas (See E. grandilobum),
grandilobum (See Phyllocactus grandilobus),
[Af 29-6], [Af 37-1], [Aj 3-1977],
guatamalense (See E. phyllanthus -v- guatamalense),
[RR], [Az], [Ad-], [Af 19-12], [Af 32-1*]
hookeri (See E. phyllanthus -v- hookeri, E. strictrum),
[DD], [RR], [Av], [Ay], [Az], [Bo-2], [Hx],
lepidocarpum.....[Bo-2],
macrocarpum (See E. phyllanthus -v- costaricensis, E. macropterum),
macropterum (See Marniera macroptera),
[RR], [Af 19-4], [Af 32-1],
nelsonii (See Chiapasia nelsonii, Disocactus nelsonii, Phyllocactus chiapensis, Phyllocactus nelsonii),

EPIPHYLLUM.....(Cont'd),

obovatus (See Zygocactus opuntioides),
[Hy],
oxypetalum.....[M*], [Az], [Bu], [Bv-1], [Cj], [Af 6-11]
[Af 32-1*], [Aj 32-2-1981], [Aw 2-1985],
[Bb-1985],
" -v- purpusii.....
phyllanthus (See Phyllocactus phyllanthoides),
[C], [I], [H], [M*], [V], [DD], [RR], [Ay],
[Be*], [Bo-2], [Af 32-1], [Gu III-1953],
" -v- boliviense.....
" -v- columbiense (columbianus).....
" -v- costaricense (See E. thomasianum -v-
costaricense).....[Az],
" -v- grandiflora.....
" -v- guatamalense (See E. guatamalense),
[Az], [Af 36:110 July-Aug.
1964], [Aw 1-1985],
" -v- hookeri (See E. hookeri, E. strictum),
[Az], [Af 36:115 July-Aug. 1964],
" -v- paraguayensis....
" -v- pittieri (See E. pittieri),
[Af 36:115 July-Aug. 1964],
" -v- rubrocoronatum.....[[I], [Ay],
" -v- thomasianum (See E. macropterum)....[Az],
pittieri (See E. phyllanthus -v- pittieri),
[RR], [QQ], [Dq*], [Af 32-1*], [Aj 11-8-1960],
pumilum (See E. caudatum),
[Az], [Hx*], [Af 17-12], [Af 32-1], [Aw 3-1979]
[Flower - Aw 2-1965], [Bb-1986],
ruestii.....
russelianum (See Schlumbergera russeliana),
" -v- gaertneri (See Hatiora gaertneri, Schlum-
bergera gaertneri),
stenopetalum.....[RR], [Bo-2], [Hx*], [Af 1-4],
[Af 32-1],
strictum (See E. phyllanthus -v- hookeri, Phyllocactus
strictus),
[Bt], [Dq], [Ad-], [Af 32-1*], [Aw 2-1966],
[Bb-1987],
thomasianum (See E. macropterum, E. phyllanthus -v-
thomasianum).....[Aw 2-1985],
" -v- costaricense (See E. phyllanthus -v-
costaricense),
[Af 37:168 Nov. Dec. 1965],
" -v- thomasianum.....[Az],
truncatum (See Schlumbergera truncata, Zygocactus
truncatus).....[Be*], [Cn],
violaceum.....[Id],
Hybrid.....[Ar], [Ay], [Gr], [Hv],
Spec. Kinchinjunga.....[Az],

EPITHALANTHA

bokei (See E. micromeris -v- bokei),
[Ar], [Ax], [Af 60-1-1988],
densispina (See E. micromeris -v- densispina),
micromeris.....[D], [L], [P], [U], [Y], [Z], [DD], [GG],
[LL], [Ar], [Av], [Ax], [Ay], [Cr], [Cx],
[Cz], [Df], [Gr], [Hu], [Hv], [Hx*],
[Hy], [Ib], [Iu], [Iz], [Ad-], [Ae-1982],
[Af 2-3], [Af 16-5], [Af 57-3], [Af 58-4],
[Af 60-1-1988, [Aj 28-10-1977], [Am 9-1981],
" -v- bokei.....[GG], [Iz], [Ae-1982], [Af 50-4],
[Af 52-5], [Af 58-5], [Aj 1-1983],
[Am 1-1980],
" -v- caespitosa.....[Hy],
" -v- densispina (See E. densispina),
[Ar], [Bo-5], [Iv 3-1954],
" -v- fungifera.....[Hy], [Im], [Hl 6-1976],
" -v- greggii.....[F], [H], [NN], [Ar], [Ay],
[Bo-5], [Hy], [Ad-], [Ae-1981],
[Af 50-4], [Aj 1-1983], [Hw-1982],
" -v- " -f. cristata.....[Ar],
" -v- longicaulis.....[In],
" -v- micromeris.....[Aj 34-1-1983],
" -v- neomexicana.....[Hl 6-1976],
" -v- pachyrhiza (See E. pachyrhiza),
[Ay], [Bo-5], [Bt], [Iz],
[Af 16-11], [Af 50-4],
[Af 58-1], [Ag 1987],
" -v- polycephala (See E. polycephala),
[Iz], [Af 50-4], [Af 53-2],
[Aj 34-1-1983], [Iv 6-1954],
" -v- rufispina.....[Bo-5], [Hl 5-1976],
[Iv 3-1954],
" -v- tuberosa.....[Cu 1-1960],
" -v- unguispina.....[Bo-5], [Hy], [Ae-1982],
[Af 50-4], [Aj 34-1-1983],
[Ba 11-1980], [Hl 5-1976],
" -v- weberi.....
" -f. cristata.....[Hn], [Hy], [Im], [Iv 6-1954],
pachyrhiza (See E. micromeris -v- pachyrhiza),
[Bo-5], [Aj 23-4-1972], [Iv 6-1954],
" -v- elongata...[Bo-5], [Hl 5-1976], [Iv 6-1954],
" -v- flavido-cylindrica.....
" -v- variegata.....[Bo-5],
polycephala (See E. micromeris -v- polycephala),
[Ar], [Bo-5],
rufispina (See E. micromeris -v- rufispina),
spinosior (See E. micromeris -v- unguispina),
unguispina (See E. micromeris -v- unguispina),
[Af 59-3],
Spec. .....[Ar],

ERDISIA

apiciflora (See Cereus apiciflorus, Corryocactus apiciflora, E. maxima),
[Ay], [Bo-2], [Hx*], [Jd],
aureispina (See Corryocactus aureispina, Corryocactus erectus, E. erecta),
[Bo-2],
ayacuchensis (See Corryocactus ayacuchensis),
" -v- fusca (See Corryocactus ayacuchensis -v- fusca),
" -v- leucacantha (See Corryocactus ayacuchensis -v- leucacantha),
erecta (See Corryocactus aureispinus, Corryocactus erectus, E. aureispina),
[Bo-2], [Hx*],
fortalezensis (See Corrycactus tenuiculus, E. tenuiculus),
maxima (See Corryocactus apiciflorus, Corryocactus maximus, E. apiciflora),
[Bo-2], [Bv-1], [Cz], [Af 31-5], [Hl 2-1971],
melanotrichus (See Corryocactus melanotrichus),
meyanii (See Corryocactus aureus),
[RR], [Ay], [Bo-2], [Hx*], [Aj 31-3-1980],
philippii (See Austrocactus philippii),
quadrangularis (See Corryocactus quadrangularis),
[DD], [Bo-2], [Hu], [Hx*],
ruthae.....[Bo-6],
sextoniana (See Borzicactus sextonianus, Loxanthocereus sextonianus),
spiniflora (See Austrocactus spiniflorus),
[RR], [Bo-2], [Hx*],
squarrosa (See Corryocactus squarrosus),
[RR], [Bo-2], [Dq], [Hx*], [Af 18-12],
tarijensis (See Corryocactus tarijensis),
tenuicula (See Corryocactus tenuiculus),
[NN], [QQ], [Bo-2], [Hx*], [Db 6-1980],
" -v- .....[Bo-2],

ERIOCACTUS

ampliocostatus (See Notocactus ampliocostatus),
[Bi 4-1986],
claviceps (See Notocactus claviceps),
[Hx*], [Hi-1], [Aj 19-1-1968], [Hl 6-1967],
[Hl 3-1981],
graessneri (See Brasilicactus graessneri, Notocactus graesssneri).....[X], [Ar],
" -v- albisetus.....
grossei (See E. schumannianus -v- grossei, Notocactus grossei).....[Ik], [Bi 4-1986], [Hl 3-1981],

ERIOCACTUS.....(Cont'd),

grossei.....(cont'd),
" -v- aureispinus (See Notocactus grossei -v- aureispinus),
" -v- grossei.....[Hi-1],
haselbergii (See Brasilicactus haselbergii, Notocactus haselbergii),
[Ar],
itapetiensis.....[Hy],
leninghausii (See Notocactus leninghausii),
[DD], [QQ], [Ar], [Av], [Ay], [Bo-3],
[Cr], [Cs], [Hi-1], [Hx], [Hy], [Jd],
[Hl 3-1981],
" -v- albilanatus.....
" -f. apelii (See Notocactus leninghausii -f. apelii),
[Bo-3], [Hl 6-1967], [Hl 3-1981],
" -v- longispina (See Notocactus leninghausii -v- longispina),
" -v- minor (See Notocactus leninghausii -v- minor),
[Hi-1],
" -f. cristata (See Notocactus leninghausii -f. cristata),
[Hy],
magnificus (See Notocactus magnificus),
[Ar], [Hi-1], [Hx*], [Hl2-1970], [Hl 2-1980]
nigrispinus (See E. schumannianus -v- nigrispinus, Malacocarpus nigrispinus, Notocactus nigrispinus, Notocactus schumannianus -v- nigrispinus),
[Bi 4-1986],
schumannianus (See Notocactus schumannianus),
[Ar], [Av], [Ay], [Bo-3], [Hx], [Hy],
[Jd], [Hl 4-1972], [Hl 3-1981],
" -v- albispinus.....[Hy],
" -v- brevispinus.....
" -v- grossei (See E. grossei, Notocactus grossei),
" -v- longispinus.....[Hy],
" -v- nigrispinus (See E. nigrispinus, Malacocarpus nigrispinus, Notocactus nigrispinus),
[Ay], [Hl 5-1967],
" -f. (extreme form).....[Ik],
warasii (See Notocactus warasii),
[Gp], [Hi-1], [Ae-1983], [Aj 31-7-1980],
[Hl 3-1981],

ERIOCEREUS

adscendens (See Harrisia adscendens).....[RR], [Bo-4],
arendtii (See Eriocereus tortuosus, Harrisia tortuosus)
" -v- uruguayensis (See E. pomanensis -v- uruguayensis, Harrisia tortuosa -v- uruguayensis),
bonplandii (See Cereus bonplandii, Harrisia bonplandii) [H], [M*], [NN], [Ay], [Bo-4], [Dn], [Hi-2], [Hv], [Hx*],
crucicentrus.....(Probably a form of Eriocereus bonplandii),
guelichii (See Harrisia guelichii),
jusbertii (See Cereus jusbertii, Harrisia jusbertii), [Ar], [Bo-4], [Hu], [Hx], [Hy], [Flower - Aj 28-8-1977], [Aj 34-1-1983],
laticostatus (See Cereus argentinensis, Cereus platygonus, E. platygonus, Harrisia platygona),
martinii (See Cereus martinii, Harrisia martinii), [Ar], [Ay], [Bo-4], [Hx*], [Hy], [Af 20-7],
" -v- perviridis.....
" -v- regelii (See Harrisia regelii),
platygonus (See Cereus argentinensis, Cereus platygonus, Harrisia platygona),
polyacanthus (See Harrisia polyacantha),
pomanensis (See Harrisia pomanensis), [Av], [Bn], [Bo-4], [Hy], [Jd],
" -v- uruguayensis (See Eriocereus arendtii -v- uruguayensis),
portoricensis (See Harrisia portoricensis),
regelii (See Harrisia regelii),
spinosissimus (See Arthrocereus spinosissimus), [Hx*], [Aj 28-3-1977],
tarijensis.....[Hi-2],
tephracanthus (See Cereus tephracanthus, Harrisia tephracanthus, Roseocereus tephracanthus, Trichocereus tephracanthus), [Ay], [Hi-2], [Af 15-2],
tortuosus (See Cereus tortuosus, Eriocereus arendtii, Harrisia tortuosus), [C], [RR], [Ay], [Bo-4], [Hy], [Jd], [Aj 21-9-1970],

ERIOSYCE

algarrobensis.....[Hi-3],
aurata (See E. ceratistes, E. sandillon), [Cu 5-1-1936], [Hl 3-1982],
ausseliana (See E. ceratistes -v- ausseliana), [Hl 3-1982],
" -v- mollesensis (See E. ceratistes -v- mollesensis).....[Ar],

ERIOSYCE.....(Cont'd),

bruchii (See Soehrensia bruchii),
[De 2-1981],
ceratistes (See E. aurata, E. sandillon),
[D], [DD], [Ar], [Ay], [Bo-3], [Bo-6], [Dq], [Hx*], [Hy], [Il], [Jd], [Aj 29-4-1978, [Aj-30-11-1979], [Hl 6-1972], [Hl 3-1982],
" -v- aurea.....[Af 50-5],
" -v- ausseliana (See E. ausseliana)....[Bo-6],
" -v- combarbalensis (See E. ihotzkyanae),
[Bo-3], [Hl 3-1982],
" -v- coquimbensis (See E. ihotzkyanae),
[Bo-3], [Hl 3-1982],
" -v- ihotzkyanae (See E. ihotzkyanae),
[Bo-6], [Af 50-5],
" -v- jorgensis (See E. ihotzkyanae),
[Bo-3],
" -v- mollesensis (mollensis) (See E. sandillon)
[Bo-3], [Hl 3-1982],
" -v- rivadavia.....[Hl 3-1982],
" -v- tranquillaensis (See E. sandillon),
[Bo-6],
" -v- vallenarensis (See E. algarrobensis or E. ihotzkyanae),
[Bo-3], [Hy], [Hl 3-1982]
" -v- zorillaensis (See E. ihotzkyanae)
[Bo-3], [Hl 3-1982],
" -f. (Copiapo).....[Au Vol. 3, 1978, Zone 3 Compendium],
combarbalensis (See E. ceratistes -v- combarbalensis),
copiapensis.....
ihotzkyanae (See E. ceratistes -v- ihotzkyanae),
[Ar], [Hi-3], [Hy], [Hl 3-1982],
" -f. cristata.....[Ar],
lapampaensis.....[Bo-6],
megacarpa (See Rodentiophila megacarpa),
[Hi-3],
rodentiophila (See Rodentiophila atacamensis),
[Hi-3],
" -v- lanata (See Rodentiophila lanata),
sandillon (See E. aurata, E. ceratistes),
[Hi-3],
spinibarbis.....[Hi-3],
tranquillaensis (See E. ceratistes -v- tranquillaensis)

ERYTHRORHIPSALIS

pilocarpa (See Rhipsalis pilocarpa),
[E], [DD], [PP], [RR], [Ay], [Bo-2], [Bv-1], [Dq*], [Gr], [Hv], [Hx*], [Jd], [Ad-], [Ag-1984], [Ba 8-1979], [Bb-1985],

ESCOBARIA

aguirreana (See Gymnocactus aguirreanus),
[G], [Z], [Hl 5-1972], [Hl 2-1981],
albicolumnaria (See Coryphantha albicolumnaris, Coryphantha orcuttii, Escobaria orcuttii, Escobaria orcuttii -v- columnaris, Mammillaria albicolumnaria),
[Y], [Df], [Aa 5-1985], [Af 60-1-1988],
bella (See Coryphantha strobiliformis, Mammillaria bella),
[RR], [Ar], [Df], [Hx*], [Iz], [Ba 6-1976], [Hl 3-1981],
bisbeeana (See Coryphantha vivipara -v- bisbeeana, E. vivipara -v- bisbeeana).....[An],
casedoi (Invalid name - See Coryphantha laradoi, E. rigida),
chaffeyi (See Coryphantha chaffeyi),
[F]. [NN], [Bo-5], [Bu], [Ih], [Il], [Iz], [Jd], [Af 60-1-1988], [Am 3-1979], [Ba 7-1976], [Cu 2-1978], [Hl 3-1981],
chihuahuensis (See Coryphantha chihuahuensis),
[Bo-5], [Hy], [Aa 5-1985], [Aj 34, 76, 120-123, 136-140, 184-188, - 1983], [Aw 3-1966], [Bi 2-1986],
chlorantha (See Coryphantha chlorantha, Coryphantha vivipara -v- desertii).....[RR],
cubensis (See Coryphantha cubensis, Neobesseya cubensis, Neolloydia cubensis),
[Aa 2-1983], [Bi 2-1986], [Bi 5-1],
dasyacantha (See Coryphantha dasyacantha, Escobesseya dasyacantha, Mammillaria dasyacantha).....[E], [U], [Y], [LL*], [RR], [Hy], [Jd], [Af 52-6], [Af 60-1-1988], [Aj 29-7-1978], [Bi 2-1986],
" -v- chaffeyi (See Escobaria chafeyi),
[Bi 2-1986],
" -v- duncanii (See Coryphantha duncanii, Escobaria duncanii, Escobesseya duncanii, Mammillaria duncanii),
" -v- varicolor (See Escobaria varicolor, Mammillaria varicolor).....[Iz],
desertii (See Coryphantha vivipara -v- desertii, Coryphantha desertii),
[E], [U], [P], [CC], [FF], [RR], [An],
duncanii (See Coryphantha duncanii, Escobesseya duncanii, Mammillaria duncanii),
[Y], [Bo-5], [Af 39-2], [Af 60-1-1988],
emskoetteriana (See E. muehlbaueriana).....[Bo-5], [Iz], [Jd], [Aa 5-1985], [Bi 2-1986],
" -v- muehlbaueriana (See E. muehlbaueriana),
" -v- runyonii (See E. runyonii, Coryphantha robertii),

ESCOBARIA.....(Cont'd),

fobei (See Escobaria lloydii),
gigantea.....
guadalupensis.....[Af 58-4],
henricksonii (See Coryphantha henricksonii),
[Af 51-3], [Aj 4-1980], [Bi 2-1986],
[Hl 2-1981], [Hm 2-1978],
hesteri (See Coryphantha hesteri, Mammillaria hesteri),
[Y], [DD], [Df], [Hx], [Aj 29-7-1978, [Ba 7-1976], [Ba 9-1980], [Bi 2-1986],
laredoi (Invalid name - See Coryphantha laredoi, Escobaria rigida).....[Iz], [Af 50-5],
leei (See Coryphantha sneedii -v- leei, Escobaria sneedii -v- leei),
[Y], [Ar], [Df], [Af 41-3], [Af 57-3], [Aj 32-12-1981],
lloydii (See Coryphantha lloydii, Mammillaria lloydii)
macdougallii (See Ortegocactus macdougallii),
[Iz], [Hl 2-1981],
minima (See Coryphantha minima, Coryphantha nellieae, Escobaria nellieae, Mammillaria nellieae),
[Iz], [Bi-2-1986],
missouriensis (See Coryphantha missouriensis, Neobesseya missouriensis),
[Iz], [Ba 2-1981], [Bi 2-1986],
" -v- asperispina (See Coryphantha asperispina, Neobesseya asperispina),
" -v- caespitosa (See Coryphantha missouriensis -v- caespitosa, Coryphantha similis, Mammillaria similis, Neobesseya similis),
" -v- marstonii (See Coryphantha marstonii, Coryphantha missouriensis -v- marstonii, Neobesseya missouriensis -v- marstonii),
" -v- notesteinii.....
" -v- robustior (See Coryphantha missouriensis -v- robustior, Neobesseya similis -v- robustior),
[Cu 2-1978],
" -v- similis (See Coryphantha similis, Coryphantha missouriensis -v- caespitosa, Coryphantha missouriensis -v- similis, Mammillaria similis, Neobesseya similis),
[Bi 2-1986],
muehlbaueriana (See Coryphantha muehlbaueriana, E. emskoetteriana),
[Bo-5], [Hx*], [Iu], [Cu 2-1978],

ESCOBARIA.....(Cont'd),

nellieae (See Coryphantha minima, E. minima, Coryphantha nellieae, Mammillaria nelieae),
[DD], [Ar], [Bo-5], [Df], [Hu], [Hx], [Hy],
[Aj 30-3-1979], [Ba 6-1976], [Hl 3-1971],

neomexicana (See Coryphantha neomexicana, Coryphantha vivipara -v- neomexicana, E. vivipara -v- neomexicana),

oklahomensis (See Coryphantha oklahomensis, Coryphantha radiosa, Coryphantha vivipara -v- radiosa),

orcuttii (See Coryphantha albicolumnaria, Coryphantha orcuttii, Escobaria albicolumnaria, Mammillaria albicolumnaria),
[Hl 2-1976],

" -v- albicolumnaria (See Coryphantha albicolumnaria, E. albicolumnaria, Mammillaria albicolumnaria),
[Iz],

" -v- duncanii (Invalid name - See Coryphantha duncanii, Escobesseya duncanii, Mammillaria duncanii),
[An],

" -v- koenigii (See Coryphantha strobiliformis -v- orcuttii),
[An],

" -v- macraxina (See Coryphantha strobiliformis -v- orcuttii),
[An],

" -v- orcuttii (See Coryphantha strobiliformis -v- orcuttii),
[Bi 2-1986],

organensis (See Coryphantha organensis, Coryphantha sneedii -v- sneedii),

peloncillensis (See Coryphantha strobiliformis, Coryphantha tuberculosa, E. strobiliformis, E. tuberculosa),

radiosa (See Coryphantha oklahomensis, Coryphantha radiosa, Coryphantha vivipara -v- radiosa),
[Aj 11-10-1960],

rigida (Invalid name - See Coryphantha laredoi, E. laredoi),
[Ar], [Bo-5], [Af 50-1],

robbinsorum (See Cochiseia robbinsorum, Coryphantha robbinsorum),
[Iz], [Bi 2-1986], [Hm 3-1979],

robertii (See Coryphantha robertii, E. runyonii),

roseana (See Gymnocactus roseanus, Neolloydia roseana, Thelocactus roseanus),
[Hv], [Hy], [Af 56-4], [Ba 7-1976], [Bi 2-1986],
[Hl 2-1981],

runyonii (See Coryphantha robertii, E. emskoetteriana -v- runyonii),
[E], [Y], [GG], [LL], [PP], [RR], [Ar], [Av], [Hx], [Hy], [Iz], [Hm 2-1979],

sandbergii (See Coryphantha sneedii -v- leei, E. leei, Mammillaria leei),
[Aa 5-1985], [Af 53-2],

shaferi.....[Bv-1], [Cz],

sneedii (See Coryphantha sneedii -v- sneedii, Mammillaria sneedii),
[E], [F], [Y], [RR], [Bo-5], [Gr], [Hx*], [Hy], [Af 32-5], [Af 57-3], [Bi 1-4],

" -v- leei (See Coryphantha sneedii -v- leei, Escobaria leei, Mammillaria leei),
[Df], [Iz], [Aa 2-1983], [At-], [Bm-1987],

" -v- sneedii (See Coryphantha sneedii -v- sneedii),
[Bi 2-1986],

strobiloides (See E. strobiliformis),
[Hy],

" -v- pubescens.....[Hy],

strobiliformis (See Coryphantha albicolumnaria, Coryphantha tuberculosa, Coryphantha strobiliformis, E. tuberculosa, Mammillaria tuberculosa),
[An], [Gr], [Aa 2-1983], [Ba 9-1981], [Bi 2-1986], [Cu 2-1978], [Hl 3-1976],

" -v- caespititia (See Coryphantha strobiliformis)
[Hy],

" -v- durispina (See Coryphantha strobiliformis -v- durispina),
[Hy], [Iz],

" -v- orcuttii (See Coryphantha strobiliformis -v- orcuttii).....[An],

" -v- strobiliformis (See Coryphantha strobiliformis -v- strobiliformis),
[Aw 4-1977],

tuberculosa (See Coryphantha strobiliformis, Coryphantha tuberculosa, E. strobiliformis, E. tuberculosa, Mammillaria tuberculosa),
[E], [P], [Y], [LL*], [RR], [Ar], [Ay], [Bo-5], [Bu], [Bz], [Cx], [Df], [Dq], [Hn], [Hv], [Il], [Af 5-2], [Af 52-3], [Aj 29-7-1978], [Am 9-1981], [Cu 1-1970],

" -v- caespititia (See Coryphantha strobiliformis)
[Ar],

" -v- durispina (See Coryphantha strobiliformis -v- durispina).....[Bo-5],

tuberculosa.....(cont'd),
" -v- gracilispina (See Coryphantha strobiliformis -v- strobiliformis),
" -v- tuberculosa.....[Af 60-1-1988],
varicolor (See Coryphantha dasyacantha -v- varicolor, Escobaria dasyacantha -v- varicolor, Mammillaria varicolor),
[Y], [Bo-5], [Df], [Af 60-1-1988],
villardii (See Coryphantha villardii),
vivipara (See Coryphantha vivipara, Mammillaria vivipara),
[Hx*], [Jd], [Aj 29-7-1978], [Bi 2-1986],
" -v- aggregata (See Coryphantha aggregata, Coryphantha vivipara -v-aggregata, Echinocereus triglochidiatus -v- melanacanthus),
" -v- alversonii (See Coryphantha alversonii, Coryphantha vivipara -v- alversonii).....[Iz],
" -v- arizonica (See Coryphantha arizonica, Coryphantha vivipara -v- arizonica, Mammillaria vivipara -v- arizonica)
" -v- bisbeeana (See Coryphantha vivipara -v- bisbeeana, Escobaria bisbeeana),
[Bi 2-1986],
" -v- borealis (See Mammillaria vivipara -v- borealis), Coryphantha vivipara -v- borealis),
" -v- buoflama (See Coryphantha vivipara -v- buoflama),
" -v- desertii (See Coryphantha desertii, Coryphantha vivipara -v- desertii, E. desertii),
" -v- kaibabensis (See Coryphantha vivipara -v- kaibabensis, Coryphantha vivipara -v- rosea),
" -v- neomexicana (See Coryphantha neomexicana, Coryphantha vivipara -v- neomexicana, Mammillaria vivipara -v- neomexicana),
" -v- radiosa (See Coryphantha radiosa, Coryphantha vivipara -v- radiosa, Escobaria radiosa, Mammillaria vivipara -v- radiosa),
" -v- rosea (See Coryphantha vivipara -v- kaibabensis, Coryphantha vivipara -v- rosea),
" -f. sonorensis (See Mammillaria sonorensis),
" -v- vivipara (See Coryphantha vivipara -v- vivipara, Mammillaria vivipara -v- vivipara).....[Iz],

ESCOBARIA.....(Cont'd),

zilzeana (See Coryphantha zilzeana, Neobesseya zilzeana),
[Iz], [Bo-6], [Aa 2-1983], [Bi 2-1986], [Cu 2-1978], [Bi Vol. 4, #6],

ESCOBESSEYA

dasyacantha (See Coryphantha dasyacantha, Escobaria dasyacantha, Mammillaria dasyacantha).....[Y],
duncanii (See Coryphantha duncanii, Escobaria duncanii, Mammillaria duncanii).....[An], [Y],

ESCONTRIA

chiotilla.....[DD], [RR], [XX], [Ay], [Az], [Bo-4], [Bv-1], [Cx], [Cz], [Dq], [Hx*], [Iu], [Jd], [Af 60-4], [Aw 1-1970],
- -f. - monstrosus.....[Aw 3-1971],
lepidantha (See Anisocereus lepidanthus, Pachycereus lepidantha),

ESPOSTOA

baumannii.....[Af 52-6],
" -v- arborescens.....
blossfeldiorum (See Cephalocereus blossfeldiorum, Thrixanthocereus blossfeldiorum).....[Jd],
calva.....[Hi-4],
cephaliomelana (See Facheiroa cephaliomelana),
chiletensis (See E. haagei -v- samnensis),
churinensis (See Pseudoespostoa melanostele -v- churinensis),
cullmannianus (See Thrixanthocereus cullmannianus),
dautwitzii (See Cereus lanatus, E. lanata, Pilocereus lanatus, Pilocereus dautwitzii),
" -f. cristata.....[L],
dybowskii (See Austrocephalocereus dybowskii, Cephalocereus dybowskii, Espostoopsis dybowskii),
gigantea (See Espostoa lanata),
guentheri (See Cephalocereus guentheri, Vatricania guentheri),
[DD], [NN], [Gr], [Jd], [Af 15-3-1964],
haagei (See E. melanostele, Pseudoespostoa melanostele).....[Hi-4],
" -v- borealis.....
" -v- rubrispina.....
" -v- samnensis.....[Hi-4],
" -v- samnensis X Haageocereus icosagonoides, (Natural hybrid).....[Hi-4],
" -f. cristata.....[Hi-4],

ESPOSTOA.....(Cont'd),

huanucoensis.....[Bo-4], [Hb], [Hi-4], [Af 33-6],
[Aj 29-6-1978], [Bi 2-3],
humboldtianus.....
hylaea.....[Hi-4], [Iz],
lanata.....[E], [H], [L], [M*], [X], [AA], [DD], [NN],
[RR], [Ar], [Av], [Ay], [Bo-4], [Bo-6],
[Cs], [Cz], [Dq], [Gr], [Hb], [Hi-4], [Hv],
[Hx], [Hy], [Il], [Jd], [Af 50-2], [Aj 33-7-1982], [Aj 33-9-1982], [Ba 11-1979], [Bi Vol. 4, #6],
" -v- atroviolacea.....
" -v- floridaensis.....[Hi-4],
" -v- gracilis (See E. nana).....[Bo-4], [Bi 2-4]
" -v- mocupensis (See E. lanata),
[H], [PP], [Hy],
" -f. procera (See E. procera),
" -v- rubrispina (See E. lanata).....[Bo-4],
" -v- ruficeps.....
" -v- sericata (See Cereus lanatus ssp. sericatus,
E. lanata, E. sericata),
[T], [DD], [Bo-4],
" " " -f. cristata.....
" -v- tenuis.....
" -f. cristata.....[Hv], [Hy], [Af 57-1],
" X Borzicactus aurivillus (Natural hybrid),
[Hi-4],
lanianuligera.....[Hi-4],
" X Matucana intertexta (Natural hybrid),
Hi-4],
laticornua (See E. lanata)..[Bo-4], [Bt], [Hl 2-1969],
" -v- atroviolacea (See E. lanata).....[Bo-4],
" -v- rubens (See E. lanata).....[Bo-4],
longispinus (See Thrixanthocereus longispinus),
melanostele (See E. haagei, Pseudoespostoa melanostele)
[D], [H], [QQ], [Ah], [Ay], [Bt], [Gr],
[Hv], [Il], [Jd],
" -v- borealis (See E. haagei -v- borealis),
" -v- canetensis (See Pseudoespostoa melanostele
-v- canetensis),
" -v- churinensis (See Pseudoespostoa melanostele
-v- churinensis, Espostoa churinensis),
" -v- inermis (See Pseudoespostoa melanostele
-v- inermis),
[Ay], [Bo-4],
" -v- longispina (See Pseudoespostoa melanostele
-v- longispina),
" -v- rubrispina (See Pseudoespostoa melanostele
-v- rubrispina),
[Iz], [Aj 9-3-1958],

ESPOSTOA.....(Cont'd),

mirabilis.....[Ay], [Hi-4], [Hy],
" -v- primigena.....[Hi-4],
" -f. cristata.....[NN], [Hy],
mocupensis (See E. haagei -v- borealis, E. lanata -v- mocupensis),
[Hb], [Ba 5-1981],
nana (See Pseudoespostoa nana),
[Av], [Hi-4], [Hu], [Au 37-3],
" -v- compacta.....
" -v- minima.....
" -f. cristata.....[NN],
" -f. monstrosus.....[Aj 23-4-1972],
" X Haageocereus chryseus (Natural hybrid),
[Hi-4],
pilosa (See Facheiroa pilosa),
primigena (See E. mirabilis -v- primigena),
procera (See E. lanata -v- procera),
[Ay], [Bo-4], [Hl 5- 1982],
ritteri.....[Bo-4], [Gr], [Hy], [Aj 27-11-1976],
" -f. cristata.....[Ar], [Hb], [Hi-4], [Hx*],
ruficeps.....[Bo-6], [Hi-4], [Aj 5-1937], [Hl 2-1966],
" -v- australis.....
senilis (See Thrixanthocereus senilis),
[Iz], [Cu 2-1970],
sericata (See Cereus lanatus ssp. sericata, E. lanata -v- sericata),
[Cr], [Hy], [Af 21-3], [Aj 1-1937], [Aj 5-1937], [Aj 15-3-1964], [Cu 6-3-1938],
" -f. cristata.....[Hy], [Af 3-8],
superba.....[NN], [Hi-4], [Aj 11-6-1960],
" -f. hylaea (See E. hylaea),
" -f. cristata.....[Hi-4],
ulei (GUERKE) (See Cephalocereus ulei, Facheiroa ulei),
[Gr],

ESPOSTOOPSIS

dybowskii (See Austrocephalocereus dybowskii, Cephalocereus dybowskii, Espostoa dybowskii),
[Hi-1],

EULYCHNIA

acida.....[Bo-2], [Hi-3], [Hy],
" -v- choapa.....
" -v- elata.....[Hi-3],
" -v- ovalle.....
" -v- rivadavia.....
" -v- procumbens.....[Hi-3],

EULYCHNIA.....(Cont'd),
aricensis (See E. cephalophora),
[Hi-3],
barquitensis (See E. longispina -v- lanuginosior, E. spinibarbis -v- lanuginosior),
[Hi-3],
breviflora (See E. procumbens, E. spinibarbis),
[Hi-3], [Hv],
" -v- taltalensis.....[Hi-3],
" -v- tenuis.....
castanea (See Philippicereus castaneus),
[Hi-3], [Af 25-6],
cephalophora (See E. aricensis),
[Hl 1-1972],
clavata (See Austrocactus spiniflorus, Erdisia spiniflora),
eburnea (See Cereus chilensis -v- eburneus),
floresiana (floresii).....[Bu], [Bv-1], [Cz], [Hy],
[Af 56-4], [Ah-1985], [Am 8-1979],
[Hl 3-1967],
iquiquensis (See Cereus iquiquensis),
[DD], [RR], [Ay], [Bo-2], [Hi-3], [Hy],
[Af 25-3], [Aj 9-8-1958], [Cu 4-1969],
" -v- pullilana.....
" -v- taltalensis (See E. breviflora -v- taltalensis),
" -f. cristata.....[Ar], [Hy],
lanuginosior (See E. barquitensis),
longispina (See E. breviflora, E. spinibarbis),
[Hy],
" -v- lanuginosior (See E. barquitensis),
morromorenoensis.....[Hi-3],
procumbens (Invalid name - See E. breviflora),
[DD], [Ay], [Af 53-4],
" -f. cristata.....
ritteri (See E. saint-pieana),
[NN], [Ay], [Hi-4], [Hv], [Hy], [Aj 9-8-1959],
" -f. cristata.....[Hi-4],
saint-pieana.....[NN], [PP], [Ar], [Bo-2], [Bt], [Hi-3],
[Hx*], [Hy], [Iz], [Jd], [Hl 4-1968],
spinibarbis (See E. breviflora, E. procumbens),
[D], [NN], [RR], [Bo-2], [Gr], [Hx*], [Hy],
[Il], [Af 57-3], [Hl 3-1967],
" -v- taltalensis.....[Av],
" -v- lanuginosior (See E. barquitensis),
[Hy],
taltalensis (See E. breviflora -v- taltalensis, E. spinibarbis -v- taltalensis),
Sp. de Arica.....

EUREBUTIA

cylindrica (See Sulcorebutia cylindrica).....
durispina (See Rebutia durispina).....
" -v- violrosea.....
glaucescens.....
graessneri (See R. minuscula -v- violaciflora).....
[Hr],
" -v- rubriflora.....
halmii.....
hahniana (See R. ritteri -v-nigricans -f. hahniana)....
longiflora.....
longispina.....
lucida.....
maximiflora (See R. maximiflora, R. pseudodeminuta).
minuscula (See R. minuscula).....
" -v- multiflora.....
muricata (See R. minuscula -v- violaciflora),
pseudograessneri.....
rigidispina.....
rubispina.....
rubrispina.....
salmonea.....[Hr],
scarlatea (See R. minuscula).....[Hr],
" -v- brevispina (See R. minuscula).....[Hr],
senilis (See R. senilis, R. minuscula -v- senilis),
" -v- calva.....
" -v- cana (See R. senilis -v- senilis).....
" -v- longispina.....
spinosissima (See Aylostera spinosissima, Rebutia spinosissima).....
" -v- similis (See R. spinosissima),
violacea (See R. violacea, R. xanthocarpa -v- violaciflora),
" -v- albispina.....
" -v- densispina.....
" -v- paucipetala.....
" -v- rosea.....
" -v- stellata.....
violbrunispina.....
violsalmonea.....

FACHEIROA

blossfeldiorum (See Espostoa blossfeldiorum, Thrixanthocereus blossfeldiorum),
cephaliomelana (See Espostoa cephaliomelana), [Ad 4-1983], [Aj 26-6-1975],
pilosa (See Espostoa pilosa),
pubiflora (See Cephalocereus ulei, Espostoa ulei, F. ulei, Pilosocereus ulei),

FACHEIROA.....(Cont'd),

ulei (GUERKE) (See Cephalocereus ulei, Espostoa ulei),
[DD], [Ay], [Bo-4], [Hi-1], [It], [Af 50-2],
[Hl 4-1970],

FEROCACTUS
----

acanthodes (See F. cylindraceus),
[D], [M*], [P], [V], [AA], [BB], [DD], [GG],
[HH], [LL], [NN], [RR], [Ar], [Ax], [Ay],
[Bo-5], [Bt], [Cb], [Cz], [Do], [Dq], [Gr],
[Hy], [Ja], [Af 1-2], [Af 2-6], [Aj 32-2-1981], [Au 37-1], [Aw 4-1965], [Bi 3-3],
[Dg-7-1], [Hh July-1983],
" -v- acanthodes.....[Q], [An], [Ax],
" -v- eastwoodiae (See F. eastwoodiae),
[Q], [An], [Ax], [Bh],
[Dg 5-4-1984],
" -v- intortus.....[Am 4-1980],
" -v- lecontei (See F. cylindraceus -v- lecontei,
F. lecontei),
[FF], [Ar], [Ax], [Bo-5], [Ca],
" -v- rostii (See F. acanthodes -v- acanthodes,
F. cylindraceus, F. rostii),
" -v- tortulospinus (See F. tortulospinus)
alamosanus (See F. pottsii -v- alamosanus),
[RR], [Bo-5], [Cx], [Hy], [Bi 3-2],
" -v- platygonus (See F. pottsii),
[N-1980], [Af 14-10], [Af 34-4],
[Af 56-2], [Cu 7-1-1938],
bicolor (See Thelocactus bicolor),
" -v- flavidispinus (See Thelocactus bicolor -v-
flavidispinus),
californicus (See F. viridescens),
[Hy],
chrysacanthus.....[Ar], [Bo-5], [Bt], [Dq], [Hy],
[Ae-1981], [At-1982-1983], [Aw 4-1965]
" -v- rubrispinus.....[Hy],
coloratus (See F. gracilis -v- coloratus),
[NN], [Ar], [Gr], [Hx], [Af 4-9-1933],
[Aw 4-1965], [Hm 2-1978], [Hs II-2-1983],
corniger (cornigerus) (See F. latispinus),
[Be*],
" -v- flavispinus (See F. latispinus),
" -v- sanguiniflorus.....
coulteri (See F. histrix, F. melocactiformis),
[Hy],
" -v- rufispinus.....[Hy],

FEROCACTUS

covillei (See F. emoryi),
[E], [M*], [P], [Q], [GG], [LL], [NN], [RR],
[An], [Ax], [Ay], [Bu], [Cb], [Cx], [Dq],
[Af 24-2], [Af 57-4], [Au 37-1],
" -f. cristata.....[Af 30-4],
crassihamatus (See Ancistrocactus crassihamatus, Glandulicactus crassihamatus, Hamatocactus crassihamatus).....[RR],
cylindraceus (See F. acanthodes)........[Hy], [Iz],
[Ct-1984],
" -v- chrysacanthus (See F. chrysacanthus),
[Hy],
" -v- eastwoodiae (See F. acanthodes -v- eastwoodiae),
" -v- johnstonianus....
" -v- lecontei (See F. acanthodes -v- lecontei,
F. lecontei),
" -v- tortulospinus (See F. tortulospinus),
diguetii.....[RR], [UU], [Ar], [Bo-5], [Dn], [Hy],
[Ia], [Im], [Af 19-12], [Ah-1985], [Af 61-1],
[Aj 24-3-1973], [Aw 4-1965], [Bi 6-2-1988],
" -v- carmenensis.....[Ar], [Af 27-6],
" -v- diguetii.....[Af 27-6],
eastwoodiae (See F. acanthodes -v- eastwoodiae, F.
cylindraceus -v- eastwoodiae),
echidne.....[RR], [Bo-5], [Bv-1], [Cx], [Hx*], [Hy],
[Af 49-6], [Aj 10-8-1959], [Hl 4-1982],
" -v- echidne.....[Cx],
" -v- victoriensis.....[Af 40-4],
electracanthus (See F. histrix),
emoryi (See Ferocactus covillei),
[G], [Ar], [Ay], [Bo-5], [Gr], [Ha], [Hy],
[Ad Dec. 1979], [Aj 32-2-1981],
" -v- covillei.....(See F. covillei).....
" -v- grandis.....
" -v- rectispinus.....[Ct-1984],
flavispinus (See latispinus -v- flavispinus),
flavovirens.....[JJ], [RR], [Ar], [Ay], [Bo-5], [Cx],
[Gr], [Jd], [Af 25-4], [Aj 2-1983],
[Ba 4-1980], [Ct-1984],
fordii.....[CC], [PP], [RR], [Ar], [Bu], [Hy], [Iz],
[Aj 28-7-1977], [Aw 4-1965], [Hh Nov. 1974],
" -v- grandiflorus (See F. chrysacanthus),
[Af 27-6], [Aw 4-1963],
gatesii.....[Ar], [Bo-5], [Gr], [Hy], [Aw 4-1963],
glaucescens (See F. pfeifferi).....[C], [X], [Z], [CC],
[Ar], [Ay], [Bd], [Bo-5], [Ha], [Hv], [Hx*],
[Hy], [Il], [Aj 23-4-1972], [Aj 29-8-1978],
[Am 5-1980], [Aw 1-1964], [Aw 4-1965], [Bm-1988], [Ct-1984], [Hw-1988],

FEROCACTUS.....(Cont'd),

glaucescens.....(cont'd),
" -v- pfeifferi (See Echinocactus pfeifferi, F. glaucescens, F. pfeifferi),
gracilis.....[CC], [Ar], [Ay], [Bt], [Gr], [Ia], [Iz], [Af 4-8], [Aw 1-1965], [Aw 4-1965],
" -v- coloratus (See F. coloratus, F. peninsulae -v- viscainensis, F. viscainensis).....[Ar],
" -v- gatesii (See F. gatesii).....
" -v- gracilis.....
" -v- rectispinus (See F. rectispinus).....
" -f. cristata.....[Ia], [Af 30-3],
guirecobensis (See F. pottsii).....
haematacanthus.....[C], [CC], [RR], [Ar], [Af 52-6], [Aj 31-1-1980], [Aj 31-5-1980], [Aw 4-1962], [Aw 2-1966], [Af 60-1-1988],
hamatacanthus (See Echinocactus longihamatus, F. longihamatus, Hamatocactus hamatacanthus), [U], [LL*], [Ax], [Bo-5], [Cx], [Df], [Gr], [Jd], [Af 6-7], [Af 57-5],
" -v- crassispinus (See Hamatocactus hamatacanthus), [Af 46:80 Mar.-Apr.-1974],
" -v- hamatacanthus.....[Ax], [Cx],
" -v- sinuatus (See Echinocactus sinuatus, Hamatocactus hamatacanthus -v- sinuatus, Hamatocactus sinuatus), [CC], [Df],
herrerae (See F. wislizenii -v- herrerae), [G], [Ar], [Ay], [Gr], [Hy], [Ad June-1979], [Ah-1985],
histrix (See F. coulteri, F. melocactiformis), [R], [Ah], [Ar], [Av], [Bd], [Bo-5], [Hx], [Af 27-6], [Af 52-1], [Aw 1-1964], [Aw 1-1986],
horridus (See F. peninsulae), [Ar], [Bo-5], [Bu], [Hy], [Aj 24-4-1973], [Bi 3-2],
" -v- brevispinus.....
" -v- rostii (See F. acanthodes, F. rostii),
ingens (See Echinocactus ingens), [Dp],
jalapensis (See F. robustus), [Ar],
johnsonii (See Echinomastus johnsonii, Neolloydia johnsonii), [CC], [LL], [RR], [Af 2-9], [Af 5-7], [Af 20-1], [Af 21-4],
johnstonianus.....[Ar], [Bo-5], [Hy], [Af 53-5], [Af 56-5], [Aw 4-1963], [Aw 4-1965], [Cu 8-3-1946],

FEROCACTUS.....(Cont'd),

latispinus.....[D], [H], [L], [X], [EE], [FF], [HH], [NN], [RR], [Av], [Ay], [Bo-5], [Cr], [Cx], [Cz], [Gr], [Hn], [Hv], [Hx}, [Hy], [Il], [Ja], [Jd], [Ae-1981], [Af 3-1], [Af 58-1], [Af 60-3], [Ah-1985], [Aj 28-11-1977], [Am 3-1980], [Hl 3-1974], [Hl 4-1982], [Hw-1989],
" -v- flavispinus.....[Hy],
" -v- greenwoodii (See F. recurvus -v- greenwoodii),
" -v- latispinus.....[Ax], [Am 5-1980],
" -v- spiralis (See Echinocactus recurvus -v- spiralis),
" -f. variegata.....[Ar], [Af 59-6],
" -f. cristata.....[Ar],
lecontei (See F. acanthodes -v- lecontei, F. cylinddraceus -v- lecontei),
[CC],
lindsayi.....[Z], [Ar], [Aw 1-1966],
longihamatus (See F. hamatacanthus),
[Ia],
" -v- sinuatus (See F. hamatacanthus -v- sinuatus)
[CC],
macrodiscus.....[Z], [NN], [RR], [Ar], [Ay], [Cx], [Hy], [Iz], [Af 54-3], [Ah-1985], [Aj 29-3-1978], [Am 3-1980], [Aw 4-1966], [Af 59-6],
" -v- decolor.....[Hy],
" -v- macrodiscus.....[Ct-1984],
" -v- multiflorus (See F. macrodiscus).....[Ar],
" -f. (Guanajuato)...[Ar Fig. 32], [Af 54-3-1982]
" -v- oaxacensis.....
" -v- septentrionalis.....[Aw 3-1987],
melocactiformis (See F. histrix].....[D], [CC], [RR], [Bo-5], [Bv-1], [Aw 3-1965],
nobilis (See F. latispinus -v- spiralis),
[CC], [RR], [Ay], [Bv-1], [Cx],
neohaematacanthus (See F. stainesii -v- haematacanthus)
[Hy],
orcuttii (See F. viridescens).....[Ar], [Bu], [Hy],
" -v- flavispinus.....
peninsulae.....[OO], [RR], [Ar], [Ay], [Bn], [Bo-5], [Bv-1], [Hy], [Aw 4-1963], [Aw 4-1965],
" -v- johnstonianus (See F. johnstonianus)
[Bi 6-3],
" -v- santa-maria (See F. townsendianus -v- santa-maria).....
" -v- townsendianus (See F. townsendianus).....
[Bi 6-3],
" -v- viscainensis (See F. gracilis -v- coloratus, F. viscainensis)..[Ar], [Aw 4-1965],

FEROCACTUS.....(Cont'd),
pfeifferi (See Echinocactus pfeifferi, F. glaucescens),
[C], [X], [Z], [CC],
pilosus (See F. stainesii),
[Af 27-6], [Af 38-4], [Af 54-2], [Aw 2-1980],
[Ct-1984], [Hl 5-1979],
" -v- pringlei (See F. pringlei, F. stainesii -
-v- pringlei),
" -v- stainesii (See F. stainesii),
pottsii (See F. alamosanus -v- platygonus),
[Bo-5], [Af 56-2],
" -v- alamosanus (See F. alamosanus),
pringlei (See F. stainesii -v- pringlei)
[RR], [Cx], [Ad Dec., 1978], [Af 27-6],
rafaelensis (See F. echidne),
[Bo-5], [Hy], [Aw 4-1971],
rectispinus (See F. emoryi -v- rectispinus),
[CC], [RR], [Ar], [Ay], [Bv-1], [Cm], [Hy],
[Iz], [Af 2-7], [Af 56-5], [Af 57-5],
[Ag-1987], [Aj 32-2-1981], [Aw 4-1965],
" -v- canipole.....
recurvus (See F. latispinus),
[CC], [JJ], [Ar], [Ay], [Hx*], [Hy], [Af 27-6]
[Af 56-5], [Aj 29-6-1978], [Ba 5-1980],
[Db 30-5],
" -v- greenwoodii.....[Ar], [Af 40-4], [Aw 4-1969]
" -v- tenuispinus.....
reppenhaagenii.....[Ar]
rhodanthus (See F. echidne).....[Ar],
robustus (See F. validus),
[E], [JJ], [RR], [Ar], [Ay], [Bj], [Bo-5],
[Bv-1], [Cx], [Gr], [Jd], [Aj 33-8-1982], [Aw
3-1970], [Ba 2-1980], [Ct-1984], [Hl 4-1980],
rostii (See F. acanthodes -v- acanthodes, F. cylind-
raceus),
[RR], [An], [Ar], [Hy],
santa-maria (See F. peninsulae -v- santa-maria, F.
townsendianus -v- santa-maria),
schwarzii.....[HH], [Ah], [Ar], [Av], [Bo-5], [Ct-1984],
[Hy], [Af 27-3], [Hl 2-1980],
setispinus (See Echinocactus setispinus, Hamatocactus
setispinus).....[Ax], [Df], [Hv], [Ib],
" -f. cachetianus.....
sinuatus.....
stainesii (See F. pilosus),
[G], [NN], [QQ], [Ar], [Av], [Bo-5], [Bt],
[Bv-1], [Cj], [Hv], [Hx*], [Hy], [Iu],
[Ad Nov., 1978], [Aj 23-4-1972], [Bm-1988],
[Hl 2-1979], [Hl 1-1982],
" -v- haematacanthus (See F. haematacanthus),
[Bo-5], [Hy],

stainesii.....(cont'd),
" -v- pilosus (See F. pilosus),
[NN], [Ay], [Bo-5], [Hy],
" -v- pringlei (See F. pringlei),
[Gr], [Iv 12-1956],
tiburonensis (See F. wislizenii -v- tiburonensis),
[Aw 4-1968],
tortulospinus (See F. acanthodes -v- tortulospina, F. cylindraceus -v- tortulospinus),
[Ar], [Bt], [Hy], [Af 4-9], [Aw 4-1965],
townsendianus (See F. peninsulae -v- townsendianus),
[I], [RR], [Ar], [Bn], [Hy], [Af 27-6], [Aw 4-1965],
" -v- santa-maria (See F. peninsulae -v- santa-maria),
[Ar], [Bo-5], [Aw 4-1965],
uncinatus (See Ancistrocactus uncinatus, Echinocactus uncinatus, Glandulicactus uncinatus, Hamatocactus uncinatus, Thelocactus uncinatus),
[CC], [RR], [Bv-3], [Cx], [Af 5-2],
" -v- wrightii (See Ancistrocactus uncinatus -v- wrightii, Echinocactus uncinatus -v- wrightii, Glandulicactus uncinatus -v- wrightii, Hamatocactus uncinatus -v- wrightii),
[G], [Y], [Iu],
validus (See F. robustus),
[Hy],
victoriensis (See F. echidne),
[Ar], [Bo-5], [Af 33-5],
viridescens.....[F], [U], [FF], [GG], [Rr], [Ar], [Ax], [Az], [Ca], [Cb], [Cd], [Ck], [Cm], [Do], [Hy], [Iz], [Ad 12-1981], [Aw 4-1965], [Ba 9-1976], [Hh Jan. 1975], [Hl 5-1971], [Hl 1-1983], [Hw-1988],
" -v- littoralis.....[Ar], [Ct-1984], [Hy], [Af 36-1],
" -v- orcuttii (See F. orcuttii),
" -v- viridescens.....[Ct-1984],
viscainensis (See F. gracilis -v- coloratus, F. peninsulae -v- viscainensis),
[Dp], [Af 4-8], [Af 16-2], [Af 41-2],
visnaga (See Echinocactus visnaga, Echinocactus platyacanthus -v- visnaga),
[Bo-5], [Cn],
wislizenii (See Echinocactus wislizenii),
[D], [M*], [P], [Q], [U], [GG], [NN], [An], [Ar], [Ax], [Ay], [Bn], [Bo-5], [By], [Df], [Dq], [Gr], [Hx*], [Hy], [Ib], [Il], [Af 22-3], [Aj 27-11-1976], [Ba 6-1981],

FEROCACTUS.....(Cont'd),

wislizenii.....(cont'd),
" -v- flaconeri.....[Hy],
" -v- herrerae (See F. herrerae),
" -v- tiburonensis (See F. tiburonensis), [Af 27-6],
" -f. cristata.....[Af 1-9],
" -f. (long-spined form).....[Ba 6-1981],
" -f. (Texas form).....[Y],
" -f. (yellow-spined form).....[Y],

FLORIBUNDA

pusilliflora (See Pilosocereus pusilliflora)....[Hi-1],

FRAILEA

alacriportana.....[Bo-3], [Aj 21-12-1970],
albiareolata.....[Hx*],
albicolumnaris.....[Hi-1], [Ba 49:124 Aug. 1970],
albifusca.....[Hi-1], [Ba 49:124 Aug. 1970], [Hl 5-1970]
ancistrophora.....[Hy],
asperispina.....[Hi-1], [Ba 49:124 Aug. 1970], [Hl 5-1970],
asterioides (See F. castanea), [H], [W], [Z], [NN], [Ay], [Bo-3], [Cr], [Hv], [Hx], [Iz], [Jd], [Em 42:6-7, 1937], [Hh March 1974], [Hl 4-1972],
" -v- backebergii.....[Hi-1],
" -v- harmoniana.....[Hi-1],
aurea (See F. pygmaea -v- aurea),
aureinitens.....[Ba 4-1976],
aureispina.....[Hi-1], [Ba 49:124 Aug. 1970],
" -v- pallidior.....[Hi-1], [Ba 49:125 Aug. 1970],
bruchii (See Gymnocalycium bruchii).....[Dr],
carminifilamentosa (Invalid name), [DD], [Aj 29-7-1878],
" -v- winkelmanniana (Invalid name).....
castanea (See F. asterioides), [DD], [NN], [QQ], [Ar], [Av], [Ay], [Bo-3], [Bt], [Gr], [Hx*], [Jd], [Ba 4-1978],
cataphracta.....[F], [NN], [Ar], [Ay], [Bo-3], [Ik], [Il], [Jd], [Af 55-6], [Aj 23-2-1972],
" -v- duchii.....[Ik],
" -v- tuyensis.....[Ik], [Am 8-1980], [Ba 50:64-65 Apr. 1971], [Ba 50:135 July, 1971],
" -f. cristata.....Iz],
cataphractoides (Invalid name).....[NN],
chicquitana (See F. pullispina), [DD], [Bo-3], [Hx*], [Aj 23-6-1972],

FRAILEA.....(Cont'd),
chrysacanthus.....[Hl 65:131 Nov.-Dec. 1965],
" -v- rubriflorus.....
columbiana.....[Bo-3], [Hu], [Hx*], [Hy], [Jd], [Iv 10-1955],
concepcionensis.....[Ik], [Aj 23-2-1972], [Ba 50:49-50, March 1971, :135 July 1971],
curvispina.....[Ay[, [Hx*],
dadakii (See F. pygmaea -v- dadakii),
deminuta.....
friedrichii.....[Hx*], [Ik], [Aj 23-2-1972], [Ba 50:25, Feb. 1971 & :135 July 1971], [Hl 3-1971]
fulviseta.....[Aj 24-8-1973],
gracillima.....[Bo-3], [Hi-1], [Hx*], [Jd], [Ba 8-1981],
grandiflora.....
grahliana.....[E], [Ay], [Bo-3], [Bv-1], [Dq*], [Gr], [Hi-1], [Hx*], [Hy], [Ik],
" -v- adustior.....
" -v- rubispina.....
horstii....[Ar], [Ay], [Hi-1], [Hx*], [Jd], [Ba 49:125, 1970], [Hl 1-1967], [Hl 5-1970],
ignacionensis.....[Hx*]. [Ik], [Aj 23-2-1972], [Ba 50:-60-64, Apr. 1971, & :125 July 1971],
knippeliana.....[Hx*], [Ik], [Hl 3-1972],
lepida.....
magnifica.....[Ar], [Ag-1981], [Ah-1980], [Cu 3-1978]
" -f. cristata.....[Ar],
mammifera.....
matoana.....[Af 43-4],
melitae (See F. cataphracta).....[Hx*],
moseriana.....[Ik],
obscura.....[Aj 29-3-1978],
perbella.....[Ba 5-1980], [Ba 8-1980],
perumbilicata.....[Ba 49:125 Aug. 1970], [Hl 5-1970],
" -v- spinosior.....[Ba 49:125 Aug. 1970],
phaeodisca (See F. pygmaea -v- phaeodisca)....[Hh March 1974], [K & L Nursery Catalog - California],
pseudogracillima (See F. gracillima),
pseudograhliana (See F. schilinzkyana),
pseudopulcherrima.....[Bo-3],
pullcherrima (See Malacocarpus pulcherrimus), [C], [Hy],
pullispina (Invalid name - See F. chiquitana), [DD], [Hx], [Aj 19-3-1968], [Cu 1-1978],
" -v- atrispina....(Invalid name).....[DD],
" -v- centrispina.....(Invalid name).....[DD],
" -v- rubrispina.....(Invalid name).....
" -v- tarapezana (See F. tarapezana),
pumila.....[RR], [Ar], [Bo-3],
" -v- alegretensis.....[RR],
" -v- atrofusca.....
" -v- catiensis.....

FRAILEA.....(Cont'd),

pumila.....(cont'd),
" -v- major.....[Hi-1],
" -v- rubrispina.....
pygmaea.....[F], [Ar], [Ay], [Bo-3], [Hy],
" -v- altigibbera.....[Hi-4],
" -v- atrofusca (Invalid name).....[Hy],
" -v- aurea (See F. aurea).....
[Aj 22-4-1971],
" -v- curvispina.....
" -v- dadakii (See F. dadakii),
[Ba 9-1979],
" -v- gracilior.....
" -v- grandiflora.....
" -v- lilalunula.....[Hi-1],
" -v- longispina.....[Hi-1],
" -v- lorencoenensis.....
" -v- major.....[Hi-1],
" -v- planicosta.....
" -v- rufida.....[Ba 8-1980],
schilinzkyana.....[Bo-3], [Hx], [Hy], [Ik], [Jd],
[Aj 17-6-1966],
uhligiana (Invalid name).....[DD],
ybatense.....[Ik], [Iz], [Aj 23-2-1972], [Ba 50:46-47,
Mar. 1971, & :135 July 1971], [Hl 4-1971],
Spec. Sant-Robore.....[Ba 1-1976],

Forma cristata.....[Ba 4-1976],

Spec. .....[Ba 1-1976],

FRIESIA

umadeave (See Pyrrhocactus umadeave),
[Hr],

FURIOLOBIVIA (See Lobivia)

GLANDULICACTUS

crassihamatus (See Ancistrocactus crassihamatus, Ferocactus crassihamatus, Hamatocactus crassihamatus, Thelocactus crassihamatus),
[Ar], [Bo-5], [Hy], [Aj 33-12-1982],
[Aw 4-1966],
uncinatus (See Ancistrocactus uncinatus, Echinocactus uncinatus, Ferocactus uncinatus, Hamatocactus uncinatus, Thelocactus uncinatus),
[NN], [Ar], [Ay], [Bo-5], [Hy], [Il], [Jd], [Aj 28-6-1977], [Hl 5-1965], [Hl 6-1978],

GLANDULICACTUS......(Cont'd),

uncinatus.....(cont'd),
" -v- wrightii (See Ancistrocactus uncinatus -v- wrightii, Echinocactus uncinatus -v- wrightii, Ferocactus uncinatus -v- wrightii, Hamatocactus uncinatus -v- wrightii),.....
[Hl 2-1967],

GRUSONIA

bradtiana (See Opuntia bradtiana),
[E], [L], [DD], [RR], [Ar], [Ay], [Cx], [Hx*], [Hy], [Iu], [Aw 3-1968],
cereiformis (See Grusonia bradtiana, Opuntia cereiformis),
grahamii (See Opuntia grahamii, Opuntia schottii -v- grahamii),
hamiltonii (See Cylindropuntia hamiltonii, Opuntia hamiltonii),
[Az], [Bo-1], [Bo-6], [Bv-1], [Dq],
rosarica (See Cylindropuntia rosarica, Opuntia rosarica)
[C],
santamaria (See Opuntia santamaria),
[Bl], [Bo-1], [Af 6-4],
schottii (See Corynopuntia schottii, Opuntia schottii),
" -v- grahamii (See Corynopuntia grahamii, Opuntia grahamii, Opuntia schottii -v- grahamii),
stanlyi (See Corynopuntia stanlyi, Cylindropuntia stanlyi, Opuntia stanlyi),
wrightiana (See Corynopuntia stanlyi -v- wrightiana, Opuntia stanlyi -v- wrightiana, Opuntia wrightiana),.....
[FF],

GYMNANTHOCEREUS

altissimus (See Browningia altisssima, Gymnocereus altissima),.....
[Hi-4],
amstutziae (See Browningia amstutziae, Gymnocereus amstutziae).....[Ay],
chlorocarpus (See Gymnocereus chlorocarpa),
[Hi-4], [Hl 1-1979], [Cu 6-3-1938],
macracanthus (See Browningia macracantha),
[Hi-4],
microspermus (See Browningia microsperma, Gymnocereus microspermus, Seticereus microspermus),
pilleiferus (See Browningia pilleifera),
[Hi-4], [Ba Aug. 1966:118],

GYMNOCACTUS

aguirreanus (See Escobaria aguirreana),
[G], [Z], [Av], [Hx*], [Af 50-5],
beguinii (Invalid name - See Echinocactus erectocentrus, Echinomastus erectocentrus, Neolloydia erectocentra, Neolloydia beguinii, Neolloydia smithii -v- beguinii),
[F], [Ar], [Bo-5], [Hx*], [Ae-1981], [Af 50-5], [Af 55-6], [Ba 8-1977], [Ba 9-1977], [Db 5-1979] [Hl 1-1973],
" -v- senilis (See Neolloydia smithii -v- senilis)
[Z], [Ar], [Bo-5], [Hx*], [Ak-1981],
" -v- smithii (See Neolloydia smithii).....[Jd],
conothelos (See Thelocactus conothelos),
gielsdorfianus (See Neolloydia gielsdorfiana, Thelocactus gielsdorfianus),
[Ar], [Bo-5], [Il], [Ae-1981], [Af 55-2], [Af 59-6, [Ah-1985], [Ba 10-1980], [Hl 6-1966], [Hl 5-1969],
" -f. cristata.....[Ar],
goldii (See Gymnocactus horripilus, Neolloydia horripila, Thelocactus goldii, Thelocactus horripilus),
[Hy],
horripilus (See Gymnocactus goldii, Neolloydia horripila, Thelocactus goldii, Thelocactus horripilus)....[Ar], [Bo-5], [Il], [Ae-1981], [Af 50-5], [Af 58-2], [Hl 6-1966],
knuthianus (See Echinocactus knuthianus, Neolloydia knuthiana, Thelocactus knuthianus, Turbinicarpus knuthianus),
[Ar], [Bo-5], [Hy], [Jd], [Af 52-1], [Ba 10-1977], [Hl 1-1973],
" -f. cristata.....[Ay],
laredoi.....
mandragora (See Neolloydia mandragora, Rapicactus mandragora, Thelocactus mandragora),
[Ar], [Ay], [Bo-5], [Hx*], [Hy], [Hl 6-1966],
roseanus (See Escobaria roseana, Thelocactus roseanus),
[Af 56-4],
saueri (See Neolloydia saueri, Thelocactus saueri),
[Ar], [Hx*], [Hy], [Af 56-6], [Af 60-4], [Aj 16-7-1965],
saussieri (See Thelocactus saussieri).....[Hl 6-1966],
smithii (See Neolloydia smithii),.....
subterraneus (See Neolloydia subterranea, Rapicactus subterraneus, Thelocactus subterraneus),
[Bo-5], [Hx*], [Hl 6-1966], [Hl 1-1973], [Hl 1-1981],
" -v- subterraneus (See Neolloydia subterranea),
[NN], [Ar], [Af 50-6], [Af 52-2],

GYMNOCACTUS.....(Cont'd),

subterraneus.....(cont'd),
" -v- zarogosae (See Neolloydia subterranea -v- zarogosae).....[Af 50-6], [Af 53-2], [Hl 1-1981], [Hm 3-1979],
valdezianus (See Echinocactus valdezianus, Neolloydia valdeziana, Normanbokea valdeziana, Pelecyphora valdeziana, Thelocactus valdezianus, Turbinicarpus valdezianus).....[Ar], [Bo-5, [Hy], [Jd], [Ba 11-1977],
" -v- albiflorus (Invalid name - See Pelecyphora valdeziana -v- albiflora, Turbinicarpus valdezianus -v- albiflorus) [Ar],
viereckii (See Neolloydia viereckii, Thelocactus viereckii),
[DD], [Ar], [Bo-5], [Hx], [Ah-1983], [Aj 6-1981], [Hl 1-1981],
" -v- major.....[Ae-1983], [Af 50-6], [Af 57-2], [Aj 32-6-1981],
" -v- viereckii (See Thelocactus viereckii -v- viereckii),
[Af 50-6],
ysabelae (See Thelocactus ysabelae),
[Bo-5], [Ba 11-1977],
" -v- brevispinus.....

GYMNOCALYCIUM

achirasense.....[Hx*], [Aj 30-2-1979....Flower], [Hl 3-1980],
alboareolatum.....
albiflorum (See G. mihanovichii -v- albiflorum),
albispinum (See G. bruchii),
[Bo-3], [Hx*], [Ba 3-1981],
albopulpa.....[See K & L nursery catalog, California - 1988],
altagraciense.....[Hl 3-1981],
ambatoense.....[Aj 31-1-1980],
andreae (See Echinocactus andreae),
[K], [V], [NN], [Bo-3], [Cm], [Hu], [Hv], [Hx], [Hy], [Il], [Iz], [Jd], [Hl 4-1968], [Hw-1986],
" -v- grandiflorum (See G. grandiflorum),
" -v- longispina.....[Ba 8-1977],
" -v- svecianum.....
anisitsii.....[RR], [Ar], [Bo-3], [Hx*], [Hy], [Ik], [Flower - Ba 1-1978], [Hl 5-1966],
antherosacos.....
antherostele.....
armatum.....[Hi-2],
ardoa.....[Cu 1-1962],

artigas (See G. uruguayense),
[EE], [KK], [Hx*], [Iv 9-1951],
" -v- rosada.....
asterium (See G. stellatum),
[Ar], [At], [Av], [Ay], [Bo-3], [Hx*], [Hy],
[Ba 1-1979],
" -v- cinereum.....[Db 2-1979],
" -v- minimum (See G. stellatum -v- minimum)
[Aj 15-2-1964], [Aj 27-9-1976],
" -v- paucicostatum.....
" -v- paucispinum (Invalid name - See G. stell-
atum -v- paucispinum),
[DD], [Ae-1982], [Aj 27-9-1976],
" -v- variispinum.....[Aj 23-9-1972],
baldianum (See G. venturianum),
[AA], [Ar], [Bo-3], [Bt], [Cr], [Cs], [Cy],
[Hb], [Hu], [Hv], [Hx], [Hy], [Il], [Iz],
[Jd], [Ag-1983], [Aj 30-6-1979], [Am 3-1982],
[Ba 3-1978], [Flower - Ja], [Hw-1982],
" -v- albiflorum.....
" -v- sanguiniflorum (See G. sanguiniflorum),
[Aj 23-9-1972],
" -v- venturianum (See G. venturianum),
[Aj 23-9-1972],
" -f. cristata.....[Hy],
bayrianum.....[KK], [Ar], [Hx*], [Ae-1981], [Ah-1x985],
[Bi 6-1], [Hl 6-1983],
bicolor (See G. hypopleurum, G. tobuschianum),
[KK], [Iz], [Aj 14-12-1963], [Hl 3-1970],
bodenbenderianum (See G. riojense),
[NN], [Ar], [Bo-3], [Bv-1], [Cm], [Hi-2]
[Hx*], [Hy], [Iu], [Iz], [Ab-1985],
[Af 56-1], [Bi 2-4], [Db 7-1979],
" -f. cristata.....[Ij],
bolivianum (See G. tudae -v- bolivianum),
borthii.....[Hx*], [Aj 27-2-1976],
bozsingianum.....[Aj 33-7-1982], [Hl 6-1976],
brachyanthum (See G. monvillei).....[Hy],
brachypetalum (See G. gibbosum).....[Bo-3], [Hy],
brevistylum (See under G. tudae),
bruchii (See G. lafaldense),
[C], [H], [AA], [BB], [KK], [Ar], [Bj], [Bo-3],
[Bv-1], [Gr], [Hu], [Hv], [Hx], [Hy], [Il],
[Jd], [Ag-1988], [Aj 15-2-1964], [Ba 7-1978],
[Bi 2-3], [Db 4-1980],
" -v- enorme.....
" -f. evolvens.....
" -v- hossei (See G. hossei).....[Bo-3], [Hy],
" -v- spinosissimum.....[Hy],
buenekeri.....[Iz], [Cu 4-1978], [Hl 5-1983],(Bi6-1*],

GYMNOCALYCIUM.....(Cont'd),

calochlorum (See G. proliferum -v- calochlorum),
[QQ], [Ar], [Bo-3], [Gr], [Hx*], [Hy],
[Ja], [Hl 4-1972], [Hw-1988],
" -v- proliferum (See G. proliferum),
[Bo-3], [Hy], [Iz], [Hl 4-1965],
" -v- roseiacanthum (See G. roseiacanthum, G.
sigelianum),
camaradense.....[Cq],
capillaense.....KK], [Hi-2], [Hy], [Iu], [Aj 1-1938],
[Aj 21-8-1970],
cardenasianum.....[KK], [NN], [Ar], [Ay], [Bo-3], [Hi-2]
[Hu], [Hv], [Hx*], [Hy], [Ae-1981],
[Af 53-2], [Af 56-5], [Ag-1984, 1988],
[Ah-1985], [Aj 30-6-1979], [Flower -
Ba 10-1980], [Bi 6-1], [Db 3-1980],
" -v- horridispinum (See G. horridispinum),
" -f. cristata.....[Ar],
carminanthum.....[Av], [Hx*], [Aj 4-1976],
castellanosii.....[H], [KK], [Bo-3], [Hx*], [Hy], [Jd],
[Hl 5-1978], [Hs I-3-1980],
chiquitanum (See G. hammerschmidii),
[DD], [KK], [Ah], [Hx*], [Ij], [Aj 30-6-
1979],
chubutense (See G. gibbosum -v- chubutense),
[Bo-3], [Flower - Dr*], [Hx*], [Hy],
[Hl 4-1983],
chuquisacanum (See G. pflanzii -v- izozogsii -f. chu-
quisacanum, G. izozogsii),
[Af 38-4],
comarapense (See G. pflanzii -v- zegarrae, G. zegarrae)
[R], [Bo-3], [Hx*],
cumingii (See Weingartia cumingii),
[I],
curvispinum (See G. valnicekianum),
[Hr], [Au Vol. 2, 1977, Zone 3 Compendium],
" -v- angusticostatum.....[Hl 3-1970],
" -v- portezuello.....[Hr],
damsii.....[M*], [S], [RR], [Bo-3], [Bu], [Bv-1], [Cp],
[Hi-1], [Hx], [Hy], [Ik], [Iz], [Af 27-4],
[Dd 3-1981],
" -v- centrispinum (Invalid name).....[DD],
" -v- rotundulum (Invalid name - See G. rotund-
ulum).....[Hx*], [Ab-1986],
" -v- torulosum (Invalid name),
[DD], [Hx], [Aj 22-4-1971],
" -v- tucavocense (Invalid name),
[DD], [Il],
" (FR-1171).....[Ah-1985],
" -f. cristata.....[Hy],

deeszianum.....[Bo-3], [Hy],
delaetianum (delaetii) - (See G. megalothelos -v- delaetianum), G. schickendantzii -v- delaetii).....[Be*], [Hl 1-1970],
denudatum.....[C], [H], [M*], [W], [Z], [NN], [RR], [Ar], [Av], [Bo-3], [Bv-3], [Cs], [Cy], [Gr], [Hi-1], [Hv], [Hx*], [Hy], [Iz], [Jd], [N-1980], [Af 3-1], [Af 58-4], [Ah-1986], [Am 3-1982], [Ba 12-1977], [Db 3-1980],
" -v- argentiensis.....[Ar], [Hy],
" -v- backebergii (See G. denudatum).....[Hi-1],
" -v- golzianum.....[NN],
" -v- multiflorum (See G. multiflorum),
" -v- paraguayense (See G. paraguayense), [Hy], [Im],
" -v- pentacantha.....[Ar],
" forma.....[Ar], [Cs],
" hybrid.....[Aj 32-11-1981],
euchlorum (See G. hybopleurum -v- euchlorum),
erinaceum (See Weingartia erinacea).....[Il],
eurypleurum.....[Ar], [Hi-1], [Hx*], [Flower - Ba 9-1978], [Hl 3-1972], [Hl 1-1975], [Bi Vol.4, &6]
eytianum (See G. pflanzii -v- eytianum, G. megatae), [Hy], [Aj 32-11-1981],
ferrarii.....[Aj 32-1-1981],
ferox (See G. gibbosum -v- ferox, G. mazanense -v- ferox, G. hybopleurum -v- ferox),
fidaianum (See Weingartia fidaiana).....[Il],
fleischerianum.....[KK], [Ar], [Ay], [Bm-], [Bo-3], [Hx*], [Hy], [Ik],
" -v- anoplum.....
" -v- andersohnianum.....[Friciana 40:11-1966],
" -v- heuschkelianum.....[Friciana 40:11-1966],
" -v- meiklejohnianum.....[Friciana 40:11-1966],
" -f. .....[Ik],
fricianum (See G. tudae), [Aj 15-6-1964], [Hl 3-1965],
friedrichii (See G. mihanovichii -v- friedrichii), [KK], [Cs], [Hx*], [Iz], [N-1980], [Am 3-1982], [Db 2-1980],
" -v- angustostriatum.....[Ik],
" -v- bolivianum.....
" -v- moserianum.....[Hx], [Ik], [Aj 23-9-1972], [Friciana 37:19- Aug. 1966], [Hh Sept., 1976],
" -v- " -f. aurea.....
" -v- " -v- rubra.....
" -v- pazoutianum.....[Hl 3-1967],
" -v- pirarettaense.....[Hx*],
" -f. (Hibotan - cultivar).....[Ar], [Hx], [Ik],

gerardii (See G. gibbosum -v- gerardii),
gibbosum.....[E], [M*], [KK], [RR], [Ar], [Bo-3], [Bv], [Gr], [Hv], [Hx*], [Hy], [Ia], [Il], [Iu], [Bi 1-2], [Hl 6-1969], [Hl 6-1977], [Hl 3-1983],
" -v- caespitosum.....[Hr], [Hl 4-1983],
" -v- " -f. intermedium.....
" -v- " -f. minimum.....
" -v- chubutense (See G. chubutense).....[Hy],
" -v- fenellii.....
" -v- ferox (See G. ferox, G. mazanense -v- ferox),
" -v- gerardii (See G. gerardii).....
" -v- kozelskyanum (See G. kozelskyanum), [Hl 4-1983],
" -v- leonensis.....
" -v- leucocanthum.....[Hy], [Hl 3-1983],
" -v- leucodictyon.....[Bo-3],
" -v- nigrum.....[E], [Ar], [Bo-3], [Hx*], [Hy], [Ba 7-1978 - Flower], [Hl 4-1983],
" -v- nobile.....[Ar], [Bo-3], [Hy], [Hl 1-1973], [Hl 3-1983], [Hl 4-1983],
" -v- schlumbergi........Hy], [Af 43-4],

glaucum.....[KK], [Hx*], [Aj 28-6-1977],
" -v- plurispinum.....
grandiflorum (See G. andreae -v- grandiflorum), [Bo-3], [Hx*], [Hy],
griseo-pallidum (Invalid name).....[FF], [KK], [Ar], [Af 57-2],
guanchinense (See G. mazanense).....[Hi-2], [Hy],
" -v- robustus.....
" -v- tinogastaense.....
guerkeanum (See G. artigas, G. uruguayense), [T], [RR], [Hx*], [Bi 6-1],
" -v- roseiflorum (See G. hamatum),
hamatum.....[DD], [Ar], [Hi-2],
hammerschmidii (See G. chiquitanum), [DD], [Ar],
hedinianum (See Weingartia hediniana).....[Il],
hennissii (hybrid).....[Db 3-1980],
heuschkelianum (See G. fleischerianum -v- heuschkelianum),
horridispinum.....[W], [DD], [KK], [Ar], [Hb], [Hu], [Hv], [Hw-1981], [Hx*], [Hy], [Iz], [Jd], [Ae-1981], [Aj 8-1982], [Ba 4-1981], [Hl 4-1971], [Hl 6-1979],
horstii.....[G], [KK], [Gr], [Ag-1988], [Aj 21:162-165 Sept. 1970], [At-1982-1983], [Cu 4-1978],

horstii.....(cont'd),
" -v- buenekeri (See G. buenekeri),
[Ay], [Aj 21:162-165 Sept. 1970],
hossei (See G. bruchii -v- hossei),
[H], [DD], [KK], [Bo-3], [Hx], [Hy], [Iu],
[Au 26-4-1971],
" -v- longispinum.....
hybopleurum (See G. bicolor, G. tobuschianus),
[E], [KK], [Ar], [Bo-3], [Hx], [Hy],
[Bi 2-4], [Db 4-1980], [Hl 6-1971],
" -v- breviflorum (Invalid name).....[DD],
" -v- centrispinum (Invalid name).....[Ar],
" -v- euchlorum (See G. euchlorum).....
" -v- ferocior (Invalid name).....[DD], [KK],
[Bi 4-1-1986*],
" -v- ferox (Invalid name).....[DD], [PP],
hyptiacanthum (See Echinocactus hyptiacanthus),
[Ar], [Bo-3], [Hx*], [Hy], [Aj 33-2-1982]
[Hl 4-1983],
" -v- citriflorum.....
immemoratum (See G. valnicekianum).....[Af 12-6],
intertextum (Invalid name).....[DD],
izozogsii (See G. pflanzii -v- izozogsii)...[Af 38-4],
joossensianum....[Bo-3], [Jd], [Au 3-1978 - Zone 3 - Compendium], [Db 4-1980],
kieslingii.....[Af 57-6],
" -f. alboareolatum.....
" -f. castaneum.....
knebelii (knebelianum) (See G. schickendantzii -v- knebelii),
kozelskyanum (See G. gibbosum -v- kozelskyanum),
[Aj 27-9-1976], [Bi 4-1985], [Hl 2-1966],
kurtzianum (See Echinocactus kurtzianus, G. mostii -v- kurtzianum, G. schickendantzii -v- kurtzianum)
[RR], [Bv-1], [Hx*], [Hl 6-1972],
lafaldense (See G. bruchii),
[RR], [Bv-1], [Cr], [Ia], [Iu], [Aj 24-8-1973], [Cu 4-1961], [Hl 1-1982],
" -v- albispinum.....[Aj 24-8-1973],
" -v- deviatum.....[Aj 24-8-1973],
" -v- enorme.....[Aj 24-8-1973],
" -v- evolvens.....[Aj 24-8-1973],
" -v- fraternum.....[Aj 24-8-1973],
" -v- hossei (See G. bruchii -v- hossei,
G. hossii).....[Aj 24-8-1973],
" -v- intermedium.....[Aj 24-8-1973],
" -v- spinosissimum....[Aj 24-8-1973], [Hl 2-1982]
lanatum (See Weingartia lanata).....[Il],
lagunillasense (See G. pflanzii -v- lagunillasense),
[Hx*], [Aj 9-2-1958],

GYMNOCALYCIUM.....(Cont'd),

leeanum.....[M*], [KK], [RR], [Ar], [Bu], [Bv-1], [Cz],
[Hb], [Hx], [Hy], [Il], [Af 26-3],
[Af 56-5], [Hl 4-1972],
" -v- brevispinum.....[Bo-3], [Aj 23-9-1972],
" -v- netrelianum (See G. netrelianum),
[Bo-3], [Hy], [Iz],
" -v- roseiflorum.....[Hy],
leonensis (See G. chubutense),
leptanthemum.....[Bo-3], [Hx*], [Hy], [Aj 30-6-1979],
loricatum (See G. spegazzinii),
[Cy], [Iu], [Af 24-5],
" -v- cachense (See G. spegazzinii),
lumbrerasense.....
malletianum.....[Iu],
marquezii (See G. pflanzii),
[Af 57-3], [Aj 9-2-1958], [Am 3-1982],
" -v- angustura.....
" -v- argentinense (Invalid name - See G. pflanzii).....[DD],
marsoneri.....[DD], [Bo-3], [Hx*], [Hy],
matoense.....[Aj 26-12-1975],
mazanense....[Ar], [Bo-3], [Hu], [Hx], [Hy], [Hl 5-1974]
" -v- breviflorum (See G. rhodantherum),
[Bo-3], [Hy],
" -v- ferox (Invalid name - See G. hybopleurum),
[Ar], [Hy], [Hl 5-1974],
" -v- polycephalum (See G. polycephalum),
megalothelos.....[X], [KK], [RR], [Hu], [Hx*], [Hy],
[Aj 23-9-1972], [Hl 1-1973],
" -v- delaetianum (See G. delaetianum),
[Friciana 40:11, 1966],
megatae (See G. tudae),
[Hx*], [Hy], [Ik], [Iz], [Hl 5-1966],
" -v- bolivianum (See G. tudae -v- bolivianum),
megatae-tudae (See G. tudae).....[Ik],
melanocarpum.....[RR],
melocactiformis (See G. mihanovichii -v- melocacti-formis),
mesopotamicum.....[Cu 2-1980],
michoga (See G. pungens).....[KK], [Hx*], [Iu],
mihanovichii.....[D], [I], [M*], [R], [X], [RR], [Av],
[Bo-3], [Bv-3], [Cz], [Gr], [Hu], [Hx],
[Hy], [Ik], [Il], [Iz], [Jd], [Af 16-6], [Ba 10-1976],
" -v- albiflorum (See G. albiflorum),
[Ik], [Hl 4-1965],
" -v- " -f. albiflora.....[Hu],
" -v- angusto-striatum.....[Aj 15-9-1964],
" -v- filadelfiense (Invalid name),
[DD], [Hi-1], [Hx*], [Hl 4-1965],

mihanovichii.....(cont'd),
" -v- fleischerianum (See G. fleischerianum), [Aj 15-9-1964],
" -v- friedrichii (See G. friedrichii), [G], [AA], [KK], [NN], [Ar], [Ay], [Bv-1], [Cr], [Cs], [Cz], [Hb], [Hy], [Af 57-2], [Ag-1985], Aj 15-9-1964], [Dg 2-4-1980-1981],
" -v- " -f. nigrum.....[Hy],
" -v- " -f. variegated.....[F],
" -v- " -f. aurea.....[Hv],
" -v- " -f. rubra.....[Ay], [Hv], [Aj 27-5-1976],
" -v- " -f. cristata.....[Hy], [Aj 23-4-1972], [Aj 24-5-1973],
" -v- " -f. aureorubro-variegatum.....
" -v- heesei.....
" -v- melocactiformis (See G. melocactiformis), [Aj 15-9-1964],
" -v- moserianum (See G. moserianum).....[Hx],
" -v- pazoutianum (See G. mihanovichii -v- pirarettaense),
" -v- pirarettaense (See G. mihanovichii -v- pazoutianum), [Aj 15-9-1964],
" -v- roseiflorum.....
" -v- rubra.....[I],
" -v- rysanekianum.....[Hl 4-1965],
" -v- stenogonum (See G. stenogonum).....[Hr], [Hx*], [Hy], [Ij], [Aj 13-11-1962] [Hl 5-1966],
" -v- stenopleurum (See G. stenopleurum),
" -v- stenostriatum (See G. mihanovichii -v- angusto-striatum),
" -v- (mutation).....[Aj 27-5-1976],

millaresii (See G. pflanzii -v- millaresii), [Af 38:144, 145, 161 - 1966],

monvillei.....[RR], [Ar], [Bo-3], [Hx*], [Hy], [Iu], [Ag-1983], [Ba 1-1974], [Bi 3-1987],

moserianum (See G. mihanovichii -v- moserianum), [KK], [Ar], [Hx*], [Ik], [Ae-1983], [Aj 23-9-1972], [Hl 1-1966],

mostii.....[F], [V], [RR], [Ar], [Bo-3], [Hx*], [Hy], [Aj 27-2-1976], [Aj 32-11-1981], [Hw-1982],
" -v- centrispinum....[Hr],
" -v- kurtzianum (See G. kurtzianum), [Ar], [Bo-3], [Hy],
" -f. cristata.....[Hl 4-1975],

GYMNOCALYCIUM.....(Cont'd),

mucidum.....[Hl 2-1983],
multiflorum.....[C], [J*], [RR], [Bl], [Bo-3], [Bv-1], [Cg], [Cm], [Hb], [Hv], [Hx], [Ba 12-1978], [Hw-1989],
" -v- albispinum.....[Bo-3], [Hy], [Bi 3-1987],
" -v- brevispinum.....[Hy], [Aj 23-9-1972], [Hl 6-1969],
" -v- hybopleurum (See G. hybopleurum).....
" -v- ourselsianum (See G. ourselsianum).....
" -v- parisiense.....[Hy],
" -v- (hybrid).....[Hu],
multispinum (See Weingartia multispina).....[Il],
neocumingii (See Weingartia neocumingii), [Ay], [Gr], [Il],
netrelianum (See G. leeanum -v- netrelianum), [E], [H], [RR],
neumannianum (See Weingartia neoneumanniana, Weingartia neumanniana).....[Ay], [Gr],
nidulans.....[Ar], [Bo-3], [Hx*], [Hy], [Ae-1981], [Aj 23-9-1972], [Hl 5-1974],
nigriareolatum.....[KK], [Bo-3], [Hx*], [Hy], [Af 56-1],
" -v- densispinum.....[Hy],
occultum.....[NN], [Hy], [Aj 27-9-1976], [Hl 4-1975],
" -v- flore-rosea.....[Hr],
ochoterenai (See G. stellatum), [Bo-3], [Cr], [Hx], [Hy], [Af 56-1], [Iv 5-1956],
" -v- cinereum.....[Bo-3],
" -v- ochoterenai.....[Bo-3],
" -v- polygonum (Invalid name), [DD], [Aj 27-9-1976], [Hl 6-1975],
" -v- tenuispinum (Invalid name).....[DD],
" -v- varispinum (Invalid name).....[DD],
odoratum.....
oenanthemum.....[D], [T], [NN], [Ar], [Bo-3], [Gr], [Hi-2], [Hx], [Iz], [Jd], [Ae-1981], [Aj 28-5-1977], [Ba 10-1980],
onychacanthum (See G. tudae).....[Hy],
optima rubra.....[Ja],
ourselianum (See G. multiflorum -v- ourselianum), [Bo-3], [Cr], [Hl 6-1969],
paediophilum.....[Hi-1], [Hx], [Aj 22-4-1971], [Hl 5-1977],
paraguayense (See G. denudatum -v- paraguayense), [Ik], [Friciana 40:11, 1966],
" -v- roseiflorum.....[Friciana 40:11, 1966],
" -v- scheidelianum.....[Friciana 40:11, 1966],
" -v- wagnerianum.....[Friciana 40:11, 1966],
" -v- weiditzianum.....[Friciana 40:11, 1966],

parisiense (See G. multiflorum -v- parisiense),
parvulum (See G. quehlianum),
penicillata.....
pflanzii.....[KK], [Ar], [Bo-3], [Hi-2], [Hx*], [Hy],
[Ik], [Iz], [Aj 32-11-1981], [Db 7-1980],
[Hl 5-1976], [Hl 3-1983], [Hh Sept, 1976],
" -v- albipulpa (See G. comarapense, G. zegarrae)
[Ar], [Hi-2], [Ag-1986],
[Ah-1985], [At-1982-'83],
" -v- comarapense (See G. comarapense, G. zegarrae),
" -v- eytianum (See G. eytianum),
[Au 26:100, 1971],
" -v- izozogsii (See G. izozogsii),
[Au 26-4-1971],
" -v- " -f. chuguisacanum (See G. chuguisacanum)...[Au 26:100, 1971],
" -v- " -f. izozogsii.....
" -v- lagunillasense (See G. lagunillasense),
[KK], [Au 26:100, 1971],
" -v- millaresii (See G. millaresii),
[Au 26:100, 1971],
" -v- riograndense (See G. pflanzii -v- albipulpa, G. riograndense),
[Au 26:100, 1971],
" -v- zegarrae (See G. comarapense, G. pflanzii -v- albipulpa, G. zegarrae),
[Au 26:100, 1971],
pilcomayoensis (See Weingartia pilcomayoensis),
[Gr],
piltziorum.....[Aj 33-7-1982],
pirarettaense (See G. mihanovichii -v- pirarettaense),
platense (See G. capillense),
[H], [M*], [RR], [Be*], [Bo-3], [Bv-1], [Cr],
[Dr], [Hx], [Hy], [Iu], [Jd], [Ad-],
" -v- nobilis.....
" -v- stellatus (G. stellatum),
[Dr],
" -v- ventanicola.....[DR],
polycephalum (See G. mazanense -v- polycephalum),
polyanthum (See G. multiflorum -v- ourselianum, G. ourselianum),
" -v- albispinum (See G. multiflorum -v- albispinum).....[Hy],
proliferum (See G. calochlorum -v- proliferum),
[Iv 9-1954],
" -v- calochlorum (See G. calochlorum),
pseudo-malacocarpus (Invalid name - See G. tudae -v- pseudo-malacocarpus, G. megatae),
[Z], [DD], [Hx*], [Af 54-6],

GYMNOCALYCIUM.....(Cont'd),

pseudoragonesii.....[Aj 27-9-1976], [Hl 6-1975],
pugionacanthum (Invalid name - See G. hybopleurum -v- ferox)....[DD], [KK], [Hx*], [Aj 24-10-1973],
pulquinense (See Weingartia pulquinensis),
pungens.....[KK], [Dp], [Hx*], [Hy], [Af 58-3], [Hl 6-1974],
" -f. cristata.....[KK],
quehlianum (See Echinocactus quehlianum), [D], [I], [J*], [AA], [LL], [NN], [Ar], [Av], [Bo-3], [Bt], [Bv-3], [Cg], [Hu], [Hv] [Hx], [Hy], [Il], [Iz], [Jd], [N-1980], [Aj 15-9-1964], [Cu 1-1968],
" -v- albispinum (Invalid name).....[Bo-3], [Hu], [Af 16-6], [Dd 12-1980],
" -v- aureispina.....
" -v- caespitosum.....[Hy],
" -v- flavispinum (Invalid name).....[Bo-3],
" -v- hennissii (See G. hennissii),
" -v- rolfianum.....[Bo-3],
" -v- zantnerianum.....[Bo-3], [Hb], [Af 21-2],
" -f. varieg. -f. cristata.....[Ar], [Hy],
querkeanum (guerkeanum).....[T],
ragonesii.....[G], [KK], [Hx], [Iz], [Aj 27-9-1976], [Aj 30-6-1979], [Am 3-1982],
rhodantherum (See G. mazanense -v- breviflorum), [Bo-3], [Hx*], [Aj 9-2-1958],
riograndense (See G. pflanzii -v- albipulpa), [Ay], [Gr], [Hx*],
riojense (See G. bodenbenderianum), [Hr], [Aj 27-9-1976], [Hl 4-1975],
ritterianum.....[Hx*], [Aj 23-7-1972],
roseiacanthum (See G. calochlorum -v- roseiacanthum, G. sigelianum),
" -v- albispinum.....
" -v- tucuman.....
rotundulum (See G. damsii -v- rotundulum).....[N-1980],
saglione (saglionis) ....[D], [H], [M*], [R], [Z], [KK], [RR], [Ar], [Ay], [Bv-3], [Cr], [Gr], [Hi-2], [Hr], [Hy], [Iu], [Iz], [Aj 30-6-1979], Aj 30-10-1979], [Db 4-1980], [Dd 3-1981], [Bi 6-3],
" -v- albispinum n.n. .....
" -v- catamarca.....[Hy],
" -v- leutispinus.....[BB],
" -v- longispinum.....
" -v- roseispinum n.n. .....
" -v- rubrispinum n.n. .....
" -v- tilcarensis (See G. saglione).....[Hy],
" -f. cristata.....[E], [F], [Hy],
sanguiniflorum (See G. baldianum),

schatzelianum.....[Hl 4-1983],
schickendantzii.....[H], [M*], [RR], [Bv-1], [Hv], [Hx*], Hy], [Iu], [Jd], [Af 59-6], [Ag-1984-1985], [Ah-1985], [Cu 1-1968], [Db 2-1980], [Hl 1-1970],
" -v- delaetii (See G. delaetii), [Bo-3], [Hx*], [Hy], [Am 3-1982],
" -v- kurtzianum (See G. kurtzianum).....
" -v- michoga.....[Hy],
" -f. cristata.....[Hy],
schroederianum (See G. uruguayense), [KK], [Bo-3], [Af 59-2], [Ah-1985], [Ba 6-1977], [Cu 2-1977],
" ssp. bayense.....[Af 59-2],
" ssp. paucicostatum.....[Af 59-2],
schuetzianum.....[Aj 28-6-1977], [Aj 32-10-1981], [Aj 33-4-1982],
sigelianum.....[Ay], [Bo-3], [Hx*], [Hy],
spegazzinii (See G. loricatum), [F], [L], [Z], [DD], [RR], [Ar], [Av], [Ay], [Bo-3], [Gr], [Hx*], [Hy], [Jd], [Af 51-4], [Ah-1985], [Aj 24-12-1973], [Aj 27-10-1976], [Aj 28-9-1977], [Hl 6-1983],
" -v- horizonthalonius (See G. spegazzinii), [Aj 28-9-1977], [Hl 6-1983],
" -v- major.....[Ar], [Bo-3], [Hy], [Hl 6-1983],
" -v- tortispinum.....[Hy],
" -f. cristata.....[Hy], [Af 51-5],
stellatum (See G. asterium), [KK], [Dr], [Aj 27-9-1976], [Hl 4-1975],
" -v- minimum (See G. asterium -v- minimum), [Aj 27-9-1976], [Hl 4-1975],
" -v- paucispinum (See G. asterium -v- pauci-spinum), [KK], [Ij], [Aj 27-9-1976], [Hl 4- 1975],
stenogonum (See G. mihanovichii -v- stenogonum), [K & L Nursery Catalog, Calif.-1988],
stenopleurum (See G. mihanovichii -v- stenopleurum), [Hi-1],
striglianum.....[Hx*], [Aj 24-12-1973],
stuckertii.....[H], [RR], [Bt], [Dr], [Ad 12-1981],
" -f. variegata.....[F],
sutterianum.....[Bo-3], [Hx*], [Hy], [Iu], [Aj 12-1937], [Hl 1-1981],
tilcarense (See Brachycalycium tilcarense, G. saglione).....[KK], [Ay], [Hu], [Hx*],
tillianum.....[KK], [Hx], [Iz], [Ad May 1980], [Af 58-5] [Ag-1987], [Ba 2-1981], [Bm-1989], [Hh July 1874], [Hl 1-1977],

GYMNOCALYCIUM.....(Cont'd),

tobuschianum (See G. valnicekianum),
" -v- polycentralis.....
tortuga (See G. megatae, M. tudae),
torulosum (See G. damsii -v- torulosum),
triacanthum (Invalid name - See G. quehlianum),
[Bo-3], [Hx*], [Aj 27-9-1976], Hl 6-1975],
tudae (See G. megatae),
[Hi-1], [Hy],
" -v- bolivianum (See G. megatae -v- bolivianum),
[Hi-2],
" -v- pseudo-malacocarpus (See G. pseudomalacocarpus),
[Au 26:100 Dec. 1971],
uebelmannianum.....[Hx*], [Ba 51:62-63, 1972],
uruguayense (See G. artigas, G. guerkeanum),
[RR], [Hi-1], [Hr], [Hy], [Jd], [Aj 33-2-1982],
[Flower - Ba 11-1977], [Flower - Db 4-1980],
" -v- roseiflorum (See G. artigas),
[Hr], [Aj 17-11-1966], [Aj 33-2-1982],
" -f.- (Schl. - 102, 104, 105, 108, 118.....)
[Aj 33-2-1982],
valnicekianum.....[Bo-3], [Hx*], [Hy], [Aj 28-2-1977],
[Aj 30-3-1979],
" -v- polycentralis.....
vatteri.....[DD], [KK], [NN], [Ar], [Bo-3], [Hu], [Hx*],
[Hy], [N-1980], [Ad March 1978], [Ae-1981],
[Aj 33-11-1982], [Flower - Ba 4-1979],
[Hl 4-1970],
venturianum (See G. baldianum),
[L], [Bv-1], [Hr], [Hy],
" -v- violaceum.....[Hr],
villamercedense.....
vorwerkii (See Neowerdermannia vorwerkii).....
[Ay], [Jd],
weissianum....[KK], [Bo-3], [Hy], [Ae-1981], [Hl 5-1974]
" -v- atroroseum.....[Bo-3], [Hy], [Iz],
" -v- cinerascens.....[Bo-3],
westii (See Weingartia westii),
[Af 29-1],
zantnerianum (See G. quehlianum -v- zantnerianum),
zegarrae (See G. comarapense, G. pflanzii -v- albipulpa, G. riograndense)......[Ar], [Av], [Ay],
[Bo-3], [Hy], [Jd], [Aj 28-11-1977],
" - f. cristata.....[Ar],
Spec. FR-1097.....[Hi-2],
Spec. (Marayes).....[KK], [Hl 1-1970],
Spec. P-222.....[Ba 12-1981],
Spec. WO-92.....[Aj 27-4-1976],
Spec. .....[Ar], [Hu], [Aj 30-6-1979],

GYMNOCEREUS

altissimus (See Browningia altissima, Gymnanthocereus altissimus),
[Bo-2], [Bo-4],
amstutziae (See Browningia amstutziae, Gymnanthocereus amstutziae),
[Ay], [Bo-2], [Bo-6], [Jd], [Aj 11-11-1960],
chlorocarpa (See Gymnanthocereus chlorocarpus),
[Hl 2-1971],
microspermus (See Browningia microsperma, Gymnanthocereus microspermus),
[DD], [Ay], [Bo-2], [Hy], [Jd],

HAAGEOCEREUS

acanthocladus (See H. multangularis)....[Bo-2], [Hx*],
acanthurus (See Borzicactus acanthurus, Loxanthocereus acanthurus),
acheatus (See H. deflexispinus, H. limensis -v- deflexispinus, H. pseudoacranthus).....
[Bo-2], [Hx*],
acranthus (See H. limensis -v- andicola, H. limensis -v- brevispinus).....[G], [PP], [QQ], [Bo-2],
[Hu], [Hx], [Hy], [Jd], [Aj 8-5-1957],
[Da 2-1980], [Hl 4-1969],
" -v- crassispinus (See H. limensis -v- metachrous),
[Av], [Bo-2], [Hx*],
" -v- fortalezensis (See H. fulvus),
[Bo-2], [Aj 17-3-1966],
" -v- metachrous (See H. acranthus -v- crassispinus, H. limensis -v- metachrous).....[Bo-2],
akersii (See H. multangularis -v- pseudomelanostele),
[Bo-2], [Hx*],
albidus (See H. albispinus),
albisetatus (See Neobinghamia multiareolata, Neobinghamia villigera, Peruvocereus albisetatus),
[Bo-3], [Hx*],
" -f. robustus....
albispinus (See H. albidus, H. chosicensis -v- albispinus, H. peniculatus, Peruvocereus albispinus),
[X], [DD], [NN], [Bo-2], [Hi-4], [Hv], [Hx*],
[Il],
" -v- floribundus (See H. albispinus, H. chosicensis -v- albispinus, Peruvocereus albispinus -v- floribundus),
" -v- roseospinus (See H. albispinus, H. chosicensis -v- albispinus),

HAAGEOCEREUS.....(Cont'd),

albispinus.....(cont'd),
" X H. multangularis (Natural hybrid - See H. chosicensis -v- rubrospinus, H. rubrospinus, Peruvocereus rubrospinus),
[Hi-4],
albus.....
ambiguus.....[Bo-2], [Bo-6],
" -v- reductus.....[Bo-2],
andinus (See Loxanthocereus montanus),
aticensis (H. subtilispinus),
aureispinus (See H. multangularis),
[Bo-2], [Hx], [Bi 5-1],
" -v- fuscispinus.....
" -v- rigidispinus.....[Bo-2],
" -v- spinosior.....
australis (See Binghamia australis, H. multicolorispinus),
[Bo-2], [Bo-6], [Hi-3],
" -v- acinacispinus.....[Bo-2], [Hx],
" -v- albiflorus (See H. platinospinus),
" -f. nanus....
" -v- rubriflorus (See Loxanthocereus clavispinus),
" -v- subtilispinus (See H. subtilispinus),
[Hi-3],
bicolor (See Borzicactus bicolor, H. dichromus, Loxanthocereus bicolor).....[Bv-1],
caespitosus.....
cephalomacrostibas (See Weberbauerocereus cephalomacrostibas),
chalaensis.....[NN], [Hi-4],
chilensis (See H. fascicularis)
chosicensis (See Binghamia multangularis, H. multangularis),
[NN], [Av], [Ay], [Bo-2], [Bv-1], [Cz],
[Hx*], [Hy], [Jd], [Aj 9-6-1958],
" -v- albispinus (See H. albispinus),
" -v- chrysacanthus (See H. chrysacanthus, H. chryseus),
" -v- dichromus (See H. dichromus, H. multangularis -v- dichromus),
" -v- marksianus (See H. multangularis -v- turbidus, H. turbidus -v- maculatus),
" -v- rubrospinus (See H. rubrospinus, Peruvocereus rubrospinus),
[Bo-2], [Hv],
chrysacanthus (See Cereus pseudomelanostele, H. akersii, H. chryseus, H. multangularis -v- pseudo-melanostele).....[X], [Ay], [Hy],

HAAGEOCEREUS.....(Cont'd),

chryseus (See H. chrysacanthus),
[Hi-4],
churinensis.....
clavatus (See H. multangularis -v- pseudomelanostele,
H. akersii, Peruvocereus multangularis),
clavispinus (See H. lachayensis, H. limensis, H.
olowinskianus),
[DD], [Bo-2],
comosus (H. albispinus X Espostoa haagei),
[Bo-2], [Hx*],
convergens (See Borzicactus convergens, Loxanthocereus
convergens),
[Hx*], [Hy],
crassiareolatus (See H. multangularis -v- dichromus,
H. dichromus),
[Bo-2], [Hx*],
" -v- smaragdisepalus (See H. multangularis -v-
dichromus),
[Bo-2],
cuzcoensis.....
" -v- tenuiarboreus.....
decumbens (See Borzicactus decumbens, H. mamillatus -
-v- brevior, H. oconia-camanensis),
[Ay], [Bo-2], [Hi-4], [Hv], [Hx*], [Hy], [Jd],
[Aj 8-12-1958],
" -v- brevispinus.....
" -v- multicolorispinus (See H. australis, H.
multicolorispinus),
[NN],
" -v- spinosior.....[Bo-2], [Bo-6],
" -v- subtilispinus (See H. australis -v-
subtilispinus),
deflexispinus (See H. limensis -v- deflexispinus, H.
achaetus, H. pseudoacranthus),
[Bo-2],
deserticola (See Borzicactus deserticola, Loxantho-
cereus deserticola),
dichromus (See H. multangularis -v- dichromus),
[Bo-2],
" -v- palidior (See H. multangularis -v- dich-
romus, H. crassiareolatus),
[Bo-2],
divaricatispinus (See H. multangularis),
[H], [Ay], [Bo-2], [Hx*], [Jd],
elegans (See H. icosagonoides, H. versicolor -v-
elegans),
" -v- heteracanthus.....
emeriticus (See Loxanthocereus emeriticus),
fascicularis.....[Hi-3],

faustianus (See Loxanthocereus faustianus),
[Hx*],
ferox (See Weberbauerocereus cephalomacrostibas),
[NN],
ferrugineus (See Borzicactus ferrugineus, Loxanthocereus ferrugineus),
fortalazensis (See H. acranthus -v- fortalazensis, H. fulvus),
fulvus (See H. acranthus -v- fortalazensis),
" -v- yautanensis.....[Hi-4],
horrens (See H. pacalaensis -v- pseudoversicolor),
[DD], [Bo-2], [Hx*],
" -v- sphaerocarpus (See H. multangularis),
[Bo-2},
huancavelicensis.....[Bo-2],
humifusus (See H. versicolor -v- humifusus),
hystrix (See Loxanthocereus hystrix, Loxanthocereus montanus),
icensis.....[Bo-6],
icosagonoides (See H. elegans, H. versicolor -v- elegans),
[Bo-2], [Hx*],
" -f. heteracanthus.....
imperialensis (See Loxanthocereus pacaranensis),
inquisivensis (See Corryocactus inquisivensis, Samaipatanus inquisivensis),
johnsonii (See Weberbauerocereus johnsonii),
lachayensis (See H. limensis -v- limensis, H. olowinskianus),
[Bo-2], [Hx*],
lanugispinus.....[Hi-4],
laredensis (See Binghamia laredensis, H. pacalaensis -v- laredensis),
[Bo-2], [Bo-6], [Bt], [Hy], [Aj 8-12-1957],
" -v- longispinus (See H. multangularis, H. pacalaensis -v- longispinus),
[Bo-2],
" -v- montanus (See H. pacalaensis -v- montanus, Loxanthocereus montanus),
" -v- pseudoversicolor (See H. pacalaensis -v- pseudoversicolor, H. pseudoversicolor),
limensis (See H. clavispinus, H. lachayensis, H. olowinskianus),
[Hi-4],
" -v- andicola (See H. acranthus, H. limensis -v- andicola),
" -v- brevispinus (See H. acranthus),
" -v- clavispinus (See H. clavispinus, H. lachayensis, H. limensis),

limensis.....(cont'd),
" -v- deflexispinus (See H. achaetus, H. deflexispinus, H. pseudoacranthus),
" -v- metachrous (See H. acranthus -v- crassispinus, H. acranthus -v- metachrous),
[Hi-4],
" -v- zonatus (See H. zonatus),
[Iz],
litoralis.....[Bo-2], [Hl 2-1982],
longiareolatus (See H. multangularis),
[Bo-2], [Hx*],
longicomus.....
mamillatus (See Borzicactus decumbens, H. decumbens),
" -v- brevior (See H. decumbens),
[Bo-2],
marksianus (See H. multangularis -v- turbidus, H. turbidus),
[Hy],
montanus (See Borzicactus montanus, Loxanthocereus montanus),
multangularis (See Binghamia multangularis, H. chosicensis, H. acanthocladus, H. divaricatispinus),
[Hi-4], [Hx*], [Hy], [Aj 8-12-1957],
" -v- dichromus (See H. chosicensis -v- dichromus, H. dichromus).....[Hi-4],
" -v- pseudomelanostele (See H. akersii, H. H. chrysacanthus, H. pseudomelanostele),
[Hi-4],
" -v- turbidus (See H. marksianus, H. turbidus),
[Hv], [Hl 1-1972],
multicolorispinus (See H. australis, H. decumbens -v- multicolorispinus),
[Bo-6], [Gr], [Hx*], [Aj 20-7-1969],
neglectus (See Loxanthocereus multifloccosus, Loxanthocereus neglectus),
oconia-camanensis (See Borzicactus decumbens, H. decumbens, H. mamillatus -v- brevior),
[Bo-2], [Bo-6],
olowinskianus (See H. lachayensis, H. limensis),
[H], [NN], [Bo-2], [Bo-6], [Hx),
" -v- repandus.....[DD], [Bo-2],
" -v- " -f. erythranthus.....
" -v- rubriflorior.....[Bo-6],
" -v- subintertextus.....[Bo-2],
pacalaensis (See H. laredensis, H. repens),
[NN], [Bo-2], [Hv], [Hx*], [Hy], [Aj 8-12-1957],

HAAGEOCEREUS.....(Cont'd),

pacalaensis.....(cont'd),
" -v- laredensis (See H. laredensis, H. pacalaensis).....[Hx*],
" -v- longispinus (See H. laredensis -v- longispinus, H. multangularis),
" -v- montanus (See H. laredensis -v- montanus),
" -v- pseudoversicolor (See H. horrens, H. pseudoversicolor),
" -v- repens (See H. repens),
pacaranensis (See Loxanthocereus pacaranensis), [Bo-2],
" -v- tenuispinus (See Loxanthocereus neglectus -v- chimbotensis),
pachycladus (See Loxanthocereus pachycladus),
pachystele (See H. multangularis), [Bo-2],
paradoxus (See Borzicactus faustianus, H. faustianus, Loxanthocereus faustianus), [Bo-2],
pectinatus.....
peculiaris (See Loxanthocereus cantaensis, Loxanthocereus peculiaris),
peniculatus (See H. albispinus, Peruvocereus albispinus).....[Bo-2], [Bo-6],
piliger (See H. multangularis -v- pseudomelanostele, H. pseudomelanostele), [Bo-2], [Bo-6],
platinospinus (See Borzicactus platinospinus, H. australis -v- albiflorus), [Bo-2], [Bo-6], [Hy], [Aj 1-1937], [Hl 2-1982], [Iv 12-1956],
" -v- pluriflorus (See H. pluriflorus),
pluriflorus (See H. platinospinus -v- pluriflorus), [Bo-2], [Hx],
pseudoacranthus (See H. deflexispinus, H. limensis -v- deflexispinus), (Bo-2],
" -v- melanostele.....[NN],
pseudomelanostele (See H. multangularis -v- pseudomelanostele, H. setosus, Peruvocereus setosus).....[Bo-2], [Bt], [Hx*],
" -v- carminiflorus (H. albispinus X Espostoa haagei).....[Bo-2],
" -v- chrysacanthus (See H. chrysacanthus),
" -v- clavatus (See H. multangularis -v- pseudomelanostele),
" -v- longicoma (See H. setosus -v- longicoma, Peruvocereus setosus -v- longicoma)
" -v- pseudomelanostele (See H. multangularis -v- pseudomelanostele),

pseudomelanostele.....(cont'd),
" -v- setosus (See H. multangularis -v- pseudomelanostele, Peruvocereus setosus),
pseudoversicolor (See H. horrens, H. laredensis -v- versicolor, H. pacalaensis -v- pseudoversicolor),
[Bo-2], [Hx*],
rauhii.....
" -v- laticornua.....
repens (See H. pacalaensis, H. pacalaensis -v- repens
[DD], [Ay], [Bo-2], [Jd],
rigidispinus (See H. multangularis),
rubrospinus (See H. chosicensis -v- rubrospinus, Peruvocereus rubrospinus),
[NN],
salmonoideus (H. multangularis X H. albispinus),
(See Peruvocereus salmonoideus), [Bo-2],
" -v- rubrispinus (H. multangularis X H. albispinus),
seticeps (H. albispinus X Espostoa haagei),
[Bo-2],
" -v- bruneispinus.....
" -v- robustispinus (H. albispinus X Espostoa haagei),
[Bo-2],
setosus (See H. multangularis -v- pseudomelanostele, Peruvocereus setosus).....
[QQ], [Ay], [Hx], [Jd], [Aj 16-2-1965], [Db 9-1980],
" -v- longicomus (See Peruvocereus setosus -v- longicoma),
[Bo-2],
seyboldianus (See Weberbauerocereus seyboldianus),
smaragdiflorus (H. albispinus X H. multangularis),
subtilispinus (See H. aticensis, Loxanthocereus aticensis),
[Hi-4],
superbus (See H. decumbens (forma)),
[NN], [Aj 8-12-1957],
symmetros (See H. multangularis -v- dichromus),
[Bo-2], [Hx*],
talarensis (See H. versicolor -v- fuscus),
[Iv 9-1957],
tapalcalaensis (See H. pacalaensis),
tenuis.....[Hi-4],
tenuispinus (See H. multangularis -v- psuedomelanostele), [Bo-2],
torataensis (See Weberbauerocereus torataensis),
turbidus (See H. multangularis -v- turbidus),
[Ay], [Bo-2], [Hx*],

HAAGEOCEREUS.....(Cont'd),

turbidus.....(cont'd),
" -v- maculatus (See H. multangularis -v- turbidus).....[Bo-2], [Bo-6],
" -f. cristata.....[NN],
variabilis (See Loxanthocereus variabilis),
versicolor.....[NN], [Av], [Ay], [Bo-2], [Bo-6], [Gr], [Hx*], [Hy], [Il], [Jd], [Au 37-1],
" -v- aureispinus (See H. versicolor)....[Bo-2],
" -v- catacanthus (See H. versicolor)....[Bo-2],
" -v- elegans (See H. icosagonoides),
" -v- fuscus (See H. talarensis),
" -v- humifusus (See H. humifusus),
" -v- lasiacanthus (See H. versicolor),
" -v- oder.....
" -v- " -f- aureispinus.....
" -v- paitanus (See H. versicolor -v- talarensis)
" -v- proliferus.....[Ay],
" -v- " -f. cristata.....[Ay],
" -v- xanthacanthus.....[Bo-2],
" -f. cristata.....[T], [Hx],
viridiflorus (See H. multangularis, Peruvianus viridiflorus),
vulpes.....[Hi-4],
weberbaueri.....(See Weberbauerocereus weberbaueri),
" -v- aureifuscus.....
" -v- horribilis.....
" -v- horridispinus.....
" -v- humilior.....
winterianus (See Weberbauerocereus winterianus), [Iz],
" -v- australis.....
xanthacanthus (See H. versicolor -v- xanthacanthus),
xylorhizus (See Loxanthocereus xylorhizus),
zangalensis.....
zehnderi.....[QQ], [Bo-2], [Hx*], [Af 51-1],
zonatus (See H. limensis -v- zonatus), [Ay], [Bo-2], [Hx*], [Jd], [Iv 9-1957],

HAMATOCACTUS

crassihamatus (See Ancistrocactus crassihamatus, Ferocactus crassihamatus, Glandulicactus crassihamatus, Thelocactus crassihamatus), [Hx],
davisii (See H. hamatacanthus -v- davisii),
hamatacanthus (See Ferocactus hamatacanthus), [L], [R], [U], [Y], [GG], [Ar], [Bo-5], [Bv-3], [Bz], [Cm], [Df], [Hx*], [Hy], [Aj 31-10-1980],

hamatacanthus.....(cont'd),
" -v- crassispinus (See Echinocactus longi-hamatus),
[Hy],
" -v- davisii (See Brittonia davisii),
[Ar], [Bo-6], [Dq],
" -v- " -f. cristata.....[Ar], [Dq], [Hy],
" -v- gracilispinus.....
" -v- insignis.....
" -v- papyracanthus.....[Hl 1-1978],
" -v- sinuatus (See Ferocactus hamatacanthus -v- sinuatus, H. sinuatus),
[Df],
setispinus (See Echinocactus setispinus, Ferocactus setispinus, Thelocactus setispinus),
[DD], [QQ], [Ar], [Av], [Ay], [Bo-5], [Bt], [Bv-3], [Bz], [Cm], [Cz], [Df], [Hb], [Hn], [Hu], [Hx], [Hy], [Il], [Flower - Aj 30-5-1979], [Dg 2-3-1980],
" -v- cachetianus.....[Iu],
" -v- hamatus.....[Y], [Df], [Aj 32-12-1981],
" -v- orcuttii.....
" -v- setaceus.....[Y], [Df], [Af 16-6], [Aj 32-12-1981],
sinuatus (See Ferocactus hamatacanthus -v- sinuatus, Hamatocactus hamatacanthus -v- sinuatus),
[Y], [Df], [Hn], [Flower - Aj 31-10-1980],
uncinatus (See Ancistrocactus uncinatus, Ferocactus uncinatus, Glandulicactus uncinatus, Thelocactus uncinatus),
[F], [GG], [NN], [Av], [Bm-], [Bv-1, [Bz], [Hn], [Hr], [Jd], [Af 37-6],
" -v- wrightii (See Ancistrocactus uncinatus -v- wrightii, Ferocactus uncinatus -v- wrightii, Glandulicactus uncinatus -v- wrightii),
[G], [Ad], [Aw 1-1967],

HARIOTA - (See Hatiora)

HARRISIA

aboriginum (See Cereus gracilis -v- aboriginum, H. gracilis -v- aboriginum),
[Ha], [Bj],
adscendens (See Cereus adscendens, Eriocereus adscendens),
[RR], [Bl], [It],
arendtii (See Eriocereus arendtii, Eriocereus tortuosus, Harrisia tortuosa),

HARRISIA.....(Cont'd),

bonplandii (See Cereus bonplandii, Eriocereus bonplandii),
[H], [I], [L], [RR], [Ay], [Bu], [Bv-1], [Hy],
" -v- brevispina.....[Bv-1], [Af 30-4],
brookii.....
crucicentra.....
deeringii (See Cephalocereus deeringii, Cereus robinii -v- deeringii, Pilosocereus deeringii),
divaricata.....[Aj 30-2-1979],
earlei.....
eriophora (See Cereus eriophorus),
[M*], [RR], [Gr], [Hx*], [Gu III-1953],
" -v- fragrans.....[M*],
fernowii.....[RR],
fimbriata (See H. nashii),
" -v- straminia (See H. nashii -v- straminia),
[Dq],
fragrans (See H. eriophora -v- fragrans),
[G], [PP], [RR], [Bv-3],
gracilis (See Cereus gracilis),
[DD], [RR], [Bo-6], [Jd], [Af 16-5],
" -v- aboriginum (See Cereus gracilis -v- aboriginum, Harrisia aboriginum),
" -v- simpsonii (See Cereus gracilis -v- simpsonii, H. simpsonii),
guelichii (See Cereus guelichii, Eriocereus guelichii)
[V], [RR], [Bl], [Bv-1], [Cm], [Af 30-4],
hahniana (See Mediocactus hahnianus),
[Af 59-2*],
hurstii.....Dq],
jusbertii (See Cereus jusbertii, Eriocereus jusbertii),
[C], [H], [L], [Bu], [Hy], [Il], [Af 57-4],
martinii (See Cereus martinii, Eriocereus martinii),
[C], [H], [M*], [RR], [TT], [Ar], [Ay],
[Bv-1], [Dn], [Hy],
nashii (Cereus nashii),
[RR], [Bo-4], [Af 15-1], [Af 56-3],
" -v- straminia (See H. fimbriata -v- straminia)
platygona (See Cereus platygonus, Eriocereus platygonus, Cereus argentinensis),
polyacantha (See Eriocereus polyacanthus),
pomanensis (See Eriocereus pomanensis),
[RR], [Hr], [Hy],
" -v- uruguayensis (See Eriocereus pomanensis -v- uruguayensis).....
portoricensis (See Eriocereus portoricensis),
[RR], [Af 38-4],
regelii (See Eriocereus regelii),
[Af 2-10],
repandus (See H. gracilis),

HARRISIA.....(Cont'd),

simpsonii (See Cereus gracilis -v- simpsonii),
[G], [RR], [Ad],
subrepanda (See Cereus gracilis, H. undata),
taetra.....
tarijensis (See Eriocereus tarijensis),
taylorii.....[RR], [Hl 2-1972],
tephracantha (See Cereus tephracanthus, Eriocereus tephracanthus, Roseocereus tephracanthus),
[Il],
tortuosa (See Cereus tortuosus, Eriocereus tortuosus),
[C], [RR], [Ay], [Bv-1], [Cz], [Hy],
[Af 30-4],
" -v- uruguayensis.....
undata (See Cereus gracilis, H. gracilis),
[DD], [RR],

HASELTONIA

columna-trajani (See Cephalocereus hoppenstedtii),
[DD], [Ar], [Ay], [Az], [Hx*], [Hy],
[Jd],

HATIORA

bambusoides.....[RR], [YY], [Bo-2], [Bv-1], [Hx*],
[Af 52-3], [Ah-1985],
clavata (See Rhipsalis clavata),
crenata (See Acanthorhipsalis crenata, Lepismium crenatum, Rhipsalis crenata),
cylindrica.....[DD], [RR], [Bo-2], [Bv-1], [Af 52-3],
epiphylloides (See Rhipsalis epiphylloides, Pseudozygocactus epiphylloides),
[Ct-1987], [Af 52-3],
" -v- bradei (See Pseudozygocactus epiphylloides -v- bradei),
[Ay], [Af 52-3],
gaertneri (See Epiphyllum russelianum -v- gaertneri, Schlumbergera gaertneri),
herminiae (See Rhipsalis herminiae),
[Ay], [Bo-2], [Bv-3], [Hv], [Hx*], [Jd],
rosea (See Rhipsalidopsis rosea, Rhipsalis rosea),
salicornioides.....[C], [H], [I], [X], [NN], [PP], [RR],
[Ar], [Ay], [Bo-2], [Bt], [Bv-1],
[Cm], [Cz], [Gr], [Hi-1], [Hv], [Hx],
[Il], [Jd], [Ab-1985], [Af 52-3],
[Ah-1985], [Aj 31-9-1980], [Ba 9-1978], [Hh 1-10-1975],
" -v- gracilis.....[Bo-2],
" -v- stricta.....[Bo-2],
" -v- villigera.....

HELIABRAVOA

chende (See Cereus chende, Lemaireocereus chende), [DD], [JJ], [Av], [Az], [Bo-4], [Gr], [Hx*], Hy], [Jd],

HELIANTHOCEREUS

andalgalensis (See Lobivia huascha -v- andalgalensis, Trichocereus andalgalensis),
antezanae (See Echinopsis antezanae, Helianthocereus orurensis, Lobivia formosa ssp. tarijensis -v- bertramiana, Trichocereus orurensis, Trichocereus tarijensis -v- orurensis), [Bo-2],
atacamensis (See Echinopsis atacamensis, Trichocereus atacamensis),
auricolor (See Helianthocereus huascha -v- auricolor, Trichocereus andalgalensis -v- auricolor, Trichocereus auricolor),
bertramianus (See Lobivia formosa ssp. tarijensis -v- bertramiana, Trichocereus bertramianus, Trichocereus herzogianus), [DD], [Bo-2],
chilensis (See Lobivia backebergii -v- chilensis, Lobivia chilensis, Trichocereus chilensis),
" -v- puritama.....
conaconensis (See Lobivia formosa ssp. tarijensis -v- bertramiana),
crassicaulis (Invalid name - See Lobivia crassicaulis, Lobivia huascha -v- crassicaulis), [DD], [Hx*],
enneacanthus (See Trichocereus enneacanthus)....[NN],
escayachensis (See Trichocereus escayachensis, Trichocereus werdermannianus),
fricii (See Lobivia fricii, Lobivia tiegeliana -v- fricii),
grandiflorus (See Echinopsis rowleyi, Lobivia grandiflora, Lobivia huascha -v- grandiflora), [DD], [Bo-2], [Hr], [Hx*], [Jd], [Aj 27-3-1976],
herzogianus (See H. orurensis, Lobivia formosa ssp. tarijensis -v- bertramiana, Trichocereus bertramianus, Trichocereus herzogianus), [Bo-2],
" -v- totorensis (See Lobivia formosa ssp. tarijensis -v- totorensis, Trichocereus herzogianus -v- totorensis, Trichocereus totorensis), [Bo-2],
huascha (See Echinopsis huascha, Lobivia huascha), [Ar], [Bo-2], [Hb], [Aj 30-5-1979],

HELIANTHOCEREUS.....(Cont'd),

huascha.....(cont'd),
" -v- albiflorus.....
" -v- auricolor (See Echinopsis huascha -v- auricolor, Lobivia huascha, Trichocereus auricolor, Trichocereus andalgalensis -v- auricolor),
[Bo-2], [Iv-7-1955],
" -v- macranthus (Invalid name - See Echinopsis huascha -v- macranthus, Lobivia huascha),
[DD],
" -v- rosiflorus (See Echinopsis huascha -v- rosiflorus, Lobivia huascha),
" -v- rubriflorus (See Echinopsis huascha -v- rubriflorus, Lobivia huascha -v- andalgalensis, Lobivia huascha -v- rubriflorus, Trichocereus andalgalensis),
[Bo-2], [Iv 7-1955],

hyalacanthus (See Helianthocereus huascha, Lobivia huascha, Lobivia hyalacantha),

narvaecensis (See Lobivia huascha ssp. narvaezensis, Trichocereus narvaezensis),

orurensis (See H. bertramianus, Lobivia formosa ssp. tarijensis -v- bertramianus, Lobivia orurensis, Trichocereus bertramianus),
[Bo-2], [Hx*],
" -v- albiflorus (See Lobivia formosa ssp. tarijensis -v- bertramianus, Trichocereus bertramianus).....[Bo-2],

pasacana (See Trichocereus pasacana),
[T], [Ar], [Ay], [Bo-2], [Hr], [Hx*], [Hy], [Af 51-4], [Aj 28-12-1977], [Hl 1-1972],
" -v- albicephalus.....
" -f. cristata.....[Hy],

pecheretianus (See Lobivia huascha),
[DD], [Bo-2],
" -v- viridior (invalid name).....[DD],

poco (See Lobivia formosa ssp. tarijensis, Trichocereus poco, Trichocereus tarijensis -v- poco),
[Bo-2], [Il], [Aw 4-1964],
" -v- albiflorus (See Lobivia formosa ssp. tarijensis Trichocereus randallii),
[Bo-2],
" -v- fricianus (See Lobivia formosa ssp. tarijensis)
[Bo-2],
" -v- sanguiniflorus (Invalid name - See Lobivia formosa ssp. tarijensis, Trichocereus poco -v- sanguiniflorus),
[DD],

HELIANTHOCEREEUS.....(Cont'd),

pseudocandicans (Invalid name - See Echinopsis pseudocandicans),
[Hx*],
" -v- flaviflorus (Invalid name).....
" -v- roseoflorus (Invalid name).....[DD],
randallii (See Helianthocereus tarijensis, Lobivia formosa ssp. tarijensis, Trichocereus randallii, Lobivia tarijensis), [Ay],
schickendantzii (See Trichocereus schickendantzii),
tarijensis (See Lobivia formosa ssp. tarijensis, Lobivia tarijensis, Trichocereus tarijensis),
[Ay],
(hybrid).....[DD],

HELIOCEREUS

amecamensis (See H. speciosus -v- amecamensis),
[Ja], [Aw 3-1973],
aurantiacus.....[Af 46:67, March, April, 1974],
bifronus (See H. speciosus).....[Hy],
cinnabarinus (See H. heterodoxus),
[Az], [Bo-4], [Hx*],
coccineus (See Mediocactus coccineus),
[Aw 2-1983],
elegantissimus.....[YY], [RR], [Az], [Cx], [Af 14-1],
[Af 34-1],
" -v- elegantissimus.....[Az],
" -v- helanae.....
" -v- stenopetalum (See Epiphyllum stenopetalum),
[Af 38-3],
heterodoxus (See H. cinnabarinus, H. speciosus -v- heterodoxus),
[Af 22-5], [Aw 2-1964],
luzmariae.....[Af 57-6],
schrankii (See H. speciosus -v- schrankii).....[Hy],
serratus (See H. speciosus -v- serratus),
speciosus.....[L], [M*], [DD], [NN], [QQ], [RR], [UU],
[Ay], Be*], [Bo-4], [Bt], [Bu], [Cm],
[Cz], [Gr], [Hv], [Hx*], [Il], [Jd],
[Af 14-4], [Bb-1983], [Hh Jan. 1974],
[Hl 6-1979],
" -v- albiflorus.....[K],
" -v- amecamensis (See H. amecamensis),
[G], [Az], [Hv], [Hh 2-3-1975],
" -v- heterodoxus (See H. heterodoxus),
[Aj 24-12-1973],
" -v- ignescens.....[Af 14-1],
" -v- schrankii (See H. schrankii),
[Aj 31-9-1980],

HELIOCEREUS.....(Cont'd),

speciosus.....(Cont'd),
" -v- serratus (See Epiphyllum anuliger, H. serratus),
" -v- speciosus.....[Az], [Cx],
" -v- superbus (See H. speciosus, H. superbus)
superbus (See H. speciosus, H. speciosus -v- superbus)

HERTRICHOCEREUS

beneckei (See Lemaireocereus beneckei, Stenocereus beneckei),
[DD], [Ar], [Av], [Ay], [Az], [Bo-4], [Bt], [Hx*], [Hy], [Ia], [Aj 30-6-1979], [Da-I-],
" -f. cristata.....[Hy],

HICKENIA

microsperma (See Parodia microsperma),
[RR], [Dl*], [Hy], [Iu], [Af 16-11], [Af 20-9],

HILDEWINTERA

aureispina (See Borzicactus aureispinus, Loxanthocereus aureispinus, Winteria aureispina, Winterocereus aureispina),
[HH], [Ay], [Bt], [Ha], [Hv], [Hx], [Il], [Jd] [Bb-1983],
" -f. cristata.....[G], [W],

HOMALOCEPHALA

texensis (See Echinocactus texensis),
[D], [H], [M*], [Y], [AA], [DD], [GG], [NN], [Ar], [Ax], [Ay], [Bz], [Hn], [Hx], [Hy], [Il], [Jd], [Ae-1981], [Af 16-9], [Af 58-5], [Aj 27-12-1976], [Aj 30-5-1979], [Dg 2-3-1980] [Dg 7-1],
" -v- gourgensii (See H. texensis -v- longispinus) [Hy],
" -v- latispinus.....
" -v- longispinus (See H. texensis -v- gourgensii)
" -f. aureovariegata.....
" -f. cristata.....[Hy], [Af 4-9],
" -f. (White flower).....[Af 58-5],
Spec. .....[Ar],

HORRIDOCACTUS

aconcaguensis (See Pyrrhocactus aconcaguensis, Pyrrhocactus horridus -v- aconcaguensis),

HORRIDOCACTUS.....(Cont'd),

aconcaguensis.....(cont'd),
" -v- orientalis (See Pyrrhocactus horridus -v- orientalis, Pyrrhocactus aconacguensis -v- orientalis),
andicolus (See Neochilenia andicola, Pyrrhocactus andicolus),
[Bo-3],
" -v- descendens (See Pyrrhocactus andicola -v- descendens),
[Bo-3],
" -v- mollensis (See Pyrrhocactus andicola -v- mollensis),
" -v- robustus (See Pyrrhocactus andicola -v- robustus),
aricensis (See Neochilenia aricensis, Neoporteria aricensis, Pyrrhocactus aricensis),
[Hl 1-1977],
armatus (See Neoporteria armata, Pyrrhocactus armatus)
[Hx*],
aspillagai (See Neochilenia aspillagai, Neoporteria aspillagai, Pyrrhocactus aspillagai),
[Hl 5-1977],
atroviridis (See Neochilenia atroviridis, Pyrrocactus atroviridis),
[Hx],
cachytayensis (cachetiensis).....[Hl-1-1977],
calderanus (See Neochilenia calderana, Neoporteria calderana, Pyrrhocactus calderanus),
carrizalensis (See Neoporteria carrizalensis, Neoporteria curvispina -v- carrizalensis, Pyrrhocactus carrizalensis),
centeterius (See Echinocactus centeterius),
[Bo-6],
chilensis (See Neochilenia chilensis, Neoporteria chilensis, Pyrrhocactus chilensis),
[Hl 5-1977],
choapensis (See Neoporteria choapensis, Pyrrhocactus choapensis),
[NN], [Bo-6], [Hx*],
combarbalensis (See H. curvispinus -v- combarbalensis, Pyrrhocactus curvispina -v- combarbalensis),
confinis (See Neochilenia confinis, Pyrrhocactus confinis),
[Bo-6],
copiapensis (See Neochilenia kunzei),
[Bo-6],
crispus (See Neochilenia crispa, Neochilenia nigriscoparia, Pyrrhocactus crispus),
[Hx],

curvispinus (See Malacocarpus curvispinus, Neoporteria curvispina, Pyrrhocactus curvispinus), [NN], [Bo-3], [Hx*], [Hl 6-1966], [Hl 5-1977],
" -v- aconcaguensis (See H. aconcaguensis, Pyrrhocactus aconcaguensis, Pyrrhocactus horridus -v- aconaconcaguensis),
" -v- combarbalensis (See H. combarbalensis, Pyrrhocactus curvispinus -v- combarbalensis), [Bo-3],
" -v- felipensis (See Pyrrhocactus curvispinus -v- felipensis).....[Bo-3],
" -v- kesselringiana (See H. kesselringianus, Neoporteria kesselringiana, Neoporteria curvispina -v- kesselringiana, Pyrrhocactus kesselringianus),
" -v- lissocarpus (See H. lissocarpus, Neoporteria curvispina -v- lissocarpus, Pyrrhocactus lissocarpus),
" -v- mollensis (See H. andicolus -v- mollensis, Pyrrhocactus andicola -v- mollensis)
" -v- santiagensis (See Pyrrhocactus curvispinus -v- santiagensis), [Hl 5-1977],

cyrtacanthus (See H. curvispinus, Pyrrhocactus curvispinus), [Hy],

dimorphus (See Neochilenia dimorpha, Pyrrhocactus dimorphus),

echinus (See Neochilenia echina, Pyrrhocactus echinus) [Hl 1-1977],
" -v- minor (See Pyrrhocactus echinus -v- minor, Pyrrhocactus floccosus -v- minor),

engleri (See Pyrrhocactus engleri), [Bo-3], [Bo-6],

eriosyzoides (See Neochilenia eriosyzoides, Neoporteria eriosyzoides, Pyrrhocactus eriosyzoides), [Bo-3],

floccosus (See Neochilenia eriocephala, Neochilenia floccosa, Horridocactus floccosus, Pyrrhocactus floccosus), [Hl 1-1977],

froehlichianus (See Pyrrhocactus froelichianus), [Bo-3], [Hx*], [Hy], [Hl 5-1977],

fusciflorus.....

fuscus.....

HORRIDOCACTUS.....(Cont'd),

garaventai (See Neoporteria subaiana, Neoporteria curvispina -v- garaventai, Pyrrhocactus garaventai, Pyrrhocactus subaiana).....[Bo-3],
geissei (See Neoporteria curvispina -v- geissei, Neoporteria geissei),
[Bo-3], [Hx],
giganteus (See Neochilenia gigantea),
glaucescens (See Neochilenia glaucescens, Neoporteria eriocephala -v- glaucescens, Pyrrhocactus glaucescens),
grandiflorus (See Neoporteria curvispina -v- grandiflora, Pyrrhocactus grandiflorus),
hankeanus (See Neochilenia hankeana, Neoporteria hankeana).....[Hl-1-1977],
heinrichianus (See Neochilenia heinrichiana, Neoporteria curvispina -v- heinrichiana, Pyrrhocactus heinrichianus).....[Bo-3],
horridus (See H. tuberisulcatus, Neoporteria horrida, Neoporteria tuberisulcata, Pyrrhocactus horridus, Pyrrhocactus tuberisulcatus),
[NN], [Aj 27-10-1976],
" -v- minor (See H. tuberisulcatus -v- minor),
intermedius (See Neochilenia intermedia, Neoporteria intermedia, Pyrrhocactus intermedius),
jussieui (See Neochilenia jussieui, Neoporteria jussieui, Pyrrhocactus jussieui),
" -v- spinosior.....
kesselringianus (See Neoporteria curvispina -v- kesselringiana, Pyrrhocactus kesselringianus),
[Hx*], [Hy], [Gx 5-1942],
" -v- minor.....
" -v- subaequalis (Invalid name) (See Neoporteria curvispinus -v- subequalis)
[Bo-3],
kunzei (See Neochilenia kunzei, Pyrrhocactus kunzei),
limariensis (See Neoporteria nigrihorrida -f. limariensis),
lissocarpus (See Neochilenia lissocarpa, Neoporteria curvispina -v- lissocarpa, Pyrrhocactus lissocarpus),
" -v- gracilis (See Neochilenia lissocarpus -v- gracilis),
" -v- luteospina (See Neochilenia lissocarpus -v- luteospina),
marksianus (See Neoporteria marksiana, Pyrrhocactus marksianus).....[Bo-6], [Hy], [Hl 6-1966],
" -v- tunensis (See Neoporteria marksiana -v- tunensis, Pyrrhocactus marksianus --v- tunensis),
minor.....[Hu], [N-1980],

HORRIDOCACTUS.....(Cont'd),

nigricans (See Neoporteria nigricans, Pyrrhocactus nigricans),
[Bo-3], [Bo-6], [Hx*], [Hy],
" -v- coquimbana (See Neoporteria coquimbana, Neoporteria nigricans -v- coquimbana),
paucicostatus (See Neochilenia paucicostata, Neoporteria paucicostata, Pyrrhocactus paucicostatus),
[Hl 1-1977],
" -v- viridis (See Neochilenia paucicostata -v- viridis, Neoporteria paucicostata -v- viridis, Pyrrhocactus paucicostata -v- viridis).....[Hl 1-1977],
pelachicus.....[Bo-3],
pulchellus (See Neochilenia pulchella, Pyrrhocactus pulchellus),
pygmaeus (See Neochilenia pygmaea, Neoporteria pilispina -f. pygmaea, Pyrrhocactus pygmaeus),
[Aj 15-1-1964], [Hl 4-1970],
reconditus (See Neochilenia recondita, Pyrrhocactus reconditus, Thelocephala recondita),
residuus (See Neochilenia residua, Pyrrhocactus residuus).....[Bo-3],
robustus (See Pyrrhocactus horridus -v- robustus, Pyrrhocactus robustus),
[Bo-3], [Hy], [Il],
" -v- vegasanus (See Horridocactus robustus, Pyrrhocactus horridus -v- robustus),
rupicolus (See Neochilenia rupicola, Neochilenia taltalensis, Neoporteria rupicola, Neoporteria taltalensis, Pyrrhocactus rupicolus, Pyrrhocactus taltalensis).....[F], [Hl 1-1977],
scoparius (See Neochilenia scoparia, Pyrrhocactus scoparius),
setosiflorus (Ritt.) (See Neochilenia setosiflora, Pyrrhocactus setosiflorus (Ritt.)),
setosiflorus (Backbg.) (See Pyrrhocactus catamarcensis, Pyrrhocactus setiflorus),
simulans (See Neochilenia simulans, Pyrrhocactus simulans).....[Bo-6],
taltalensis (Hutch.) (See H. rupicolus, Neochilenia rupicola, Neochilenia taltalensis, Neoporteria rupicola, Neoporteria taltalensis, Pyrrhocactus rupicolus, Pyrrhocactus taltalensis),
taltalensis (Ritt) (See Pyrrhocactus neohankeanus).
transitensis (See Neochilenia transitensis, Pyrrhocactus transitensis)....[Bo-6], [Aj 15-3-1964],

HORRIDOCACTUS.....(Cont'd),

trapichensis (See Neochilenia trapichensis, Pyrrho-
cactus trapichensis),
truncatipetalus (See Pyrrhocactus truncatipetalus)
tuberisulcatus (See H. horridus, Pyrrhocactus
horridus, Pyrrhocactus tuberisulcatus),
[DD], [Ar], [Bo-3], [Bo-6], [Hx*], [Hy],
" -v- minor.....[DD], [Hy], [Hl 6-1966], [Hl 5-
1977],
" -v- spiralis.....[Hy],
vallenarensis (See Neochilenia vallenarensis, Neo-
porteria vallenarensis, Pyrrhocactus
vallenarensis),
[DD], [Bo-3], [Hx*], [Hy],
vexatus (See Neochilenia vexata, Pyrrhocactus vexatus)
" -v- horridus (See Neochilenia vexata -v-
horrida),
wagenknechtii (See Neochilenia wagenknechtii, Neopor-
teria wagenknechtii, Pyrrhocactus
wagenknechtii),
Spec. (de Ovalle).....[Ar],

HYLOCEREUS

antiguensis.....[RR], [Bo-6],
bronxensis (broxensis).....(Af 27-4),
calcaratu.....[HH], [RR], [Bl], [Bo-2], [Af 19-7],
[Af 51-6], [Hh Jan. 1974],
compressus (See Cereus triangularis, H. triangularis),
costaricensis.....[RR], [Al*], [Bo-2], [Af 20-12],
cubensis.....[G], [PP], [RR],
escuintlensis.....[Af 56-4],
estebanensis.....[Bo-2],
extensus (See Cereus extensus),
[Ft - Curtis's Bot. Mag. 70:Pl. 4066, 1873],
guatamalensis (See Acanthocereus guatamalensis)
[RR], [Ay],
lemairei.....[RR], [Bo-2], [Bu], [Bv-1],
microcladus (See Wilmattea minutiflora).....[Bo-2],
minutiflorus (See Wilmattea minutiflora),
monacanthus.....[T], [RR],
napoleonis (See H. triangularis),
[Af 56-3], [Aw 1-1985],
ocamponis.....[M*], [RR], [Az], [Cx], [Hy], [Aj 31-1-
1980], [Aj 31-12-1980],
peruvianus.....[Ay], [Bo-2],
polyrhizus.....[DD], [Bo-2], [Hx*],
purpusii.....[Az], [Bo-2], [Cx], [Hx*], [Aj 32-5-1981],
scandens.....
schomburgkii.....
stenopterus.....[M*], [RR],

HYLOCEREUS.....(Cont'd),

triangularis.....[RR], [Bj], [Hx*], [Ia], [Af 24-4],
tricostatus (See Hylocereus undatus),
[Fg], [Hy], [Im], [Af 21-1],
trigonus.....[LL*], [RR], [Hy],
trinitatensis (See H. lemairei),
undatus (See Cereus undatus, H. tricostatus),
[G], [V], [LL*], [RR], [Av], [Az], [Bb], [Bo-2]
[Bt], [Bu], [Bv-1], [Cx], [Cz], [Dq], [Fl],
[Gr], [Hv], [Hx*], [Il], [Jd], [Ad Dec. 1978],
[Aj 29-1-1978], [Bm-1987], [Gu III-1953],
[Hl 3-1979],
venezuelensis.....[I], [Ay], [Bo-2], [Hl 2-1981],

HYMENOLOBIVIA

buiningiana (See Lobivia buiningiana, Lobivia
jajoiana).....[Hi-2],
miniatinigra (See Lobivia miniatinigra),
[Hi-2],
purpureo-miniata (See Lobivia purpureo-miniata),
[Hi-2],
schuldtii (See L. klusacekii),
[Hr],
tuberculosa (See Lobivia tuberculosa),
[Hi-2],

HYMENOREBULOBIVIA

albicentra (See Lobivia albicentra),
arachnoidea (See Lobivia haematantha ssp. densispina)
cabradai (See Lobivia cabradai, Lobivia haematantha
ssp. densispina),
[Hr],
" -v- aureiflora.....[Hy],
" -f. cristata.....[Hy],
carnea (See Lobivia haematantha ssp. densispina),
carneopurpurea (See Lobivia haematantha ssp.
densispina),
cerasiflora (See Lobivia haematantha ssp. densispina),
citriflora (See Lobivia haematantha ssp. densispina),
cordipetala (See Lobivia haematantha ssp. densispina),
crispa (See Lobivia haematantha ssp. densispina),
gigantea (See Lobivia haematantha ssp. densispina),
kavinai (See Lobivia haematantha ssp. densispina,
Lobivia kavinai),
[Hy],
kreuzingeri (See Lobivia haematantha ssp. densispina,
Lobivia kreusingeri),
maresii (See Lobivia haematantha ssp. densispina),
melanea (See Lobivia haematantha ssp. densispina),

HYMENOREBULOBIVIA.....(Cont'd),

minima grandiflora (See Lobivia haematantha ssp. densispina, Lobivia minima -v- grandiflora),
multiflora (See Lobivia haematantha ssp. densispina),
nigra (See Mediolobivia pygmaea -v- flavovirens),
nivosa (See Lobivia haematantha ssp. densispina),
paucipetala (See Lobivia haematantha ssp. densispina),
pectinata (Backbg.) (See Rebutia pygmaea),
pectinata (Fric) (See Lobivia haematantha ssp. densispina),
purpurea (See Lobivia purpurea),
robusta sanguinea (See Lobivia haematantha ssp. densispina),
sanguiniflora (See Lobivia famatimensis -v- haematantha)
spinosissima (See Lobivia haematantha ssp. densispina),

HYMENOREBUTIA

albolanata (See Lobivia albolanata),
aurea (See Echinopsis aurea, Lobivia aurea, Pseudolobivia aurea),
" -v- callochrysea (See Echinopsis callochrysea, Echinopsis aurea -v- fallax, Lobivia aurea -v- fallax, Pseudolobivia aurea -v- fallax), [Hi-2],
" -v- catamarcensis (See Echinopsis catamarcensis, Trichocereus catamarcensis),
" -v- cylindrica (See Lobivia cylindrica),
" -v- depressicostata.....[Hi-2],
" -v- lariojensis (See Lobivia aurea -v- fallax), [Hi-2],
chlorogona (See Lobivia haematantha ssp. densispina -v- rebutioides),
chrysantha (See Lobivia chrysantha).....[Hi-2],
cintiensis (See Lobivia cintiensis, Lobivia lateritia) [Hi-2],
" -f. cristata.....[Hi-2],
densispina (See Lobivia famatimensis -v- densispina)
drijveriana (See Lobivia drijveriana, Lobivia haematantha ssp. kuehnrichii),
kreuzingeri (See Lobivia haematantha ssp. densispina),
kuehnrichiana (See Lobivia haematantha ssp. kuehnrichii, Lobivia kuehnrichii),
leucomalla (See Lobivia aurea -v- leucomalla, Lobivia leucomalla),
napina (See Lobivia haematantha -v- kuehnrichii ?), [Aj 15-7-1964],
nealeana (See Echinopsis nealeana, Lobivia nealeana, Lobivia saltensis -v- nealeana),

HYMENOREBUTIA.....(Cont'd),

pectinifera (See Lobivia haematantha ssp. densispina -v- pectinifera, Lobivia pectinifera),
pseudocachensis (See Echinopsis nealeana, Lobivia nealeana, Lobivia pseudocachensis, Lobivia saltensis -v- nealeana),
[Cu 1-1961],
pusilla (See Lobivia pusilla, Lobivia tiegeliana -v- pusilla),
[Hi-2],
" -v- flaviflora (See Lobivia tiegeliana -v- flaviflora),
[Hi-2],
" -f. pusilla.....[Hi-2],
quinesensis (See Echinopsis aurea -v- quinesensis, Lobivia aurea -v- quinesensis),
[Hi-2],
rebutioides (See Lobivia haematantha ssp. densispina -v- rebutioides, Lobivia rebutioides),
" -v- citriniflora (See Lobivia haematantha ssp. densispina -v- rebutioides),
[Aj 33-3-1982],
" -v- kraussiana (See Lobivia haematantha ssp. densispina -v- rebutioides),
[Aj 33-3-1982],
" -v- rebutioides (See Lobivia haematantha ssp. densispina -v- rebutioides),
[Aj 33-3-1982],
" -v- sublimiflora (See Lobivia haematantha ssp. densispina -v- sublimiflora, Lobivia sublimiflora),
scoparia (See Lobivia haematantha ssp. densispina, Lobivia scoparia),
tiegeliana (See Lobivia tiegeliana),
[Hi-2],
" -v- dimorphipetala.....[Hi-2],
" -v- distefanoiana (See Lobivia tiegeliana -v- distefanoiana, Lobivia tiegeliana).....[Hi-2],
" -v- ruberrima (See Lobivia -v- ruberrima),
torataensis.....[Hi-2],
torreana.....[Hi-2],

ISLAYA

aticensis (See H. subtilispinus, Loxanthocereus aticensis),
bicolor (See Neoporteria bicolor),
[G], [Ar], [Bo-3], [Hi-4], [Hx*], [Hy], [Jd], [Hs II-2-1981],

ISLAYA.....(Cont'd),

brevicylindrica (See Neoporteria islayensis -f. brevicylindrica),
[K], [Ay], [Bo-3], [Hi-4], [Hx*],
[Aj 16-7-1965],
chalaensis (See I. copiapoides -v- chalaensis),
copiapoides (See I. islayensis -v- copiapoides, Neoporteria islayensis -v- copiapoides),
[Ar], [Bo-3], [Hy], [Jd], [Ae-1981],
[Flower -Aj 28-9-1977],
" -v- copiapoides.....[Hi-4],
" -v- chalaensis (See I. chalaensis).....[Hi-4],
" -v- echinata.....
" -v- pseudomollendensis.....[Hi-4],
divaricatiflora (See Neoporteria islayensis -v- divaricatiflora),
[G], [DD], [Hi-4], [Af 52-2],
flavida.....[J], [Hi-4], [Hy], [Af 60-2], [Hl 6-1976],
grandiflorens (See I. maritima),
[Ar], [Av], [Hx*], [Hy], [Jd], [Ae-1981],
[Cu-1-1976], [Hl 3-1982],
" -v- spinosior (See I. maritima),
[Bo-3], [Af 52-2], [Af 52-3],
[Aj 27-12-1976],
" -v- tenuispina (See I. maritima),
grandis (See Neoporteria islayensis -v- grandis),
[Bo-3], [Hx*], [Aj 9-7-1958],
" -v- brevispina (See I. grandis),
[Bo-3],
" -v- neglecta (See Neoporteria grandis -v- brevispina),
intermedia (See Neoporteria intermedia, Pyrrhocactus intermedius),
islayensis (See I. mollendensis, Neoporteria islayensis)
[Bo-3], [Hx*], [Hy], [Hl 3-1982],
" -v- aticensis (See Haageocereus subtilispinus, Loxanthocereus aticensis),
" -v- compacta.....
" -v- copiapoides (See I. copiapoides),
" -v- " -f. pseudomollendensis.....
" -v- minor (See I. minor, Neoporteria islayensis -f. minor),
[Hi-4], [Hl 5-1968],
" -f. mollendensis (See I. mollendensis, I. islayensis),
" -v- nigra.....
krainziana (See Neoporteria krainziana),
[G], [Ar], [Bo-6], [Hi-3], [Hy], [Il],
" -v- longispina.....[Hl 3-1982],
" -f. cristata.....[Hi-3],

ISLAYA.....(Cont'd),

maritima (See I. grandiflorens),
[Hi-4], [Hu], [Hy], [Hl 3-1982],
" -f. cristata.....[Hy],
minor (See I. islayensis -v- minor),
[Av], [Bo-3], [Gr], [Hu], [Hx*], [Hy], [Af 52-4],
[Hl 3-1982],
minuscula.....[Hi-4],
mollendensis (See I. islayensis).....[NN], [Ay],
[Bo-3], [Hx], [Hy], [Iv 12-1956],
omasensis.....[Aj 34-3-1983],
paucispina (See I. copiapoides -v- chalaensis, I. paucispinosa, Neoporteria lindleyi),
[Ay], [Hx*], [Jd], [Af 52-1], [Aj 11-12-1960],
" -v- copiapoides (See I. copiapoides),
" -v- curvispina.....
paucispinosa (See I. paucispina, Neoporteria lindleyi -f- lindleyi).....[T], [QQ], [Bo-3], [Hx],
roseiflora (See I. divaricatiflora),
solitaria.....[G], [PP],
unguispina.....[Hl 3-1982]

Spec. .....[Bo-6],

ISOLATOCEREUS

dumortieri (See Lemaireocereus dumortieri, Stenocereus dumortieri),
[DD], [Bo-4], [Bt], [Hx*], [Hy], [Aj 23-9-1972],

JASMINOCEREUS

galapagensis (See Cereus thouarsii, J. thouarsii),
[RR], [Hx*], [Hy],
howellii (See Cereus thouarsii, J. thouarsii),
[Af 34-3], [Aj 18-8-1967],
" -v- delicatus (See J. thouarsii -v- delicatus),
[DD], [Af 34-3], [Af 39-5],
sclerocarpus (See J. thouarsii -v- sclerocarpus),
[DD], [Af 39-5],
thouarsii (See Cereus thouarsii),
[Ay], [Bo-2], [Jd],
" -v- chathamensis.....[DD], [Af 34-3],
" -v- delicatus.....[Ch],
" -v- sclerocarpus.....[G], [DD], [Af 34-3],
" -v- thouarsii.....[Ch], [Af 34-3],

KRAINZIA

guelzowiana (See Mammillaria guelzowiana),
[T], [NN], [Ar], [Bo-5], [Bt], [Hy], [Jd],
[Ae-1981], [Ba 3-1977], [Ba 1-1978],
" -v- comocephala.....[Hy],
" -v- inermis.....[Hy],
" -v- splendens (See Mammillaria guelzowiana -v- splendens).....[Hy],
longiflora (See Mammillaria longiflora),
[DD], [NN], [Ar], [Ay], [Bo-5], [Hy], [Jd],
[Ae-1981], [Aj 26-10-1975], [Ba 3-1977],
" -v- stampferi.....

LASIOCEREUS

fulvus.....[Ay], [Hi-4], [Ba Aug. 1966:119],
rupicola.....[DD], [Bo-6], [Hi-4], [Hx*],

LEMAIREOCEREUS

aragonii (See Marshallocereus aragonii, Stenocereus aragonii),
[RR], [Bv-1], [Hy], [Af 5-3], [Af 58-1],
[Aj 24-5-1973],
beneckei (See Hertrichocereus beneckei, Stenocereus beneckei),
[RR], [Ay], [Bv-1], [Cx], [Il], [Jd], [Af 5-3],
[Af 16-2],
chende (See Cereus chende, Heliabravoa chende),
[DD], [JJ], [RR], [Av], [Bv-1], [Cx], [Hy],
[Jd], [Aj 11-8-1960],
chichipe (See Cereus chichipe, Polaskia chichipe),
[D], [V], [Bv-1], [Cx], [Jd], [Af 21-4],
[Af 36-5],
clavatus.....[Af 19-9],
cumengei (See Machaerocereus gummosus),
deficiens (See Cereus deficiens, Ritterocereus deficiens),
[Iu],
dumortieri (See Isolatocereus dumortieri, Stenocereus dumortieri)
[G], [L], [RR], [Bv-1], [Cx], [Il], [Af 1-11],
[Ah-1985],
eichlamii (See Ritterocereus eichlamii, Stenocereus eichlamii),
[Ay], [Az],
" -v- longispina.....[Af 37-3],
euphorbioides (See Neobuxbaumia euphorbioides, Rooksbya euphorbioides).....[X], [Az],
" -v- olfersii.....
eruca (See Machaerocereus eruca).....[Il],

LEMAIREOCEREUS.....(Cont'd),

godingianus (See Armatocereus godingianus)
[RR],
griseus (See Cereus griseus, Ritterocereus griseus, Stenocereus griseus),
[RR], [Az], [Bl], [Cz], [Dq], [Hy], [Af 57-3], [Aj 26-11-1975], [Fg], [Hl 2-1981],
gummosus (See Machaerocereus gummosus, Stenocereus gummosus),
hollianus (See Cereus hollianus, Pachycereus hollianus)
[DD], [JJ], [RR], [Bo-4], [Bv-1], [Cx], [Hy], [Af 22-6], [Ba 4-1980], [Db 3-1979],
" -f. cristata.....
humilis.....[RR],
hystrix (See Cereus hystrix, Ritterocereus hystrix, Stenocereus hystrix),
[U], [RR], [Af 4-5], [Af 56-3], [Gu III-1953], [Hl 3-1972],
laetus (See Armatocereus laetus),
litoralis (See L. thurberi -v- litoralis, Stenocereus thurberi -v- litoralis),
[Af 3-1], [Af 30-4],
longispina (See Cereus laevigatus, Ritterocereus laevigatus, Stenocereus laevigatus),
[RR], [Bo-4], [Bv-1], [Dq],
marginatus (See Marginatocereus marginatus, Pachycereus marginatus, Stenocereus marginatus),
[WW*], [Bv-1], [Cz], [Dq], [Af 14-4], [Af 55-6]
" -v- gemmatus.....[Bv-1],
" -v- marginatus.....[Az],
" -f. monstrosus.....[Af 21-4]
martinezii (See Cereus martinezii, Stenocereus martinezii),
[Bv-1], [Af 21-4], [Af 57-4]
matucanensis (See Armatocereus ghiesbreghtii, Armatocereus matucanensis, Cereus ghiesbreghtii
mixtecensis (See Lemaireocereus chichipe, Polaskia chichipe),
montanus (See Cereus montanus, Stenocereus montanus),
[Cx], [Af 10-10], [Af 15-4], [Af 34-4],
olivaceus (See Lemaireocereus hystrix, Ritterocereus hystrix).....[Flower - Hy*],
pruinosus (See Cereus pruinosus, Ritterocereus pruinosus, Stenocereus pruinosus)...[E], [X], [Bv-1], [Cx], [Hy], [Il], [Af 14-4],
" - mutant.....[Bi Vol. 4, #6],
queretaroensis (See Cereus queretaroensis, Pachycereus queretaroensis, Ritterocereus queretaroensis, Stenocereus queretaroensis),
[Fg], [Hy], [Af 21 -1, & 2], [Iv 12-1946],
quevedonis (See Stenocereus quevedonis),

LEMAIREOCEREUS.....(Cont'd),

schottii (See Lophocereus schottii),
[Af 56-6],
schumannii (See Stenocereus schumannii),
setispinus (See Neobuxbaumia scoparia),
standleyi (See Ritterocereus standleyi, Stenocereus standleyi),
[Aw 1-1966], [Aw 2-1970],
stellatus (See Ritterocereus stellatus, Stenocereus stellatus, Cereus stellatus),
[RR], [Ay], [Bu], [Bv-1], [Cx], [Cz], [Jd],
thurberi (See Cereus thurberi, Marshallocereus thurberi, Pachycereus thurberi, Stenocereus thurberi),
[D], [I], [M*], [BB], [GG], [LL], [RR], [An], [Cx], [Cz], [Hy], [Il], [N-1980], [Am 3-1982], [Au 37-1], [Aw 3-1972],
" -f. cristata.....[Ay], [Cx], [Af 1-9], [Aw 3-1972],
" -v- litoralis (See Lemaireocereus litoralis, Stenocereus thurberi -v- litoralis)
[Af 41-5],
treleasei (See Stenocereus treleasei),
[RR], [Bu], [Bv-1], [Cx], [Af 9-1], [Af 21-3]
[Aw 2-1964],
weberi (See Cereus weberi, Pachycereus weberi, Stenocereus weberi),
[Ay], [Bl], [Bt], [Bv-1], [Cx], [Hy], [Im], [Af 21-3], [Af 25-5], [Aw 2-1962], [Aw 2-1964],
L. -f. cristata.....[Af 30-4],

LEOCEREUS

bahiensis.....[DD], [RR], [Ay], [Bo-4], [Bo-6], [Hi-1], [Hx*], [Jd],
" -v- debilispinus.....
" -v- palmierensis.....
glaziovii.....[Bo-4],
melanurus.....[It],
mello-barretoi (See Arthrocereus mello-barretoi),
paulensis.....
serpens.....[Bo-2],
urandianus.....[Hi-1],

LEPIDOCORYPHANTHA

macromeris (See Coryphantha macromeris),
[Ar], [Bo-5], [Hx*],
runyonii (See Coryphantha macromeris -v- runyonii, Coryphantha runyonii, Mammillaria runyonii),
[DD], [NN], [Ar], [Bo-5], [Il],

LEPISMIUM

aculeatum (See Rhipsalis aculeata).....[Ct-1987],
anceps (See L. cruciforme -v- anceps, Rhipsalis anceps)
brevispinum (See Acanthorhipsalis brevispinum, Rhipsalis brevispina),
cereoides (See Rhipsalis cereoides),
chrysanthum (See Rhipsalis chrysanthum),
commune (See L. cruciforme),
crenatum (See Acanthorhipsalis crenata),
cruciforme (See Rhipsalis cruciforme),
[T], [Ay], [Bo-6], [Bu], [Bv-3], [Cz], [Gr], [Jd],
" -v- anceps (See Rhipsalis anceps),
[QQ], [RR], [Bo-2], [Bo-6], [Hi-1], Hy*],
" -v- bolivianum.....
" -v- cavernosum.....[Bv-1],
" -v- commune (See L. cruciforme),
" -v- cruciforme.....[Bv-1],
" -v- knightii (See L. knightii),
" -v- marnieranum (See L. marnieranum),
" -v-megalanthum (See L. megalanthum).....[DD],
" -v- myosurus (See L. myosurus, Rhipsalis myosurus),
" -v- vollii (See L. vollii),
dissimile (See L. pacheco-leonii, Rhipsalis dissimilis).....[Hy*],
" -v- setulosa (See Rhipsalis disimilis),
epiphyllanthoides (See Rhipsalis epiphyllanthoides),
floccosum (See Rhipsalis floccosa),
" -v- gibberula (See Rhipsalis gibberula),
gibberulum (See Rhipsalis gibberula),
grandiflorum (See Rhipsalis grandiflora).....[Hy*],
houletianum (See Rhipsalis houletiana).....[Ct-1987],
" -v- regnellii (See Rhipsalis regnellii),
ianthothele (See Pfeiffera ianthothele, Rhipsalis ianthothele),
incachacanum (See Rhipsalis incachacana),
knightii (See L. cruciforme -v- knightii),
lorentzianum (See L. cruciforme -v- anceps, Rhipsalis boliviana, Rhipsalis lorentziana),
lumbricoides (See Rhipsalis aculeata, Rhipsalis lumbricoides),
marnieranum (Invalid name - See L. cruciforme -v- marnieranum, Rhipsalis marnieriana)...[DD],
megalanthum (See L. cruciforme -v- megalanthum, Rhipsalis megalantha)...[Bo-2], [Bo-6], [Bv-1],
micranthum (Vaupel) (See Cereus micranthus (Vaupel)),
miyagawae (See Pfeiffera miyagawae),
monacanthum (See Acanthorhipsalis monacantha, Rhipsalis monacantha).....[Ct-1987],

LEPISMIUM.....(Cont'd)

myosurus (See L. cruciforme -v- myosorus),
[E],
neves-armondii (See Rhipsalis neves-armondii),
[Bv-1], [Aj 24-1-1973],
pacheco-leonii (See L. dissimile, R. pacheco-leonii,
R. dissimilis),
paradoxum (See Rhipsalis paradoxa),
[Bo-2], [Gr], [Hy], [Il],
paranganiense (See Acanthorhipsalis paranganiensis,
Rhipsalis paranganiensis),
pittieri (See Rhipsalis pittieri),
[Bt],
pulvinigerum (See Rhipsalis pulvinigera),
[Ay],
puniceo-discus (See Rhipsalis puniceo-discus),
[Bv-1], [Im],
" -v- chrysocarpum.....
rigidum (See Rhipsalis neves-armondii, Rhipsalis
rigida),
saxatile (See Rhipsalis saxatile),
[Au 4:Dec. 1965],
trigonum (See Rhipsalis trigona),
[Bo-2], [Bv-1],
tucumanense (tucumanensis) (See Rhipsalis tucumanensis),
[Bv-1],
vollii (See L. cruciforme -v- vollii),
warmingianum (See Rhipsalis warmingiana),

LEPTOCEREUS

arboreus.....[RR], [Hy*],
assurgens (See Cereus assurgens),
[RR], [Hx*], [Af 9-1],
ekmannii.....
grantianus.....[DD], [B0-4], [Dq],
" -v- leonii (See Leptocereus leonii),
[U], [RR],
leonii (See L. grantianus -v- leonii),
[LL], [RR], [Bo-4], [Gu III-1953],
maxonii.....[RR],
prostratus.....
quadricostatus.....[RR], [Ay], [Af 5-5], [Ct-1988],
sylvestris.....[RR],
weingartianus.....[RR], [Af 56-3],
wrightii.....[Bo-4],

LEPTOCLADODIA

elongata (See Mammillaria elongata).....[Hx],

LEPTOCLADODIA.....(Cont'd)

elongata.....(cont'd)
" -v- rufrocrocea.....[Bt],
leona (See Mammillaria pottsii).....[NN],
microhelia (See Mammillaria microhelia),
[NN], [QQ], [Ay], [Cr], [Ig],
" -v- droegeana (See Mammillaria droegeana, Mammillaria microhelia -v- droegeana)..[NN]
" -v- microheliopsis (See Mammillaria microhelia -v- microheliopsis, Mammillaria microheliopsis)....[NN],
sphacelata (See Mammillaria sphacelata),

LEUCHTENBERGIA

principis.....[E], [I], [T], [V], [X], [AA], [DD], [NN], [Ar], [Av], [Ay], [Be*], [Bt], [Bv-3], [Bz], [Cm], [Cr], [Ct], [Cx], [Cz], [Gr], [Hr], [Hu], [Hv], [Hx], [Hy], [Il], [Iz], [Jd], [Ae-1983], [Aj 28-5-1977], [Au 37-4], [Ba 2-1978], [Ct-1983], [Da 4-1979],
" -v- variegata.....[Bu],

LEUCOSTELE

rivierei (See Echinopsis rivierei),
[DD], [Bo-2], [Hy], [Aj 15-7-1964], [Hl 1-1977,

LOBIERA

macdougallii (See Nopalxochia macdougallii),
[DD], [Bo-2], [Hx*], [Hy], [Af 16-12], [Af 20-10],

LOBIVIA

abrantha (See L. caineana, L. muschii),.....[Hy],
acanthoplegma (invalid name - See Cinnabarinea acanthoplegma, L. cinnabarina ssp. acanthoplegma, Pseudolobivia acanthoplegma),
[DD], [Cv-1], [Hx*],
" -v- oligotricha (See L. oligotricha, L. neocinnabarina, L. pseudocinnabarina),
" -v- patula (See L. cinnabarina).....[Cv-1],
" -v- pilosa (See L. cinnabarina ssp. acanthoplegma, L. oligotricha -v- pilosa),
[Jc], [Aj 31-8-1980],
" -v- roseiflora (See L. cinnabarina ssp. acanthoplegma -v- roseiflora),
[Ha], [Jc],

LOBIVIA.....(Cont'd)

aculeata (See L. pentlandii -v- hardeniana, L. rossii),
adpressispina (See L. pugionacantha -v- cornuta), [Hi-2],
aguilari (See Echinopsis obrepanda -v- aguilari, L. calorubra -v- mizquensis).....[Cv-1],
akersii (See L. tegeleriana -v- akersii), [Cv-1], [Hx*], [Aj 24-2-1973],
albicentra (See Hymenorebulobivia albicentra), [Hy],
albolanata (See Hymenorebutia kreuzingeri, L. haematantha ssp. densispina -v- pectinifera, L. kreuzingeri, L. pectinifera -v- albolanata, L. rebutioides).....[Cv-2], [Hy], [Aj 10-9-1959],
allegraiana (See L. backebergii ssp. hertrichiana), [Bo-3], [Cv-1], [Hx], [Hy], [Cu 4-1960],
amblayensis (See Echinopsis amblayensis. L. haematantha ssp. kuehnrichii -v- amblayensis).....[Cv-2], [Flower - Hu], [Ae-1981], [Aj 23-3-1972],
" -v- albispina (See L. haematantha ssp. kuehnrichii -v- amblayensis), [Cv-2], [Aj 23:68 March, 1972],
andalgalensis (Br. & R.)1922 (See Helianthocereus crassicaulis, L. huascha -v- crassicaulis, Trichocereus catamarcensis(Ritt.)),

andalgalensis (Web.) (See L. huascha -v- rubriflora),
andina (See L. rossii),
antezanae (See Trichocereus antezanae),
apurimacana (See L. cruciaureispina, L. maximiliana ssp. westii -v- intermedia), [Cv-1],
arachnacantha (See Echinopsis ancistrophora ssp. arachnacantha, Echinopsis arachnacantha), [X], [EE], [NN], [Ar], [Ay], [Bo-3], [Cv-3] [Hu], [Hx], [Ag-1982], [Ag-1988], [Am 1-1981], [Am 4-1982], [Hl 4-1967],
" -v- densiseta (See Echinopsis ancistrophora ssp. arachnacantha -v- densiseta, Echinopsis arachnacantha -v- vallegrandensis).....[Cv-3], [Jc], [Af 42-4], [Af 19:49 March, 1968],
" -v- minor (See Echinopsis ancistrophora ssp. arachnacantha -v- minor),
" -v- spinosior (See Echinopsis ancistrophora ssp. arachnacantha -v- spinosior),
" -v- sulphurea (See Echinopsis ancistrophora ssp. arachnacantha -v- sulphurea) [Hu], [Ba 6-1974],

LOBIVIA.....(Cont'd)

arachnacantha.....(cont'd)
" -v- torrecillasensis (See Echinopsis ancistrophora ssp. arachnacantha -v- torrecillasensis, Echinopsis torrecillascensis, L. torrecillasensis), [Cv-3], [Il], [Jc],
" -v- vallegrandensis (See Echinopsis ancistrophora ssp. arachnacantha -v- vallllegrandensis, Echinopsis vallegrandensis).....[Cv-3],
argentea (See L. wegheiana) Natural hybrid L. ferox X L. pentlandii, [Cv-1], [Gr], [Hy], [Aj 10-9-1959],
astranthema (See L. haematantha ssp. densispina -v- pectinifera).....[Cu 3-1962],
atrovirens (See Digitorebutia atrovirens, Mediolobivia atrovirens, Rebutia pygmaea -f. atrovirens, Mediolobivia pectinata -v- atrovirens), [Jc],
" -v- haefneriana (See Mediolobivia haefneriana), [Jc],
" -v- huasiensis (See Aylostera huasiensis)
" -v- pseudoritteri (See Rebutia ritteri),
" -v- raulii (See Digitorebutia raulii),
" -v- ritteri (See Lobivia ritteri),
" -v- yuncharasensis.....
" -v- yuquinensis (See Digitorebutia yuquinensis),
" -v- zecheri (See Aylostera zecheri),

aurantiaca (Invalid name - See L. pentlandii -v- hardeniana),
aurea (See Echinopsis aurea, Hymenorebutia aurea, L. cylindrica, Pseudolobivia aurea).....[Bu], [Bv-3], [Cr], [Cv-3], [Cz], [Dq], [Gr], [Af 16-12], [Cu 3-1961], [Hl 3-1979], [Bi Vol. 4, #6],
" -v- albiflora (See Pseudolobivia aurea -v- albiflora).....[Jc], [Ba 7-1979],
" -v- calochrysea (See Hymenorebutia aurea -v- calochrysea),
" -v- catamarcensis (See Hymenorebutia aurea -v- catamarcensis, [Jc],
" -v- cylindrica (See L. aurea, L. cylindrica, Salpingolobivia cylindrica), [Bn],
" -v- depressicostata.....
" -v- dobeana (See Lobivia dobeana, Cereus huascha -v- rubriflorus)..... [Cv-3], [Jc],

aurea.....(cont'd)
" -v- fallax (See Pseudolobivia aurea -v- fallax, Hymenorebutia aurea -v- lariojensis), [Cv-3],
" -v- leucomalla (See Echinopsis aurea -v- leucomalla, Lobivia leucomalla), [Iz],
" -v- quinesensis (See Echinopsis aurea -v- quinesensis),
" -v- robustior (See Lobivia aurea -v- fallax),
" -v- shaferi (See Echinopsis aurea -v- shaferi, L. cylindracea, L. shaferi), [Jc],
" -v- sierragrandensis.....[Jc],
" -v- tortuosa.....[Jc],
aureolilacina (See L. ferox -v- longispina, Lobivia longispina, Pseudolobivia wilkeae), [Af 33-4],
aureosenilis (See L. maximiliana ssp. westii, L. pampana).....[Cv-1],
backebergiana (See L. multicostata), [Hx*], [Hy], [Jc],
" -v- capinerensis.....
" -v- fleischeriana.....
backebergii (See Echinopsis werdermannii, Pseudolobivia backebergii), [F], [M*], [Bo-3], [Bv-1], [Cv-1], [Cy], [Dq] [Gr], [Hb], [Hu], [Hx], [Hy], [Jc], [Af 32-1] [Hl 2-1976],
" -v- larae (See Lobivia larae).....[Jc],
" -v- oxyalabastra (See Lobivia oxyalabastra), [Cv-1],
" ssp. hertrichiana (See L. hertrichiana, L. allegraiana, L. binghamiana, L. echinata, L. incaica), [Cv-1], [Iz], [Jd],
" " " -v- lauii (See L. lauii)....[Cv-1],
" " " -v- simplex (See L. simplex), [Cv-1],
" ssp. schieliana (See L. schieliana)..... [Cv-1],
" ssp. wrightiana (See L. wrightiana)..... [Cv-1], [Hl 1-1980],
" " " -v- chilensis (See L. chilensis),
" " " -v- winteriana (See L. winteriana), [Iz],
" ssp. zecheri (See L. zecheri),
binghamiana (See L. backebergii -v- hertrichiana), [Bo-3], [Bu], [Bv-1], [Cv-1], [Hx], [Hy], [Af 32-1], [Cu 4-1960],

boedekeriana (See Cinnabarinea boedekeriana, L. rossii, L. pugionacantha -v- rossii),
boliviensis (See L. carminantha, L. pentlandii, L. raphidacantha),
[RR], [Ar], [Bo-3], [Cv-1], [Flower - Gr], [Hx], [Hy], [Ae-1982],
" -v- croceantha (See L. varians -v- croceantha), [Hy],
" -v- robustiflora.....
" -v- violaciflora.....
brachyantha (See Mediolobivia brachyantha, Rebutia brachyantha, Rebutia steinmannii -v- brachyantha),
breviflora (See L. sanguiniflora -f. breviflora), [Bo-3], [Cv-2], [Hy],
" -v- chereauniana (See Lobivia cinnabarina, L. chereaunianus).....[NN],
bruchii (See L. formosa ssp. bruchii, Echinopsis bruchii, Soehrensia bruchii),
[Bu], [Bv-1], [Cv-3], [Ad March 1978], [Ad May - 1978], [Af 59-1], [Aj 28-12-1977],
" -v- kieslingii (See L. formosa ssp. bruchii -v- kieslingii, L. kieslingii). [Cv-3],
" -v- nivalis (See L. formosa ssp. bruchii -v- nivalis, Soehrensia bruchii -v- nivalis).....[Cv-3],
brunneo-rosea (See L. boliviensis, L. pentlandii), [Bo-3], [Cv-1],
buiningiana (See Hymenolobivia buiningiana), [Hi-2], [Ae-1981], [Af 54-4],
cabradai (See Hymenorebulobivia cabradai, L. haematantha ssp. densispina, L. rebutioides),
" -v- aureiflora.....[Hy],
" -f. cristata.....[Hy],
cachensis (See L. saltensis, Pseudolobivia cachensis, L. haematantha -v- kuehnrichii).....
[RR], [Bo-3], [Cv-3], [Hx*],
caespitosa (See L. maximiliana ssp. caespitosa), [E], [Bo-3], [Hx], [Hy], [Iu],
" -v- caespitosa.....[Hi-2],
" -v- altiplani.....
" -v- rinconadensis (See L. maximiliana ssp. caespitosa -v- hermanniana, L. hermanniana)..... [Hi-2],
" -v- violacea.....[Aj 30-7-1979],
caineana.....[NN], [Cv-2], [Hx*], [Jc], [Af 53-3], [Aj 10-3-1959],
" -v- albiflora.....[Cv-2], [Af 24-6],

calochrysea (See L. aurea -v- fallax).....
[Cv-3],
callianthus (See L. huascha, Trichocereus callianthus)
[R], [Cv-3],
callichroma (See Echinopsis obrepanda -v- purpurea),
[Cv-1],
calorubra (See Echinopsis calorubra, Echinopsis obrepanda -v- calorubra, Pseudolobivia carmineoflora).....[Cv-1],
" -v- grandiflora (See Echinopsis rauschii -v- grandiflora, L. pojoensis -v- grandiflora),
" -v- megalocephala (See Echinopsis rauschii -v- megalocephala),
" -v- mizquensis (See L. aguilari, L. mizquensis)
" -v- pojoensis (See Echinopsis rauschii, L. pojoensis).....[Jc],
camargoensis (See Aylostera camargoensis)
camataquiensis (See Hymenolobivia cintiensis, L. cintiensis, L. lateritia).....[Cv-2],
campicola (See L. culpinensis).....[Hi-2],
cardenasiana (See Echinopsis ancistrophora ssp. cardenasiana, Echinopsis cardenasiana),
[Cv-3], [DD-472], [Jc], [Aj 23-2-1972],
cariquinensis (See L. maximiliana),
[Cv-1], [Iv 12-1959],
carminantha (See L. pentlandii),
[V], [Bo-3], [Cj], [Cv-1], [Hx*],
carmineoflora (See Echinopsis obrepanda -v- calorubra, Pseudolobivia carmineoflora),
[Cv-1],
cerasiflora (See Hymenorebulobivia cerasiflora, L. haematantha ssp. densispina),
charazanensis (See L. maximiliana ssp. caespitosa -v- charazanensis),
[Bo-3], [Cv-1], [Hx*], [Am 4-1983], [Iv 12 -1957],
charcasina (See L. cinnabarina).....[Cv-1],
chereaunianus (See L. breviflora -v- chereauniana, L. cinnabarina),
[Cv-1], [Hx],
chilensis (See L. backebergii ssp. wrightiana -v- chilensis).....[Cv-1],
chionantha (See Acanthocalycium chionanthum, Acanthocalycium thionanthum -v- chionanthum),
[RR],
chiquitana (See L. rigidispina),
chlorogona (See Hymenorebutia chlorogona, L. haematantha ssp. densispina -v- rebutioides, L. rebutioides -v- chlorogona),

chorrillosensis (See L. haematantha ssp. chorrillosensis).....[Cv-2], [Hx*],
chrysantha (See Hymenorebutia chrysantha, L. dragai, L. janseniana, L. klusacekii, L. polaskiana, L. staffenii),
[T], [Bo-3], [Cv-2], [Jd], [Hx*], [Hy],
" -v- hossei (See L. hossei),
" -v- hypocyrta (See L. hypocyrta),
[Jc], [Aj 23-11-1972],
" -v- janseniana (See L. janseniana).....[Hy],
" -v- klusacekii (See L. klusacekii),
" ssp. jajoiana (See L. jajoiana),
[Cv-2], [Iz], [Jd],
" " " -v- caspalasensis.....
" " " -v- fleischeriana (See L. chrysantha)
" " " -v- paucicostata (See L. jajoiana -v- paucicostata).....
[Cv-2],
" " " -v- vatteri (See L. vatteri).....
[Cv-2],
" ssp. marsoneri (See L. marsoneri).....
[Cv-2],
" " " -v- rubescens (See L. rubescens),
[Cv-2],
chrysochete (See L. hystrix).....
[Bo-3], [Cq], [Cv-2], [Hx*], [Il], [Jc], [Aj 16-12-1965],
" -v- chunchullensis (See L. chrysochete),
" -v- markusii (See L. markusii),
[Ay], [Cv-2], [Af 54-1],
" -v- minutiflora.....[Cv-2], [Jc], [Aj 28-4-1977],
" -v- subtilis.....[Ba 3-1980],
" -v- tenuispina (See L. tenuispina).....
[Cv-2],
churinensis (See L. tegeleriana -v- akersii).....
[Cv-1],
cinnabarina (See Cinnabarinea cinnabarina, Echinopsis chereauniana, L. charcasina),
[L], [DD], [Bo-3], [Cv-1], [Gr], [Hx*], [Hy], [Iu], [Jc], [Jd], [Bm-...],
" -v- grandiflora (See L. cinnabarina ssp. prestoana, L. prestoana).....
[Jc], [Aj 23-11-1972],
" -v- oligotricha (See Lobivia oligotricha),
" -v- walterspielii (See Cinnabarinea walterspielii -v- sanguiniflora, L. walterspielii).....
[Cv-1],

cinnabarina.....(cont'd),
" -v- zudanensis (See L. zudanensis).....[Cv-1],
" ssp. acanthoplegma (See L. acanthoplegma, Pseudolobivia acanthoplegma), [Cv-1], [Iz],
" " " -v- neocinnabarina (See L. neocinnabarina, L. oligotricha),
" " " -v- oligotricha, (See L. oligotricha) [Cv-1],
" " " -v- patula.....[Cv-1],
" " " -v- roseiflora.....[Cv-1],

" ssp. prestoana.....[Cv-1],
" " " -v- draxleriana (See L. draxleriana), [Cv-1],
" -f. cristata.....[Hy],
cintiensis (See Hymenolobivia cintiensis, L. lateritia) [Cv-2], [Iv 12-1959],
" -v- elongata (See L. lateritia),
" -v- grandis.....
claeysiana (See L. ferox -v- longispina), [Bo-3], [Cv-2], [Aj 16-12-1965], [Iv 7-1956],
corbula (See L. maximiliana -v- corbula), [RR], [Bv-1], [Cv-1], [Hx*], [Af 16-8], [Af 32-1],
cornuta (See L. pugionacantha -v- cornuta), [DD:473], Hx*], [Hy],
cotagaitensis (See L. lateritia -v- cotagaitensis), [Cv-2],
crassicaulis (See L. huascha -v- crassicaulis, L. andalgalensis).....[Cv-3], [Dr],
cruciaureispina (See L. maximiliana ssp. westii), [Cv-1], [Hi-4],
culpinensis (See L. pugionacantha -v- culpinensis), [Cv-2], [Hi-2], [Hu], [Hx*], [Ba June 1966]
cumingii (See Sulcorebutia neocumingii, Weingartia neocumingii),
cylindracea (See L. aurea -v- shaferi, Salpingolobivia cylindracea).....[Cv-3],
cylindrica (See L. aurea, Hymenorebutia aurea, Salpingolobivia cylindrica), [Cv-3], [Hx*], [Am 4-1981], [Iv 7-1956],
deesziana (See L. haematantha ssp. densispina -v- pectinifera),
densispina (See L. haematantha ssp. densispina), [E], [L], [NN], [UU], [Cv-2], [Hv], [Hx], [Hy], [Aj 28-6-1977],
" -v- albiflora (See L. pectinifera -v- albiflora),

densispina.....(cont'd),
" -v- albolanata (See L. haematantha ssp. densispina -v- pectinifera),
" -v- aurantiaca (See L. famatimensis -v- aurantiaca),
" -v- blossfeldii (See L. haematantha ssp. densispina),
" -v- haematantha (See L. haematantha),
" -v- kreuzingeri (See Hymenorebulobivia kreuzingeri),
" -v- rebutioides (See L. haematantha ssp. densispina -v- rebutioides), [Hv],
" -v- " -f. chlorogona (See Hymenorebutia chlorogona),
" -v- setosa (See L. famatimensis -v- setosa),
" -f. cristata.....[W], [Ah-1986],
diersiana (See Mediolobivia diersiana, Rebutia diersiana,
digitiformis (See Mediolobivia pectinata -v- digitiformis, Rebutia pygmaea).....[Bu],
dislefolia.....
distafandiana.....[K & L Nursery Catalog],
divaricata (See L. backebergii ssp. hertrichiana, Neolobivia divaricata),
dobeana (See Cereus huascha -v- rubriflora, L. aurea -v- dobeana),
dragai (See L. chrysantha -v- klusacekii), [Hr],
draxleriana (See L. cinnabarina ssp. prestoana -v- draxleriana), [DD:474], [Hx*], [Ba 50:193-194 Oct. 1971],
drijveriana (See Hymenorebutia drijveriana, L. haematantha ssp. kuehnrichii), [H], [DD], [Bo-1], [Bo-3], [Cg], [Hy], [Ia],
" -v- astranthema (See L. haematantha ssp. kuehnrichii).....[Bo-3],
" -v- aurantiaca (See L. haematantha ssp. kuehnrichii).....[Hy],
" -v- nobilis (See L. haematantha ssp. kuehnrichii)....... [C], [Bo-3],
ducis-paulii (See L. ferox -v- longispina, Pseudolobivia ducis-paulii),
" -v- rubriflora (See L. ferox -v- longispina),
" -v- longispina (See L. ferox -v- longispina),
duursmaiana (See L. sanguiniflora -v- duursmaiana),
echinata (See L. backebergii ssp. hertrichiana, Neolobivia echinata), [Hx*], [Il], [Aj 24-8-1973],

einsteinii (See Cylindrorebutia einsteinii, Mediolobivia einsteinii, Rebutia einsteinii, Rebutia schmiedcheniana -v- einsteinii)..... [Jc],
" -v- atrospinosa.....
" -v- aureiflora (See Mediolobivia aureiflora, Rebutia aureiflora).....[Jc],
" -v- elegans (See Mediolobivia elegans),
" -v- gonjianii (See Rebutia gonjianii)..... [Jc],
emmae (See L. saltensis -v- nealeana)..... [Bo-3], [Hx*],
" -v- brevispina (See L. saltensis -v- mealeana),

euanthema (See Mediolobivia euanthema, Rebutia euanthema, Rebutia oculata)..... [Hy], [Jc],
" -v- tilcarensis.....[Jc],
eucaliptana (See Rebutia eucaliptana),
famatimensis (BACKBG.) (See L. haematantha -v- densispina -subvar.- pectinifera), [Cv-2],
famatimensis (SPEG.) (See Echinocactus famatimensis (SPEG.), Rebutia famatimensis (SPEG.), Reicheocactus pseudoreicheanus (BACKBG.)), [F], [H], [L], [M*], [DD:366a], [Ay], [Bo-3], [Bt], [Bu], [Bv-3], [Cm], [Cr], [Cs], [Cv-3], [Cy], [Hi-2], [Hv], [Hx*[, [Hy], [Iq], [Iz], [Jc], [Jd], [Af 59-1], [Aj 10-8-1959], [Au 37-4], [Cu 1-1968], [Cu 4-1971], [Hl 2-1970],
" -v- albiflora.....[T],
" -v- " -sv- sufflava....[Flower Aj 28-2-1977] [Hl 2-1974],
" -v- aurantiaca.....[E], [H], [Bo-3], [Hy],
" -v- densispina (See L. densispina, L. haematantha ssp. densispina, Salpingolobivia densispina), [Ar], [Bo-3], [Hx*], [Hy],
" -v- " -sv- blossfeldii.....[Hb],
" -v- famatimensis (SPEG.) Br. & R. ...[DD 366b],
" -v- haematantha (See L. haematantha), [D], [Bo-3], [Hu], [Il], [Ad-...],
" -v- jachalensis.....[Cv-3], [Jc],
" -v- leucomalla (See L. aurea -v- leucomalla, L. leucomalla), [Bo-3],
" -v- " -f. rubrispina.....[Hb],

famatimensis.....(cont'd),
" -v- oligacantha.....[Gr],
" -v- sanjuanensis.....[Cv-3], [Jc],
" -v- setosa.....[Bo-3],
" -f. cristata.....[Hy],
ferox (See Echinopsis ferox, Echinopsis lecoriensis, L. ducis-paulii, Pseudolobivia ferox, Soehrensia ferox).....[X], [RR], [Ck], [Cv-2], [Hi-2], [Hx*], [Hy], [Af 56-3],
" -v- camargensis (See Lobivia interjuncta)..... [Hi-2],
" -v- horrida.....
" -v- longispina (See L. ducis-paulii -v- rubriflora, L. longispina), [Cv-2], [Iz], [Jc],
" -v- nigra (See L. ferox -v- longispina),
" -v- potosina (See Echinopsis cerdana, Echinopsis potosina, L. potosina, Pseudolobivia potosina)..... [Cv-2], [Jc],
formosa (See L. oreopepon, Soehrensia formosa, Soehrensia oreopepon), [Ay], [Bv-1], [Cv-3], [Af 17-12], [Bm-...), [Jc],
" ssp. bruchii (See Eriocyse bruchii, L. bruchii, Soehrensia bruchii), [Cv-3], [Iz], [Jc], [Jd],
" " " -v- kieslingii (See L. kieslingii), [Cv-3], [Jd],
" " " -v- nivalis (See L. korethroides, Soehrensia bruchii -v- nivalis, Soehrensia koreth- roides).....[Cv-3],
" ssp. grandis (See Echinopsis grandis, L. gran- dis, Soehrensia grandis, Soehrensia ingens), [Bo-3], [Jc]?,
" " " -v- pinchasensis.....
" ssp. tarijensis (See Echinopsis tarijensis, L. tarijensis, Helianthocereus tari- jensis, Helianthocereus poco, Trichocereus poco, Trichocereus tarijensis)..... [Cv-3], [Jc],
" " " -v- bertramianus (See L. formosa ssp. tarijensis -v- bertramianus, Mediolobivia orurensis, Tricho- cereus bertramianus, Tricho- cereus orurensis)..... [Jc],

formosa.....(cont'd)
" -v- tarijensis.....(cont'd),
" " " -v- totorensis (See Trichocereus herzogianus -v- totorensis),
" -v- amaichensis.....
" -v- ingens (See Soehrensia ingens),
" -v- hyalacantha (See L. hyalacantha, L. huascha?)
" -v- kieslingii (See L. formosa ssp. bruchii -v- kieslingii),
" -v- randallii (See L. formosa ssp. tarijensis, Trichocereus randallii),
" -v- rosarioana (See Lobivia rosarioana)...[Jc],
" -v- uebelmanniana (See Soehrensia uebelmanniana, Trichocereus uebelmanniana),
" -v- velascoensis.....
frankii (See Echinopsis obrepanda),
fricii (See L. tiegeliana -v- fricii),
[Cv-2], [DD-475], [Aj 24-10-1973],
" -v- ruberrima (See L. tiegeliana -v- ruberrima),
[Cv-2],
glauca (See L. chrysantha ssp. jajoiana -v- paucicostata),
[Cv-2], [Hx*], [Il], [Ae-1981], [Am 4-1982],
[Au 37-2], [Ba 50:168 Sept. 1971],
" -v- paucicostata (See L. chrysantha ssp. jajoiana -v- paucicostata),
[Cv-2], [DD-476], [Aj 24-8-1973],
glaucescens.....[Hi-4],
grandiflora (See Echinopsis rowleyi, L. huascha -v- grandiflora),
[C], [Cv-3], [Jc],
" -v- crassicaulis (See Helianthocereus crassicaulis, L. andalgalensis, L. crassicaulis, Trichocereus catamarcensis).....[Jc],
" -v- lobivioides (See Helianthocereus grandiflorus. L. grandiflora).....
[Jc],
" -v- longispina.....[Jc],
" -v- pumila.....[Jc],
grandis (See Echinopsis grandis, L. formosa ssp. grandis, Soehrensia grandis).....[RR],
graulichii (Invalid name - See Echinopsis ancistrophora, Pseudolobivia graulichii).....[Cv-3],
" -v- cinnabarina (See L. fricii, L. tiegeliana -v- fricii),
gregeri (See L. chrysantha ssp. jajoiana -v- vatteri),
haageana (See L. chrysantha ssp. marsoneri -v- rubescens, L. rubescens),
[F], [H], [NN], [Hx], [Hy], [Ia],

haageana.....(cont'd)
" -v- albihepatica (See L. chrysantha ssp. marsoneri -v- rubescens)....[Bo-3],
" -v- bicolor (See L. chrysantha ssp. marsoneri -v- rubescens),
" -v- chrysaantha (See L. chrysantha)....[Hi-2],
" -v- cinnabarina (See L. chrysantha ssp. marsoneri -v- rubescens).....
" -v- croceantha (See L. chrysantha ssp. marsoneri -v- rubescens)..... [Hl 2-1974],
" -v- durispina (See L. chrysantha ssp. marsoneri -v- rubescens).....
" -v- grandiflora-stellata (See L. chrysantha ssp. marsoneri -v- rubescens), [Bo-3], [Hx*],
" -v- leucoerythrantha (See L. chrysantha ssp. marsoneri -v- rubescens)....[Bo-3],
haagei (See Digitorebutia haagei, Mediolobivia haagei, L. neohaageana, Pygmaeolobivia haagei, Rebutia haagei).....[Jc],
" -v- canacruzensis (See Digitorebutia canacruzensis),
" -v- crassa.....[Jc],
" -v- elegantula.....
" -v- eos (See Digitorebutia eos, Mediolobivia rosalbiflora, Rebutia rosalbiflora),
" -v- mudanensis (See Digitorebutia mudanensis), [Jc],
" -v- nazarenoensis (See Digitorebutia nazarenoensis).....[Jc],
" -v- orurensis (See Digitorebutia orurensis, L. orurensis, Mediolobivia orurensis, Mediolobivia pectinata -v- orurensis, Rebutia orerensis, Rebutia pygmaea).....[Jc],
" -v- pallida (See Digitorebutia pallida),
" -v- pelzliana.....[Jc],
" -v- violascens (See Rebutia violascens),
haematantha (See Echinocactus haematanthus), [Cv-2], [Dr], [Hx*],
" -v- amblayensis (See Echinocactus amblayensis, L. amblayensis, L. haematantha ssp. kuehnrichii -v- amblayensis)
" ssp. chorrillosensis (See L. chorrillosensis), [Cv-2], [Jc],
" ssp. densispina (See L. densispina, L. scoparia).....[Cv-2],
" " " -v- pectinifera (See L. famatimensis (Backbg.), L. pectinifera), [Cv-2],

haematantha.....(cont'd),
" ssp. densispina.....(cont'd),
" " " -v- rebutioides (See L. rebutioides),
" " " -v- sublimiflora (See L. sublimiflora).....[Cv-2],
" -v- elongata (See L. elongata).....[Cv-2],
" -v- fechseri (See L. hualfinensis -v- fechseri) [Cv-2],
" -v- hualfinensis (See L. hualfinensis), [Cv-2], [Jc],
" -v- jasimanensis.....[Jc],
" ssp. kuehnrichii (See L. kuehnrichii)..... [Cv-2], [Jc], [Jd],
" " " -v- amblayensis (See L. amblayensis) [Cv-2],
" -v- rebutioides (See Lobivia rebutioides, Hymenorebutia rebutioides -v- kraussiana),
" -v- viridis.....[Jc],
hardeniana (See L. pentlandii -v- hardeniana),
hastifera (See L. ferox -v- longispina).....[Hy],
hermanniana (See L. maximiliana ssp. caespitosa -v- hermanniana).....[Hx*], [Hy],
" -v- breviorior (Invalid name - See L. maximiliana ssp. caespitosa -v- hermanniana),
hertrichiana (See L. backebergii -v- hertrichiana, L. allegraiana, L. binghamiana, L. incaica), [B], [H], [M*], [S], [Bn], [Bo-3], [Bv-1], [Ck], [Cv-1], [Hv], [Hx*], [Hy], [Jc], [Jd], [Af 16-8], [Af 32-1], [Hl 2-1975],
" -v- lauii (See Lobivia lauii),
" -v- simplex (See Lobivia simplex),
higginsiana (See L. pentlandii, L. schneideriana), [Bo-3], [Cv-1], [Hx*], [Hy], [Cu 2-2-1933],
" -v- carnea.....[Hy],
hirsutissima (See L. tiegeliana -v- pusilla),
hoffmanniana (Invalid name - See Sulcorebutia hoffmanniana).....[Bo-3],
horrida (See L. lateritia -v- kupperiana).....[Hi-2],
" -v- sanguiniflora (See L. lateritia -v- sanguiniflora, L. sanguiniflora),
hossei (See L. chrysantha -v- hossei).....[Hx], [Hy],
hualfinensis (Invalid name - See L. haematantha -v- hualfinensis).....[Ae-1981], [Aj 19:67 April 1968], [Aj 21:45 March 1970],
" -v- fechseri (See L. haematantha -v- fechseri), [G], [Ba 51-123, 1972],
huariensis (See L. pugionacantha -v- rossii),

LOBIVIA.....(Cont'd),
huascha (See Cereus huascha -v- flaviflorus, Echinopsis huascha, Helianthocereus huascha, Helianthocereus pecheretianus, Trichocereus huascha, Trichocereus vatteri).....[Bu], [Bv-1], [Cv-3], [Dq], [Jc],
" -v- andalgalensis (See L. huascha -v- rubriflora, Trichocereus andalgalensis).....[Cv-3],
" -v- calliantha (See Trichocereus callianthus),
" -v- crassicaulis (See L. crassicaulis)...[Cv-3],
" -v- grandiflora (See L. grandiflora, Trichocereus grandiflora).....[Cv-3],
" -v- purpureaminiata (See Hymenolobivia purpureaminiata, L. purpureamineata),
" -v- robusta.....
" -v- rubriflora (See L. huascha -v- andalgalensis, Helianthocereus huascha -v- rubriflorus, Trichocereus andalgalensis, Trichocereus auricolor), [C], [H], [Bu], [Bv-3], [Af 17-4],
" ssp. narvaezensis (See Trichocereus narvaezensis),
" " " -v- caspalasensis (See L. chrysantha ssp. jajoiana -v- caspalasensis)
" " " -v- pajonalensis.....
" -v- walteri (See Lobivia walteri),
huilcanota (See L. backebergii ssp. hertrichiana), [Bo-3], [Hx*],
hyalacantha (See L. huascha).....[Dr],
hystrichacantha.....[Hy],
hystrix (See L. chrysochete), [Hi-2], [Ba June 1966],
hypocyrta (See L. chrysantha -v- hypocyrta),
importana (See Hymenolobivia cintiensis, L. cintiensis, L. lateritia, L. scopulina).....[Hx*],
" -v- brevispina.....
" -v- elongata.....
incaica (See L. backebergii ssp. hertrichiana, Neolobivia incaica).....[Hy], [Ip],
incuiensis (See L. tegeleriana -v- incuiensis)....[Ay],
ingens (See L. formosa, Soehrensia ingens),
interjuncta (See L. ferox -v- camargensis),
intermedia (See L. maximiliana ssp. westii -v- intermedia, L. westii -v- intermedia), [Aj 23-9-1972],
iridescens (See L. chrysantha ssp. marsoneri)....[Hx*],
jajoiana (See Hymenolobivia buiningiana, L. chrysantha ssp. jajoiana), [F], [J*], [T], [NN], [TT], [Ar], [Bo-3], [Bv-1], [Cs], [Hv], [Jd],
" -v- aurata.....
" -v- carminata (See L. chrysantha ssp. jajoiana),
" -v- caspalasensis.....[Jc],

jajoiana.....(cont'd),
" -v- elegans.....
" -v- fleischeriana (See L. chrysantha ssp. jajoiana -v- fleischeriana), [Ar], [Hr], [Hy], [Aj 19-9-1968], [Aj 23-3-1972], [Hl 1-1969],
" -v- gregeri (See L. chrysantha ssp. jajoiana -v- vatteri).....[Ba 3-1974],
" -v- longispina.....[Hy],
" -v- nidularis.....
" -v- nigristoma (See L. chrysantha ssp. jajoiana -v- vatteri, L. nigristoma), [NN], [Av], [Er], [Jc], [Bm-1989],
" -v- paucicostata (See L. glauca -v- paucicostata, L. chrysantha ssp. jajoiana -v- paucicostata).....[Jc],
" -v- pungens.....[Jc],
" -v- splendens (See L. chrysantha ssp. jajoiana)
" -v- striatipetala.....[Hy],
" -v- vatteri (See L. chrysantha ssp. jajoiana -v- vatteri, L. vatteri),
" -f. cristata.....[Hy],
janseniana (See L. chrysantha -v- janseniana, L. chrysantha).....[Dq*],
" -v- leucacantha (See L. chrysantha).....[H],
johnsoniana (See L. boliviensis, L. pentlandii), [Bo-3], [Hx*],
jujuyana (jujuensis) (See L. chrysantha ssp. jajoiana -v- vatteri, L. vatteri),
kavinai (See L. haematantha ssp. densispina, Hymenorebulobivia kavinai).....[Hy],
kermesina (See Echinopsis kermesina, Echinopsis mamillosa -v- kermesina, Pseudolobivia kermesina).....[Av],
kieslingii (See L. formosa ssp. bruchii -v- kieslingii [Aj 28-11-1977],
klimpeliana (See Acanthocalycium klimpelianum)....[Jd],
klusacekii (See L. chrysantha -v- klusacekii, L. polaskiana).....[Hr],
" -v- roseiflora (See L. chrysantha -v- klusacekii),
korethroides (See Echinopsis korethroides, L. formosa ssp. bruchii -v- nivalis, Soehrensia korethroides),
kratochviliana (See Echinopsis ancistrophora -v- kratochviliana, Echinopsis kratochviliana, Pseudolobivia kratochviliana)...[Af 51-5],
kreuzingeri (See Hymenorebutia kreuzingeri, L. haematantha ssp. densispina), [Ar], [Hy],

kuehnrichii (See Hymenorebutia kuehnrichii, L. haematantha ssp. kuehnrichii),
[Bo-3], [Hr], [Hv], [Jd], [Af 56-3],
" -v- molinensis (See L. drijveriana, L. haematantha ssp. kuehnrichii)
" -v- penca-poma (See L. penca-poma, L. haematantha ssp. kuehnrichii)...[Hu],
kupperiana (Invalid name - See L. lateritia -v- kupperiana).....[Bo-3], [Hi-2], [Hx*],
" -v- rubriflora (See L. lateritia -v- kupperiana),
larabei (See L. backebergii ssp. hertrichiana),
[Bo-6], [Bu], [Hx*],
larae (See L. pentlandii -v- larae),
[Hx*], [Ae-1981], [Af 36-...],
lateritia See Echinopsis lateritia, Hymenorebutia cintiensis, L. camataquiensis, L. cintiensis, L. scopulina),
[L], [RR], [Cv-2], [Hy], [Iz], [Jc], [Bm-1988],
" -v- borealis (See L. lateritia, L. kupperiana -v- rubriflora),
" -v- citriflora.....[Flower - Ba 2-1980], [Jc],
" -v- cotagaitensis (See L. cotagaitensis),
[Cv-2],
" -v- kupperiana (See L. horrida, L. kupperiana,
[Jc],
" -v- rubriflora (See L. kupperiana -v- rubriflora, L. lateritia -v- borealis)
" -v- sanguiniflora (See L. sanguiniflora),
" -v- schneideriana (See L. pentlandii).....
[H], [Cv-1],
lauii (See L. backebergii ssp. hertrichiana -v- lauii),
[Af 51-6], [Hh Jan. 1974],
lauramarca (See L. maximiliana -v- corbula),
[Bo-3], [Hx*],
leptacantha (See L. maximiliana ssp. quiabayensis -v- leptacantha),
[DD-477], [Hx*], [Aj 33-12-1982],
leucomalla (See L. aurea -v- leucomalla, Salpingolobivia leucomalla),
[Bv-1], [Af 32-1], [Cu 3-1961],
" -v- rubrispina (See L. aurea -v- leucomalla),
leucorhodon (See L. brunneo-rosea, L. pentlandii),
[Bo-3], [Hx*],
leucosiphus.....[Af 38-4],
leucoviolacea (See L. omasuyana, L. pentlandii),
[Bo-3], [Hx*],
longispina (See L. ferox -v- longispina, Pseudolobivia longispina).....[Av], [Hi-2], [Hu], [Hx*],
[Ae-1982], [Cu 8-1-1939],

longispina.....(cont'd),
" -v- nigra (See Echinopsis nigra, Pseudolobivia longispina -v- nigra).....
[Hy],
luteiflora (See Echinopsis aurea, L. aurea, Pseudolobivia aurea, Pseudolobivia luteiflora),
[Hy],
markusii (See L. chrysochete -v- markusii),
[Aj 17-7-1966],
marsoneri (See L. chrysantha ssp. marsoneri, L. haageana, L. rubescens).....[Bu], [Hx*], [Jc],
" -v- iridescens (See L. iridescens, L. muhriae, L. uitewaaleana).....
[Jc],
" -v- rubescens (See L. chrysantha ssp. marsoneri -v- rubescens, L. rubescens),
maximiliana (See Echinopsis maximiliana, L. cariquinensis, L. pseudocariquinensis).....
[Cv-1], [Hi-2], [Hx*], [Iz], [Jc], [Ad Jan.-1978], [Aj 33-12-1982], [Bm-1986], [Hl 2-1971],
" ssp. caespitosa (See L. caespitosa, Pseudolobivia caespitosa).....
[Jc],
" " " -v- altiplani (See L. caespitosa -v- altiplani),
" " " -v- charazanensis (See L. charazanensis).....[Cv-1],
" " " -v- hermanniana (See L. hermanniana)
[Hl 4-1976],
" " " -v- miniatiflora (See L. miniatiflora)...[Cv-1], [Db 2-1979],
" " " -v- rincondensis (See L. caespitosa -v- rincondensis),
" " " -v- violacea (See L. caespitosa -v- violacea),
" -v- charazanensis (See L. charazanensis),
[Jc],
" -v- corbula (See L. lauramarca).....
[Cv-1], [Jc], [Jd],
" -v- durispina.....
" -v- hermanniana (See L. hermanniana, L. caespitosa -v- rinconadensis)...[Jc],
" -v- intermedia (See L. intermedia),
" -v- miniatiflora (See L. miniatiflora).....
[Jc],
" ssp. quiabayensis (See L. quiabayensis),
[Cv-1], [Aj 31-12-1980],
" " " -v- leptacantha (See L. leptacantha)
[Cv-1],

maximiliana.....(cont'd),
" -v- sicuaniensis (See L. sicuaniensis),
" -v- violacea (See L. caespitosa -v- violacea)
[Jc],
" ssp. westii (See L. westii, L. cruciaureispina)
[Cv-1], [Jc],
" " " -v- intermedia (See L. intermedia),
[Cv-1],
megacarpa (See Echinopsis megacarpa, Echinopsis ayo-
payana),
megatae.....
microthele (microthelis) (See Cinnabarinea pseudocinn-
abarina -v- microthelis, L. cinnabarina ssp.
acanthoplegma -v- oligotricha),

miniatiflora (See L. maximiliana ssp. caespitosa -v-
miniatiflora)......
[Bt], [Hi-2], [Hu], [Hx*],
miniatinigra.....[Hi-2],
minima.....
" -v- grandiflora (See L. haematantha ssp. densi-
spina).....[Hy],
minuta (See L. backebergii ssp. hertrichiana, Neo-
lobivia minuta),
[H], [Bo-6], [Af 51-6], [Aj 27-1-1976],
mirabunda.....[Bo-3], [Hx*],
mistiensis (See L. pampana),
[T], [Bo-3], [Cr], [Gr], [Hy], [Ig*],
[Aj 16-7-1965],
" -v- leucantha (See Echinopsis leucantha, L.
pampana).....[Hx*],
mizquensis (See Echinopsis obrepanda -v- mizquensis),
[DD-478], [Hx*], [Aj 23-6-1972],
muhriae (Invalid name - See L. chrysantha ssp. marson-
eri, L. muhriae, L. iridescens).....[DD],
" -v- flaviflora (Invalid name - See L. chrysantha
ssp. marsoneri),
multicostata (Invalid name - See L. saltensis -v-
multicostata, L. backebergiana),
muschii (See L. abrantha, L. caineana),
napina (See Hymenorebutia napina),
narvaezensis (See L. huascha ssp. narvaezensis),

nealeana (See Hymenorebutia nealeana, Echinopsis neal-
eana, L. saltensis -v- nealeana, L. pseudo-
cachensis),
[Ar], [Bo-3], [Hx], [Hy], [Jd],
" -v- grandiflora.....[Ar],
" -v- purpureiflora.....

neocinnabarina (Invalid Name - See Cinnabarinea oligotricha, L. cinnabarina ssp. acanthoplegma -v- oligotricha),
[DD], [Hy],
neohaageana (See Mediolobivia haagei, Mediolobivia pygmaea, Rebutia pygmaea).....[Ia],
" -v- flavovirens (See Mediolobivia pygmaea -v- flavovirens),
nigra (See Hymenorebulobivia nigra, Echinopsis nigra, Lobivia longispina),

nigricans (See Digitorebutia nigricans, Mediolobivia nigricans, Rebutia ritteri -v- nigricans),
" -v- albispina.....[Jc],
" -v- carmeniana (See Digitorebutia carmeniana, Rebutia carmeniana).....[Jc],
" -v- peterseimii (See Digitorebutia peterseimii)

nigrispina (See L. haematantha ssp. kuehnrichii),
[Bo-3], [Bt], [Hx*],
" -v- rubriflora (See L. haematantha ssp. kuehnrichii).....[Bo-3], [Hy],
nigrostoma (See L. chrysantha ssp. jajoiana -v- vatteri, L. jajoiana -v- nigrostoma),
[Aj 5-1-1954], [Aj 5-4-1954], [Hl 1-1971],

nuquensis.....[Aj 23:151, 1972],
ochroleuca.....
oculata (See Mediolobivia euanthema -v- oculata, Rebutia euanthema -v- oculata),

oligotricha (See Cinnabarinea oligotricha, L. cinnabarina ssp. acanthoplegma -v- oligotricha, L. neocinnabarina),
[Il],
" -v- pilosa (See L. cinnabarina ssp. acanthoplegma -v- oligotricha),
omasuyana (See L. leucoviolacea, L. pentlandii),
[Aj 16-2-1965],
oreopepon (See Acanthocalycium oreopepon, L. formosa, Soehrensia formosa, Soehrensia oreopepon),
[Dr],
orurensis (See L. formosa ssp. tarijensis -v- bertramiana, Mediolobivia orurensis, Trichocereus bertramianus, Trichocereus tarijensis -v- orurensis).....[C],
otukae (See L. famatimensis).....
oxyalabastra (See L. backebergii -v- oxyalabastra),
[Aj 17:76 April 1966],

LOBIVIA.....(Cont'd),

oyonica (See L. churinensis, L. tegeleriana -v- akersii),
pachyacantha (See Furiolobivia pachyacantha).....[Hy],

pampana (See L. aureosenilis, L. mistiensis, Lobivia glaucescens),
[Bo-3], [Cv-1], [Hx*], [Jc], [Aj 33-4-1982],
" -v- borealis.....[Jc],
" -v- sanguiniflora (See L. sanguiniflora),

paucartambensis.....
peclardiana (See L. tiegeliana),
[Bo-3], [Bv-1], [Cr], [Hx*], [Af 32-1], [Hl 2-1972],
" -v- albiflora (See L. tiegeliana),
" -v- winteriae (winteriana) (See L. tiegeliana),

pectinata (See Hymenorebulobivia pectinata, Rebutia pygmaea),
pectinifera (See L. haematantha ssp. densispina -v- pectinifera),
[NN], [Hy], [Aj 32-7-1981], [Hh Sept. 1974],
" -v- albiflora.....[Hy],
" -v- albolanata (See L. haematantha ssp. densispina -v- pectinifera),
" -v- cinnabarina (See L. famatimensis -v- cinnabarina),
" -v- citriflora (See L. haematantha ssp. densispina -v- pectinifera),
" -v- eburnea.....

" -v- haematantha (See L. haematantha),
" -v- oligacantha (See L. famatimensis -v- oligacantha).....[Hy],
" -v- subcarnea.....
" -v- sufflava.....
penca-poma (See L. haematantha ssp. kuehnrichii),
[RR],
pentlandii (See L. argentea, L. boliviensis, L. brunneo-rosea, L. caraminantha, L. higginsiana, L. johnsoniana, L. leucorhodon, L. leuco-violacea, L. omasuyana, L. raphidacantha, L. rossii -v- bustilloensis, L. schneideriana, L. titicacensis, L. varians, L. wegheiana),
[E], [T], [DD], [Bo-3], [Bv-3], [Cq], [Cv-1], [Gr], [Hi-2], [Hr], [Hv], [Hx], [Hy], [Iz], [Jc], [Jd], [N-1980], [Af 52-6], [Hl 2-1981],

" -v- achatina (hybrid).....(See L. pentlandii)

LOBIVIA.....(Cont'd),

pentlandii.....(cont'd),
" -v- albiflora.....[Bo-3], [Hx], [Hy],
" -v- atrocarnea.....[Hy],
" -v- forbesii.....
" -v- grandiflora.....[Hy],
" -v- hardeniana (See L. hardeniana),
" -v- larae (See L. larae).....[Cv-1],
" -v- longispina (See L. boliviensis)...[Bo-3],
" -v- maximiliana (See L. maximiliana).....
[NN], [Hr],
" -v- " -f. cristata.....[NN],
" -v- ochroleuca (See L. pentlandii),
" -v- scheeri (See L. pentlandii),
" -f. cristata.....[I], [QQ],
peterseimii (See Rebutia ritteri -f. peterseimii),
pictiflora (See L. ferox -v- longispina, Pseudolobivia pictiflora).....[Hi-2], [Ba June 1966],

pilifera (See Mediolobivia ritteri -v- pilifera, Rebutia costata -f. pilifera),

planiceps (See L. backebergii ssp. hertrichiana), [Bo-3], [Hb], [Hx*], [Hy],

pojoensis (See Echinopsis ancistrophora ssp. pojoensis, Echinopsis pojoensis),
[Aj 19:8 Jan. 1968], [Aj 21:45 March 1970],
[Hl 3-1974],
" -v- grandiflora (See Echinopsis ancistrophora ssp. pojoensis -v- grandiflora, Echinopsis pojoensis),
" -v- megalocephala (See Echinopsis ancistrophora ssp. pojoensis -v- megalocephala),

polaskiana (See L. chrysantha -v- klusacekii), [Bo-3], [Hx*], [Hy],

polycephala (See L. sanguiniflora -v- duursmaiana), [Bo-3], [Hy],

potosina (See Echinopsis potosina, L. ferox -v- potosina, Pseudolobivia potosina), [L], [Hx*], [Hy],

prestoana (See Cinnabarinea prestoana, L. cinnabarina ssp. prestoanas)......[Af 42:185, 187, July, Aug. 1970],

prolifera (See L. backebergii -v- hertrichiana),

LOBIVIA.....(Cont'd),

pseudocachensis (See Echinopsis nealeana, Hymenorebutia nealeana, L. nealeana, L. saltensis -v- nealeana),
[Bo-3], [Bu], [Dq], [Hy], [Iv 9-1954],
" -v- cinnabarina (See L. saltensis -v- nealeana)
[Bo-3], [Hy],
" -v- salynii.....[D],
" -v- sanguinea (See L. saltensis -v- nealeana),
[Bo-3],
pseudocariquinensis (See L. maximiliana).....[Af 33-4],
pseudocinnabarina (Invalid name - See Cinnabarinea pseudocinnabarina, L. cinnabarina ssp. acanthoplegma),
[DD],
pugionacantha (See L. pugionacantha -v- flaviflora),
[Bo-3], [Cv-2], [Hx*], [Hy], [Jc],
" -v- bustilloensis (See L. pugionacantha -v- rossii, L. rossii -v- pugionacantha),
" -v- carminea.....
" -v- cornuta (See L. adpressispina, L. cornuta)
[Jc],
" -v- culpinensis (See L. campicola, L. culpinensis).....[Cv-2], [Jc], [Ad Jan. 1978], [Am 9-1978],
" -v- corrugata.....
" -v- flaviflora (See L. pugionacantha),
" -v- haemantha.....[Cv-2], [Jc],
" -v- rossii (See Cinnabarinea boedekeriana, L. stollenwerkiana, L. rossii).....
[Cv-2], [Jd],
" -v- salitrensis (See L. salitrensis),
[Cv-2], [Jc],
" -v- salmonea (See L. pugionacantha -v- rossii, L. rossii -v- salmonea),
" -v- sayariensis (See L. rossii -v- sayariensis)
" -v- versicolor (See L. versicolor).....[Jc],
purpurea (See Hymenorebulobivia purpurea).....[Hy],
purpureaminiata (See L. huascha -v- purpureaminiata).
[Hi-2],
pusilla (See Hymenorebutia pusilla, L. tiegeliana -v- pusilla, Mediolobivia hirsutissima),
[DD-479], [Hu], [Af 57-2], [Ba June 1966],
" -v- flaviflora (See Hymenorebutia pusilla -v- flaviflora, L. tiegeliana -v- flaviflora),
[DD-480], [Hu], [Ba June 1966],
pygmaea (See Lobivia digitiformis, Mediolobivia digitiformis, Mediolobivia pectinata -v- digitiformis, Rebutia pygmaea).....[Jc],

pygmaea.....(cont'd),
" -v- colorea (See Rebutia colorea),
" -v- diersiana (See Digitorebutia diersiana, Rebutia diersiana, Rebutia rutiliflora).....[Jc],
" -v- friedrichiana (See Digitorebutia friedrichiana, Mediolobivia friedrichiana, Rebutia friedrichiana), [Jc],
" -v- iscayachensis (See Digitorebutia iscayachensis, Rebutia iscayachensis),
" -v- knizeii.....[Jc],
" -v- minor (See Digitorebutia diersiana -v- minor, Rebutia diersiana -v- minor),
" -v- nigrescens (See Digitorebutia diersiana -v- atrovirens, Rebutia diersiana -v- atrovirens).....[Jc],
" -v- polypetala.....[Jc],
" -v- tafnaensis.....[Jc],
" -v- violaceostaminata.....[Jc],
quiabayensis (See L. maximiliana -v- quiabayensis), [Hx*], [Aj 33-12-1982],
quinesensis (See Echinopsis aurea -v- quinesensis, L. aurea -v- quinesensis),
raphidacantha (See L. boliviensis, L. pentlandii), [Bo-3], [Hx*], [Af 19-8],
rauschii (Friedr.) (See Echinopsis ancistrophora ssp. pojoensis, Echinopsis rauschii, L. pojoensis).....[Cv-3],
rauschii (Zecher).....[Cv-2], [Hx*], [Jc],
rebutioides (See L. haematantha ssp. densispina -v- rebutioides), [EE], [Ar], [Bo-3], [Gr], [Hy],

" -v- chlorogona (See Hymenorebutia chlorogona, L. haematantha ssp. densispina -v- rebutioides).....[Hx],
" -v- citriniflora (See L. haematantha ssp. densispina -v- rebutioides) [E], [BO-3],
" -v- kraussiana (See L. haematantha ssp. densispina -v- rebutioides), [Bo-3],
" -v- multiflora (See Hymenorebulobivia multiflora, L. haematantha ssp. densispina),
" -v- sublimiflora (See L. haematantha ssp. densispina -v- sublimiflora, L. sublimiflora)...[T], [Bo-3],

rigidispina (Invalid name).....[DD], [Hx],
ritteri (See Mediolobivia ritteri, Rebutia ritteri),
rosaricana.....[Aj 30-12-1979],
" -v- rubriflora.....[Aj 30-12-1979],
rossii (See L. pugionacantha -v- rossii),
[Ar], [Bv-1], [Gr], [Hx*], [Hy], [Jd],
[Af 32-1],
" -v- boedekeriana (See Cinnabarinea boedekeriana, L. pugionacantha -v- rossii),
" -v- bustilloensis (See L. pugionacantha -v- bustilloensis).....[Hi-2],
" -v- carminata (See L. pugionacantha -v- rossii).....[Bo-3],
" -v- rossii (See L. pugionacantha -v- rossii), [Hi-2],
" -v- salmonea (See L. pugionacantha -v- rossii)....[Bo-3],
" -v- sanguinea (See L. pugionacantha -v- rossii).....[Bo-3], [Hy],
" -v- sayariensis (See L. pugionacantha -v- rossii).....[Hi-2],
" -v- stollenwerkiana (See L. pugionacantha -v- rossii, L. stollenwerkiana) [Hy],
" -v- walterspielii (See Cinnabarinea walter-spielii, L. cinnabarina -v- walterspielii).....[Bo-3],
ruberrima (See L. tiegeliana -v- ruberrima), [Hy],
rubescens (See L. chrysantha ssp. marsoneri -v- rubescens, L. haageana),
[D], [T], [Hx], [Af 24-5], [Iv 5-1956],
" -v- tenuispina (See L. chrysantha ssp. marsoneri -v- rubescens, L. haageana), [Bo-3],
salitrensis (See L. pugionacantha -v- salitrensis), [Hx*], [Ba 4-1974],
" -v- flexuosa (See L. pugionacantha -v- salitrensis).....[Ba 4-1974],
saltensis (See L. cachensis, Pseudolobivia cachensis), [RR], [Bt], [Cv-3], [Aj 28-12-1977],
" -v- multicostata (See L. backebergiana, L. multicostata).....[Jc],
" -v- nealeana (See L. nealeana, L. pseudocachensis).....[Cv-3], [Jd],
" -v- pseudocachensis (See L. emmae, L. nealeana, L. saltensis -v- nealeana, L. pseudocachensis),
" -v- schreiteri (See L. schreiteri), [Cv-3], [Aj 28-12-1977],

saltensis.....(cont'd),
" -v- stilowiana (See L. stilowiana),
[Cv-3], [Iz],
" -v- zapallarensis.....
sanguiniflora (See L. sanguiniflora -v- pseudolater-
itia).....[NN], [Bo-3], [Cv-2], [Hx*],
[Hy], [Jd], [Jc],
" -v- breviflora (See L. breviflora),
[Cv-2],
" -v- duursmaiana (See L. duursmaiana, L. poly-
cephala).....[Cv-2],
" -v- pseudolateritia (See L. sanguiniflora),
[Bo-3],
" -v- violaciflora.....[NN],
scheeri (See L. pampana).....
schieliana (See L. backebergii ssp. schieliana, L.
schieliana -v- albescens),
[Bo-3], [Hi-2], [Hx], [Jc], [Aj 33-12-1982],
" -v- albescens (Invalid name - See L. backe-
bergii ssp. schieliana, L. schiel-
iana).....[DD],
" -v- leptacantha (See L. leptacantha).....[Jc],
" -v- quiabayensis (See L. quiabayensis, L.
maximiliana ssp. quiabayensis),
schmiedcheniana (See Mediolobivia schmiedcheniana,
Rebutia einsteinii -f. schmiedcheniana),
schneideriana (See L. boliviensis, L. higginsiana, L.
pentlandii),
[Bo-3], [Bo-6], [Hx*], [Hy],
" -v- carnea (See L. pentlandii),
[Bo-3], [Hl 5-1967],
" -v- cuprea (See L. pentlandii).....[Bo-3],
schreiteri (See L. saltensis -v- schreiteri),
[T], [Bo-3], [Hu], [Hx], [Jc], [Aj II 3, 4 -
1930], [Cu 2-1975],
" -v- riolarensis.....[Jc],
" -v- stilowiana (See L. saltensis -v- stilowiana,
L. stilowiana).....[Jc],
schuldtii (See L. chrysantha -v- klusacekii, L. klus-
acekii, L. polaskiana, Hymenolobivia schuldti),
scoparia (See Hymenorebutia scoparia, L. haematantha
ssp. densispina),
[Hx], [Aj 10-10-1959],
scopulina (Invalid name - See L. lateritia),
[DD], [Bo-6],
shaferi (See Echinopsis aurea -v- shaferi, L. aurea
-v- shaferi, Lobivia cylindracea),
[Bo-3], [Gr], [Hb], [Hx*],
sicuaniensis (See L. maximiliana -v- corbula),
[DD-481], [Ba 50:229 Dec. 1971],

LOBIVIA.....(Cont'd),

silvestrii (See Cereus silvestrii, Chamaecereus silvestrii, Echinopsis silvestrii, Echinopsis chamaecereus),
[UU], [Ay], [Hb], [Hv], [Jd], [Au 22: 68 Sept. -1967], [Dg 7-1],
simplex (See L. backebergii ssp. hertrichiana -v- simplex].....[Ba 51:134, 1972],
spiniflora (See Echinopsis spiniflora, Acanthocalycium spiniflorum),
" -v- klimpeliana (See Acanthocalycium klimpelianum, Acanthocalycium spiniflorum -v- klimpelianum),
" -v- macrantha.....[Jc],
" -v- peitscheriana (See Acanthocalycium peitscherianum, Acanthocalycium spiniflorum -v- peitscherianum)
" -v- violacea (See Echinopsis violacea, Acanthocalycium violaceum),
spinosissima (See L. ferox -v- longispina),
staffenii (See L. chrysantha, L. janseniana),
[Bo-3], [Hr],
" -v- lagunilla (See L. chrysantha),
steinmannii (See Aylostera steinmannii, Mediolobivia eucaliptana, Mediolobivia steinmannii, Rebutia steinmannii).....[Hc],
" -v- applanata.....
" -v- brachyantha (See L. brachyantha, Rebutia brachyantha).....[Jc],
" -v- camargoensis (See Aylostera camargoensis, Rebutia camargoensis),
" -v- christinae (See Digitorebutia christinae, Rebutia christinae),
" -v- cincinnata (See Digitorebutia cincinnata, Rebutia cincinnata).....[Jc],
" -v- costata (See Rebutia costata).....[Jc],
" -v- leucacantha.....
" -v- major.....[Jc],
" -v- melanocentra.....
" -v- parvula.....[Jc],
" -v- rauschii (See Digitorebutia rauschii, Mediolobivia rauschii),
" -v- tuberculata.....[Jc],
stilowiana (See L. saltensis -v- stilowiana),
[T], [Hx], [Af 56-2],
stollenwerkiana (See L. pugionacantha -v- rossii, L. rossii).....[Hx],
sublimiflora (See L. haematantha ssp. densispina -v- sublimiflora),
[Hy], [Ia], [Aj 10-10-1959],

taratensis (See Cinnabarinea acanthoplegma, L. cinnabarina ssp. acanthoplegma, L. acanthoplegma), [Af 38:141 July, Aug., 1966],
" -v- leucosiphus (See L. cinnabarina ssp. acanthoplegma), [Af 38:142, 161,- 1966],
tarijensis (See L. formosa ssp. tarijensis), [Ay],
tegeleriana (See Acantholobivia tegeleriana, Acantholobivia ayacuchensis), [DD], [Ay], [Bo-3], [Cv-1],
" -v- akersii (See L. akersii, L. tegeleriana -v- akersii), [Cv-1], [Jc],
" -v- eckardiana (See L. tegeleriana),
" -v- incuiensis (See Acantholobivia incuiensis, L. incuiensis).....[Jc],
" -v- medingiana (See L. tegeleriana),
" -v- plominiana (See L. tegeleriana),
" -v- puquiensis.....[Hi-4],
" -v- tegeleriana.....[Hi-4],
tenuispina (See L. chrysochete -v- tenuispina), [Hi-2],
thionantha (See Acanthocalycium catamarcense, Acanthocalycium thionanthum).....[Dr],
" -v- aurantiaca (See Acanthocalycium aurantiaca)
" -v- brevispina (See Acanthocalycium brevispinum),
" -v- catamarcensis (See Acanthocalycium catamarcense),
" -v- chionantha (See Acanthocalycium chionanthum, Acanthocalycium thionanthum -v- chionanthum, L. chionantha,
" -v- erythranta.....[Jc],
" -v- ferrarii (See Acanthocalycium ferrarii),
" -v- glauca (See Acanthocalycium glaucum),
" -v- munita.....[Jc],
" -v- variiflora (See Acanthocalycium variiflorum, Acanthocalycium thionanthum -v- variiflorum),
tiegeliana (See L. peclardiana), [T], [DD-482], [Ar], [Au], [Ay], [Bo-3], [Cv-2], [Hu], [Hv], [Hx], [Hy], [Il], [Jc], [Jd], [Bi 1-2],
" -v- albiflora (See L. tiegeliana),
" -v- cinnabarina (See Cinnabarinea graulichii, L. fricii, L. tiegeliana -v- fricii, L. graulichii -v- cinnabarina).....[Jc],
" -v- dimorphipetala.....

tiegeliana.....(cont'd),
" -v- distefanoiana (See Hymenorebutia tiegeliana -v- distefanoiana, L. tiegeliana).....[NN], [Hu], [Hx*],
" -v- flaviflora (See Hymenorebutia pusilla -v- flaviflora, L. pusilla -v- flaviflora).....[Cv-2],
" -v- fricii (See L. tiegeliana -v- cinnabarina, L. fricii),
[DD-475], [Cv-2],
" -v- peclardiana (See L. peclardiana),
" -v- pusilla (See Hymenorebutia pusilla, L. pusilla, Mediolobivia hirsutissima).
" -v- ruberrima (See Hymenorebutia tiegeliana -v- ruberrima, L. ruberrima),
[DD-482], [Cv-2], [Jc], [Ba 51:102, 1972],
" -v- tiegeliana.....[Af 60-2],
" -v- uriondoensis (See L. tiegeliana -v- pusilla)
[Ba 10-1979],
" -v- winteriana (See L. tiegeliana),
titicacensis (See L. johnsoniana, L. pentlandii),
[Hl 2-1975], [Iv 12-1959],
torataensis.....
torreana.....
torrecillasensis (See Echinopsis torrecillasensis, Echinopsis ancistrophora ssp. arachnacantha -v- torrecillasensis, Pseudolobivia torrecillasensis).....[X], [Bo-2], [Hy],
tuberculosa (See Hymenolobivia tuberculosa)....[Hi-2],
uitewaaleana (See L. chrysantha ssp. marsoneri),
[Bo-3], [Hx*],
vanwurkiana (Invalid name).....[Hx*],
varians (See L. pentlandii),
[Bo-3], [Hx*], [Hy], [Ah-1985],
" -v- croceantha (See L. pentlandii).....[Bo-3],
" -v- rubro-alba (See L. pentlandii).....[Bo-3],
variispina.....[Hi-2],
vatteri (See L. chrysantha ssp. jajoiana -v- vatteri, L. jajoiana -v- nigristoma, L. nigristoma),
[Bo-3], [Af 21-4],
" -v- robusta (See L. chrysantha ssp. jajoiana -v- vatteri).....[Bo-3],
versicolor (See L. pugionacantha -v- versicolor),
[Hx*],
vilcabambae (See L. backebergii ssp. hertrichiana -v- simplex, Neolobivia vilcabambae),
violacea (See Acanthocalycium violaceum).....[Jd],
walterii (See L. huascha -v- walteri, Soehrensia walterii),

LOBIVIA.....(Cont'd),

walterspielii (See Cinnabarinea walterspielii, L. cinnabarina -v- walterspielii),
" -v- sanguiniflora (See Cinnabarinea walterspielii -v- sanguiniflora, L. cinnabarina -v- walterspielii),
watadae (hybrid)?
" -v- salmonea (See L. rossii -v- salmonea)

wegheiana (Natural hybrid - L. ferox X L. pentlandii) [Bo-3], [Hx*], [Hy], [Af 53-2],
" -v- leucantha (See L. pentlandii),
wegneriana (See L. hertrichiana),.....[Hx*],
wessneriana (See Hymenorebutia kreuzingeri, L. haematantha ssp. densispina -v- sublimiflora, L. rebutioides -v- sublimiflora),
westii (See L. maximiliana ssp. westii), [H], [Bo-3], [Hx], [Il], [Af 26-3],
" -v- intermedia (See L. intermedia, L. maximiliana ssp. westii -v- intermedia), [Hi-4],
wilkeae (See L. ferox -v- longispina, Pseudolobivia wilkeae),
" -v- carminantha.....(See L. ferox -v- longispina),
winteriana (winteriae) (See L. backebergii ssp. wrightiana -v- winteriana), [Av], [Bt], [Hv], [Hx], [Ae-1981], [Aj 31-12-1980], [Flower - Aj 34-3-1983],
wrightiana (See L. backebergii ssp. wrightiana, L. chilensis, Neolobivia wrightiana), [D], [H], [J*], [NN], [Bo-3], [Hx*], [Hy], [Jc], [Ag-1986], [Aj 8-5-1957], [Aj 28-10-1977], [Am 4-1982], [Cu 7-2-1938], [Hl 2-1972], [Hl 4-1972],
" -v- brevispina (See L. backebergii ssp. wrightiana).....[Hl 5-1967],
" -v- winteriana (See L. winteriana),
zecheri (See L. backebergii ssp. zecheri), [DD-483], [Jc], [Ba 50:146 Aug. 1971],
" -v- fungiflora (See L. backebergii ssp. zechneri, L. zecheri), [Stachelpost 9:138 Sept. 1973],
zudanensis (See Cinnabarinea zudanensis, L. cinnabarina -v- zudanensis), [Af 42:34 Jan., Feb., 1970],
Spec. Famatima (Ritter).....[Hu],
Spec. FR - 82.....[Bo-6],
Spec. FR - 99.....[Bo-6],
Spec. Lau - 331.....[Af 54-2],

LOBIVIA.....(Cont'd),

Spec. Lau - 352.....[Af 54-4],
Spec. Lau - 1004.....[Aj 31-12-1980],

LOBIVIOPSIS

ducis-pauli (See L. ferox -v- longispina), [Hr], [Hy],
graulichii (See L. ancistrophora, Pseudolobivia graulichii).....[Hr],

LOPHOCEREUS

australis (See L. schottii -v- australis), [CC], [Af 2-9], [Af 3-8], [Af 21-3],
gatesii.....[Az], [Bo-4], [Bv-1], [Cx], [Af 5-10], [Af 16-2], [Af 35-6], [Af 39-6], [Af 41-5],
mieckleyanus (See Cereus schottii -f. monstrosus, L. schottii -f. mieckleyanus, L. schottii -f. monstrosus).....[DD], [Bo-4],
sargentianus (See Cereus schottii, Lophocereus schottii),....[CC], [Bo-4], [Hy], [Af 3-8],
schottii (See Cereus schottii) [M*], [U], [GG], [HH], [LL], [RR], [UU], [An], [Ar], [Ay], [Bo-4], [Bt], [Cz], [Dq], [Gr], [Ha], [Hv], [Hx*], [Hy], [Il], [Iz], [Jd], [Af 1-1], [Af 56-6], [Aw 2-1976],
" -v- australis (See L. australis), [Az], [Af 7-8], [Af 35-6],
" -v- mieckleyanus (See L. mieckleyanus, L. schottii -f. monstrosus), [Ay], [Bo-4], [Gr], [Af 20-8], [Af 35-6],
" -v- obesus.....[Af 20-8],
" -v- tenuis.....[Ay], [Az], [Af 35-6],
" -v- schottii.....[Az], [Cx], [Aw 1-1966],
" -f. monstrosus.....[C], [R], [U], [AA], [LL], [Ar], [Ay], [Az], [Be*], [Bu], [Cz], [Hy], [Af 35-6], [Af 51-4], [Af 60-2], [Aw 1-1972], [Db 10-1980],
" -f. spiralis (espiralda).....[Az], [Aw 1-1966],

LOPHOPHORA

diffusa (See L. echinata -v- diffusa), [G], [Av], [Gg], [Aw 1-1967], [Hl 5-1969], [Hl 5-1972], [Hl 1-1979], [Hw-1989],
echinata (See L. williamsii -v- echinata), [DD], [Ar], [Bo-5],
" -v- diffusa.....(See L. diffusa).....[Bo-5], [Hx*], [Ae-1982], [Ba 3-1980],

LOPHOPHORA.....(Cont'd),

echinata.....(cont'd),
" -f. cristata.....[Hn],
fricii.....[Hl 6-1974], [Hl 3-1980],
jourdaniana (See L. williamsii -v- jourdaniana)
[Au 37-4], [Hl 5-1969], [Hl 1-1975],
lewinii (See L. williamsii -v- lewinii, L. williamsii),.....[Cc], [Hl 5-1969],
lutea (See L. ziegleri),
[Ar], [Bo-5], [Gr], [Hl 5-1969],
" -v- texana (Invalid name).....[Bo-5],
" -f. fricii (See L. fricii),
williamsii.....[I], [L], [M*], [T], [LL*], [NN], [QQ], [TT], [UU], [Ar], [Av], [Ax], [Ay], [Be*], [Bt], [Bv-3], [Bz], [Cg], [Cm], [Cq], [Cr], [Cs], [Cx], [Cz], [Df], [Dq], [Gr], [Hn], [Hu], [Hv], [Hx], [Hy], [Il], [Iz], [Jd], [Af 16-10], [Af 43-2], [Aj 29-7-1978], [Aj 31-2-1980], [Ba 2-1980], [Bi 1-3], [Bi 4-3], [Da 1-1980], [Hl 3-1983], [Hw-1982],
" -v- caespitosa.....[Ar], [Hy], [Ba 3-1980], [Hl 4-1969], [Hl 3-1975],
" -v- decipiens.....[Ar], [Bo-5], [Hy], [Ae-1983], [Hl 4-1969], [Hl 3-1983],
" -v- " -f. cristata.....[Ar],
" -v- echinata (See L. echinata),
[Y], [Df], [Aw 12:12 1967],
" -v- heptagona.....[Hy],
" -v- jourdaniana (See L. jourdaniana)..... [Bm-1985],
" -v- lewinii (See L. lewinii, L. williamsii).
" -v- pellotinica.....
" -v- pentagona.....[Bo-5], [Hy], [Hl 4-1969],
" -v- peyotl-zacatensis.....
" -v- pluricostata.....[Hr], [Hy], [Hl 4-1969], [Hl 3-1975],
" -v- texana.....
" -v- williamsii.....[Df],
" -f. cristata.....[G], [K], [Bo-5], [Hn], [Hy], [Ij], [Iu], [Af 2-1], [Aw 4-1962], [Cu 2-4-1934], [Db 1-1980], [Hh May-June, 1977],
" -f. (spiral).....[Hl 3-1975],
" -f. (typica).....[Hl 3-1975],
ziegleri (See L. lutea),
[TT], [Ck], [Cm], [Cu 3-1962], [Da 1-1980], [Hl 5-1969],
" -v- albilanata.....
" -v- diagnonalis.....
" -v- flavilanata.....

LOXANTHOCEREUS

acanthurus (See Borzicactus acanthurus, Borzicactus eriotrichus, Haageocereus acanthurus), [Bo-2], [Hy], [Hx*], [Jd], [Af 60-5],
" -v- ferox (See L. acanthurus), [DD], [Bo-2],
aticensis (See B. gracilis -v- aticensis, L. gracilis),.....[Bo-2], [Bo-6],
aureispinus (See Borzicactus aureispinus, Hildewintera aureispina, Winterocereus aureispina), [Gr],
brevispinus (See L. peculiaris),
bicolor (See Haageocereus bicolor).....[Hi-4],
camanaensis (See Borzicactus gracilis -v- camanaensis), [Bo-2], [Bo-6], [Hi-4], [Hx*],
canetensis (See L. pacaranensis).....[Bo-2],
cantaensis (See Haageocereus peculiaris, L. peculiaris) [Hx*],
casmaensis (See L. pacalaensis).....[Iv 6-1957],
clavispinus (See Haageocereus pacalaensis, Haageocereus laredensis, Haageocereus repens),
convergens (See Haageocereus convergens), [Hi-4], [Aj 34-3-1983],
crassiserpens (Natural hybrid - Borzicactus aurivillus X Borzicactella serpens)-(See Borzicactus crassiserpens, Cleistocactus crassiserpens) [QQ], [Ay], [Bo-6], [Db 9-1980],
cullmannianus (Invalid name - See Borzicactus cullmannianus).....[Bo-6], [Hy],
deserticola (See Borzicactus deserticola, Haageocereus deserticola).....
erectispinus.....[Bo-2], [Bo-6],
eremiticus (See Haageocereus eremiticus),
erigens (See Loxanthocereus bicolor),
eriotrichus (See Borzicactus acanthurus, Borzicactus eriotrichus, L. acanthurus), [Hy], [Da 2-1980], [Gy 1937 No. 11, p. 4],
eulalianus (eulaliensis) (See Haageocereus faustianus, L. faustianus).....[Bo-2],
faustianus (See Borzicactus faustianus, L. kellerbadensis, L. eulalianus, L. paradoxus), [Bo-2], [Hx], [Hl 4-1971],
ferrugineus (See Borzicactus ferrugineus, Haageocereus ferrugineus).....[Bo-2],
formosus (See Borzicactus formosus, Matucana formosa),
gracilis (See Borzicactus gracilis, L. aticensis, L. nanus, Maritimocereus gracilis), [NN], [Bo-2], [Hi-4], [Hu], [Hx*], [Ba 7-1978],
" -v- aticensis (See L. aticensis),
" -v- camanaensis (See L. camanaensis),
gracilispinus (See L. neglectus).....[Bo-2],

LOXANTHOCEREUS.....(Cont'd),

granditesselatus (See L. tesselatus -v- spinosior), [Bo-2], [Hi-4],
hoffmannii.....
hystrix (See L. montanus).....[Bo-2],
" -v- brunnescens.....
jajoianus (See Borzicactus jajoianus).....[Bo-2],
keller-badensis (See Borzicactus keller-badensis, L. faustianus).....[Dt 2:22 Fig. 1948],
madisoniorum (See Borzicactus madisoniorum),
montanus (See Borzicactus montanus, Haageocereus andinus, Haageocereus hystrix, Haageocereus montanus),
multifloccosus (See L. neglectus)....[Bo-2], [Bo-6],
nanus (See Borzicactus gracilis, L. gracilis, Maritimo-cereus gracilis, Maritimocereus nanus)....[Bo-6],
neglectus (See Haageocereus neglectus, L. gracili-spinus).....[Hi-4],
" -v- chimbotensis (See Haageocereus pacaranensis -v- tenuispinus),....[Hi-4],
otuscensis.....[Hi-4],
pacaranensis (See Haageocereus pacaranensis)...[Hi-4],
pachycladus (See Haageocereus pachycladus, L. yauyosensis).....[Bo-2],
paradoxus (See Haageocereus paradoxus, Haageocereus paradoxus, L. paradoxus),
parvitesselatus.....[Hi-4],
peculiaris (See Haageocereus peculiaris, L. cantaensis) [Bo-2], [Bo-6], [Hi-4],
piscoensis (See Oreocereus piscoensis)...[DD], [Bo-2],
pullatus (See Borzicactus acanthurus), [Bo-6],
" -v- brevispinus (See Borzicactus acanthurus), [Bo-2],
" -v- fulviceps (See Borzicactus acanthurus), [Bo-2],
puquiensis.....[Hi-4],
rhodoflorus.....[Bo-2],
riomajensis.....[Bo-2],
sextonianus (See Borzicactus sextonianus, Erdisia sextonianus).....[Bv-1], [Hi-4], [Hl 4-1982],
splendens (See Borzicactus gracilis, L. gracilis), [Bo-2],
sulcifer (See Borzicactus sulcifer), [Ay], [Bo-2], [Hx],
" -v- longispinus.....[Bo-2],
tesselatus (See Borzicactus fieldianus -v- tesselatus, Borzicactus tesselatus, Clistanthocereus tesselatus),
" -v- spinosior (See L. granditesselatus),
trujilloensis.....[Hi-4],

LOXANTHOCEREUS.....(Cont'd),

variabilis (See Haageocereus variabilis), [Hi-4],
xylorhizus (See Haageocereus xylorhizus)....[Hi-4],
yauyosensis (See L. pachycladus),

Spec. FR-181.....(Bo-2],
Spec. FR-319.....[Hi-4],
Spec. ....[Bo-2],
Spec. Arequipa.....[Af 52-3],

LYMANBENSONIA

micrantha (See Acanthorhipsalis micrantha, Rhipsalis micrantha).....[Af 56-3],

MACHAEROCEREUS

eruca (See Cereus eruca, Stenocereus eruca), [I], [DD], [NN], [OO], [RR], [Ay], [Az], [Bt], [Cz], [Gr], [Hx*], [Hy], [Jd], [Af 52-2], [Ah-1985], [Aj 33-7-1982], [Aw 3-1985], [Bi 1-4], [Cu-2-1972], [Dg 7-3-1985],
gummosus (See Cereus gummosus, Stenocereus gummosus), [D], [CC], [OO], [RR], [Ay], [Az], [Bl], [Cx], [Dq], [Gr], [Hy], [Af 14-7], [Af 60-2], [Aw 3-1962],
" -v- tortuosus.....[Af 37-6],
" -f. cristata.....[Af 2-1], [Af 30-3], [Aw 1-1965] [Aw 3-1973], [Aw 4-1986], [Cu 6-2-1937],

MAIHUENIA

albolanata.....
" -v- viridulispina.....
brachydelphys.....[Bk], [Bo-1],
cumulata.....
latispina.....
patagonica (See M. tehuelches), [Bo-1], [Hy], [Ij],
philippii [Bo-1], [Bt], [Af 59-1],
poeppigii (Opuntia poeppigii), [D], [RR], [Ay], [Bl], [Bo-1], [Bt], [Dq], [Hv], [Hx*], [Ij], [Jd], [Af 55-6], [Bi 1-4], [Bi 4-1],
tehuelches (See M. patagonica), [RR], [Bl], [Iu], [Af 6-7],
valentinii.....[DD], [RR], [Hi-2],

MAIHUENIOPSIS

albomarginata.....
archiconoidea (See Tephrocactus conoideus),
[Hi-3],
atacamensis (See Opuntia atacamensis, Tephrocactus atacamensis, Tephrocactus chilensis),
[Hi-3],
boliviana.....[Af 59-1],
camachoi (See Opuntia camachoi, Tephrocactus camachoi),
colorea (See Tephrocactus colorea),
conoidea (See Tephrocactus conoideus),
crassispinus.....[Hi-3],
darwinii (See Opuntia darwinii, Tephrocactus darwinii),
domeykoensis (See Opuntia domeykoensis),
[Hi-3],
grandiflora.....[Hi-3],
hypogaea (See Opuntia hypogaea,
[Hi-2],
leoncito (See Opuntia leoncito, Tephrocactus leoncito),
[Hi-3],
leptoclada (See Opuntia leptoclada),
mandragora (See Tephrocactus heteracanthus, Tephrocactus mandragora).....[Hi-2],
molfinoi (See Opuntia molfinoi),
[DD], [Bo-1], [Dr*], [Hx*],
minuta (See Tephrocactus mandragora).....[Af 59-1],
neuquensis (See Opuntia neuquensis, Tephrocactus neuquensis),
ovallei (See Opuntia ovallei, Tephrocactus ovallei),
ovata (See Opuntia russelii, Tephrocactus russelii),
" -f. calva.....
" -f. sterilis (See Opuntia grata).....[Hi-3],
rahmeri (See Opuntia rahmeri),
tarapacana (See Opuntia tarapacana, Tephrocactus tarapacana),
wagenknechtii (See Opuntia wagenknechtii),

MALACOCARPUS

aciculatus (See Wigginsia erinacea),
apricus (See Notocactus apricus),
arechavaletai (See Notocactus arechavaletai, Wigginsia arechavaletai, Wigginsia maldonadensis),
[E], [Bo-3], [Dq*], [Hr], [Ba 6-1977],
beltranii (See Wigginsia beltranii),
bezrucii (See Wigginsia bezrucii)....[Hr],
calispina (See Malacocarpus arechavaletai, Notocactus arechavaletai, Wigginsia arechavaletai),
catamarcensis (See Pyrrhocactus catamarcensis, Wigginsia catamarcensis, Pyrrhocactus setiflorus).....[RR],

MALACOCARPUS.....(Cont'd),

camantii (See Wigginsia camantii),
concinnus (See Notocactus concinnus),
corynodes (See Notocactus corynodes, Wigginsia corynodes),
[E], [PP], [Bo-3], [Bv-1], [Hx*], [Af 29-5],
" -v- erinaceus (See M. erinaceus, Wigginsia acuata, Wigginsia erinacea),
courantii (See Wigginsia courantii, Wigginsia tephracantha -v- courantii),
curvispinus (See Horridocactus curvispinus, Pyrrhocactus curvispinus),
erinaceus (See Echinocactus erinaceus, Wigginsia acuata, Wigginsia erinacea),
fricii (See Wigginsia fricii),
[Bo-3], [Hr],
hennisii (See Notocactus vorwerkianus, Wigginsia vorwerkiana),
islayensis (See Islaya islayensis),
kovaricii (See Notocactus kovaricii, Wigginsia kovaricii),
langsdorfii (See Wigginsia langsdorfii, Wigginsia polyacantha),
leprosorum (See Notocactus leprosorum, Wigginsia leprosorum),
[Hu],
leucocarpus (See Wigginsia leucocarpus),
linkii (See Notocactus linkii),
[RR], [Af 4-10],
longispinus (See Notocactus longispinus, Wigginsia longispina),
maassii (See Echinocactus maassii, Parodia maassii, Wigginsia maassii),
macrocanthus (See M. sellowii -v- macrocanthus, Wigginsia macrocanthus, Wigginsia sellowii -v- macrocanthus),
[DD], [Bo-3],
macrogonus (See Wigginsia macrogona),
maldonadensis (See M. arechavaletai, Wigginsia maldonadensis, Wigginsia arechavaletai),
mamillarioides (See Neoporteria mamillarioides, Wigginsia mamillarioides),
mammulosus (See Notocactus mammulosus),
[RR], [Dl*],
" -v- pampeanus (See Notocactus mammulosus -v- pampeanus),
martinii (See Wigginsia sessiliflora -v- martinii),
napina (See Neoporteria napina, Thelocephala napina),

nigrispinus (See Eriocactus nigrispinus, Eriocactus schumannianus -v- nigrispinus, Notocactus nigrispinus),
[Af 6-7],
orthacanthus (See Echinocactus orthacanthus, Notocactus orthacanthus, Wigginsia orthacantha),
ottonis (See Notocactus ottonis).....[Dl*], [Af 4-10],
pauciareolatus (See Wigginsia sessiliflora -v- pauciareolata),
[F], [Af 16-6], [Iv 4-1957],
polyacanthus (See M. erinaceus, Wigginsia langsdorfii, Wigginsia polyacantha),
prolifera (See Notocactus prolifera, Wigginsia prolifera),
" -v- longispina (See Notocactus prolifera -v- longispina, Wigginsia prolifera -v- longispina),
pulcherrimus (See Frailea pulcherrima, Wigginsia pulcherrimus),
[RR],
reichei (See Neochilenia reichei, Neoporteria reichei, Thelocephala reichei),
[RR], [Hy], [If],
rubricostatus (See Wigginsia rubricostata),
scopa (See Notocactus scopa).....[Dn],
sellowii (See Echinocactus sellowii, Notocactus sellowii, Wigginsia sellowii),
[Bv-1], [Dq], [N-1980], [Af 29-5],
" -v- courantii (See M. courantii),
" -v- macrocanthus (See M. macrocanthus),
" -v- macrogonus (See M. macrogonus),
" -v- turbinatus (See M. turbinatus),
" -v- vorwerkiana (See M. vorwerkiana, Notocactus vorwerkianus, Wigginsia vorwerkiana),
sessiliflora (See Notocactus sessiliflorus, Wigginsia sessiliflora).....[EE], [Bo-3], [Jd],
" -v- martinii (See Wigginsia sessiliflora -v- martinii),
" -v- pauciareolata (See M. pauciareolata, Wigginsia sessiliflora -v- pauciareolata),
stegmannii (Invalid name - See Wigginsia stegmannii),
[Bo-3], [Hx*],
strausianus (See Neoporteria strausiana, Pyrrhocactus strausianus),
submammulosus (See Notocactus submammulosus),
[Cu 6-1-1937],
" -f. cristata.....[Bl],

MALACOCARPUS.....(Cont'd),

tephracanthus (See Notocactus acuatus, Wigginsia acuata, Notocactus tephracanthus, Wigginsia tephracantha).....[RR], [Iu],
" -v- courantii (See M. courantii, Wigginsia courantii),
" -v- depressus (See Wigginsia tephracantha -v- depressus),
tuberisulcatus (See Horridocactus horridus, Pyrrhocactus horridus, Pyrrhocactus tuberisulcatus),
turbinatus (See Wigginsia turbinata),
vorwerkianus (See Notocactus vorwerkianus, Wigginsia vorwerkiana)....[Bo-3], [Bv-1], [Cz], [Af 29-5],

MAMILLOPSIS

diguettii (See M. senilis, Mammillaria diguettii, Mammillaria senilis),
senilis (See Mammillaria senilis).....[D], [X], [DD], [HH], [NN], [PP], [RR], [Ar], [Ay], [Cx], [Hr], [Hx], [Il], [Jd], [Ae-1981], [Af 57-4], [Af 50-2], [Ag-1981], [Aj 28-5-1977], [Aj 30-8-1979], [Aj 31-1-1980], [Au 35-2], [Bm-1987], [Bi 6-2],

MAMMILLARIA

acanthoplegma (See M. elegans, M. haageana).....[Hy],
" -v- abducta (See M. haageana),
affinis (See M. polythele).....[Bv-1], [Io*],
alamensis (See M. sheldonii, M. pseudoalamensis), [VV], [Hl 5-1981],
alba (See M. albiflora, M. rhodantha -v- alba), [Hy],
" -f. cristata.....[Hy],
albata.....[Jb],
" -v- longispina.....
" -v- sanciro.....[Jb],
albescens (See Dolichothele albescens, M. camptotricha -v- albescens),
[D], [VV], [Bv-1], [Ah-1985], [Cu 8-3-1946],
albiarmata (See M. coahuilensis -v- albiflora, Porfiria coahuilensis -v- albiflora),
[VV], [Aq], [Hx*],
albicans.....[RR], [Aq], [Ar], [Av], [Gr], [Aa 5-1980], [Aa 4-1985], [Aj 29-5-1978],
albicolumnaria (invalid name - See Coryphantha albicolumnaris, Escobaria albicolumnaria),
[Y], [Df],
albicoma.....[Ct-1983], [Af 58-1], [Af 59-2],
albidula (Invalid name - See M. haageana),
[Bo-5], [Ct-1983],

MAMMILLARIA.....(Cont'd),

albiflora (See M. herrerae -v- albiflora, M. alba),
[Ar], [Hl 6-1973],
albilanata (See M. fauxiana),
[D], [VV], [Aq], [Ar], [Bo-6], [Bv-1],
[Ct-1983], [Hy], [N-1980], [Aa 4-1980],
[Ae-1981], [Af 55-2], [Db 6-1979], [Db 11-1979],
" -v- reppenhagenii (See M. reppenhagenii),
" -v- tegelbergiana (See M. tegelbergiana),
albiseta.....
aljibensis (See M. perbella),
[VV], [Bv-1], [Hy], [Ba 9-1980],
alpina (See M. kraehenbuehlii),
amoena (See M. discolor),
[RR], [VV],
ancistroides (See M. glochidiata, or M. jaliscana),
anebeliana.....
angelensis (See M. verhaertiana),
[VV], [Aq], [Ar], [Aa 1-1983], [Aw 4-1963],
" -v- estebanensis (See M. estebanensis),
angularis (See M. compressa),
anniana.....[Ab-1985], [Af 53-2], [Aj 33-4-1982],
[Hs III 4-1982],
apamensis......[Jb],
" -v- pratensis.....
apozolensis.....[Jb].
" -v- saltensis.....[Jb],
applanata (Invalid name - See M. heyderi -v- applanata)
[J*], [U], [LL*], [RR], [Bv-1], [Cx], [Hy],
[Aw 2-1964],
arida (See M. baxteriana),
[CC], [RR], [Aa 6-1985], [Af 51-2], [Bi 6-2-1988],
arizonica (See Coryphantha vivipara -v- arizonica),
[Af 14-1],
armatissima (See M. gigantea).....[VV],
armillata.....[A], [G], [RR], [VV], [Aq], [Ay], [Bv-1],
[Ct-1983], [Hx*], [Hy], [Aa 3- 1981],
[Aa 1-1986], [Aj 10-12-1959],
" -v- bullardiana (See M. bullardiana),
" -v- capensis (See M. capensis),
" -v- cerralboa (See M. cerralboa),
" -v- fraileana (See M. fraileana),
ascensionis (See M. glassii -v- ascensionis),
[Aj 30-3-1979], [Hl 5-1982], [Hm 2-1979],
" -v- cerroviejo.....
" -v- dulcis.....
atroflorens (Invalid name - See M. mystax).....[Bo-6],
aureiceps (See M. rhodantha -v- sulphureus, M. rhodantha -v- aureiceps)....[RR], [Aq], [Aa 2-1987], [Aw 1-1976], [Hl 3-1977],

aureilanata (See Coryphantha pycnacantha),
[NN], [VV], [Ac], [Aq], [Ar], [Av], [Hv],
[Hx], [Hy], [Il], [Ae-1981]. [Af 53-6],
[Af 56-5], [Ba 4-1979], [Db 26th Anniv.
Edition],
" -v- alba (See M. aureilanata).....[Ar],
aureoviridis (See M. aurihamata),
[Du 1937:56 Descrip.], [Hx*],
auriareolis (See M. parkinsonii),
[Dv 48:412, 1933 Descrip.], [Iv 4-1959],
auricantha (See M. canelensis),
auricoma (See M. spinosissima),
aurihamata (See M. aureoviridis, M. boedekeriana),
[VV], [Aq], [Bv-1], [Ct-1983], [Cz], [Hu],
[Hx*], [Hy],
aurisaeta (Invalid name - See M. picta).....[Bo-6],
aurispina (See M. rekoi -v- aurispina)
auritricha (See M. canelensis).....[VV],
autumnalis (See M. mystax),
avila-camachoi (Invalid name - See M. cadereytensis -v-
avila-camachoi, M. parkinsonii)....[Bo-6],
aylostera (See Dolichothele beneckei, M. beneckei),
bachmannii (See M. caput-medusae, M. sempervivi),
backebergiana.....[NN], [Ay], [Ff], [Il], [Au 21:47 June
1966], [Aw 3-1966], [Db 4-1979], [Hl 6-1982],
" -v- backebergiana.....[Aq],
" -v- ernestii (See M. ernestii),
[Aq], [Af 51-3], [Hl 6-1982],
balsasensis (See M. beneckei)......[Dw 3: 121, 1931],
balsasoides (See M. beneckei).....[VV], [Aw 3-1967],
bambusiphila.....
" -v- parva.....
barbata (See M. orestera, M. garessii, M. chavezii,
M. morricalii, M. viridiflora),
[VV], [RR], [Aq], [Iz], [Aa 2-1980], [Aj 29-7-1978]
barkeri (Invalid name - See M. beneckei, M. guiengo-
lensis).....[Bo-6], [Cu 4-1962],
baumii (See Dolichothele baumii).....[E], [Aq], [Ay],
[Ct-1983], [Hv], [Hw-1981], [Hx], [Iz], [Af 59-
6], [Ba 5-1981], [Dg 7-1],
" -v- radiaissima.....
baxteriana.....[VV], [Aq], [Bv-1], [Hy], [Aa 6-1985],
[Af 6-1],
" -v- marshalliana (See M. marshalliana),
beiselii.....[F], [VV], [Ac], [Aq], [Ct-1983], [Af 52-1]
[Af 59-3], [Aj 30-3-1979],
" -v- albispina.....
bella (Backbg.).....[Aq], [Ar], [Bt], [Ct-1983], [Hx],
[Hy], [Il], [Af 56-5], [Db 6-1979],
bella (Br. & R.) Weniger, -(See Escobaria bella)..[Df],

MAMMILLARIA.....(Cont'd),

bellacantha (See M. canelensis).....[VV],
bellisiana (See M. sonorensis or M. tesopacensis)..[VV],
beneckei (See M. aylostera, M. balsasensis, M. balsasoides, M. barkeri, M. colonensis, M. guiengolensis, M. nelsonii, Dolichothele beneckei), [VV], [Ag], [Aq], [Av], [Ct-1983], [Iz], [Aa 5-1982], [Af 55-2], [Ag-1986], [Ba 4-1979],
" -v- multiceps.....(See M. guiengolensis),
bergii (See M. ortegae), [Fig. in Linnaea 12:10 t.2 - Fig. 7-1838],
berkiana (berkeana) (Lau 1245).....
bernalensis......[Jb],
bicolor (See M. geminispina),
" -v- longispina.....
" -v- nivea (See M. nobilis).....[Hy],
" -v- " -f. cristata.....[Hy],
" -v- nobilis (See M. nobilis).....
" -f. cristata.....[Hy],
bicornuta (Invalid name - See M. bucareliensis, M. magnimamma).....[Bt],
blossfeldiana (See M. shurliana), [F], [VV], [Ac], [Aq], [Ar], [Bt], [Bu], [Ct-1983], [Hx*], [Hy], [Iz], [Af 4-11], [Af 55-1], [Ba 10-1976], [Hh July 1974],
" -v- shurliana (See M. shurliana).....[N], [Af 13-5], [Af 24-3], [Ba 9-1980],
bocasana.....[C], [I], [R], [AA], [NN], [VV], [Aq], [Ar] [Ay], [Be*], [Bu], [Bv-1], [Cm], [Cr], [Cy] [Cz], [Gr], [Hx*], [Hy], [Il], [Jd], [Af 5-10], [Af 51-6], [Af 52-2], [Hw-1989],
" -v- inermis (See M. bocasana).....[VV], [Ah],
" -v- kunzeana (See M. kunzeana).....[VV],
" -v- longicoma (See M. longicoma).....
" -v- multilanata.....[Ah],
" -v- murivora.....[Aj 28-4-1977],
" -v- rosea.....[Ba 7-1977],
" -v- roseiflora.....
" -v- sericata.....
" -v-splendens.....[TT], [Ae-1981], [Aj 23-6-1972]
" -f. (Ed Hummel).....[VV], [Ar], [Bu],
" -f. ISI - 1424.....[Af 56-2],
" -f. Hunt - 9447.....[Ct-1983],
" -f. cristata.....[Ab-1985], [Af 27-3],
" -f. mutation.....[Aj 28-11-1977],
bocensis (See M. neoschwarzeana, M. rubida), [E], [VV], [Aq],
" -v- movasana.....[Jb].
" -v- rubida (See M. rubida),

MAMMILLARIA......(Cont'd),

boedekeriana (See M. aurihamata),
[C], [RR], [VV],
bogotensis (See M. columbiana -v- bogotensis),
[VV], [Bv-1],
bombycina.....[C], [H], [M*], [W], [NN], [RR], [TT],
[VV], [Aq], [Ar], [Av], [Ay], [Bv-1],
[Cr], [Cs], [Gr], [Hu], [Hv], [Hx], [Hy],
[Iz], [Jd], [Aj 28-7-1977], [Bi 1-4],
[Bm-1986],
" -v- flavispina.....[Be*],
" -f. cristata.....
bonavitii (Invalid name - See M. rhodantha),
[Bo-5], [Ia], [Jb], [Aj 33-10-1982], [Ba 12-1980], [Db - 26th Anniv. Edition),
boolii.....[G], [UU], [Aq], [Av], [Ay], [Bt], [Ff], [Hb]
[Hx], [Il], [Iz], [Jd], [Aa 1-1985], [Ad Jan. 1980], [Ag-1986], [Hl 3-1982],
bosshardii (Hybrid).....[Iv 6-1950],
brandegeei.....[VV], [Aq], [Ct-1983], [Cu 3-1968],
" -v- gabbii (See M. gabbii),
[Hy], [Af 41-2],
" -v- glareosa (See M. glareosa),
[Aq], [Bo-5],
" -v- lewisiana (See M. brandegeei).....[Aq],
" -v- magdalensis.....
brauneana.....[K], [VV], [Aq], [Ar], [Ct-1983], [Hx*],
[Hy], [Ia], [Il], [Hl 3-1968],
bravoae (See M. hahniana),
[E], [VV], [Aq], [Bv-1], [Af 52-2], [Cu 8-4-1946],
brevicrinita.....[Jb],
brunispina (See M. perbella),
" -v- major.....[Iv 10-1960],
bucareliensis (See M. magnimamma),
[F], [VV], [Ar],
" -v- bicornuta (See M. bicornuta),
" -v- multiflora.....[Jb],
" -v- tamaulipa.....[Jb],
buchenaui (See M. crucigera).....[DD], [Hl 6-1982],
bullardiana (See M. hutchisoniana),
[Ar], [Hy], [Il], [Af 6-1], [Ba 8-1977],
[Cu 1-1973], [Hl 2-1975],
buxbaumeriana (buxbaumiana) - (See M. boolii)....[Jb],
cadereytana (See M. pygmaea),
cadereytensis (See M. parkinsonii),
[VV - Fig. 222], [Aq], [Bv-1], [Ct-1983],
[Hy], [Ae-1981], [Af 55-4],
" -v- avila-camachoi (See M. avila-camachoi),
" -v- quadrispina (See M. parkinsonii).....[VV],

caerulea (See M. chionocephala,.....
[VV],
calacantha (See M. rhodantha -v- calacantha),
[TT], [VV], [Aq], [Ct-1983], [Hx*], [Hy],
" -v- rubra.....[Hy],
californica (See M. grahamii -v- arizonica),
calleana (See M. wildii -f. calleana),
[Aq], [Bo-5], [Iv 3-1952],
" -v- albescens.....[Aq],
" -v- schwarzii (See M. schwarzii).....[Aq],
camptotricha (See Dolichothele albescens, Dolichothele camptotricha, M. albescens, M. decipiens -v- camptotricha),
[E], [AA], [NN], [RR], [VV], [Aq], [Ay],
[Bv-1], [Cz], [Hv], [Hx], [Jd], [Af 6-7],
[Af 53-2],
" -v- albescens (See M. albescens),
[Hl 4-1974],
" -v- albicans.....
" -v- subinermis.....[Af 57-2], [Iv 11-1947],
" -f. (Mme Marnier).....[Ah-1981], [Af 53-2],
candida (See Mammilloydia candida),
[E], [H], [L], [M*], [NN], [RR], [VV], [Aq],
[Ar], [Av], [Ay], [Bt], [Bv-1], [Cz], [Hx],
[Hy], [Il], [Jd], [Ae-1981], [Aj 31-1-1980],
" -v- caespitosa.....[AA], [Af 42-6],
" -v- candida.....
" -v- estanzuelensis (See M. ortiz-rubiona),
" -v- ortiz-rubiona (See M. ortiz-rubiona, Mammilloydia ortiz-rubiona),
" -v- rosea.....[TT], [Bt], [Hy], [Aj 27-12-1976],
[Hl 5-1972],
" -v- senilis (See Mamillopsis senilis, M. senilis),
[TT],
canelensis (See M. standleyi),
[VV], [Ac], [Hy], [Jd], [Aa 4-1980], [Ae-1981], [Ah-1985], [Aj 31-6-1980], [Bi 2-3],
capensis.....[G], [VV], [Aq], [Ar], [Hy], [Iz],
[Aa 5-1984], [Aa 3-1986], [Af 4-11],
" -v- pallida.....[Ar],
caput-medusae (See M. sempervivi).....[Bv-1], [Gr],
carmenae.....[Aq], [Bv-1], [Ct-1983], [Hx*], [Il], [Iz],
[Jd], [Aa 6-1983], [Ad 3-1983], [Af 50-2],
[Af 50-4], [Af 52-2], [Ag-1985, 1988],
[Aj 28-12-1977], [Aw 4-1988],
" -f. cristata.....[Ab-1985], [Af 59-4], [Ah-1986]
carnea.....[RR], [VV], [Aq], [Av], [Bv-3], [Cx], [Gr],
[Hu], [Hv], [Hy], [Ia], [Ba 4-1980],

carnea.....(cont'd),
" -v- aeruginosa (See M. carnea),
" -v- cirrosa (See M. carnea),
" -v- orcuttii (See M. orcuttii),
" -v- robustispina (See M. carnea),
" -v- subtetragona.....[Hx*],
carretii (See M. saffordii),
[CC], [NN], [RR], [VV], [Ac], [Aq], [Ar],
[Fg], [Hx*], [Hy], [Aa 1-1981], [Aa 6-1981],
[Ba 12-1981],
casio (See M. mystax).....[JJ], [Bo-5], [Ae-1981],
" -v- longispina.....[Bo-5],
celsiana (See M. muehlenpfordtii, M. neopotosina),
[C], [H], [R], [NN], [RR], [VV], [Ay], [Bv-1],
[Ck], [Cx], [Cz], [Dq], [Gr], [Hb], [Hx*],
[Hy], [Aw 4-1987],
" -v- guatamalensis (See M. ruestii),
" -v- longispina.....
" -f. cristata.....[Hy],
centralifera.....[Jb],
centraliplumosa (See M. spinosissima -v- centrali-
plumosa).....[Ct-1983], [Il], [Aw 2-1971],
[Hl 6-1982],
centricirrha (See under M. magnimamma),
[J*], [EE], [QQ], [Ay], [Be*], [Cr], [Cs],
[Hx], [Hy], [N-1980],
" -v- bockii (See M. magnimamma),
[Hx*], [Aj 18-10-1967],
" -v- divergen (See M. magnimamma).....[Hy],
" -v- gladiata (See M. magnimamma).....
" -v- glauca (See M. magnimamma).....
" -v- inermis.....[Hy],
" -v- krameri (See M. magnimamma).....
" -v- macracantha (See M. magnimamma).....[Bk],
" -v- neumanniana.....
" -v- pentacantha.....
" -v- recurva.....
" -v- versicolor.....
cephalophora (See M. aureilanata),
cerralboa (See M. armillata -v- cerralboa),
[RR], [Aq], [Hy], [Aa 5-1980], [Af 53-1],
chapinensis (See M. voburnensis),
chavezei (Invalid name - See M. viridiflora)....
[Au 18:8 Mar. 1963], [Hl 6-1972],
chica (See M. viescensis).....[Jb],
chionocephala (See M. caerulea),
[E], [S], [T], [RR], [VV], [Ar],
[Bv-1], [Hx*], [Hy], [Aa 1-1982],
[Ae-1981], [Af 3-3], [Af 52-5],
[Af 56-6], [Aw 3-1987],

MAMMILLARIA.....(Cont'd),

cirrhifera (See M. compressa),
" -v- longispina.....[Af 19-7],
claviformis.....[Jb],
coahuilensis (See M. schwartzii, Porfiria coahuilensis, Porfiria schwartzii),
[Aq], [Hx*], [Jd], [Ae-1982], [Af 56-4],
" -v- albiflora (See M. albiarmata, Porfiria coahuilensis -v- albiflora),
" -v- albiarmata (See M. albiarmata, M. coahuilensis -v- albiflora),
collina (See M. haageana, M. elegans).....[L], [JJ], [RR], [VV], [Bo-6], [Cx], [Dq], [Hy], [Jd],
collinsii (See M. voburnensis -v- collinsii).....[NN], [RR], [TT], [VV], [Ac], [Aq], [Bv-1], [Cx], [Hu], [Hy], [Jd], [Af 6-5], [Aw 3-1986], [Ba 7-1978],
" -v- chiapas.....[Cu 3-1963],
colonensis (See M. beneckei).....[VV],
columbiana (See M. soehlemannii, M. hennisii, M. tamajonis (tamayonis)),
[MM], [VV], [Aq], [Ay], [Cz], [Hy], [Bo-6],
" -v- albescens (Invalid name).....[Bo-6],
" -v- bogotensis (See M. bogotensis),
[Bo-5], [Bo-6], [Bv-1], [Hx*],
" -v- ruestii (See M. ruestii),
" -v- yucatanensis (See M. yucatanensis),
compacticaulis.....[Jb],
compressa.....[E], [R], [T], [RR], [TT], [VV], [Aq], [Ar], [Ax], [Ay], [Bv-1], [Ct-1983], [Cx], [Cz], [Gr], [Ha], [Hx], [Iz], [N-1980], [Ad April 1979], [Ae-1981], [Ag-1984],
" -v- fulvispina.....
" -v- longiseta.....[Dq],
" -v- longispina.....[Ih],
" -v- rubrispina.....[Bv-1],
" -v- triacantha.....
" -f. (long-spined form) - (See M. cirrhifera) [Af 19-7],
" -f. cristata.....[Am 8-1984],
confusa (See M. karwinskiana)....[RR], [Ay], [Bv-1], [Cz], [Hy], [Aw 2-1962], [Cu 3-1963],
" -v- centrispina.....[VV], [Hy],
" -v- conzattii (See M. conzattii),
[VV], [Ba 7-1981],
" -v- longispina.....[Aw 2-1962],
" -v- robustispina.....[VV], [Hy],
conoidea (See Neolloydia conoidea).....[Io],
conspicua (See M. elegans, M. haageana),
[R], [VV], [Ct-1983], [Hu], [Hx], [Ba 4-1980]

conspicua.....(cont'd),
" -v- vaupelii (See M. vaupelii).....
conzattii (See M. karwinskiana).....[RR], [Cx],
coronaria (See M. spinosissima),
[CC], [Be*],
cowperae (See M. moelleriana),
[H], [Ac], [Ar], [Bt], [Hx*], [Hy], [Aa 2-1980], [Hl 2-1973], [Cu 21:58, 1959],
craigii.....[G], [VV], [Jd], [Ad Dec. 1979], [Af 14-8],
crassa.....[Jb],
crassimamillis.....[Jb],
crassior.....[Jb],
crassispina (See M. fuscata, M. rhodantha).....[Bv-1],
criniformis (See M. wildii, M. crinita)....[F], [Io*],
crinita (See M. wildii -f. crinita, M. criniformis),
[VV], [Aq], [Io*], [Ab-1985],
" -v- pauciseta.....
crispiseta (See M. mystax).....[VV],
crocidata (See M. obconella),
[E], [H], [RR], [VV], [N-1980],
" -v- quadrispina.....[Aw 4-1966],
crucigera (See M. buchenaui).....[G], [Z], [JJ], [NN],
[VV], [Aq], [Ar], [Av], [Cx], [Hv], [Iz],
[Jd], [Ae-1981], [Af 6-1], [Aj 31-2-1980],
[Hl 6-1983], [Hm 1-1978],
" -f. cristata.....[VV],
cylindrica (See M. hamata),
dasyacantha (See Coryphantha dasyacantha, Escobaria dasyacantha).....[Y], [Df],
dawsonii (See M. glareosa),
[VV], [Ay], [Hx*], [Af 32-6], [Cu 3-1961],
dealbata (See M. elegans, M. haageana, M. dealbata),
[RR], [Ar], [Av], [Cx], [Hb], [N-1980], [Af 6-7],
decipiens (See Dolichothele decipiens),
[C], [RR], [VV], [Aq], [Bk], [Iu], [Af 1-9],
" -v- albescens.....
" -v- camptotricha (See Dolichothele camptotricha, M. camptotricha).....[Iz],
deherdtiana.....[Ar], [Av], [Hx*], [Ae-1981], [Af 41-1],
[Ba 9-1978], [Ba 1-1981], [Db 4-1979],
[Hm 2-1978],
" -v- deherdtiana.....[Aq], [Iz],
" -v- dodsonii (See M. dodsonii),
[Aq], [Ar], [Hv], [Af 51-3],
deitrichae (deitrichiana) (See M. parkinsonii),
[Cm], [Af 8-2],
deliusiana (See M. bella),
densispina.....[T], [NN], [RR], [VV], [Aq], [Ar], [Dq],
[Hy], [Il], [Jd], [Ag-1982],
" -v- mieheana (See M. mieheana).....[Iz],

MAMMILLARIA.....(Cont'd),

denudata (See M. lasiacantha -v- denudata),
[Bv-3], [Bz], [Df], [Hy], [Af 52-1],
[Hl 4-1973],
diguetti (See Mamillopsis diguettii, Mamillopsis
senilis, M. senilis),
dioica.....[L], [M*], [LL], [RR], [VV], [Aq], [Ar],
[Ax], [Ca], [Cb], [Ct-1984], [Hx*], [Hy],
[Ij], [Aa 3-1984], [Af 2-2], [Af 54-3],
[Flower - Ba 10-1980], [Cu 1-1971], [Db -
26th Anniv. Edition),
" -v- estebanensis (See M. estebanensis),
" -v- incerta.....[Do],
" -v- verhaertiana.....
" -f. cristata.....[Af 34-5],
discolor.....[VV], [Aq], [Ar], [Bo-6], [Hu], [Hy], [Jd],
" -v- esperanzaensis (See M. esperanzaensis),
" -v- longispina.....[Jb],
" -v- multispina.....[Jb],
" -v- ochoterenae.....
" -v- schmollii.....
" -f. cristata.....[K],
dixanthocentron.....[DD], [JJ], [NN], [Aq], [Ar],
[Ct-1984], [Hu], [Hv], [Hx*],
[Aa 1-1980], [Ae-1981], [Af 57-3],
[Af 60-3], [Aw 2-1964], [Db 11-
1979], [Hl 2-1976],
" -v- flavicentra (See M. flavicentra).....[Jb],
" -v- robusta.....
" -f. rubra.....
" -f. cristata.....[Hn],
dodsonii (See M. deherdtiana -v- dodsonii),
[Z], [Ar], [Av], [Ct-1984], [Hx*], [Af 42-6],
[Af 51-3], [Aj 27-11-1976], [Au 36-4],
[Aw 1-1970], [Aw 15:36 Jan-Mar. 1971],
[Ba 9-1978], [Ba 1-1981],
dolichocentra (See M. obconella, M. tetracantha),
[Bk], [Bu], [Bv-3],
donatii (See M. collina, M. haageana),
[VV], [RR], [Iq], [Hl 1-1970],
droegeana (See M. microhelia),
[VV], [Hx*], [Jb], [Ba 4-1981],
dumetorum (See M. schiedeana),
[NN], [Ar], [Af 39-2], [Aj 8-7-1957],
duncanii (Invalid name - See Coryphantha duncanii,
Escobaria dasyacantha -v- duncanii, Escobaria
duncanii, Escobesseya duncanii)....[Y], [Df],
duoformis.....[Aq], [Af 55-3], [Aw 1-1966], [Ba 4-1979],
" -v- rectiformis.....[Bo-5], [Dx No. 2:60 Tab.19,
Fig. 5, 1948],
" -v- xuchiapensis.....[Jb],

durangicola.....[Jb],
durispina (See M. polythele, M. calacantha, M. moglicherweise).....[VV], [Aq], [Bv-1], [Ct-1984] [Hy], [Jd], [Aa 6- 1983],
duwei (See M. nana).....[Ag-1988], [K&L Nursery catalog -1988],
dyckiana (See M. elegans, M. haageana),
ebenacantha (See M. karwinskiana).....[Bo-6],
echinaria (See M. elongata -v- echinaria), [Ay], [Bv-1], [Ct-1984], [Cx], [Cz],
" -v- rufrocrocea (See M. elongata -v- rufrocrocea).....[RR], [VV],
echinata.....(See M. elongata),
echinus (See Coryphantha echinus, Coryphantha cornifera -v- echinus).....[Y], [Df],
egregia (See M. lasiacantha).....[Aq], [Bo-5], [Ff], [Hx], [Aa 3-1980], [Hl 4-1973],
" -f. cristata.....[Hy], [Aa 5-1981],
ehreteana (See M. magnimamma).....[Af 30-2],
eichlamii (See M. voburnensis -v- eichlamii)....[RR], [VV], [Aq], [Ay], [Aa 4-1981], [Af 57-2],
" -v- albida (See M. eichlamii),
ekmannii (See M. mammillaris),
elegans (See M. dealbata, M. haageana, M. dyckiana, M. sanangelensis), [C], [I], [NN], [RR], [VV], [Aq], [Ar], [Av], [Ay], [Be*], [Bv-1], [Cr], [Cx], [Hu], [Hv], [Hw-1981], [Hx*], [Jb], [Jd], [Aw 1-1964], [Aw 3-1986], [Ba 2-1981], [Db 7-1979],
" -v- dealbata (See M. dealbata, M. haageana) [Dq],
" -v- donatii (See M. donatii).....[NN],
" -v- longicaudata.....[Jb],
" -v- lupina.....[Jb],
" -v- oaxacensis.....[Cu 1-1963],
" -v- schmollii (See M. haageana), [VV], [Ba 2-1981],
" -v- supertexta (See M. supertexta)....[VV],
" -v- teyuca.....[Jb],
" -f. cristata.....[Ah-1985],
elongata (See Leptocladodia elongata, M. echinaria, M. echinata).....[C], [M*], [R], [X], [AA], [RR], [VV], [Aq], [Ay], [Bv-1], [Cm], [Ct-1984], [Cx], [Cz], [Gr], [Hv], [Hx*], [Il], [Ja], [Jd], [Af-6-5], [Af -51-6], [Ah-1986], [Am 8-1984], [Ba 8-1977],
" -v- anguinea (See M. echinaria).....
" -v- densa (See M. mieheana).....[Bo-5],
" -v- echinaria (See M. echinaria)..[Ar], [Bo-5],
" -v- echinata.....(See M. elongata),

MAMMILLARIA.....(Cont'd),

elongata.....(cont'd),
" -v- intertexta.....[Ar], [Bo-5],
" -v- minima.....[Bv-1], [Dq],
" -v- obscuriosa.....
" -v- rufrocrocea.....[Bo-5], [Bt], [Il],
" -v- stella-aurata.....[H], [CC], [Bo-5],
" -v- straminea.....[Bo-5],
" -v- subcrocea.....[Ar], [Bo-5],
" -v- tenuis (See M. fuscata).....[Bo-5],
" -v- viperina.....[Bo-5],
" -v- .....[Ar],
" -f. cristata.....[Ar], [N-1980], [Ah-1983],
eppeliana.....[Cu 1-1962],
erectacantha (See M. wiesingeri -v- erectacantha)
[NN], [Aq], [Aw 1-1968],
erectohamata (See M. aurihamata),
[VV], [Bv-1], [Aj II-9-1930],
eriacantha.....[RR], [Aq], [Ct-1984], [Hy], [Jd],
[Aw 3-1986], [Hl 3-1978],
" -v- flavispina.....
ernestii (See M. backebergiana -v- ernestii),
[Ac], [Hx*], [Il], [Aw 2-1971], [Aw 1-1978],
[Ba 11-1978], [Ba 9-1979],
" -v- duocentralis.....
erythra.....[Jb],
erythrocalix (erythrocalyx) - (See M. duoformis),
[Aw 1-1966],
" -v- robusta.....
erythrosperma (See M. multiformis),
[C], [NN], [VV], [Aq], [Ar], [Cj], [Cr],
[Af 52-2],
eschanzieri (See M. monancistracantha)
[Ar], [Hr], [Hl 2-1979],
eshauzieri.....
esperanzaensis (See M. discolor)....[JJ], [Ct-1984],
esseriana (See M. karwinskiana)
[VV], [Bo-6], [Hx*], [Dy 3-289 Fig. 1928],
estanzuelensis (See M. candida).....[Bo-5],
estebanensis (See M. dioica -v- estebanensis)
[G], [Aq], [Af 39-1], [Hl 5-1982],
euthele (See M. melanocentra).....[Hx*],
evermanniana.....[RR], [VV], [Aq], [Hy], [Ae-1981],
falsicrucigera (See M. buchenaui, M. crucigera),
[Bo-6], [Aj 15-10-1964],
fasciculata (Invalid name - See M. thornberi),
[A], [B], [D], [P], [CC], [LL], [NN], [RR],
[VV], [An], [Ad Sept. 1979], [Af 3-11],
[Ba 6-1977],
felicis (See M. nejapensis),
felipensis (See M. bocasana).....[Jb],

fera-rubra (See M. rhodantha -v- fera-rubra),
[R], [VV], [Aq], [Ct-1984], [Aw 2-1976],
fertilis (See M. backebergiana).....[Bo-5],
fischeri (See M. karwinskiana).....[Cu-1-1970],
fittkaui.....[Z], [Aq], [Hx*], [Il], [Iz], [Af 52-2],
[Ag-1985, 1988], [Aj 23-3-1972], [Hl 6-1977],
flava (See M. spinosissima),
flavescens (Haw.) .....(See M. nivosa)
[NN], [Bo-5], [Af 52-1], [Ba 3-1978],
" -v- nivosa (See M. nivosa).....Hx*], [Il],
flavicentra (See M. dixanthocentron -v- flavicentra)..
[Hx*], [Jd], [Af 60-3],
flavihamata (See M. gilensis).....[Bo-6],
flavovirens (Discard name - See M. magnimamma group),
[VV], [Cx], [Ah-1980],
floresii (See M. canelensis).....[Bo-5], [Iv 9-1952],
" -v- hexacentra.....
formosa.....[NN], [TT], [VV], [Aq], [Bt], [Ct-1984],
[Aa 3-1983], [Ae-1982], [Af 56-1], [Hl 4-1979],
" -v- brauneana.....
" -v- chionocephala.....[Hl 4-1979],
foveolata (See M. seitziana).....[Aw Vol. 27:85],
fragilis (See M. gracilis -v- fragilis, M. gracilis),
[S], [RR], [VV], [Bv-1], [Cz], [Af 6-5],
fragrans (Invalid name - Coryphantha fragrans, Coryphantha radiosa, Coryphantha vivipara -v-radiosa).....[Y], [Df],
fraileana.....[G], [CC], [NN], [VV], [Aq], [Cr],
[Ct-1984], [Gr], [Hx*], [Iz], [Aa 1-1981],
[Aa 3-1986], [Af 55-4], [Bm-1987],
freudenbergi.....
fuauxiana (See M. albilanata),
[Ay], [Hy], [Af 56-5], [Db 6-1979], [Iv 9-1952],
fuliginosa (See M. obconella, M. polythele, M. tetracantha).....[VV], [Aj 23-4-1972],
fuscata (See under M. aureiceps, M. rhodantha -v-sulphureus).....[Bv-1], [Hy], [Hw-1988],
" -v- russea.....[Bo-5],
" -v- sulphurea (See M. rhodantha -v- sulphureus),
fuscata-esperanza (See M. esperanzaensis),
fuscohamata (See M. jaliscana),
[DD], [Bo-6], [Ff], [Aj 15-1-1964],
gabbii (See M. brandegeei -v- gabbii).....
galeottii (See M. tetracantha -v- galeottii,
M. obconella -v- galeottii).....[Hy],
garessii (See M. barbata),
[Ct-1983], [Iz], [Aa 2-1980], [Af 42:115-117, 1971],

MAMMILLARIA.....(Cont'd),

gasseriana (See M. stella-de-tacubaya),
[VV], [Jd], [Ae-1981], [Aj 31-10-1980], [Hl 4-1973],
gasterantha (See M. spinosissima -v- virginis, M. virginis).....[Aj 31-5-1980],
gatesii (See M. petrophila -v- gatesii),
[VV], [Aa 2-1981], [Af 8-7], [Af 52-3],
gaumeri (See M. heyderi -v- gaumeri, M. heyderi -v- meiacantha),
[RR], [VV], [Aq], [Ct-1984], [Cx], [Hx*],
[Aa 3-1984], [Ae-1981], [Af 41-2], [De 4-1982],
geminispina.....[C], [M*], [N], [R], [V], [X], [NN],
[RR], [VV], [Aq], [Ar], [Bt], [Bu],
[Bv-1], [Cx], [Cz], [Hv], [Hx], [Iz],
[N-1980], [Af 16-3], [Af 55-6],
[Af 58-1], [Ag-1987], [Bi 1-2], [Bi 2-1]
[Bi 2-3],
" -v- brevispina.....[Ay],
" -v- infernillensis.....
" -v- longispina.....[Ad-1983],
" -v- nivea (See M. geminispina)
[VV], [Bu], [Bv-1],
" -v- " -f. cristata.....[Bl]
" -v- nobilis (See M. geminispina)
[Bo-5], [Ae-1981], [Hl 1-1965],
" -f. cristata.....[E], [NN], [Ia], [Af 16-3],
[Af 30-2],
gigantea.....[NN], [RR], [VV], [Aq], [Bv-1], [Ct-1986],
[Hv], [Hx*], [Hy], [Iz], [Aa 2-1981],
gilensis (See M. rettigiana),
[H], [VV], [Ba 11-1979],
ginsaumae (See M. discolor),
gladiata (See M. centricirrha -v- gladiata, M. magnimamma),
glareosa (See M. brandegeei -v- glareosa),
[Aq], [Bo-5], [Ct-1984], [Hy],
glassii.....[G], [Z], [VV], [Aq], [Ar], [Ay], [Ct-1983],
[Hx*], [N-1980], [Af 56-2], [Ah-1985],
[Aw 1-1971],
" -v- ascensionis.....[Af 52-2], [Ag-1982],
" -v- nominis-dulcis (See M. schwarzii),
[Af 57-5], [Iz],
" -v- nova (Cerro del Viejo).....[Ah-1985],
" -v- nova (Dulces Nombres),
[Ag-1988],
" -v- siberiensis (Lau 1322) (See M. schwarzii),
[Af 57-5],
" -f. (Lau-1186).....[Af 57-5],
glochidiata.....[F], [RR], [VV], [Ac], [Ar], [Hu], [Hx],
(cont'd) [Hy],

glochidiata.....(cont'd),
" -v- xiloensis.....[Jb],
glomerata (See M. ekmannii),
goldii (See M. saboae -f. goldii),
[G], [Z], [Aq], [Ar], [Ay], [Ha], [Hx],
[Aa 5-1981], [Ad Nov. 1980], [Af 51-1],
[Af 57-4], [Aw 4-1969], [Ba 10-1977], [Bi 2-1],
" -f. cristata.....[Aa 5-1981],
goodridgei (goodrichii).....[VV], [RR], [Aq], [Ar],
[Aa 2-1986], [Af 2-4], [Af 23-3],
" -v- hutchisoniana (See M. hutchisoniana),
" -v- rectispina.....[Af 23-3],
gracilis (See M. fragilis)
[C], [HH], [Aq], [Cg], [Gr], [Hx*], [Il],
" -v- fragilis (See M. fragilis).....[Ar],
" -v- gracilis.....[Bo-5],
" -v- pulchella.....[E], [H], [Aq], [Ay],
" -f. cristata.....[Ct-1984], [Aa 5-1981],
[Aj 8-12-1957],
" -f. monstrosus.....[Af 55-3],
graessneriana (See M. columbiana),
[DD], [RR], [VV], [Aq], [Hx*], [Hy],
grahamii (See Coryphantha grahamii, M. microcarpa)
[Q], [U], [An], [Bz], [Df], [Ds], [Ib],
" -v- grahamii.....[An], [Ax],
" -v- oliviae (See M. oliviae),
[Q], [An], [Ax], [Am 8-1983],
" -v- (Gl. & F. - 729).....[Af 56-1],
greggii (See Epithalantha micromeris -v- greggii)..[Fg],
grusonii (See M. mexicensis).....[Aq],
" -v- lewisiana.....
" -v- zeyeriana.....
gueldemanniana (See M. sheldonii).....
[Aq], [Gr], [Dz 1941:57 Fig.)
" -v- guirocobensis (See M. quirocobensis).[Bo-5],
guelzowiana.....[G], [L], [R], [NN], [UU], [VV], [Aq],
[Bt], [Cr], [Cu], [Cx], [Dh-1], [Hb],
[Hv], [Hx], [Il], [Iz], [Ja], [Jd], [Ag-1981], [Aj 33-12-1982], [Ba 1-1978],
[Flower - Ba 11-1978], [Bi Vol. 4, #6],
" -v- robustior.....
" -v- splendens.....
" -v- superba.....
guerreronis.....[BB], [VV], [Aq], [Ay], [Cx], [Hx*],
[Hy], [Jd], [Af42-3], [Af 55-2], [Aj 8-5-1957], [Db 6-1979], [Db - 26th Anniv. Edition],
" -v- subhamata.....[Dm], [Hy],
" -v- zapilotensis (See M. zapilotensis),

guiengolensis (See M. beneckei),
guillauminiana (See M. mercadensis -v- guillauminiana).. [Bo-5], [Aa 3-1983], [Iv 9-1952], [Bi Vol. 4, #6],
guirocobensis (See M. gueldmanniana -v- guirocobensis), [VV],
gummifera (See M. heyderi)...[D], [M*], [VV], [Ct-1984], [Hx*], [Hy], [Af 60-5], [Am 8-1983],
" -v- applanata (See M. applanata, M. heyderi -v- applanata).....[U], [An], [Ib],
" -v- gaumeri (See M. gaumeri),
" -v- hemisphaerica (See M. hemisphaerica, M. heyderi -v- hemisphaerica), [Dg 2-3-1980],
" -v- macdougallii (See M. heyderi -v- macdougallii, M. macdougallii), [Q], [An], [Af 52-1], [Au 37-1],
" -v- meiacantha (See M. heyderi -v- meiacantha, M. meiacantha) (M*], [U], [An], [Ib],
" -v- petrophila (See M. petrophila),
haageana (See M. elegans, M. dealbata, M. collina) [C], [RR], [VV], [Aq], [Ar], [Hv], [Hx*], [Hy], [Af 16-12], [Af 60-3], [Aw 4-1970], [Aw 4-1987],
" -v- abducta.....
" -v- acanthoplegma (See M. acanthoplegma),
" -v- conspicua......[Aq],
" -v- schmollii (See M. elegans -v- schmollii), [Aq],
haasii.....
haehneliana (See M. knebeliana).....[VV], [Ah-1985],
hahniana (See M. mendeliana, M. woodsii)....[C], [H], [I], [L], [T], [X], [AA], [NN], [Tt], [VV], [Aq], [Ar], [Ay], [Bv-1], [Cs], [Ct-1984], [Cx], [Cz], [Gr], [Hv], [Hx], [Hy], [Il], [Jd], [Af 1-7], [Af 55-1], [Ag-1985], [Aj 29-3-1978], [Ba 10-1978],
" -v- albiflora (See M. albiflora),
" -v- bachmannii (See M. bachmannii),
" -v- bravoae (See M. bravoae),
" -v- chionocephala.....
" -v- giseliana (See M. tarijensis)...[F], [VV], [Dm], [Ad-], [Hl 1-1973],
" -v- hahniana.....[Iz],
" -v- haseltonii.....[VV], [Bv-1],
" -v- klissingiana.....
" -v- mendeliana (See M. mendeliana).....[VV],
" -v- superba.....[Ah-1985], [K & L Nursery Catalogue, California],
" -v- werdermanniana (See M. werdermanniana), [VV], [Dm], [Hy], [Ad-1983],

MAMMILLARIA.....(Cont'd),

hahniana.....(cont'd),
" -v- woodsii (See M. woodsii),
" -f. cristata.....[Ia],
halbingeri.....[VV], [Hy], [Ea 1933:9 Fig.],
halei (See Cochemiea halei),
[Aq], [Bk], [Hv], [Iz], [Jd], [Aa 3-1981], [Aa 1-1984],
hamata (See M. rossiana),
[RR], [Bo-5], [Ct-1984], [Hx], [Aj 15-2-1964], [Ba 3-1978], [Ba 4-1979], [Cu 38:38 1976],
hamiltonhoytea (See M. petterssonii)....[VV], [Cx], [Hx*], [Hy], [Af 20-12], [Af 22-3],
" -v- fulvaflora.....[Hu],
hamta.....(See M. crucigera),
hastifera (See M. gigantea).....[Hx*],
haudeana (See M. saboae -f. haudeana),
[Ay], [Ad Nov. 1980], [Af 51-1], [Aj 29-11-1978], [Am 4-1980],
heeriana (See M. duoformis),
heeseana (See M. petterssonii).....[Ic],
heidiae (See Dolichothele heidiae),
[Aq], [Hv], [Iz], [Aa 3-1984], [Ae-1981], [Aj 26-10-1975], [Aj 27-10-1976], [Hl 4-1982],
hemisphaerica (See M. heyderi -v- hemisphaerica),
[CC], [LL], [PP], [RR], [VV], [Ar], [Bu], [Bv-1], [Bz], [Cx], [Hn], [Hy],
" -v- walteri (See M. heyderi -v- walteri, M. walteri).....[VV],
hennisii (See M. columbiana).....[Aj 31-6-1981],
hernandezii (See M. napina).....[Dq], [Af 55-1], [Af 56-3], [Af 58-5], [Af 60-3],
herrerae.....[D], [I], [VV], [Ac], [Aq], [Ar], [Av], [Ay], [Bt], [Cr], [Cx], [Hx], [Hy], [Il], [Iz], [Af 16-11], [Af 53-6], [Ag-1985], [Aj 29-2-1978],
" -v- albiflora (See M. albiflora, M. alba), [Ar], [Hy],
hertrichiana (See M. standleyi),
[VV], [Ac], [Aq], [Hu], [Af 59-2],
" -v- robustior.....
heyderi.....[A], [P], [NN], [RR], [An], [Ar], [Be*], [Cx], [Df], [Dh-1], [Hn], [Hy], [Af 52-6], [Af 55-6], [Am 2-1979],
" -v- applanata (See M. applanata).......[An], [Ct-1984], [Df], [Af 57-3], [Af 60-1-1988],
" -f. bensonensis.....
" -v- bullingtoniana.....[Ae-1981],
" -v- gaumeri (See M. gaumeri),
" -v- gummifera (See M. gummifera), [Aq], [Aa 4-1986],
MAMMILLARIA.....(Cont'd),

heyderi.....(cont'd),
" -v- hemisphaerica (See M. hemisphaerica, M. gummifera -v- hemisphaerica), [Y], [LL], [Aq], [Ar], [Az], [Df], [Aa 4-1986], [Ae-1982],
" -v- heyderi (See M. applanata), [LL*], [Aq], [Ax], [Df], [Aa 4-1986],
" -v- macdougallii (See M. gummifera -v- macdougalllii, M. macdougallii),
" -v- meiacantha (See M. gummifera -v- meiacantha, M. meiacantha).....[Aq], [Ax], [Df],
" -v- waltheri (See M. hemisphaerica -v- waltheri, M. waltheri),
hidalgensis (See M. polythele), [CC], [EE], [VV], [Bu], [Bv-1], [Hx*],
hildmanniana (See M. hidalgensis).....Hy],
hirsuta (See M. bocasana, M. aureiviridis, M. erecto-hamata, M. longicoma, M. mollihamata.....[Hx*],
" -v- grandis.....[Jb],
hoffmanniana (See M. obconella), [VV], [Hy], [Cu 3-1964]
huajuapensis (See M. mystax).....[Hx*],
hubertmulleri.....[Ct-1984],
huiguerensis.....[Jb],
huitzilopochtli.....[Ac], [At], [Iz], [Aa 37-1], [Af 52-2], [Ag-1987, 1988], [Ah-1985], [Au 35-2], [Db 1-1980], [Db 2-1980], [Hl 4-1980], [Hl 1-1983],
humboldtii.....[VV], [Aq], [Ar], [Av], [Hx*], [Hy], [Iz], [Jd], [Aa 3-1980], [Ab-1985], [Ad-1983], [Ae-1981], [Af 50-6], [Af 52-1], [Ag-1988], [Ah-1985], [Aj 9-6-1958], [Ba 1-1979], [Db 9-1979],
" -v- caespitose.....[See K & L Nursery Catalog],
" -f. .....[Ar], [Af 50-6],
hurtiana.....
hutchisoniana (See M. bullardiana), [I], [VV], [Aq], [Hx], [Af 6-1], [Af 57-2], [Aw 3-1972],
" -v- albissima.....[Aa 37-1],
" -v- hutchisoniana.....[Iz],
" -v- louisiae (See M. louisiae).....[Iz],
icamolensis (See M. carretii ? - Should be discarded), [VV], [Bo-5], [Ct-1984], Hx*],
ignota.....[Jb],
igualensis.....[Jb],
" -v- palmillensis.....[Jb],
inaiae (See M. sheldonii).....[VV], [Hy],
infernilensis (See M. parkinsonii).....[VV],
ingens (See M. obconella).....[Bo-5],
MAMMILLARIA.....(Cont'd),

insularis (See Bartschella insularis),
[VV], [Aq], [Ar], [Hu], [Hx*], [Hy], [Il],
[Aa 1-1983], [Af 10-2], [Af 51-6], [Af 57-1],
[Ba 9-1977], [Bm-1985],
irregularis.....[Bo-5],
isotensis.....[Jb],
jalapensis.....
jaliscana (See M. flavihamata, M. posseltiana, M. rettigiana).....[VV], [Aq], [Ar], [Hy], [Aj 31-2-1980], [Ba 11-1979],
johnstonii.....[RR], [Aq], [Bv-1], [Hy], [Jd], [Ah-1980], [Aj 29-5-1978],
" -v- guaymensis.....[VV], [Aa 4-1983],
" -v- johnstonii.....[Aa 4-1983],
" -v- sancarlensis.....[VV], [Bv-1], [Hy], [Aa 4-1983], [Ae-1981],
" -v- typica.....[VV],
joosseniana (See M. bella).....[Bo-5],
jozef-bergeri (See M. collinsii),
karwinskiana...[C], [H], [RR], [VV], [Aq], [Bv-1], [Ct-1984], [Cx], [Fg], [Hu], [Hx], [Hy], [Dh-1],
" -v- centrispina.....
" -v- nejapensis (See M. nejapensis),
" -f. cristata.....[Hy],
kelleriana (See M. durispina).....[VV], [Bv-1], [Dm], [Cu 8-3-1946],
kewensis (See M. polythele).....[RR], [VV], [Aq], [Bv-1], [Di*], [Gr], [Hy], [Ia],
" -v- craigiana (See M. polythele)..... [VV], [Af 33-5],
kladiwae (See M. nunezii, M. spinosissima),
klissingiana.....[VV], [Aq], [Ar], [Bv-1], [Cx], [Hy], [Il], [Iz], [Ae-1981], [Af 56-1], [Db - 26th Anniv. Edition],
knebeliana (See M. leucantha, M. haehneliana), [VV], [Dw 4:52 Fig. 1932],
knippeliana.....[VV], [Aq], [Ct-1984], [Af 56-2],
kraehenbuehlii.....[Ac], [Aq], [Hx*], [Iz], [Aa 4-1982], [Ab-1985], [Aj 22-5-1971], [Aj 24-6-1973], [Hl 6-1982],
krasuckae.....[Jb],
kuentziana (See M. vetula),
kunthii (See M. lanata).....[Bo-5], [Hx], [N-1980], [Aj 15-11-1964], [Hl 2-1976],
kunzeana.....[RR], [VV], [Ar], [Bv-1], [Cr], [Dm], [Hu], [Hx*], [Hy], [Il],
laetescens (See M. magnimamma),
laeta (See M. uberiformis),
lanata (See M. supertexta, M. albilanata), [CC], [RR], [VV], [Bv-1], [Hx], [Hy],
laneusumma (See M. canelensis)....[VV],

MAMMILLARIA.....(Cont'd),

lanifera (See M. rhodantha).....[Jb],
" -v- juxtlahuacensis.....[Jb],
lasiacantha (See M. denudata, M. egregia)...[B], [D], [M*], [P], [U], [LL*], [VV], [An], [Aq], [Ar],[Ax], [Hx], [Iz], [Ae-1982], [Af 57-3], [Af 58-4], [Af 58-5], [Af 60-1-1988], [Dg 5-4-1984], [Hl 4-1973],
" -v- denudata (See M. denudata)....[U], [Y], [An], [Df], [Af 52-5], [Af 56-3],
" -v- lasiacantha.....[Y], [Df], [Af 56-3],
" -v- minor.....
laui.....[Aa 22-2-1982],
" -f. dasyacantha (See M. subducta -v- dasyacantha) [Aq], [Aa 2-1982], [Af 52-6], [Af 56-5], [Af 57-6], [Ag-1985], [Hs -III 4-1982],
" -f. laui ...[Iz], [Aa 2-1982], [Ae-1981], [Af 52-2], [Af 56-1],
" -f. subducta (See M. subducta -v- subducta).[Aq], [Aa 2-1982], [Ae-1983], [Af 52-2],
" -f- (L-777).....[Af 50-2],
leei (Invalid name - See Coryphantha sneedii -v- leei, Escobaria leei, Escobaria sneedii -v- leei), [Y], [Df],
lengdobleriana (See M. lasiacantha, or M. magallanii), [Bo-5], [Ct-1985], [Hl 4-1973],
lenta.....[VV], [Aq], [Ar], [Gr], [Hx], [Hy], [Iu], [Iz], [Af 53-6], [Af 57-5], [Aj 27-10-1976], [Hw-1982],
leona (See Leptocladodia leona, M. pottsii), [Hx*], [Il], [Hl 5-1983],
" -v- similis.....[N],
leptacantha (See M. rekoi -v- leptacantha)
leucantha....[VV], [Aq], [Ct-3-1985], [Ee 1933:233 Fig.]
leucocentra (See M. geminispina group)...[Bo-5], [Bo-6], [Ct 3-1985], [Dm], [Hy], [Cu 8-3-1946],
leucocephala (See M. acanthoplegma).....[Bo-6],
lewisiana (See M. brandegeei -v- lewisiana)....[Bv-1], [Ff], Hx*], [Hy], [Aa 2-1986], [Af 27-6],
limonensis (See M. variabilis),
lindsayi (See M. standleyi), [VV], [Aq], [Aa 2-1981], [Af 12-11],
" -v- cobrensis.....
" -v- robustior (See M. standleyi),
lloydii (See M. uncinata -v- lloydii), [RR], [VV], [Aq], [Dm], [Hx*], [Hy], [Aa 4-1980],
longicoma (See M. bocasana -v- longicoma), [C], [QQ], [RR], [Aq], [Ar], [Ay], [Bv-1], [Dq], [Hy], [Af 5-10],
" -v- flavispina.....[Hy],
" -v- multiflora.....[Hy],

longiflora (See Krainzia longiflora),
[G], [K], [QQ], [RR], [TT], [VV], [Aq], [Hv]
[Hx*], [Ia], [Il], [Jd], [Af 57-5], [Af 57-6], [Ag-1981], [Aj 16-11-1965], [Aj 28-4-1977], [Au Vol. 3], [Ba 9-1978], [Bi 1-4],
[Hh May-June, 1977],
" -v- albispina.....[Iz],
" -v- breviflora.....[Af 51-3],
" -v- longiflora.....[Aq], [Iz], [Af 54-1],
" -f. stampferi.....[Aq], [Ct-1987], [Af 52-2],
[Af 54-1], [Cu 4-1979],
" -f. (White-spined).....[Aq], [Aa 3-1980],
longimamma (See Dolichothele longimamma).....[Av],
[Ay], [Dq], [Hv], [Jd], [Flower - Ba 9-1979],
" -v- gigantothele (See Dolichothele longimamma -v- gigantothele),
" -v- longimamma (See Dolichothele sphaerica, M. sphaerica).....[Ax], [Df],
" -v- uberiformis (See Dolichothele uberiformis, M. uberiformis),
[Aq], [Ij], [Aa 5-1982],
longiseta (See M. compressa v- longiseta).....[C],
louisiae (See M. hutchisoniana -v- louisiae),
[G], [X], [NN], [UU], [Aq], [Ff], [Hx*],
[Aa 1-1983], [Ae-1981], [Af 32-6], [Af 57-2],
[Ba 8-1978], [Ba 9-1980],
" -f. .....[Aa 37-1],
ludwigii (See M. phymatothele).....[Hy],
macdougallii (See M. gummifera -v- macdougallii, M. heyderi -v- macdougallii).....[A], [E], [RR],
[VV], [Ac], [Hy], [Aj 28-11-1977],
macracantha (See M. centricirrha, M. magnimamma),
[M*], [RR], [Bo-5], [Bu], [Bv-1], [Cg], [Hy],
" -v- retrocurva (See M. macracantha).....[Bo-5],
macromastos (See M. magnimamma).....[Hy],
macromeris (See Coryphantha macromeris, Lepidocoryphantha macromeris).....[Y], [Df],
magallanii (See M. roseocentra),
[VV], [Aq], [Bo-6], [Hv], [Hx], [Hy],
[Iz], [Ba 8-1976], [Ba 9-1980],
[Hl 4-1973],
" -v- hamatispina (Invalid name - See M. stella-de-tacubaya).....[Bo-6],
magneticola (See M. vetula)....[Ct-1985], [Ba 11-1977],
magnifica.....[VV], [Ac], [Ad], [Aq], [Ar], [Ct-1985],
[Aa 1-1980], [Ae-1982], [Af 55-2],
[Af 56-1], [Aj 5-1978], [Aj 33-1-1982],
[Au Vol. 2-1977, Zone 3 Compendium],
[Aw 1-1967],
" -v- aguafria.....

magnifica.....(cont'd),
" -v- magnifica.....[Aq], [Iz],
" -v- minor.....[Aq],
magnimamma.....[RR], [TT], [VV], [Aq], [Ar], [Av], [Be*], [Bv-1], [Cx], [Cz], [Hv], [Aa 6-1983], [Ag-1987], [Aw 4-1966], [Db 2-1979], [Hw-1982],
" -v- arietina (See M. magnimamma),
" -v- bockii (See M. magnimamma),
" -v- divergens (See M. magnimamma),
" -v- krameri (See M. magnimamma),
" -v- lutescens (See Coryphantha pycnacantha),
" -v- recurvus (See M. magnimamma),
" -v- rubriflora.....
" -f. vagaspina.....[Af 55-6],
" -f. monstrosus.....[Af 59-6], [Ah-1985], [Af 59-6],
mainiae.....[M*], [P], [Q], [CC], [NN], [RR], [VV], [Ac], [An], [Aq], [Ax], [Bt], [Cx], [Hx], [Hy], [Aa 1-1982], [Af 55-1], [Af 57-3], [Am 9-1981],
mammillaris (See M. ekmanii, M. glomerata, M. simplex, M. pseudosimplex), [F], [RR], [VV], [Aq], [Iz], [Af 57-3],
" -v- albidispina.....
maritima (See Cochemiea maritima, M. pondii -v- maritima)....[Aq], [Hv], [Aa 1-1984], [Ba 1-1982],
marksiana....[Aq], [Ar], [Hu], [Hx], [Hy], [Iz], [Aa 3-1982], [Af 56-3], [Aw 3-1977], [Db 2-1980],
marnieriana (See M. oliviae, M. microcarpa).[Ar], [Gr],
marshalliana (See M. baxteriana -v- marshalliana), [Af 6-1],
martinezii.....[D], [H], [VV], [Ar], [Bv-1], [Dm], [Hy], [Cu 8-3-1946],
mathildae.....[Aq], [Hx*], [Af 50-2], [Aj 24-12-1973],
matudae.....[E], [Aq], [Ar], [Ff], [Hx*], [Il], [Jd], [Ae-1981], [Ag-1985], [Aj 23-3-1972], [Bi 5-4-1987],
" -f. duocentralis.....
" -v- matudae.....[Aq], [Iz],
" -v- serpentiformis.....[Aq], [Aw 2-1971],
maycobaensis.....[Aj 27-2-1976],
mayensis (See M. canelensis).....[VV],
mazatlanensis (See M. occidentalis, M. patonii)... [RR], [VV], [Aq], [Bv-1], [Gr], [Hx], [Hy], [Jd], [Af 56-2], [Aj 27-5-1976],
" -v- monocentra (See M. occidentalis -v- monocentra).....[VV],
" -v- patonii (See M. patonii),
" -v- sinaloensis (See M. patonii -v- sinaloensis, M. occidentalis -v- sinaloensis),

MAMMILLARIA.....(Cont'd),

meiacantha (See M. heyderi -v- meiacantha),
[U], [Y], [LL*], [RR], [Ar], [Av], [Bz],
[Ct-1984], [Df], [Hy], [Af 57-3], [Af 58-4],
[Af 60-1-1988], [Am 1-1980], [Cu 1-1970],
[Cu 1-1971],
meissneri (See M. elegans -v- schmollii, M. haageana
-v- schmollii).....[Bo-5], [Hx],
melaleuca (See Dolichothele melaleuca),
[L], [Aq], [Cr], [Af 50-3],
melanocentra.....[RR], [VV], [Ag], [Bv-1], [Hu], [Hx*],
[Hy], [Ad 12-1982], [Hl 6-1969],
" -v- rosea.....
" -v- runyonii (See M. melanocentra),
" -v- typica.....[Aw 4-1986],
melispina (See M. winterae, M. zahniana),
mendeliana (See M. hahniana -v- mendeliana).......[H],
[NN], [VV], [Cx], [Hx], [Hy], [Af 9-3-1958],
mercadensis ...[G], [NN], [RR], [VV], [Ac], [Aq], [Ar],
[Bo-6], [Ct-1985], [Hv], [Hx*]. [Jd],
[Aa 4-1982], [Aa 1-1985], [Ad Nov.
1980], [Af 42-3], [Af 56-1], [Ba 10-
1977], [Db 2-1979],
" -v- guillauminiana (See M. guillauminiana).....
meridio-rosei (See M. wrightii -v- wilcoxii),
[Ae-1981], [Af 50-4],
mexicensis (See M. grusonii or M. zeyerana),
[VV], [Bo-5], [Aa 5-1984], [Cu 4-1961],
meyranii.....[Aq], [Af 55-1], [Aw 3-1968], [Aw 4-1983],
" -v- michoacana.....[Aw 4-1969],
microcarpa (See M. milleri, M. grahamii, M. mar-
nierana, M. oliviae),
[B], [D], [M*], [Q], [R], [Y], [GG], [HH],
[LL*], [RR], [VV], [Ac], [Aq], [Ar], [Ay],
[Bt], [Bz], [Df], [Gr], [Hn], [Hy], [Jd],
[Ad Sept. 1980], [Af 13-9], [Af 52-1],
[Aj 27-7-1976], [Am 2-1979], [Au 37-1],
" -v- auricarpa.....[F], [P], [Aq], [Aa-],
" -v- grahamii (See M. grahamii),
[Aq], [Bo-5], [Hu], [Ba 9-1980],
" -v- gueldemanniana (See M. gueldemanniana),
" -v- microcarpa (See M. microcarpa),
[Aq], [Ax],
" -v- milleri (See M. milleri),
[Bn], [Cu 4-1975],
" -v- oliviae (See M. grahamii -v- oliviae, M.
oliviae),
[Aq] [Aa -],
" -v- swinglei.....

MAMMILLARIA.....(Cont'd),

microhelia (See Leptocladodia microhelia, M. microheliopsis),
[H], [K], [NN], [VV], [Aq], [Bv-1], [Cx], [Hu], [Hv], [Hx*], [Hy], [Il], [Jd], [Af 53-6], [Aj 26-8-1975], [Am 5-1985], [Au Vol. 3 - 1978 Zone 3 Compendium], [Ba 3-1978], [Db 11-1979],
" -v- droegeana (See M. droegeana).....[NN],
" -v- microheliopsis (See M. microheliopsis), [Ar], [Bo-5], [Hx], [Hy], [Ba 8-1977], [Ba 9-1979],
" -v- rubrispina.....[Af 4-11],
microheliopsis (See M. microhelia -v- microheliopsis), [C], [H], [TT], [VV], [Aq], [Ar], [Bv-1] [Cm], [Hu], [Am 1-1984],
microthele.....[Aq], [Ar], [At], [Jd], [Ad -], [Af 59-2], [Ag-1987, 1988],
" -v- superfina.....[Ae-1981], [Ba 8-1980],
miegeana (See M. johnstonii)
[Aq], [Hx*], [Jd], [Af 57-4], [Aj 24-9-1973], [Hl 6-1974],
mieheana (See M. densispina -v- mieheana).....[Aq], [Dm], [Aa 2-1982], [Aa 3-1984], [Cu 8-4-1946],
" -v- globosa (See M. esaussieri),
mihavoinandis.....[Bo-5],
milleri (See M. microcarpa -v- milleri, M. microcarpa), [P], [CC], [RR], [Iz], [Aa 3-1981],
mitlensis (See M. rekoi).....[Bo-5], [Aw 1:86 Fig. 61 & 62, 1956], [Hl 5-1982],
mixtecensis (See M. casio, M. mystax),
[Hx*], [Eb 25:534, Fig. 28, 1954],
moelleriana (See M. cowperae),..... [VV], [Aq], [Hx*], [Hv], [Hy], [Iu], [Iz], [Jd], [Aa 2-1980], [Aa 4-1984], [Ae-1981], [Af 57-3], [Aj 28-2-1977], [Aj 32-12-1981], [Hw-1986], [Hw-1989],
" -v- cowperae (See M. cowperae, M. moelleriana),
moglicherweise (See M. durispina),
mollendorffiana (See M. rhodantha -v- mollendorffiana), [Aq], [Ff], [Hy], [Jd], [Ad 5-1978], [Ad 4-1982], [Af 55-2],
mollihamata (See M. pygmaea).....[Ar], [Cu 7-1960],
monancistracantha (Invalid name - See M. eschanzieri, M. nana).....[Bo-6], [Il], [Hl 2-1979],
monancistria (See M. erythrosperma),
monocentra (See M. montensis -v- monocentra),
montensis.....[VV],
" -v- monocentra.....
" -v- quadricentra.....
monticola.....[Jb],

MAMMILLARIA.....(Cont'd),

morganiana (See M. parkinsonii),
[VV], [Aq], [Cm], [Ia], [Ah-1985],
morricalii (See M. barbata).....[Ct-1983], [Iz],
[Aa 2-1980], [Af 41-5], [Hl 6-1972],
movensis (See M. craigii).....[VV], [Af 56-5],
muehlenpfordtii.....[Aq], [Ct-1985], [Aa 4-1982],
[Af 52-1],
" -v- brevispina.....[Hy],
" -v- hexispina.....
" -v- heyderi.....[U],
" -v- longispina.....[Bo-5], [Hy],
multiceps (See M. prolifera -v- multiceps, M. prolifera
-v- texana, M. pusilla).....[C], [U], [Y],
[LL*], [RR], [VV], [Ar], [Bv-1], [Cz], [Df],
[Af 52-1], [Am 5-1985], [Ba 10-1977],
" -v- humilis.....
" -c- perpusilla (See M. prolifera).....[Bo-5],
multidigitata..[G], [Aq], [Hy], [Aa 1-1981], [Af 19-10],
[Af 34-4], [Af 52-6], [Cu 4-1974], [Db 12-1980],
multiformis (See M. erythrosperma).....[RR], [VV],
multihamata.....[RR], [VV],
multiseta (See M. karwinskiana).....[Ct-1985],
mundtii (See M. wiesingeri)...[RR], [VV], [Hx*], [Hy],
[Aj 24-9-1973], [Aj 31-1-1980], [Ba 6-1980],
mystax (See M. atroflorens, M. crispiseta, M. muta-
bilis, M. neomystax),
[CC], [NN], [RR], [VV], [Aq], [Ar], [Bm-
1985], [Bv-1], [Cx], [Hx], [Hy], [Il], [Jd],
[Ae-1981], [Af 56-5], [Af 57-4], [Aw 4-1962],
[Ba 6-1980], [Db 5-1979], [Hw-1982],
nagliana.....[Jb],
nana (See M. monancistracantha)
[Ah], [Ct-1986], [Hb], [Aa 6-1980], [Aa 4-1982],
[Af 52-1], [Af 58-2], [Af 59-2], [Af 60-2],
" -v- chrysacantha.....
" -v- monancistracantha.....[Aq],
napina.....[Z], [JJ], [NN], [VV], [Ac], [Aq], [Ct-1986],
[Hx*], [Iz], [Aa 1-1980], [Af 56-6], [Aj 8-
11-1957], [Aw 2-1969], [Bm-1988], [Cu 4-1981]
[Hl 1-1980], [Bi 6-3],
" -v- centrispina.....
nazasensis (See M. pennispinosa -v- nazasensis).....
nealeana (See M. muehlenpfordtii),
nejapensis (See M. karwinskiana),
[VV], [Ah], [Aq], [Ar], [Av], [Ay], [Bv-1],
[Cz], [Ff], [Hu], [Hx], [Hy], [Aa 5-1987],
[Ab -1985], [Af 51-6], [Hl 4-1982],
" -v- aureispina.....
" -v- brevispina.....[Bo-5], [Hy],
" -v- elegans.....[Hy],

nejapensis.....(cont'd)
" -v- longispina.....[Bo-5], [Hy], [Ae-1981], [Aj 23-3-1972],
" -f. cristata.....[Bu],
nellieae (See Coryphantha minima, Coryphantha nellieae, Escobaria minima, Escobaria nellieae), [Y], [Df],
nelsonii (See Dolichothele nelsonii, M. beneckei, Oehmea nelsonii).....[Jd],
neobertrandiana (See M. magallanii, M. lengdobleriana), [Bo-5], [Iv 9-1952],
neoconspicua (See M. conspicua),
neocoronaria (See M. coronaria, M. spinosissima), [Hy], [Il],
neocrucigera (Invalid name - See M. parkinsonii),
neoelegans (See M. elegans).....[Hy],
neoformosa (See M. formosa).....[Hy],
neomystax (See M. karwinskiana), [Ar], [Ay], [Hy], [Ba 2-1978],
" -v- rhodacantha.....
" -f. cristata.....[Hn],
neopalmeri.....[VV], [Aq], [Hx*], [Aa 3-1982], [Aa 5-1987], [Ae-1981], [Af 51-6], [Af 58-1], [Ag-1987], [Aj 31-1-1980], [Hl 1-1971],
" -v- multidigitata.....
neophaeacantha (See M. kewensis, M. polythele), [Bo-5], [Bv-1],
neopotosina (See M. muehlenpfordtii), [Af 52-1], [Aw 3-1965],
" -v- brevispina.....
" -v- hexispina.....
neoschwarzeana (See M. bocensis, M. tesopacensis), [Bo-5], [Ff], [Iv 3-1952],
neotetracantha (See M. tetracantha).....[Hy],
nigricans (See M. phaeacantha).....
nivosa (See M. flavescens -v- nivosa, Coryphantha nivosa)..[RR], [VV], [Aq], [Bv-1], [Hx], [Ae-1981], [Au Vol. 2-1977 Zone 3 Compendium], [Hl 5-1972],
nobilis (See M. bicolor -v- nivea, M. geminispina),
nolascana (See M. tayloriorum),
noureddineana.....[Jb-Abb-39],
nunezii (See M. bella, M. solisii, M. wutheniana), [NN], [RR], [VV], [Aq], [Ar], [Dm], [Hy], [Aa 4-1981],
" -v- bella (See M. bella),
" -v- solisii (See M. solisii),
obconella (See M. dolichocentra, M. tetracantha), [Aq], [Ar], [Hy], [Af 58-2],
" -v- galeottii (See M. tetracantha -v- galeottii] [Ar],

obscura (See M. hamiltonhoytea, M. gigantea),
[VV], [Bo-5], [Hl 4-1977],
" -v- heeseana.....
" -v- wagneriana.....
" -v- " -f. tortulospina (See M. wagneriana
-v- tortulospina)....[VV],
obvallata (See M. fuliginosa),
ocamponis (See M. mercadensis),
[Cu 2-1-1933], [Ec 2:355 Fig. #1, 1918],
occidentalis.....[RR], [VV], [Aq], [Gr], [Hy], [Af 6-1],
[Af 50-4],
" -v- monocentra (See M. mazatlanensis -v-
monocentra).....[Bo-5],
" -v- patonii (See M. mazatlanensis -v- patonii,
patonii, M. patonii),
" -v- sinaloensis (See M. mazatlanensis -v-
sinaloensis).....[VV],
ochoterenae (See M. discolor, M. esperanzaensis, M.
pachyrhiza).....[VV], [Cx], [Hy],
ocotillensis (See M. gigantea).....[VV], [Bv-1],
" -v- brevispina.....[VV],
" -v- longispina.....[VV],
olivacea (See M. tentaculata, M. fuscata),
oliviae (See M. grahamii -v- oliviae, M. microcarpa
-v- oliviae),
[P], [RR], [VV], [Hx], [Aa 3-1981], [Ba 9-1980],
oraticeps.....[Hy],
orcuttii.....[CC], [VV], [Aq], [Ay], [Ct-1986], [Hy],
[Ae-1982], [Aj II-12-1930], [Aj 16-11-1965]
orestera (See M. chavezei, M. viridiflora),
[Hx*], [Ae-1981],
ortegae (Should be discarded).....[RR], [VV],
ortiz-rubiona (See M. candida, Mammilloydia candida),
[D], [NN], [VV], [Bv-1], [Ct-1983], [Cx], [Hy],
oteroi.....[Z], [Ah], [Aq], [Ct-1986], [Ae-1981], [Af-
55-1], [Ba 7-1977], [Hl 5-1982], [Hm 2-1978],
pachycylindrica (See M. grusonii)
[Aq], [Aa 3-1983], [Cu 21:82, 1959],
pachyrhiza (See M. discolor),
[VV], [Aj 31-12-1980],
pacifica (See M. baxteriana, M. petrophila),
[VV], [Bv-1], [Af 6-1],
painteri (See M. pygmaea).....
[RR], [VV], [Aq], [Af 41-3],
palmeri (See M. neopalmeri).....[RR],
papasquiarensis (See M. tesopacensis -v- papasquiar-
ensis, M. wagneriana).....[Jb],
papyracantha (See Pediocactus papyracantha),
paradensis (See M. stella-de-tacubaya),
parensis (See M. pringlei).....[VV:316],

MAMMILLARIA.....(Cont'd),

parkinsonii (See M. auriareolis, M. avila-camachoi, M. rosensis, M. tiegeliana, M. vonwyssiana), [D], [I], [L], [S], [AA], [NN], [RR], [VV], [XX], [Ac], [Aq], [Ar], [Av], [Ay], [Bu], [Bv-1], [Cg], [Cm], [Cx], [Cz], [Dq], [Hv], [Hx], [Hy], [Iz], [Aa 4-1981], [Ae-1981], [Af 6-7], [Af 55-6], [Ai-1978-1979],
" -v- brevispina.....[VV], [Hy], [Cu 1-1964],
" -v- dietrichae.....[Bo-5],
" -v- tiegeliana.....[Bv-1],
" -f. cristata...[Bu], [Hy], [Af 2-7], [Af 31-3],
parrasensis.....[Jb].
parriana (See M. mixtecensis, M. mystax), [Aj 23-6-1972],
parvimamma (See M. mammillaris).....[Hy],
patonii (See M. mazatlanensis, M. occidentalis -v- patonii).....[Av], [Cx], [Aw 1-1988],
" -v- sinaloensis (See M. mazatlanensis, M. occidentalis -v- sinaloensis), [Ar], [Bo-5], [Aj 27-4-1976],
peacockii (See M. dealbata).....[Bo-5],
pectinifera (See Solicia pectinata), [G], [I], [NN], [RR], [VV], [Ac], [Aq], [Ar], [Av], [Ct-1986], [Hv], [Il], [Iz], [Jd], [Aa 3-1982], [At-1982], [Ba 4-1979], [Ba 10-1980], [Hl 2-1980], [Hw-1986],
" -f. solisioides.....[Ct-1987],
" -f. cristata.....
peninsularis.....[VV], [Ac], [Aq], [Hy], [Jd], [Aa 3-1983], [Aa 3-1986], [Af 58-6], [Aj 28-7-1977], [Db 6-1979],
pennispinosa.....[G], [Z], [Aq], [Ar], [Bu], [Hu], [Hv], [Hx], [Hy], [Iz], [Jd], [Aa 6-1980], [Aa 3-1983], [Ad Nov. 1980], [Ae-1981], [Ah-1986], [Aj 33-10-1982],
" -v- nazasensis (See M. nazasensis).....[G], [Aq], [Ay], [Ct-1986], [Iz], [Aa-], [Cu 4-1979], [Hl 5-1982],
" -v- pennispinosa.....[Aq],
" -f. cristata.....[Ar],
pentacantha (See M. magnimamma).....[Bo-5],
perbella.....[C], [CC], [VV], [Ac], [Aq], [Ct-1983], [Cx], [Hx*], [Jd], [Ae-1981], [Au Vol. 3, 1978, Zone 3 Compendium],
" -v- lanata.....[Ae-1981],
perezdelarosae.....[Aw 4-1985],
petrophila.....[VV], [Ac], [Aq], [Hx*], [Aa 1-1982], [Af 41-4], [Af 55-3],

petrophila.....(cont'd),
" -v- gatesii (See M. gatesii)....[Aa 37-1],
petterssonii.....[RR], [VV], [Aq], [Ct-1986], [Hy], [Aa 3-1987], [Ad June 1978], [Ad Aug. 1978], [Ba 11-1979],
-v- longispina (See M. heeseana).....
pfeifferi (See M. rhodantha),
phaeacantha.....[RR], [VV], [Bv-1], [Hy],
phellosperma (See Phellosperma tetrancistra, M. tetrancistra).....[Bv-1], [Dq],
phitauiana (See M. verhaertiana).....[V], [Aq], [Aa 6-1985], [Af 2-11], [Aj 11-2-1960],
phymatothele (See M. magnimamma).....[RR], [Hy],
" -v- trohartii.....[VV],
picta (See M. auriseta, M. schieliana), [VV], [Aq], [Bo-5], [Hy], [Aj 8-9-1957],
" -v- viereckei (See M. picta).....[Hl 5-1979],
pilcayensis (pitcayensis) (See M. spinosissima -v- pilcayensis).....[Aq], [Bt], [Aa 5-1987], [Af 59-6], [Db 7-1979],
" -v- chrysodactyla.....[Ae-1981],
" -v- crocidata.....
" -f. (Yellow-spined form).....[Aq],
pilensis (See M. petterssonii).....[Bo-5],
pilispina (See M. subtilis, M. sanluisensis), [Aq], [Ar], [Av], [Hx*], [Hy], [Ia], [Jd], [Aa 5-1987],
plumosa.....[C], [M*], [X], [AA], [HH], [NN], [RR], [VV], [Aq], [Ar], [Ay], [Bt], [Bv-1], [Cm], [Cr], [Gr], [Hu], [Hy], [Iz], [Jd], [N-1980], [Aa 6-1980], [Aa 1-1985], [Ab-1985], [Af 5-10], [Af 58-2], [Ag-1985], 1988], [Aj II-12-1930], [Aw 4-1986], [Bi 2-3], [Hw-1981],
" -f. (Golfball form].....[Aq], [Iz],
polycentra (See M. spinosissima).....[Hy],
" -v- auricoma (See M. spinosissima -v- auricoma) [Hy],
" -v- brunnea (See M. spinosissima -v- brunnea), [Hy],
" -v- pretiosa.....[Hy],
polyedra.....[AA], [RR], [VV], [Aq], [Ct-1986*], [Hx], [Hy], [Aj 18-9-1967], [Cu 3-1971],
polythele (See M. affinis, M. dolichocentra, M. kelleriana, M. subdurispina).....[RR], [VV], [Aq], [Ct-1986], [Hx*], [Aa 6-1983], [Ba 1-1978],
" -v- affinis (See M. polythele).....[Bu],
pondii (See Cochemiea pondii).....[Aq],
" -v- maritima (See Cochemiea maritima, M. maritima),

pondii.....(Cont'd),
" -v- setispina (See Cochemiea setispina, M. setispina).....[Iz],
poselgeri (See Cochemiea poselgeri),
[Aq], [Ct-1986], [Iz], [Jd], [Aa 1-1980], [Aa-1-1984], [Aa 1-1986], [Cu 1-1980], [Hl 5-1974],
posseltiana (See M. rettigiana).....[VV], [Aa 4-1984],
potosina (See M. muehlenpfordtii, M. neopotosina)..... [Bv-1], [Cz],
pottsii (See Coryphantha pottsii, Leptocladodia leona, M. leona),
[F], [M*], [Y], [Z], [CC], [GG], [RR], [VV], [Ac], [Aq], [Ax], [Bu], [Bv-1], [Df], [Hy], [Iu], [Iz], [Af 52-6], [Af 60-1-1988], [Ag-1987],
" -v- gigas.....[Jb],
" -v- multicaulis.....[Jb],
praelii (See M. karwinskiana),
[NN], [VV], [Bo-5], [Hx], [Iv 4-1959],
" -v- viridis.....
priessnitzii.....[Jb],
pringlei (See M. rhodantha -v- pringlei),
[F], [H], [T], [NN], [RR], [VV], [Aq], [Bv-1], [Cx], [Hv], [Hy], [Aa 2-1987], [Ah-1985], [Aw 2-1976],
" -v- columnaris.....[VV],
" -v- longicentra (Invalid name).....[DD],
prolifera.....[C], [H], [S], [EE], [Qq], [RR], [VV], [Aq], [Ar], [Gr], [Hx*], [Hy], [Jd], [Af 56-3], [Ag-1985], [Aj 31-11-1980], [Gu III-1953],
" -v- arachnoidea.....[Aq], [Aa 6-1986], [Cu 1-1978],
" -v- haitiensis (See M. prolifera)...[F], [H], [Aq], [Ar], [Hy], [Aa 6-1986], [Aj 30-1-1979], [Ba 2-1978],
" -v- mexicana.....[T],
" -v- minima.....
" -v- multiceps (See M. prolifera -v- texana), [Ar],
" -v- perpusilla.....
" -v- prolifera.....[Aq], [Aa 6-1986],
" -v- pusilla (See M. prolifera, M. pusilla), [Bv-1],
" -v- texana (See M. multiceps)...[M*], [Aq], [Ar], [Ax], [Ct-1986*], [Df], [Hy], [Aa 6-1986], [Hw-1989],
" -f. (small form).....[Aq],
" -f. cristata.....[Hy],
pseudoalamensis (See M. microcarpa -v- oliviae. M. marnieriana, M. alamensis),
[Hx], [Ba 8-1977], [Iv 9-1953],

pseudocrucigera (Invalid name - See M. sempervivi),
[W], [VV], [Bv-1],
pseudoperbella.....[I], [NN], [RR], [QQ], [VV], [Ac],
[Aq], [Ar], [Cx], [Gr], [Hx], [Hy],
[Ae-1981], [Bi 5-4-1987],
" -f. cristata.....[Aq], [Aa 5-1981],
pseudorekoi (See M. rekoi).....(Hx*],
pseudoschiedeana (See M. dumetorum, M. schiedeana -v-
dumetorum).....[Bo-5],
pseudoscrippsiana (See M. scrippsiana).....Bo-5],
pseudosimplex (Invalid name - See M. ekmanii, M. glomerata, M. simplex).....[Bo-6],
puberula.....[Jb], [Ct-1986],
pubigera (See M. puberula),
pubispina (See M. cadereytensis, M. mollihamata, M. pygmaea),
[VV], [Bo-5], [Ct-1986], [Aj II 3&4, 1930],
pulchra.....[Ct-1986*],
pullihamata (See M. rekoi),
[NN], [Ag-1982], [Ar], [Bo-6], [Hx*], [Jb],
purpursii (See Pediocactus simpsonii).....[Bo-5],
pusilla (See M. prolifera).....[Bk], [Fg],
" -v- texana (See M. prolifera -v- texana),
[Di*],
pygmaea.....[VV], [Aq], [Be*], [Bv-1], [Ct-1986], [Iu],
[Af 55-4], [Cu 2-1961],
" -v- pubispina (See M. pubispina),
pyrrhocephala.....[RR], [VV], [Bo-1], [Hy],
" -v- confusa (See M. confusa),
" -v- donkelaeri.....
" -v- fulvolanata.....
" -v- malletiana.....
" -f. cristata.....
queretarica (See M. infernillensis).....[VV],
quevedoi (See M. bravoae or M. woodsii),
[Dm], [Cu 8-4-1946],
quirocobensis.....[Hl 2-1975],
radiaissima (See M. baumii -v- radiaissima)...[VV],
radiosa (See Coryphantha vivipara -v- radiosa,
Escobaria radiosa, M. vivipara -v- radiosa),
" -v- alversonii (See Coryphantha vivipara -v-
alversonii).....[Bk],
" -v- neomexicana (See Coryphantha vivipara -v-
neomexicana).....[Di*],
ramillosa (See Coryphantha ramillosa).....[Y], [Df],
rayonesii.....
rectispina.....
recurva (See M. magnimamma).....[VV], [Bv-1],
recurvispina (See Coryphantha recurvispina, Coryphantha recurvata).....[RR], [An],

MAMMILLARIA.....(Cont'd),

rekoi (See M. mitlensis, M. pseudorekoi, M. rekoina), [RR], [VV], [Aq], [Bv-1], [Af 30-2],
" (affinity) (Lau 1314).....[Aq],
" -v- aurispina (See M. aurispina).....[Aq], [Aw 1-1983], [Hs III-4-1982],
" -v- leptacantha (See M. leptacantha)....... [Af 55-2], [Ag-1986],
" -v- mitlensis.....[RR], [VV],
" -v- suprema.....
rekoiana (Invalid name - See M. rekoi with straight spines).....[VV],
reppenhagenii (See M. albilanata -v- reppenhagenii, M. tegelbergiana)....[Aq], [Ae-1982], [Af 49-5], [Aj 28-6-1977], [Cu 2-1978], [Hl 1-1983],
rettigiana.....[VV], [Aq], [Bo-6], [Ad Dec. 1979], [Aj II-5-1930], [Dw 2-98, 1930],
rhodantha (See M. aureiceps, M. fuscata, M. pfeifferi),
rhodantha.....[AA], [RR], [TT], [VV], [Aq], [Ar], [Be*], [Bv-1], [Ct-1986], [Cx], [Cz], [Hb], [Hv], [Hy], [Aj 9-1976], [Aw 2-1969], [Aw 4-1975], [Ba 1-1978],
" -v- alba (See M. alba).....
" -v- andreae (See M. rhodantha -v- sulphureus),
" -v- aureiceps (See M. aureiceps, M. rhodantha -v- sulphureus),
" -v- calacantha (See M. calacantha),
" -v- crassispinus (See M. crassispina).....[H],
" -v- chrysantha (See M. rhodantha),
" -v- droegeana (See M. droegeana),
" -v- fera-rubra (See M. fera-rubra),
" -v- flavispina.....[Ia],
" -v- mollendorffiana (See M. mollendorffiana),
" -v- pfeifferi.....[If],
" -v- pringlei (See M. pringlei).....[Iz],
" -v- rubens.....[Ar], [Bo-5],
" -v- ruberrima.....
" -v- rubescens.....[Hy],
" -v- rubra.....[Bo-5], [Ba 1-1978],
" -v- sulphureus (See M. aureiceps)...[CC], [Hy], [Aa 2-1987], [Iv 4-1959],
" -f. (gold-spined form).....[Ar],
rioverdense.....[Jb],
ritteriana (See M. chionocephala).....[VV], [Hy], [Af-56-5], [Ah-1986], [Hl 1-1970],
" -v- quadricentralis (See M. formosa), [VV], [Bv-1],
roberti (Invalid name - See Coryphantha roberti, Escobaria runyonii), [Y], [Df], [Hm 2-1979],
rosensis (See M. parkinsonii complex).....[VV],

roseo-alba.....[VV], [Aq], [Ar], [Ct-1986], [Hx*], [Hy], [Iu], [Dw I:87, 1929], [Iv 4-1949],
" -v- winteriae.....
roseocentra.....[R], [Ar],
rossiana (See M. duoformis).....[Bo-5], [Aj 9-8-1958],
ruber.....
rubra (See M. crinita),
rubida (See M. bocensis -v- rubida)..... [NN], [Bo-5], [Hy], [Iv 3-1952],
rubrograndis.....[Aq], [Ae-1982], [Aj 30-4-1979],
ruesticina.....
ruestii (See M. columbiana -v- ruestii), [RR], [VV], [Aq],
runyonii (Br. & R.) Boed. (See M. melanocentra), [RR],
runyonii (Br. & R.) Weniger (See Coryphantha macromeris -v- runyonii).....[Y], [Df],
rupestris.....
rutila (See M. rhodantha).....[NN],
saboae.....[G], [VV], [Ay], [Ff], [Hx*], [Af 51-1], [Ah-1986], [Am 4-1980], [Aw 3-1966], [Ba 9-1978], [Hl 4-1971],
" -v- goldii (See M. goldii).....[Ar], [Iz], [Ag-1981],
" -f. cristata.....
" -f. haudeana (See M. haudeana), [Aq], [Hv], [Iz], [Af 51-3], [Ah-1985], [Au 36-4],
" -f. saboae.....[Aq],
saetigera (See M. brauneana), [T], [NN], [VV], [Bv-1], [Il], [Cu 3-1964],
" -v- quadrangularis.....[VV],
saffordii (See M. carretii).....[RR],
" -f. cristata.....[Hy],
salm-dyckiana (See Coryphantha salm-dyckiana),
saint-pieana (See M. gigantea or M. petterssonii), [DD],
san-angelensis.....[Aw 1-1981],
sancarlensis (See M. johnstonii -v- sancarlensis),
sanjuanensis.....[Jb],
sanluisensis (See M. pilispina), [Ct-1987], [Cu 11-1957], [Cu 3-1968],
santacarlensis (See M. barbata), [Ct-1983], [Iz], [Aa 2-1980], [Af 41-6], [Hl 6-1972],
sartorii.....[RR], [VV], [Aq], [Cx], [Hy], [Iu], [Jd], [Aa 4-1983], [Aa 2-1986], [Ah-1980],
" -v- brevispina.....
" -v- longispina.....
saxicola.....[Jb],

MAMMILLARIA.....(Cont'd),

scheeri (See Coryphantha scheeri),
[Y], [Df],
scheidweileriana (See M. erythrosperma),
[H], [L], [VV], [Hx*],
schelhasii (See M. bocasana or M. longicoma),
[VV], [Be*], [Dm], [Hy],
" -v- aurihamata (See M. aurihamata),
schiedeana.....[C], [H], [L], [T], [NN], [QQ], [RR],
[VV], [Aq], [Ar], [Ay], [Bt], [Bv-1],
[Ck], [Cm], [Gr], [Hx], [Hy], [Il],
[Jd], [Aa 2-1985], [Af 5-10], [Af 52-4],
[Ag-1988], [Am 2-1985], [Aw 3-1969],
" -v- dumetorum (See M. dumetorum),
[Af 53-3],
" -v- nigra.....[Ia],
" -v- plumosa (See M. plumosa),
[Ar], [Bo-5], [Aa 2-1985],
[Ae-1981], [Am 2-1985],
" -v- schiedeana.....[Iz], [Af 53-3],
" -f. cristata.....[Ar],
schieleana (See M. picta, M. viereckii),
[Ee 3:27 Fig. 1949],
schmollii (See M. discolor),.....[Cx],
schmuckerii (See M. kelleriana).....[Bo-6],
schulzeana (See M. graessneriana).....[Bo-5],
schumannii (See Bartschella schumannii),
[G], [Ac], [Aq], [Ar], [Ay], [Bv-1], [Dq],
[Ha], [Il], [Iz], [Jd], [Aa 1-1986], [Aa 6-1987], [Af 50-4], [Cu 2-3-1934],
" -v- insularis (See Bartschella insularis,
M. insularis),
schwartzii (Invalid name - See M. coahuilensis, Porfiria coahuilensis, Porfiria schwartzii),
[R], [Ar], [Aa-], [At-1982], [Cu 11:17],
schwarziana (See Coryphantha schwarziana),
schwarzii.....[Aq], [Ar], [Ct-1987], [Ff], [Iz], [Aa 1-1980], [Aa 2-1985], [Af 60-2], [Ah-1985],
[Am 2-1985], [Hw-1988],
" -f. (red-spined form).....[Aq],
scolymoides (See Coryphantha scolymoides)..[Y], [Df],
scrippsiana.....[Z], [RR], [ZZ], [VV], [Aq], [Hy],
[Ad-], [Ae-1981], [Aj 31-7-1980],
-v- armeria.....[Jb],
" -v- autlanensis.....[Af 20-9],
" -v- pseudoscrippsiana (See M. pseudoscrippsiana),
" -v- rooksbyana.....[Bo-5], [Iv 3-1952],
seideliana (See M. pygmaea, M. kunzeana),
[RR], [VV], [Il], [Ed 21:154, 1911],
[Hl 4-1981],

seitziana (See M. compressa),
[NN], [RR], [VV], [Bu], [Bv-1],
" -v- tolantongensis.....[Jb],
sempervivi.....[D], [NN], [QQ], [RR], [VV], [Ac], [Aq],
[Ar], [Ay], [Bv-1], [Cz], [Gr], [Il],
[Jd], [Af 14-7], [Af 55-4], [Af 59-2],
[Aj 31-1-1980],
" -v- caput-medusa (See M. caput-medusa),
[Ar], [Aj 31-7-1980],
" -v- pseudocrucigera (See M. pseudocrucigera),
[Iz],
" -v- rosa.....
" -v- tetracantha (See M. tetracantha)...[VV],
senilis (See Mamillopsis diguettii, Mamillopsis senilis).....[Aq], [Hv], [Iz], [Jd],
setispina (See Cochemiea setispina, M. pondii -v- setispina).....
[Aq], [Iz], [Jd], [Aa 1-1984], [Hl 5-1974],
sheldonii.....[J*], [P], [RR], [VV], [An], [Aq], [Ar],
[Av], [Ay], [Cs], [Ct-1987], [Cx], [Dp],
[Hx], [Aa 6-1987], [Ba 9-1980], [Flower - Ba 3-1981],
shurliana (See M. blossfeldiana -v- shurliana)
[VV], [Ar], [Bt], [Bv-1], [Hh July 1974],
[Hl 2-1975],
shurlyi (Invalid name - See M. schwarzii).....[Be*],
silvatica.....[Jb],
similis (See Coryphantha similis, Escobaria missouiensis -v- similis, Neobesseya similis),
[Y], [Df],
simplex (See M. mammillaris),
[Bo-6], [Hx], [Io], [Aj 11-4-1960], (Aj 33-7-1982], [Ba 5-1979],
" -v- albidispina.....
sinaloensis (See M. mazatlanensis),
sinistrohamata.....[VV], [Ac], [Aq], [Aa 4-1984],
[Af 42-3], [Aj 29-2-1978],
slevinii.....[Aq], [Cx], [Aa 5-1980], [Aj 27-6-1976],
sneedii (See Coryphantha sneedii, Escobaria sneedii),
[Y], [Df],
soehlemannii (See M. columbiana)......[Hx],
solisii (See M. nunezii -v- solisii),
[H], [RR], [VV], [Hy],
solisioides (See M. pectinifera -f. solisiodes),
[Aq], [Ar], [Av], [Ay], [Ff], [Iz], [Ae-1981], [Aj 30-6-1979], [Au Vol. 2, 1977 Zone 3 Compendium], [Bi Vol. 4 - 3], [Db 5-1980], [Hl 2-1979],
sonorensis.....[Aq], [Dq], [Ad 12-1981], [Ae-1981],
[Af 44:201, 1972], [Cu 4-1970],

sonorensis.....(cont'd),
" -v- bocensis.....
" -v- brevispina.....[VV], [Af 12:156, 1940],
" -v- canelensis.....
" -v- evermanniana.....
" -v- gentri.....[VV], [Af 12:10, 1940],
" -v- hiltonii.....[VV], [Af 12-156, 1940],
" -v- longispina.....[VV], [Af 12:156, 1940],
" -v- maccartyi.....[VV], [Af 12:10, 1940],
" -v- marksiana.....
" -v- tayloriorum.....
sphacelata.....[C], [H], [JJ], [NN], [VV], [Aq], [Ar], [Ay], [Ct-1987], [Cx], [Gr], [Hx*], [Hy], [Aa 1-1981], [Aa 6-1987], [Af 56-5], [Aw 2-1966], [Ba 4-1980], [Hl 1-1983],
" -v- viperina (See M. viperina).....[Iz],
sphaerica (See Dolichothele sphaerica, M. longimamma -v- sphaerica) [Df], [Aa 5-1982], [Af 55-6], [Ba 5-1979],
spinosissima (See M. auricoma, M. flava, M. kladiwae), [F], [H], [L], [X], [BB], [NN], [QQ], [RR], [VV], [Aq], [Ar], [Bv-1], [Gr], [Hu], [Hv], [Hx], [Il], [Jd], [N-1980], [Ab-1985], [Af 43:200 Fig. 138, 1971], [Af 58-2], [Aw 2-1983],
" -v- auricoma (See M. polycentra -v- auricoma),
" -v- brunnea (See M. polycentra -v- brunnea),
" -v- centraliplumosa (See M. centraliplumosa),
" -v- eastaneoides.....
" -v- flavida (See M. spinosissima -v- auricoma),
" -v- pilcayensis (See M. pilcayensis).....[Iz],
" -v- rubens.....[E],
" -v- " -f. cristata.....[Hy],
" -v- rubra.....[Dm],
" -v- sanguinea.....
" -v- spinosissima.....[Iz],
" -v- virginis (See M. virginis),
stampferi (See M. longiflora f. stampferi), [Aj 30-8-1979],
standleyi.....[G], [CC], [RR], [VV], [Aq], [Cx],
" -v- lindsayi.....
" -v- robustispina (See M. standley).....[VV],
stella-de-tacubaya.....[Aq], [Ct-1987],
" -v- aurata.....
" -v- paradensis (See M. stella-de-tacubaya),
stenocephala (See M. polythele).....[Af 30-2],
strobilina.....(See M. collinsii or M. karwinskiana - form).....[Af 42:182, 1980], [Dv 48: 329, 367, 1933],
subangularis (See M. compressa),
" -v- rubrispina.....[Hy],

subcrocea (See M. elongata).....[Hy],
" -v- echinata (See M. elongata -v- echinata),
[Hy],
" -f. cristata.....[Hy],
subducta -v- subducta (See M. laui -v- subducta).....
" -v- dasyacantha (See M. laui -v- dasyacantha),
subdurispina (See M. kewensis)
[Bo-5], [Ff], [Hy], [Iv 9-1952],
subpolyedra (See M. polyedra).....[RR],
subtetragona (See M. carnea).....
subtilis (See M. pilispina),
[Ac], [Bo-5], [Ia], [Cu 12:81, 1950],
subcata (See Coryphantha sulcata).....[Y], [Df],
supertexta (See M. elegans -v- supertexta),
[Il], [Aa 6-1981], [Am 6-1980],
[Cu 39:97-99, 1977],
" -v- leucostoma (Invalid name),
[Bo-5], [Aj 26-8-1975],
supraflumen.....[Jb],
suprema.....[Aa 6-1984],
surculosa (See Dolichothele surculosa),
[D], [H], [NN], [VV], [Aq], [Ay], [Hv], [Iz],
[Jd], [Af 55-6], [Af 60-2], [Ag-1985],
[Aj 29-11-1978], [Am 2-1979], [Hl 4-1974],
[Hw-1986],
swinglei (See M. inaiae),
[R], [EE], [VV], [Aq], [Ar], [Hx], [Jd],
[Aa 6-1987], [Aj 9-7-1958],
" -v- gueldemanniana (See M. gueldemanniana),
" -v- sheldonii.....
tamayonis (tamajonis) (See M. columbiana, M. hennissii),
[Ef 12 No. 38:62 Fig. 1949],
tarajensis (See M. hahniana -v- giseliana),
[Bo-6], [Cu 8-3-1946],
tayloriorum.....[Z], [Aq], [Ct-1987], [Aa 4-1980],
[Af 47:173, 1975], [Aw 2-1979],
tegelbergiana (See M. albilanata -v- tegelbergiana),
[VV], [Aq], [Bu], [Bv-1], [Ff], [Hy],
tenampensis (See M. sartorii),
[RR], [VV], [Bv-3], [Hx*], [Af 6-5],
tentaculata.....
tenuis (See M. elongata -v- tenuis).....[Io*],
tenunicaulis (See M. gracilis).....[Hy],
tesopacensis (See M. sonorensis).....
[VV], [Hx*], [Af 51-5],
" -f. brevispina.....
" -v- papasquiarensis (See M. wagneriana)
[Af 52-5], [Aw 4-1966], [Hl 4-1977],
" -v- rubriflora (See M. sonorensis),
[Ad June - 1979],

tetracantha (See M. polythele, M. dolichocentra, M. hidalgensis),
[RR], [VV], [XX], [Bv-1], [Cm], [Cz], [Aj 29-6-1978],
" -v- galeottii (See M. galeottii, M. obconella -v- galeottii).....[Bv-1],
" -v- longispina......[Ia],
tetracentra (See M. sempervivi),
tetrancistra (See Phellosperma tetrancistra),
[B], [M*], [P], [BB], [GG], [LL*], [RR], [An], [Aq], [Ax], [Ca], [Do], [Iz], [Jd], [Aa 5-1984], [Au 37-2], [Db 4-1979], [Hl 4-1982],
theresae.....[G], [I], [Z], [NN], [Aq], [Ar], [Av], [Ay] [Bt], [Ct-1987], [Ff], [Gr], [Hu], [Hv], [Hx], [Iz], [Jd], [Ae-1982], [Af 39:239, 1967], [Af 51-1], [Aj 33-1-1982], [Au 37-4], [Au Vol. 2 - 1977 Zone 3 Compendium], [Flower - Ba 2-1978],
" -f. cristata.....[Cu 3-1975],
thornberi......[GG], [Ac, [An], [Aq], [Ax], [Iz], [Aa 3-1982], [Af 60-5],
" -v- yaquensis (See M. yaquensis),
tiegeliana (Invalid name - See M. perbella)
[D],
tlalocii.....[Jb],
tolimensis (Form of M. compressa),
[H], [Hy], [Ad Jan. 1978], [Ag-1986], [Aw 2-1963],
" -v- brevispina.....[VV], [Hy],
" -v- longispina.....[VV], [Bv-1], [Ia], [Hl 6-1976],
" -v- subuncinata.....
tonalensis.....[Aq], [Ct-1987], [Iz], [Aa 6-1981], [Af 52-2], [Am 6-1981], [Hl 6-1982],
trichacantha (See M. nana),
[NN], [RR], [TT], [VV], [Ar], [Bv-1], [Hy], [Aa 6-1980], [Aj 28-6-1977],
trohartii (See M. magnimamma, M. phymatothele),
[Gr],
tropica.....[Jb],
tuberculosa (See Coryphantha tuberculosa, Escobaria strobiliformis, Escobaria tuberculosa),
[Y], [Df],
tulensis-stachellos.....
uberiformis (See M. longimamma -v- uberiformis),
[Hl 4-1974],
umbrina (See M. ernestii, M. virginis)....[VV], [Hx],
uncinata.....[E], [L], [VV], [Ac], [Aq], [Ar], [Cx], [Hv], [Hy], [Il], [Af 54-6], [Af 56-5],

uncinata.....(cont'd),
" -v- bihamata.....
" -v- biuncinata.....
" -v- lloydii (See M. lloydii).....
unihamata.....[Z], [Ad -], [Ag-1988], [Aj 26-7-1975], [Ea 1937:40, Fig.],
vagaspina (M. magnimamma complex), [VV], [Ar], [Bv-3], [Hy], [Af 56-5], [Hl 5-1976],
" -f. cristata.....[Af 56-5],
vallensis.....[Jb],
" -v- brevispina.....[Jb],
variabilis (See M. limonensis),
varicolor (See Coryphantha dasyacantha -v- varicolor, Escobaria varicolor), [Y], [Df],
varieaculeata.....[Aq], [Iz], [Aa 1-1982], [Aa 6-1984], [Aj 4-1966], [Aj 29-6-1978], [Aj 29-8-1978],
vaupelii (See under M. haageana), [C], [CC], [VV], [Cz], [Hx], [Hw-1981], [Hy], [Ia], [Af 7-21, 1935], [Bi-1],
" -v- flavispina (See M. tegelbergiana).....[VV],
" -f. cristata.....[Bu], [Bv-1], [Cz],
velthuisiana (See M. petterssonii).....[Bo-V],
verhaertiana.....[NN], [VV], [Ar], [Bo-6], [Hy],
verticealba.....[Jb],
" -v- zacatecasensis.....
vetula (See M. kuentziana, M. magneticola, M. radiaissima).....[RR], [VV], [Aq], [Bo-6], [Jd],
viereckii (See M. picta -v- viereckii)....[NN], [VV], [Aq], [Bv-1], [Ck], [Ct-1987], [Dq], [Hx*], [Il], [Jd], [Af 5-10], [Aj 9-4-1958],
" -v- brunispina.....[Bo-5],
" -f. cristata.....[D],
viescensis.....[Ae-1982], [Hl 4-1973],
villifera (See M. polyedra).....[Hx*],
viperina (See M. elongata -v- viperina), [DD], [JJ], [NN], [VV], [Ay], [Ct-1987], [Cx], [Aa-], [Jd], [Af 17-1],
virginis (See M. spinosissima -v- virginis), [Aq], [Hx*],
" -v- robusta.....
viridiflora (See M. wrightii -v- viridiflora), [Aq], [Ax], [Hl 6-1972], [Hl 6-1983],
vivipara (See Coryphantha vivipara, Escobaria vivipara)
" -v- arizonica (See Coryphantha vivipara -v- arizonica).....[Y], [Df],
" -v- borealis (See Coryphantha vivipara -v- borealis).....[Y], [Df],

vivipara.....(cont'd),
" -v- neomexicana (See Coryphantha vivipara -v- neomexicana).....[Y], [Df],
" -v- radiosa (See Coryphantha vivipara -v- radiosa).....[Y], [Df],
" -v- vivipara (See Coryphantha vivipara -v- vivipara).....[Y], [Df],
voburnensis (See M. woburnensis), [RR], [VV], [Aq], [Ar],
" -v- beiselii (See M. beiselii),
" -v- collinsii (See M. collinsii),
" -v- eichlamii (See M. eichlamii).....[Iz],
" -v- quetzalcoatl.....[Jb],
vonwyssiana (See M. auriareolis, M. avila-camachoi, M. rosensis, M. tiegeliana, M. parkinsonii), [Eg 1945-1946, 171 Fig.], [Hx*],
wagneriana.....[Aq], [Ct-1987], [Hy], [Dw 4:199, Fig.-1932], [Hl 4-1977],
" -v- tortulospina (See M. obscura -v- wagneriana -f. tortulospina), [Ae-1981], [Af 58-2], [Ah-1985],
waltheri (See M. heyderi -f. waltheri), [Ct-1987], [Dy 3:72 Fig. 1927], [Iv 4-1949],
weingartiana (See M. unihamata).........[VV], [Aq], [Ct-1987], [Iz], [Dw 4:219 Fig. 1932],
werdermanniana (See M. hahniana -v- werdermanniana), [Bv-1], [Cu 8-4-1946], [Ea 204:1933],
wiesingeri.....[VV], [Aq], [Ct-1987], [Aw 3-1987], [Ea 204 Fig. 1933],
" -v- erectacantha (See M. erectacantha)...[Iz],
wilcoxii (See M. wrightii -v- wilcoxii, M. meridiorosei, M. santaclarensis), [Y], [DD], [RR], [VV], [An], [Av], [Bn], [Df], [Hx*], [Ae-1982], [Af 14-5], [Aj 27-2-1976], [Ba 5-1979],
" -v- viridiflora (See M. wrightii -v- viridiflora).....[VV],
" -v- wolfii (See M. wrightii -f. wolfii), [Aj 1-1976],
wildii.....[C], [H], [S], [T], [RR], [VV], [Ac], [Aq], [Ar], [Be*], [Bu], [Bv-1], [Cm], [Cr], [Hu], [Hx*], [Hy], [Il], [Jd], [Aa 4-1984], [Aj 24-6-1973], [Am 2-1981],
" -f. calleana (See M. calleana)
" -f. crinita (See M. crinita),
" -f. cristata.....[Cr], [Dn], [Ds], [Hn], [Ia], [Ig], [AJ 23-4-1972],
winterae.....[CC], [VV], [Aq], [Ar], [Bv-1], [Cm], [Ct-1987], [Hy], [Iu], [Iz], [Aa 2-1981], [Af 58-2], [Aj 30-11-1979], [Aw 3-1968],

MAMMILLARIA.....(Cont'd),

woburnensis (See M. voburnensis)
[RR], [VV], [Gr], [Hy],
wohlschlageri sp. nov. .....[Jb],
wolfii (See M. wrightii -v- wolfii).....[Ae-1981],
woodsii (See M. hahniana, M. mendeliana),
[NN], [QQ], [VV], [Ar], [Bm], [Bv-1], [Cr],
[Hx], [Hy], [Ia], [Il], [N-1980], [Ae-1981],
[Af 15-3],
wrightii....[G], [P], [Y], [Q], [GG], [RR], [VV], [An],
[Ar], [Ax], [Bz], [Df], [Dq], [Hx*], [Jd],
[Aa 1-1985], [Af 6-2], [Af 14-5], [Aj 27-2-1976], [Hl 6-1972],
" -v- viridiflora (See M. viridiflora, M. wilcoxii -v- viridiflora),
[B], [P], [An],
" -v- wilcoxii (See M. wilcoxii).....[P], [An],
[Aq], [Ax], [Df], [Iz], [Aa 2-1986],
" -f. wolfii (See M. wolfii),
[Aq], [Ct-1987], [Cu 4-1979],
[Hl 5-1982], [Hm 3-1979],
" -v- wrightii.....[Aq], [Ax], [Df],
wuthenauiana (See M. nunezii).....[Bo-5:3444-1961],
[Aa 15:25 - 1975], [Ff 25-1973],
xaltianguensis.....[Aq], [Ae-1981], [Aj 29-10-1978],
[Am 3-1979], [Hl 1-1983],
" -v- aguilensis.....[Jb],
xanthina.....[RR], [VV],
xochipilli.....[Jb],
yaquensis.....[X], [VV], [Aq], [Ay], [Hx], [Iz],
[Jd], [Aa 15:30-1975], [Af 57-4], [Aj 34-1-1983], [Am 7-1983], [Db 6-1979],
yucatanensis (See M. columbiana).....[RR], [VV], [Aq],
zacatecasensis.....[NN], [Aq], [Ct-1987], [Aa 4-1984],
[Cu 22:51, 1960], [Ff 26 Fig. 1973],
zahniana (See M. winterae).....[VV], [Ct-1987], [Jd],
[Aa-], [Dw 1:120 Fig. 1929],
zaltiaguensis.....
" -v- aguilensis.....
zapilotensis (See M. guerreronis -v- zapilotensis,
M. guerreronis).....[VV],
zeilmanniana.....[H], [T], [X], [EE], [QQ], [TT], [VV],
[Aq], [Ar], [Av], [Ay], [Bv-1], [Cs], [Gr],
[Hu], [Hv], [Hw-1981], [Hx*], [Hy], [Il],
[Iz], [Jd], [Ad 3-1983], [Af 60-6], [Aj 30-3-1979], [Ba 10-1978], [Dg 7-1],
" -v- alba (See M. albiflora, M. herrerae -v- albiflora).....[Hv], [Il],
" -v- robusta.....
" -f. cristata.....[Ab-1985], [Aj 9-7-1958],
[Aj 27-2-1976],

MAMMILLARIA.....(Cont'd),

zephranthoides (See Dolichothele zephranthoides), [S], [NN], [RR], [VV], [Ac], [Aq], [Hu], [Hv], [Hx], [Ae-1982], [Ah-1980], [Bm-1988],
zeyeriana.....[CC], [NN], [VV], [Ad], [Aq], [Ar], [Hy], [Iu], [Jd], [Aa 4-1981],
zubleri.....[Jb],
zuccariniana.....[F], [RR], [VV], [Aq], [Ct-1987], [Hx*], [Hy], [Ij], [Ad Oct. 1980], [Ai 1978-1979],
" -v- ochroleuca.....
" (3 varieties).....[Hy],

Spec. Ab - G-78-029 (Pueblo Nuevo).....[Af 56-5],
Spec. BK-100.....[Aj 30-2-1979],
Spec. de Cuicatlan.....[Ae-1981],
Spec. East of Aramberri, N. L. Mexico (Hinton s.n. 1987) [Af 59-3],
Spec. FR-759.....[Db - 26th Anniv. Edition],
Spec. GS-7, GS-15, GS-35], GS-36, GS-50, GS-52..... [Aj 33-7-1982],
Spec. Lau-060.....[Ae-1981],
Spec. Lau-066.....[Aj 27-3-1976],
Spec. Lau-086.....[Ae-1981],
Spec. Lau-694.....[Aj 30-10-1979],
Spec. Lau-777.....[Af 50-2], [Aj 27-8-1976], [Ba 9-1978]
Spec. Lau-1044.....[Af 2-1980],
Spec. Lau-1045.....[Ae-1981],
Spec. Lau-1095.....[VV],
Spec. Lau-1108.....[Ae-1981],
Spec. Lau-1171.....[Af 50-4],
Spec. Lau-1186 & 1186A.....[Am 6-1981],
Spec. Lau-1219.....[Af 50-4],
Spec. Lau-1222.....[Af 50-4],
Spec. Lau (Tepalcingo).....[Aa-],
Spec. Rep.-841 (Tepalcingo).....[Aa 3-1980],
Spec. Tenango del valle.....[Ae-1981],

MAMMILLOYDIA

candida (See Mammillaria candida).....[NN], [Jd],
ortiz-rubiona (See Mammillaria candida, Mammillaria ortiz-rubiona).....[NN],

MARENOPUNTIA

marenae (See Opuntia marenae), [DD], [Ay], [Bo-1], [Hx*], [Hy], [Eh 22:27, 1950],

MARGINATOCEREUS

lomatopus (See M. marginatus).....[Hy],
marginatus (See Cereus marginatus, Lemaireocereus marginatus, Pachycereus marginatus, Stenocereus marginatus),
[W], [X], [DD - Flower], [Ar], [Ay], [Bo-4], [Bo-6], [Bt], [Hx*], [Jd], [Af 51-6], [Bm-1988],
" -v- gemmatus.....[Bo-4],
" -v- oaxacensis.....[Bo-6],

MARITIMOCEREUS

gracilis (See Borzicactus gracilis, Borzicactus sextonianus, Loxanthocereus gracilis, Loxanthocereus sextonianus),
[Aj 10-9-1959],
nanus (See Borzicactus nanus, Loxanthocereus nanus),
[Bu],
sextonianus (See Borzicactus sextonianus pro parte, Erdisia sextonianus, Loxanthocereus sextonianus,
[Bv-1],

MARNIERA

chrysocardium.....(See Epiphyllum chrysocardium),
[E], [DD], [QQ], [Ay], [Az], [Bo-2], [Bt], [Bu], [Hx*], [Jd], [Ad-], [Aj 5-1976], [Ba 12-1976],
macroptera (See Epiphyllum macropterum),
[DD], [Bo-2], [Bo-6], [Hx*], [Hy], [Af 23:17-1951],
" -v- kimnachii (See Epiphyllum crenatum -v- kimnachii),

MARSHALLOCEREUS

aragonii (See Cereus aragonii, Lemaireocereus aragonii)
[Hx*], [Hy], [Af 23:121-1951],
chacalapensis (See Ritterocereus chacalapensis, Stenocereus chacalapensis),.....[Bo-4],
thurberi (See Cereus thurberi, Lemaireocereus thurberi, Pachycereus thurberi, Stenocereus thurberi),
[An], [Ar], [Bo-4], [Hu], [Hx*], [Hy],
" -v- litoralis (See Lemaireocereus thurberi -v- litoralis, Stenocereus thurberi -v- litoralis),
[Bo-4],
" -f. cristata.....[Ay],

MATUCANA

aurantiaca (See Borzicactus aurantiacus, Submatucana aurantiacus),
[F], [I], [K], [Bo-6], [Bt], [Jd], [Cu 1-1971], [Hi-4],
" -v- calvescens (See M. aurantiaca -v- densispina).....
" -v- densispina.....[Hi-4],
aureiflora (See Borzicactus aureiflorus, Submatucana aureiflora),
[Av], [Ay], [Hi-4], [Hv], [Ab-1985], [Ag-1983] [Ah-1985], [Aj 28-4-1977], [Hl 2-1982],
" -v- elata.....
blancii (See Borzicactus haynei -v- perplexus),
[DD], [Bo-2], [Af 51-1],
" -v- nigriarmata (See Borzicactus haynei -v-perplexus),
[Bo-2], [Au 11:71, 1956],
breviflora (See Borzicactus haynei -v- breviflorus, M. multicolor -v- breviflora).....Bo-2],
bruneispina.....[Hl 2-1982],
brunnescens.....[Iv 12-1956],
calliantha (See Borzicactus callianthus),
[Hi-4], [Hl 1-1974], [Hw-1981],
" -v- prolifera.....[Hi-4],
calocephala (See Borzicactus calocephalus).....[Hx*],
calvescens (See Borzicactus calvescens, Borzicactus aurantiacus -v- calvescens, Matucana aurantiaca, Matucana calvescens, Submatucana aurantiacus, Submatucana calvescens),
celendinensis (See Borzicactus intertextus -v-celendinensis, Submatucana celendinensis),
[Hi-4],
cereoides (See Borzicactus haynei, M. haynei,
[Bo-2], [Il], [Da I],
colorisplendida (See M. yanganucensis),
" -v- grandiflora (See M. yanganucensis -v-grandiflora, M. yanganucensis),
" -v-setosa (See M. yanganucensis -v- setosa),
comacephala (See Borzicactus comacephalus),
[Gr], [Hi-4], [Af 50-6], [Aj 22-3-1971], [Hs II-4-1981],
" -f. (Type DM 28 from Llameltin).....
[Hh Sept. 1973],
" -f. (Type Lau 185 from Huaytani).....
[Hh Sept. 1973],
" -f. (Ecotype FR-587 from E. Cord. Blanca).....
[Hh Sept. 1973],
crinifera (See Borzicactus criniferus, B. haynei -v-perplexus).....Bo-6], [Hi-4], [Hv], [Hy],
[Af 50-6], [Aj 22-3-1971],

currundayensis (See Submatucana currundayensis, Borzicactus aurantiacus -v- calvescens, Borzicactus calvescens),
[Bo-2], [Hi-4], [Aj 28-11-1977], [Hl 6-1981],
elongata (See M. haynei -v- elongata),
[Bo-2], [Hx*], [Aj 22-3-1971],
formosa (See Borzicactus formosus),
[I], [Ay], [Hi-4], [Ad Oct. 1979], [Hl 1-1974],
" -v- minor (See Borzicactus formosus, Borzicactus formosus -v- minor),
[Hi-4],
" -v- paucicostatus (See Borzicactus paucicostatus, Submatucana paucicostata),
" -v- tuberculatus (See Borzicactus tuberculatus),
fruticosa (See Borzicactus fruticosus).....[Hi-4],
hastifera.....[Hi-4],
haynei (See Borzicactus haynei, M. cereoides, M. variabilis).....[F], [Z], [QQ], [RR], [Ar], [Av], [Ay], [Bl], [Bo-2], [Dq], [Hi-4], [Hx*], [Hy], [Jd], [Af 51-1], [Ag-1988], [Aj 23-2-1972], [Hl 6-1971], [Hl 1-1974],
" -v- atrispina (See Borzicactus haynei -v- atrispina, M. elongata, M. multicolor),
" -v- blancii.....[Hw-1989],
" -v- elongata (See M. elongata).....[Hi-4],
" -v- erectipetala (See M. haynei).....[Bo-2],
" -v- gigantea (See M. multicolor).....
" -v- hystrix (See M. hystrix, M. multicolor -v- hystrix, M. robusta),
[Aj 28-11-1977],
" -v- multicolor (See M. multicolor, M. haynei -v- gigantea),
" -v- pectinifera.....[Hy],
" -v- perplexa (See Borzicactus perplexus, M. herzogiana -v- perplexa, M. yanganucensis)
herzogiana (See Borzicactus haynei -v- perplexus, M. haynei -v- perplexa, M. yanganucensis),
[Bo-2], [Au 11:71, 1956],
" -v- perplexa (See Borzicactus haynei -v- perplexus, M. haynei -v- perplexa),
[Au 11:71, 1956],
huagalensis (See Borzicactus huagalensis, Submatucana huagalensis),
huarinensis.....[Hh Sept, 1973],
humboldtii (See Borzicactus humboldtii, Seticereus humboldtii, Seticereus neohumboldtii),

MATUCANA.....(Cont'd),

hystrix (See Borzicactus hystrix, M. multicolor -v- hystrix, M. robusta),
[NN], [Bo-2], [Hx*],
" -v- armillata.....
" -v- atrispina (See Borzicactus haynei -v- atrispinus, M. haynei -v- atrispina),
" -v- nigrispina.....
" -v- umadeavoides.....
icosagona (See Binghamia aurivillus, Borzicactus aurivillus, Borzicactus icosagonus, Seticereus icosagonus),
intertexta (See Borzicactus intertextus, Submatucana intertexta).....[Hi-4], [Hw-1988],
krahnii (See Borzicactus krahnii, Submatucana krahnii),.....[Av],
madisoniorum (See Borzicactus madisoniorum, Eomatucana madisoniorum, Submatucana madisoniorum),
[Ay], [Jd], [Ab-1985], [Af 56-5], [Ag-1985], [Ah-1985],
" -v- asterium.....
" -v- madisoniorum (See Borzicactus madisoniorum, Submatucana madisoniorum),
[Aj 28-4-1977], [Aj 31-6-1980],
" -v- paucicostata (See Borzicactus paucicostatus, Submatucana paucicostata),
" -v- pujupatjii (See Borzicactus madisoniorum -v- pujupatjii),
[Hl 1-1974],
megalantha (See Borzicactus aurantiacus -v- megalanthus, M. herzogiana -v- perplexa, Matucana yanganucensis -v- suberecta),
[Hi-4],
mirabilis (See Arequipa mirabilis, Borzicactus mirabilis),
multicolor (See M. haynei -v- gigantea),
[I], [Bo-2], [Bo-6], [Hi-4], [Hx*], [Db 2-1979], [Hs II-4-1981],
" -v- armillata.....
" -v- breviflora (See M. breviflora),
" -v- hystrix (See Borzicactus haynei -v- atrispinus, M. hystrix),
[Hi-4],
" X Oreocereus ritteri (Natural hybrid),
[Hi-4],
myriacantha (See Borzicactus myriacanthus),
[Hi-4], [Hv], [Af 50-2], [Aj 28-10-1977 - Flower], [Aj 30-9-1979], [Hs II-4-1981],
oreodoxa (See Borzicactus oreodoxa, Eomatucana oreodoxa).....[Ag-1985],

MATUCANA.....(Cont'd),
pallarensis (See Borzicactus weberbaueri -v- pallarensis),
[Hi-4], [Hv], [Aj 22-11-1971],
paucicostata See Borzicactus paucicostatus, Submatucana paucicostata),
polzii.....
purpureo-alba......[Hi-4],
rarissima (Natural hybrid - Oreocereus ritteri X M. multicolor).....[Bo-6], [Aj 22-3-1971],
ritteri (See Borzicactus ritteri, Submatucana ritteri),
[Ay], [Bo-2], [Hi-4], [Hl 2-1982],
robusta (See M. hystrix, M. multicolor -v- hystrix)
Bo-6], [Aj 23-2-1972],
roseoalba.....[Hh Sept. 1973], [Ae-1981], [Af 50-3],
[Hw-1982],
setosa (See M. yanganucensis -v- setosa)
supertexta.....(See M. variabilis).....[Hi-4],
tuberculata (See Borzicactus tuberculatus)
tuberculosus (See Borzicactus tuberculosa),
[Hi-4], [Hl 1-1974],
turbiniformis (See Submatucana turbiniformis),
variabilis (See Borzicactus variabilis, Borzicactus supertextus, M. supertexta),
[Bo-2], [Hx*], [Aj 22-3-1971], [Hs II-4-1981],
" -v- fuscata.....[Bo-2],
" -v- mirabilis (See Borzicactus mirabilis),
variicolor (See Arequipa variicolor, Oreocereus variicolor, Submatucana variicolor),
[Aj 22-3-1971],
weberbaueri (See Borzicactus weberbaueri),
[Gr], [Hi-4], [Hu], [Hv], [Af 50-2],
[Aj 30-9-1979],
" -v- blancii (See M. yanganucensis -v- albispina)
[Iv 12-1956],
" -v- flammea (See Borzicactus weberbaueri -v- flammeus),
[Ay], [Af 50-2], [Aj 30-9-1979],
winteri (winteriana).....[Bo-6], [Hi-4],
yanganucensis (See Borzicactus yanganucensis, Matucana herzogiana),
[T], [Av], [Ay], [Bo-2], [Gr], [Hi-4],
[Hy], [Hs II-4-1981],
" -v- albispina (See M. yanganucensis),
[Bo-2],
" -v- fuscispina (See M. yanganucensis),
[Bo-2],
" -v- grandiflora (See M. colorisplendida -v- grandiflora, M. yanganucensis)
" -v- longistyla (See M. yanganucensis),
" -v- parviflora (See M. yanganucensis)...[Bo-2],
" -v- salmonea (See M. yanganucensis).....[Bo-2],

MATUCANA......(Cont'd),

yanganucensis.....(Cont'd),
" -v- santiensis.....
" -v- setosa (See M. colorisplendida -v- setosa),
[Hs II-4-1981],
" -v- suberecta (See M. megalantha),
[Bo-2],
" -v- yanganuca.....

Spec. ......[Bo-2],

MEDIOCACTUS

coccineus (See Cereus setaceus, Heliocereus coccineus),
[RR], [Bo-2], [Dq*], [Hx*], [Hy*], [Jd],
[Aj 15-10-1964],
" -v- xanthocarpus.....
hahnianus (See Harrisia hahniana),
[Bo-2], [Bo-6],
hassleri (See Selenicereus hassleri),
[Bo-2],
" -f. monstrosus.....[Bo-2],
lindmannii (See Selenicereus lindmannii),
[L], [Bo-2, 798, 1959],
megalanthus (See Selenicereus megalanthus)
[MM], [RR],
pomifer (See Selenicereus pomifer).....
setaceus (See Cereus setaceus, M. coccineus),

MEDIOLOBIVIA

albilongiseta (See M. aureiflora -v- albilongiseta,
M. aureiflora, Rebutia aureiflora),
albopectinata (See Digitorebutia albopectinata, Rebu-
tia albopectinata),
atrovirens (See M. pectinata -v- atrovirens, Rebutia
pygmaea -v- atrovirens),
auranitida (See Rebutia auranitida),
[Bo-3], [Hx*], [Dt 1:18, 1947], [Hl 2-1969],
" -v- flaviflora (See Rebutia auranitida),
[Bo-3],
" -v- gracilis (See Rebutia auranitida -f. gra-
cilis),
aureiflora (See Rebutia aureiflora).....[C], [J*], [T],
[EE], [NN], [QQ], [Bo-3], [Cm], [Hx], [Hy],
[Jd], [Bi 2-3], [Cu 7-2-1938],
" -v- albilongiseta (See Rebutia aureiflora),
[Hr],
" -v- a;biseta (See Rebutia aureiflora),
[Hy],
" -v- albispina (See Rebutia aureiflora),

aureiflora.....(Cont'd),
" -v- boedeckeriana (See M. boedeckeriana, Rebutia aureiflora -f. boedeckeriana) [Bo-3], [Hy],
" -v- duursmaiana (See Rebutia duursmaiana, Rebutia aureiflora -f. duursmaiana), [Hx*], [Hy], [Aj 31-2-1980],
" -v- leucolutea (See Rebutia aureiflora),
" -v- lilacinostoma (See Rebutia aureiflora),
" -v- longiseta (See Rebutia aureiflora -f. longiseta).....[Hr], [Hx*],
" -v- rubelliflora (See M. rubelliflora, Rebutia aureiflora -f. rubelliflora), [Bo-3], [Hb], [Hr],
" -v- rubriflora (See Rebutia aureiflora -f. rubriflora), [Bo-3], [Hx*], [Aj 24-9-1973],
" -v- sarothroides (See M. sarothroides, Rebutia aureiflora -f. sarothroides),

blossfeldii (See M. aureiflora -v- rubriflora, Rebutia aureiflora -f. rubriflora),
" -v- compactiflora (See M. aureiflora -v- rubriflora, Rebutia aureiflora -f. rubriflora),
" -v- nigrilongiseta (See M. aureiflora -v- rubriflora, Rebutia aureiflora -f. rubriflora),

boedeckeriana (See M. aureiflora -v- boedeckeriana, Rebutia aureiflora -f. boedeckeriana), [Cr], [Ig*],

brachyantha (See M. costata -v- brachyantha, Rebutia brachyantha).....[EE], [Bo-3], [Hb],

bruneoradicata (See Rebutia bruneoradicata),

brunescens (See Digitorebutia brunescens, Rebutia brunescens).....[DD-484],

canacruzensis (See Digitorebutia canacruzensis, Rebutia canacruzensis),

carmeniana (See Digitorebutia carmeniana, Rebutia carmeniana),

christinae (See Digitorebutia christinae, Rebutia christinae),

cincinnata (See Digitorebutia cincinnata, Rebutia cincinnata),

columnaris (See M. conoidea -v- columnaris, Rebutia einsteinii -f. columnaris),

conoidea (See Rebutia conoidea, Rebutia einsteinii -f. conoidea).....[Ee 1:19-1947],
" -v- columnaris (See M. columnaris, Rebutia einsteinii -f. columnaris),

costata (See Rebutia costata),
[Bo-3], [Bp], [Bt], [Hx*],
" -v- brachyantha (See M. brachyantha, Rebutia brachyantha),
" -f. eucaliptana (See M. eucaliptana, Rebutia costata -f. eucaliptana, Rebutia eucaliptana),
diersiana (See Rebutia diersiana),
digitiformis (See Digitorebutia digitiformis, M. pectinata -v- digitiformis, Rebutia digitiformis, Rebutia pectinata -v- digitiformis, Rebutia pygmaea).....[Bu],
duursmaiana (See M. aureiflora -v- duursmaiana, Rebutia aureiflora -f. duursmaiana, Rebutia duursmaiana),
einsteinii (See M. schmiedcheniana -v- einsteinii, Rebutia einsteinii, Rebutia schmiedcheniana -v- einsteinii).....[Hy], [Hl 5-1968],
elegans (See Rebutia aureiflora -v- elegans),
[V], [Bo-3], [Hy], [Ia],
eos (See Digitorebutia eos, Rebutia eos),
[G], [DD-485], [Hx*],
euanthema (See Acantholobivia euanthema, Rebutia euanthema).....[Bo-3], [Hy],
" -v- fricii (See R. euanthema -f. fricii),
[DD], [Bo-3], [Hx],
" -v- oculata (See Rebutia euanthema -f. oculata, Digitorebutia euanthema -v- oculata),
[T], [Bo-3],
" -v- pygmaea (See M. neopygmaea, Rebutia euanthema -f. neopygmaea),
" -v- ritteri (See Digitorebutia ritteri, M. ritteri, Rebutia ritteri),
eucaliptana (See M. costata -f. eucaliptana, Rebutia costata -f. eucaliptana, Rebutia eucaliptana)
[Hx], [Hy], [Jd],
friedrichiana (See Digitorebutia friedrichiana, Rebutia friedrichiana),
fuauxiana (See Rebutia fuauxiana),
[Bo-3], [Hx*],
haagei (See Digitorebutia haagei, M. pygmaea, Rebutia haagei, Rebutia pygmaea),
[T], [Bp], [Cr], [Hy], [Ig*],
haefneriana (See Rebutia pygmaea -v- haefneriana),
[T], [Hx*], [Aj 6-2-1955], [Aj 6-3-1955],
hirsutissima (See Lobivia tiegeliana -v- pusilla, Rebutia hirsutissima),
[Hx*],
iscayachensis (See Digitorebutia iscayachensis, Rebutia iscayachensis),

ithyacantha (See Rebutia fiebrigii -v- vulpes, Rebutia ithyacantha),
[Hb], [Af 42-1],

kesselringiana (See M. aureiflora -v- rubriflora, Rebutia aureiflora -f. rubriflora, Rebutia kesselringiana),
[NN], [Cp],

mudanensis (See Digitorebutia mudanensis, Rebutia mudanensis),

neopygmaea (See M. euanthema -v- pygmaea, Rebutia euanthema -f. neopygmaea),
[Bo-3], [Hy],

nigricans (See Rebutia ritteri -v- nigricans),
[Dk], [Hx*], [Hy],

" -v- peterseimii (See Rebutia peterseimii, Rebutia ritteri -f. peterseimii),

orurensis (See Digitorebutia orurensis, M. pectinata -v- orurensis, Rebutia orurensis, Rebutia pectinata -v- orurensis, Rebutia pygmaea),
[Bp], [Hl 4-1978],

padcayensis (See M. pygmaea, Rebutia padcayensis, Rebutia pygmaea),

patericalyx (See Rebutia patericalyx).....

" -v- odontopetala (See Rebutia odontopetala, Rebutia patericalyx),

pectinata (See Digitorebutia pectinata, Rebutia pectinata, Rebutia pygmaea,
[Bo-3], [Bp], [Gr], [Hr], [Hy], [Jd],

" -v- atrovirens (See Rebutia pygmaea -f. atrovirens).....[Hx], [Hy],

" -v- digitiformis (See Digitorebutia digitiformis, Rebutia pectinata -f. digitiformis, Rebutia pygmaea),
[T], [Bo-3], [Bp], [Hx*],

" -v- neosteinmannii (See M. pygmaea, Rebutia neosteinmannii, Rebutia pygmaea, Rebutia pygmaea -f. neosteinmannii),

" -v- orurensis (See Digitorebutia orurensis, M. orurensis, Rebutia orurensis, Rebutia pygmaea).....[Bp], [Hy],

pygmaea (See Digitorebutia pygmaea, Rebutia haagei, Rebutia pygmaea),
[T], [EE], [Bo-3], [Hx*],

" -v- flavovirens (See Rebutia pygmaea -f. flavovirens),

raulii (See Digitorebutia raulii, Rebutia raulii),

rauschii (See Digitorebutia rauschii, Lobivia rauschii (Zecher), Rebutia rauschii),
[Ay], [Cv-2],

MEDIOLOBIVIA.....(Cont'd),

ritteri (See Digitorebutia ritteri, Rebutia ritteri,
[J*], [EE], [NN], [Bo-3], [Hy], [Il], [Hl 2-1972],
" -v- peterseimii (See Rebutia peterseimii, Rebutia ritteri -f. peterseimii)
" -v- pilifera (Invalid name - See Rebutia ritteri),
rosalbiflora (See Rebutia rosalbiflora),
" -v- amblypetala (See Rebutia rosalbiflora -v- amblypetala),
rubelliflora (See M. aureiflora -v- rubelliflora, Rebutia aureiflora -f. rubelliflora),
[F],
rubriflora (See M. aureiflora -v- rubriflora, Rebutia aureiflora -f. rubriflora),
sarothroides (See M. aureiflora -v- sarothroides, Rebutia aureiflora -f. sarothroides),
[NN],
schmiedcheniana (See Rebutia einsteinii -f. schmiedcheniana, Rebutia schmiedcheniana),
[Bo-3], [Hx],
" -v- einsteinii (Invalid name - See Rebutia einsteinii).....[Bo-3], [Hb],
" -v- karreri (Invalid name - See Rebutia karreri, Rebutia einsteinii -f. rubriviridis),
" -v- rubriviridis (Invalid name - See Rebutia einsteinii -f. rubriviridis, Rebutia schmiedcheniana -v- rubriviridis).....[Jd],
" -v- steineckeri (Invalid name - See Rebutia einsteinii -f. steineckeri),
spiralisepala (See Rebutia aureiflora - cultivated form).....Hl 3-1975],
steinmannii (See Rebutia steinmannii),
[RR], [Bu],
tarvitaensis (See Rebutia tarvitaensis)
Spec. FR-1118.....[Hb],

MELOCACTUS

aciculosus (See M. salmianus -v- aciculosus, M. macracanthus),
" -v- aductus (See M. macranthus),
acispinosus (See M. jacobinensis, M. rectispinus),
actinacanthus (Cuba).....
acuatus (See Echinocactus acuatus, Wigginsia acuata),
[Gm*],
acunai.....[Av], [Bo-4], [Hx*], [Aj 22-7-1971], [Cu 7-3-[1939], [Gu III-1953], [Hl 1-1971], [Hs IV-3-1983],

MELOCACTUS.....(Cont'd),

acunai.....(cont'd),
" -v- flavispinus.....
" -v- lagunaensis.....
aequatorialis.....
X albicephalus (M.pruinosus X M. oreas ssp. cremnophilus -f. erythracanthus (Braun 1981b, & Taylor 1982) ....[DD-486], [Jd], [Ba 4-1974],
albispinus (See M. macracanthus),
alpescens (albescens ?).....[Hl 6-1982],
amethystinus (See M. brederooianus -v- amethystinus),
ammotrophus (See M. lensselinkianus -f. ammotrophus, M. amethystinus.....[Fig.11 in Af 58-1 plant as M. longicarpus belongs here],
amoenus.....[Ar], [Bo-4], [Hy], [Ba 11-1976],
" -v- rubrispinus.....
amstutziae (See M. peruvianus -v- amstutziae), [Hb], [Ba 11-1976],
andicolus.....
angusticostatus (See M. macracanthus),
antonii (See M. intortus -v- antonii),
appropinquatus (See M. macracanthus),
applanatus (See M. peruvianus -v- applanatus),
approximatus (See M. macracanthus),
arachnites (See M. diersianus, M. levitestatus -v- diersianus),
aracutuensis.....
arcuatus (arcuatispinus) (See M. macracanthus),
argenteus (See M. macracanthus),
armatus (See M. macracanthus),
auriazureus.....
axiniphorus (See M. pruinosus).....[Ba 10-1976],

azulensis (See M. oreas ssp. ernestii)..... [Aj 28-7-1977], [Aw 2-1981],
azureus.....[DD-487], [Ar], [Ht], [Hv], [Hx*], [Jd], [Am 1-1981], [Au Vol. 2, 1977, Zone 3 Compendium], [Bi 4-3-1986], [Bi 5-4-1987], [Hl 1-1977], [Ct-1988],
" -v- ferreophilus (See M. ferreophilus).
" -v- krainzianus (See M. krainzianus),
" -v- pachyacanthus (See M. pachyacanthus),
bahiensis .....[G], [XX], [Ar], [Bv-1], [Cz], [Gr], [Hb], [Hi-1], [Hu], [It], [N-1980],
" -v- ernestii.....[Db 5-1980],
baarsianus (See M. macracanthus),
bargei (See M. macracanthus),
bellavistensis (See M. stenogonus), [I], [Ay], [Bo-4], [Hi-4], [Hx*], [Af 50-1], [Aj 15-12-1964], [Hl 6-1982],

MELOCACTUS.....(Cont'd),

bellavistensis.....(cont'd),
" -v- minor.....
" -v- multicostata.....[Gm*],
benkellinsianus (See M. salinensis),
besleri (See Discocactus besleri, Discocactus insignis -v- besleri, Discocactus placentiformis)..[Gm*],
borhidii.....[Da 2-1977], [Da 3-1978], [Da 4-1978],
bradleyi (See M. intortus),
brasiliensis.....
brederooianus (See M. salvadorensis), [Hx*], [Aj 24-4-1973], [Hl 5-1975],
" -v- amethystinus (See M. amethystinus),
" -v- inconcinnus (See M. inconcinnus)
broadwayii.....[G], [RR], [Af 51-2], [Hh May, 1974],
brongnartii (See M. communis),
brumadoensis.....
buysianus (See M. macracanthus),
caesius.....[RR], [Av], [Bo-3], [Hx], [Hy], [Il], [Af 2-6], [Af 13-12], [Aj 23-12-1972], [Ba 11-1976], [Hl 1-1981], [Hl 3-1981], [Hh Vol. 2, #9, 1977],
caititensis.....
canitiensis.....
canescens (See M. zehntneri -v- canescens ?, M. inconcinnus).....[Hi-1],
" -v- montealtoi (See M. zehntneri -v- canescens),
canudoensis.....
capillaris (See M. macracanthus),
cephalonoplus (See M. caesius),
chagualensis.....
chalensis.....
chiclayensis.....
chotensis.....
citrispinus.....
communis..(Cactus melocactus), See RR Vol. 3, p. 224 for more synonyms......[G], [K], [DD], [Ay], [Bo-4], [Cm], [Cn], [Gm*], [Hx*], [Hy], [Aj 23-1-1972], [Ba 11-1976], [Da 2-1977], [Da 1-1980],
" -f. cristata.....[Da 1-1980],
concinnus (See M. pruinosus -v- concinnus) [DD-488], Gr], [Hv], [Ae-1983], [Hl 1-1976], [Hs I-4-1980],
conoideus (See M. uebelmannianus).....[Ht], [Ae-1981], [Cu 4-1976],
conquistaensis.....[Hw-1982],
contortus (See M. macracanthus),
coronatus.....[Ba 6-1980],
crassicostatus (See M. obtusipetalus),
cremnophilus (See M. oreas ssp. cremnophilus), [DD-489], [Hx*], [Af 44-1972], [Aj 24-9-1973],

MELOCACTUS.....(Cont'd),

curvicornis (See M. zehntneri -v- curvicornis).....
[Aj 23-2-1972],
curvispinus (See M. neryi)..[Bo-4], [Hy], [Ba 11-1976],
dawsonii.....[Bc], [Ae-1982], [Aj 31-7-1980], [Aw 2-1965], [Aw 2-1970], [Aw 2-1986], [Hl 3-1983],
deinacanthus.....[Hx*], [Aj 24-10-1973],
" ssp. florschuetzianus (See M. florschuetzianus)
" ssp. longicarpus (See M. longicarpus)
" " " -f. mulequensis (See M. mulequensis),
delessertianus (See M. salvador).....[UU], [Av], [Bo-4], [Bt], [Jd], [Ae-1982], [Aw 2-1986], [Hl 3-1978], [Hl 3-1983],
depressus (See M. melocactoides -v- depressus), [Aj 31-11-1980], [Cu 6-4-1938],
diersianus (See M. arachnites, M. levitestatus -v- diersianus)....[Hi-1], [Aj 26-8-1975], [Ba 11-1976], [Hl 6-1981],
" -f. diersianus.....[Ct-1988],
" -f. rubrispinus.....[Ct-1988],
dichroacanthus (See M. intortus).....[Is*],
disciformis (See M. salvadorensis)....[Hl 6-1976],
douradaenis (See M. zehntneri)
ernestii (See M. oreas ssp. cremnophilus)..[G], [Bo-4], [Hb], [Ba 11-1976], [Hl 1-1983], [Hh Vol. 3, #3&4, 1978], [var. - Ae 1982]
erythracanthus (See M. oreas -v- erythracanthus ?), [DD-490], [Hx*], [Am 1-1981], [Ba 4-1974],
evae.....[Da 4-1978],
ferox (See M. macracanthus),
ferreophilus (See M. azureus -v- ferreophilus), [DD-491], [Hx*],
florschuetzianus (See M. deinacanthus -v- florschuetzianus)....[Ba 6-1981], [Hh Sept., 1975],
fortalezensis (See M. peruvianus).....[Bo-4], [Hy],
" -v- schoenii.....
giganteus......[Jd], [Bi 4-3*],
glaucescens (See M. pruinosus.....[DD-492], [Hx*], [Il], [Af 44-1972], [Ba 4-1974], [Bm-1985],
glauxianus (See M. amethystinus, M. griseoloviridis),
goniodacanthus (goniacanthus) (See M. melocactoides),
gracilis (See M. macracanthus).....[Aj 23-1-1972],
griseus (See M. caesius).....
griseoloviridis (See M. lensselinkianus),
guaricensis.....[Bo-4], [Bt], [Gr],
guatamalensis (See M. maxonii, M. ruestii -v- maxonii),
guitarti.....[Bo-4], [Af 51-2], [Ba 11-1976], [Cu 7-4-1939], [Hl 1-1971], [Hs I-1-1980],
harlowii.....[RR], [Av], [Bo-4], [Hl 1-1971], [Hl 6-1981], [Hl 4-1983], [Hs IV-2-1983],

MELOCACTUS.....(Cont'd),

harlowii.....(cont'd),
" -v- harlowii.....[Aj 22-7-1971],
" -v- siguaensis.....[Hl 1-1971],
havannensis (See M. guitarti, M. commumis -v- havannensis, M. intortus),
helvolilanatus (See M. zehntneri group).....
[Ba 12-1976], [Hs III-2-1982],
hispaniolicus (See Melocactus lemairei).....[Bo-4],
holguinensis (See M. jakusii).....[Aw 2-1987],
horridus (See M. zehntneri group)....[Ei 12:227, 1934],
horstii.....
huallancaensis (See M. peruvianus),
[Bo-4], [Hs IV-4-1983],
huancensis.....
iacuensis.....[Bv-3],
inconcinnus (See M. brederooianus -v- inconcinnus),
[Hi-1], [Aj 26-9-1975], [Aw 2-1981],
" -v- brederooianus (See M. brederooianus),
indiae occidentalis (See M. intortus).....[Ct-1984*],
intermedius (See M. macracanthus),
interpositus (See M. oreas ssp. rubrisaetosus)
intortus.......[G*], [M*], [LL*], [RR], [Bo-4], [Bu],
[Bv-1], [Hy], [Af 1-11], [Af 4-1], [Af 52-1], [Af 60-6], [Am 4-1985],
" -v- antonii (See M. antonii),
" -v- inaquensis.....[Bv-1],
" -f. cristata.....[Af 4-1], [Af 4-10],
" -f. monstrosus.....[Af 4-1],
intricatus (See Echinocactus intricatus, Malacocarpus arechaveletai, Wigginsia arechaveletai),
[Gm*],
irecensis (See M. krainzianus, M. azureus -v- krainzianus),
itaberapensis.....[Ar],
itaobimensis.....
jacobinensis (See M. acispinosus),
jakusii (See M. holguinensis).....[Da 4-1978],
jansenianus (See M. peruvianus),
[Bo-4], [Iv 5-1955],
kodwijkianus.....
krainzianus (See M. irecensis, M. azureus -v- krainzianus),
lanssensianus.....
lanuginosus (See Mammillaria mammillaris)....[Ct-1984]*
lehmannii (See M. macracanthus),
[Di*], [Is*],
lemairei (See M. hispaniolicus),
[RR], [Af 56-3], [Am 2-1981],

MELOCACTUS.....(Cont'd),

lensselinkianus (See M. griseoloviridis, M. amethystinus)
[Ht], [Ba 4-1974], [Cu 4-1982],
" -f. .....(As M. longicarpus in [Af 58-1],)
" -f. (de Itaobim).....[Ht],
" -f. (de Janio Quadros e de Santana).....[Ht],
levitestatus.....[G], [Ct-1988],
" -v- ammotrophus (See M. ammotrophus),
" -v- diersianus (See M. arachnites, M. diersianus),
" -v- rubrispinus (See M. rubrispinus),
" -v- securituberculatus (See M. securituberculatus),
lobelii.....[Bo-4], [Hl 3-1981],
loboguerreroi.....[Af 55-1], [Aj 33-4-1982], [Aw 3-1967]
[Ba 9-1977], [Hl 6-1983],
longicarpus (See M. deinacanthus ssp. longicarpus).....
[Ah-1983], [Ba 7-1980],
longispinus (Buin. & Bred.), (See M. oreas -v- longispinus.....[Ae-1981], [Ba 6-1977], [Bi 4-3],
" -v- barocensis.....
longispinus (Ritt.), (See M. florschuetzianus, M. nitidus, M. oreas -v- florschuetzianus),
macracanthoides (See M. intortus).....
macracanthus (See M. lehmannii) and see RR Vol. 3, page 222 for more synonyms.........[G], [Ar],
[Bo-4], [Gm*], [Ha], [Hy], [Is*], [Ba 11-1976], [Ba 5-1979], [Cu 2-1981],
" -v- elegans.....
macrocanthus (See M. intortus),
macrochapensis .....
macrodiscus (See M. zehntneri ssp robustispinus),
[Bo-4], [Ht], [Af 41-1], [Au Vol. 3, 1978, Zone 3 Compendium], [Hl 6-1981], [Hs III-2-1982],
" -v- macrodiscus.....
" -v- minor (See M. zehntneri -v- macrodiscus),
" -f. (Barra de Santa Rosa).....[Ht],
" -f. (Currais Novos).....[Ht],
" -f. (de Brumada).....[Ht],
" -f. (de Nossa Senhora da Gloria).....[Ht],
" -f. (Dois Riachos, AL).....[Ht],
" -f. (do Raso da Catarina).....[Ht],
" -f. (Taua, CE).....[Ht],
margaritaceus (Invalid name).....[Ht],
" -f. disciformis.....
" -f. salvadoranus.....
" -f. (Paco das Mocas).....[globose - Ht],
" -f. (Santa Amaro das Brotas).....[oblong - Ht],

matanzanus.....[G], [R], [S], [AA], [Ar], [Av], [Bo-4], [Bt], [Bv-3], [Dq], [Hv], [Hx], [Iz], [Ab-1985], [Ae-1981], [Af 6-8], [Aj 30-2-1979],
" -f. varieg. .....
maxonii (See M. ruestii ssp. maxonii).....[G], [NN], [RR], [UU], [Ar], [Ay], [Bo-4], [Bt], [Bv-1], [Hv], [Hy], [Ig*], [Ba 6-1979], [Hl 1-1969],
mazelianus.....[Aj 32-9-1981], [Hl 5-1981],
melocactoides (See M. violaceus).....[G], [M*], [Z], [PP], [RR], [Bv-3], [Hi-1], [Hy], [Au 37-2], [Hl 6-1976],
" -v- capensis.....
" -v- depressus.....[Ht],
" -v- " -f. depressus.....
" -v- " -f. sanctaritae.....[Ht],
" -f. exsanguis (See M. melocactoides -v- natalensis -f. exsanguis).....
" -v- itaipuassuensis.....[Ht],
" -v- melocactoides.....[Ht],
" -v- " -f. melocactoides.....
" -v- " -f. perspinosus.....
" -v- natalensis.....[Ht],
" -v- " -f. exsanguis.....[Ht],
" -v- " -f. natalensis.....[Ht],
" -v- violaceus (See M. violaceus).....
" -v- " -f. capensis.....
" -v- " -f. civitatis.....[Ht],
" -v- " -f. violaceus.....
meonacanthus (See M. communis).....[Gm*], [Im],
microcarpus (See M. macracanthus),
microcephalus (See M. macracanthus).....[Is*],
minimus (See Mammillaria prolifera),
miquelii (See M. intortus).....[Is*],
montanus (See M. florschuetzianus, M. deinacanthus ssp. florschuetzianus),
mulequensis (See M. deinacanthus ssp. longicarpus)........[Ba 3-1976],
" -v- saxicola.....(See M. saxicola),
multiceps.....[Bi 4-3]*,
nagyi.....[Da 4-1978],
neglectus (See M. bahiensis),
" -v- diamantinensis.....
neryi (See M. schulzianus, M. smithii)..... [RR], [Ar], [Bo-4], [Bu], [Hx], [Hy], [Jd], [Aj 11-6-1960],
nitidis (See M. azulensis).....[Hi-1],
oaxacensis (See M. ruestii ssp. oaxacensis), [EE], [RR], [Bo-4], [Af 24-2], [Af 26-3], [Ag-1988], [Hh 3-3 & 4, 1978], [Hl 3-1983], [Hs I-3-1980],

MELOCACTUS.....(Cont'd),

obtusipetalus (See M. crassicostatus)...[S], [T], [AA],
[Hx], [Hy], [Ah-1983], [Aw 3-1968],
ocujalius.....[Hl 1-1971],
onychacanthus.....[Hi-4], [Hv], [Iz], [Aj 22-7-1971],
" -v- albescens.....[Hi-4],
" -v- conicus.....[Hi-4],
" -v- onychacanthus.....
oreas (See M. ernestii).....
[Bo-4], [Ht], [Hy], [Cu-3-1980], [Hl 1-1969],
" -f. azulensis (See M. azulensis, M. nitidus)...
" ssp. bahiensis (See M. bahiensis).....[Ht],
" " " -v- bahiensis.....(See M. bahiensis),
[Ht],
" ssp. cremnophilus (See M. cremnophilus).....
" " " -f. erythracanthus (See M. erythracanthus).....[Ct-1988]
" ssp. ernestii (See M. ernestii).....
" " " -f. azulensis (See M. azulensis,
M. nitidus).....[Ct-1988],
" -v- florschuetzianus (See M. florschuetzianus),
" -v- rubrisaetosus (See M. rubrisaetosus, M.
interpositus).....[Ct-1988],
" -v- submunitus.....[Ht],
" -f. (de Itaberaba).....[Ht],
" -f. (de Jequie).....[Ht],
" -f. (de Milagres).....[Ht],
" -f. gigante (de Jequie).....[Ht],
" -f. (Itaobim).....[Ht],
" -f. (Minas Gerais).....[Ht],
" -f. (Sergipe).....[Ht],
orthacanthus (See Echinocactus orthacanthus, Notocactus
orthacanthus, Wigginsia orthacanthus)...[Gm*],
pachyacanthus (See M. azureus -v- pachyacanthus),
.....[Aj 27-1-1976], [Ct-1988],
parthoni (See M. violaceus),
paucispinus (See M. pruinosus).....[Bi 4-3],
permutabilis (See M. salvadorensis),
peruvianus (See M. huallancaensis, M. townsendii),)
[NN], [QQ], [Ay], [Bo-4], [Hi-4], [Hy],
[Da 2-1980], [Hl 4-1969], [Bi Vol. 4, #6],
" -v- amstutziae (See M. amstutziae),
" -v- applanatus.....(See M. applanatus),
" -v- canietensis (See M. peruvianus),
[Ar], [Bo-4],
" -v- lurinensis (See M. peruvianus)...[Bo-4],
" -f. cristata .....[Hi-4],
platyacanthus (See Echinocactus platyacanthus) [Gm*],
poliacanthus (See Echinocactus poliacanthus, Malacocarpus erinaceus, Malacocarpus polycanthus),
[Gm*],

portoricensis (See M. intortus),
pruinosus (See M. glauscens).....[Ct-1988],
" -v- concinnus (See M. concinnus),
purpureus (See M. lemairei),
pygmaeus.....[Hl 6-1976],
pyramidalis (See M. macracanthus).....[Gm*],
radoczii.....[Da 4-1978],
rectispinus (See M. acispinosus),
recurvus (See Echinocactus recurvus, Ferocactus recurvus),
[Gm*], [N-1980],
robustispinus (See M. zehntneri ssp. robustispinus),
[Ba 5-1977], [Ba 3-1980],
rotula (See M. macracanthus),
rubens (See M. communis),
rubrisaetosus (See M. oreas -v- rubrisaetosus),
[Ba 7-1977],
rubrispinus (See M. diersianus -v- rubrispinus),
ruestii.....[RR], [Hy], [Af 34-2], [Aw 2-1983],
" -v- alba.....
" ssp. cintalapensis.....[Aw 3-1986],
" ssp. maxonii (See M. maxonii),
" ssp. oaxacensis (See M. oaxacensis),
" ssp. ruestii.....
" ssp. sanctae-rosae.....
rufispinus (See M. communis),
salinensis (See M. benkellinsianus).
salmianus (See Echinocactus salmianus, M. macracanthus)
[Gm*],
salvador (See M. delessertianus).....[RR],
salvadorensis (See M. brederooianus)....[Ht], [Aj 32-6-1981], [Ba 11-1976], [Cu 3-1980], [Iv 9-1952],
" -f. (de Ipira).....[Ht],
" -f. (de Itatinho).....[Ht],
" -f. (de Jequie).....[Ht],
" -f. (de Mangarito.....[Ht],
" -f. (do Piaui.....[Ht],
sanguiniflorus (See M. levitestatus),
saxicola (See M. mulequensis -v- saxicola),
schatzlii.....[Aj 33-4-1982], [Aj 33-10-1982],
schulzianus (See M. neryi).....[Ba 5-1974],
seabrasensis.....
securituberculatus (See M. levitestatus -v- securituberculatus),
sellowii (See Echinocactus sellowii, Malacocarpus sellowii, Wigginsia sellowii).....[Gm*],
smithii (See M. neryi).....,
spatanginus (See M. macracanthus),
stenogonus (See M. bellavistensis),

stramineus (See M. macracanthus),
tenuispinus (See Echinocactus tenuispina).....[Gm*],
tephracanthus (See Echinocactus tephracanthus)...[Gm*],
townsendianus (townsendii) (See M. peruvianus),
[Bo-4], [Bv-1],
trigonaster (See M. macracanthus),
trujilloensis (See M. peruvianus).....[Bo-4],
" -v- schoenii (See M. peruvianus),.....[Bo-4],
uebelmannianus (uebelmannii).......[Ar], [Hl 2-1970],
[Hl 1-1980],
unguispinus (See M. peruvianus),
[Bo-4], [Hx], [Aj 9-9-1958], [Aj 16-4-1965]
" -v- sechura.....
" -v- talara.....
violaceus (See M. parthoni, M. melocactoides -v- violaceus).....[E], [Bo-4], [Bv-3], [Hy], [Iz], [Ba 11-1978], [Cu 3-1980], [var. - W],
warasii.....[Af 57-6], [Hs I-3-1980], [Ct-1988],
zehntneri.....[RR], [Bo-4], [Ht], [Jd], [Bm-1985],
[Cu 3-1967], [Hl 6-1981], [Hs III-2-1982],
" ssp. zehntneri -v- curvicornis (See M. curvicornis)....
" ssp. " -v- viridis.....
" ssp. canescens (See M. canescens).....
" ssp. robustispinus (See M. robustispinus).....
" -v- ananas (de Acu).....[Ht],
" -v- ananas (de Aiuaba).....[Ht],
" -f. amosa, magna.....[Ht],
" -v- macrodiscus (See M. macrodiscus),
" -f. (de Diamante).....[Ht],
zuccarinii (See M. macracanthus).....[Di*], [Is*],
Spec. .....[Hv], [Af 57-3], [Db 4-1980], [Aj 33-12-1982], [Aj 34-3-1983],
Spec. (Crisnejas).....[Ar],
Spec. (De Melon di Deroe).....[Ba 5-1979],
Spec. H/B-79.....[Ba 7-1980],
Spec. WK-24.....[Aj 31-7-198
Spec. HU-214.....[Aj 23-1-1972],
Spec. HU-217.....[Aj 23-1-1972],
Spec. HU-219.....[Aj 23-1-1972],

(Melocactus from Venezuela)

GS-16.....[Aj 33-10-1982],
GS-31.....[Aj 33-10-1982],
GS-38.....[Aj 33-10-1982],
GS-39.....[Aj 33-10-1982],
GS-69.....[Aj 33-10-1982],
GS-83.....[Aj 33-10-1982],
GS-88.....[Aj 33-10-1982],
GS-103.....[Aj 33-10-1982],
GS-107.....[Aj 33-10-1982],

MICRANTHOCEREUS

albicephalus (See Austrocephalocereus albicephalus, Coleocephalocereus albicephalus),
aureispinus (See Arrojadoa aureispina, Austrocephalocereus albicephalus
auri-azureus (See Arrojadoa auri-azurea), [Hx*], [Ad-], [Aj 33-12-1982], [Au Vol.-3, 1978, Zone 3 Compendium], [Ct-1988],
densiflorus (See Arrojadoa densiflora), [Hu], [Af 52-4], [Af 56-6], [Ba 4-1974],
dolichospermaticus (See Austrocephalocereus dolichospermatic),
estevesii (See Austrocephalocereus estevesii, Coleocephalocereus estevesii),
flaviflorus (See M. uilianus).....
haematocarpus (See Cephalocereus haematocarpus, Austrocephalocereus purpureus (Guerke)Backeberg)
monteazulensis (See Cephalocereus monteazulensis, Austrocephalocereus monteazulensis)....[Hi-1],
polyanthus (See Cephalocereus polyanthus), [DD], [HH], [Ay], [Bo-4], [Hi-1], [Hu], [Hx*], [Hy], [Jd], [Af 50-2],
" -v- rufispinus.....
purpureus (See Austrocephalocereus purpureus (Guerke) Buxb. 1970, Austrocephalocereus lehmannianus, (Werd.) Backbg. 1960, Cephalocereus purpureus, (Guerke) 1908, Cephalocereus lehmannianus (Werd.) 1933)........[Hi-4],
ruficeps (See Cephalocereus macrocephalus, Mitrocereus ruficeps, Neobuxbaumia macrocephala),
rufispinus.....[NN],
streckeri (See M. ruficeps).....[Ct-1988],
uilianus (See Cephalocereus uilianus, M. flaviflorus),
violaciflorus (See Cephalocereus violaciflorus), [Hx*], [Aj 20-7-1969], [Aj 33-12-1982],

MICROPUNTIA

barkleyana (See Corynopuntia barkleyana, Opuntia barkleyana)....[Bo-1], [Hy], [Ej 36:662, 1946],
brachyrhopalica (See Corynopuntia pulchella, Opuntia brachyrhopalica, Opuntia pulchella), [Bo-1], [Ej 36:661, Fig. 1, 1946],
gracilicylindrica (See Opuntia gracilicylindrica)..... [DD], [Bo-1],
pygmaea (See Opuntia pygmaea).....[DD], [Bo-1], [Hx*],
spectatissima (See Opuntia spectatissima), [Bo-1], [Ej 36:662, Fig. 3, 1946],
tuberculosirhopalica (See Opuntia tuberculosirhopalica).....[DD], [Bo-1],
wiegandii (See Opuntia wiegandii).....[DD], [Bo-1],

MICROSPERMIA

cruci-albicentra.....[Hr],
gigantea (See Parodia gigantea),
" -v- jujuyana.....[Hr],
microsperma (See Hickenia microsperma, Parodia
microsperma),
nivosa (See Parodia nivosa),.....[Hr],
" -v- golgata.....[Hr],
rigidispina (See Parodia rigidispina),
rigidissima (See Parodia rigidissima),
" -v- rubriflora.....
sanagasta (See Parodia sanagasta),
sanguiniflora (See Parodia sanguiniflora,
" -f. cristata.....[Hr],

MILA
----

albisaetacens.....[Bo-2], [Hx*], [Hh Vol 3, #3&4, 1978],
alboareolata (See M. caespitosa).....[E], [Bo-2],
australis.....(See M. caespitosa ssp. nealeana -v-
nealeana),
breviseta (See M. nealeana -f. senilis),
[Bo-2], [Bo-6], [Hi-4], [Hx*], [Hy],
caespitosa (See M. albo-areolata).....[DD], [NN], [QQ],
[Ay], [Bo-2], [Bv-1], [Bv-3], [Cz], [Hv], [Jd],
[Hh Vol. 3, #3&4,1978], [Hl 6-1969], [Hl 2-1971]
" -v- grandiflora.....[Hi-4],
" -f. kubeana (See M. kubeana),
cereoides (See M. nealeana)......[Bo-2], [Bo-6], [Hx*],
[Hh Vol. 3, #3 & 4, 1978],
churinensis.....[Hh Vol. 3, # 3 & 4, 1978],
" -v- longispina.....
colorea.....[Hi-4],
cuprifera.....
densiseta (See M. nealeana -f. senilis)...[Bo-2], [Bo-6]
eulaliana.....
fortalezensis (See M. nealeana).....[Bo-2], [Bo-6],
[Hh Vol. 3, #3 & 4, 1978],
kubeana (See M. caespitosa -f. kubeana)....[E], [Bo-2],
[Dq], [Hi-4], [Hx*], [Hh Vol.3, #3 & 4, 1978],
longispina.....
lurinensis (See M. nealeana).....[Bo-2], [Bo-6],
maritima.....[Hh Vol. 3 #3 & 4, 1978],
maxima.....
nealeana .....[Ay], [Bo-2], [Bo-6], [Dq], [Hi-4], [Hx*],
[Il], [Iz], [Jd], [Ad Mar., 1979], [Hl 3-1971],
[Hh Vol. 3 #3 & 4, 1978],
" ssp. nealeana.....
" -v- albiflora.....
" -f. senilis (See M. densiseta),.....
" -v- tenuior (See M. tenuior)...(Bo-2], [Bo-6],

MILA.....(Cont'd),
palspitosa.....
papyracantha.....
pugionacantha.....
pugionifera.....[Bo-2], [Bo-6], [Ad March-1979], [Af 51-1], [Hh Vol. 3 # 3&4, 1978],
senilis (See M. nealeana -f. senilis).....[Bo-6], [Hh Vol. 3, #3 & 4, 1978],
sublanata (See M. nealeana)...[Bo-2], [Hh 3,#3&4, 1978]
" -v- pallidior (See M. nealeana -f. senilis),
Spec. .....[Bo-6],

MIQUELIOPUNTIA

miquelii (See Cylindropuntia miquelii, Opuntia miquelii).....[Hi-3],

MIRABELLA

albicaulis (See Acanthocereus albicaulis).....[Hi-1],
minensis.....

MITROCEREUS
chrysomallus (BACKBG.) Br. & R. (See Mitrocereus fulviceps, Cephalocereus fulviceps),
columna-trajani (DAWSON) (See M. fulviceps),
fulviceps (WEB.) Backbg. (See Cephalocereus fulviceps, Pachycereus chrysomallus, Pseudomitrocereus chrysomallus, Pachycereus columna-trajani (DAWSON)),....[DD], [JJ], [Ay], [Az], [Bo-2], [Hx*], [Af 27-1], [Bi 2-1], [Hl 6-1980],
militaris (See Backebergia militaris, Cephalocereus chrysomallus, Pilosocereus chrysomallus), [Aj 16-3-1965],
ruficeps (See Cephalocereus macrocephalus, Cephalocereus ruficeps, Neobuxbaumia macrocephala), [Av], [Bo-2],

MONVILLEA

adelmarii (See Cereus adelmarii).....
alticostata (See Cereus alticostatus).....[Hi-1],
amazonica (See Praecereus amazonicus).....[Bo-4],
anisitsii (See M. marmorata, M. spegazzinii),
apoloensis (See Praecereus apoloensis).....[Af 33-3],
ballivianii.....[Iv 7-1959],
breviflora (See Brasilicereus breviflorus),
brittoniana (See M. maritima),
calliantha (Invalid name - See M. rhodoleucantha), [Bo-4], [Jd],
campinensis (See Praecereus campinensis).....[Bo-4],
" -v- piedadensis (See M. piedadensis).....

MONVILLEA......(Cont'd),
cavendishii.....[H], [M*], [RR], [Bo-4], [Bv-1], [Dq*], [Hy],
" -v- lauterbachii (See M. cavendishii, M. lauterbachii, M. paxtoniana),
" -v- paxtoniana (See M. lauterbachii, M. paxtoniana),
" -v- rhodoleucantha (See M. calliantha, M. rhodoleucantha),
" -v- saxicola (See M. saxicola),
chacoana.....[Hi-2],
diffusa (See Praecereus diffusus), [RR], [Bo-4], [Bv-1], [Jd],
ebenacantha.....[Hi-2],
euchlora.....[Bo-4],
haageana.....[DD], [NN], [Bo-4], [Hx], [Jd],
" -f. cristata.....[Am 5-1981],
insularis (See Cereus insularis),
jaenensis (See Praecereus jaenensis), [Bo-4], [Hi-4],
" -v- columbiana.....[Ay],
" -v- paucispina.....[Bo-4],
lauterbachii (See M. cavendishii -v- lauterbachii, M. cavendishii -v- paxtoniana), [Al], [Bo-4],
leucantha.....
lindenzweigiana (See Cereus spegazzinii -v- hassleri, Piptanthocereus lindenzweigiana),
maritima (See Praecereus maritima), [Ay*], [Bo-4], [Jd], [Af 34-3],
" -v- spinosior.....[Dt 2:54, 1948],
marmorata (See M. anisitsii, M. spegazzinii),
parapetiensis.....
paxtoniana (See M. cavendishii -v- paxtoniana, M. lauterbachii).....[Ay], [Bo-4],
" -v- borealis.....[Hi-1],
phaeacantha (See Brasilicereus phaeacanthus, Cephalocereus phaeacanthus),
phatnosperma.....[Hy*],
piedadensis.....[Hi-1],
pugionifera.....[Hi-1],
pucarensis (See M. jaenensis, Praecereus jaenensis),
rhodoleucantha (See M. calliantha, M. cavendishii -v- rhodoleucantha).....[Jd],
saddiana (See Cereus saddianus).....
saxicola (See M. cavendishii -v- saxicola).....[PP],
smithiana (See Cephalocereus smithianus, Cereus smithianus, Praecereus smithianus).........[Bo-4], [Jd],
spegazzinii (See M. anisitsii).....[E], [H], [M*], [RR], [Ay], [Bj], [Bo-4], [Dq], [Gr], [Hv], [Hy], [Ik], [Il], [Jd], [Aj 28-1-1977],

MONVILLEA.....(Cont'd),

spegazinii.....(cont'd),
" -v- boliviana.....
" -v- vargasiana (See Cereus vargasianus),
" -f. cristata.....[R], [Hy],

MORANGAYA

pensilis (See Cereus pensilis, Echinocereus pensilis), [Az], [Aw 2-1977], [Hh Jan.-1974],

MORAWETZIA

doelziana (See Oreocereus doelzianus), [DD], [QQ], [Ar], [Ay], [Bl], [Bo-2], [Bt], [Cy], [Hi-4], [Hx], [Hy], [Il], [Jd], [Af 51-4], [Ag-1988], [Aj 10-11-1959], [Cu 5-1-1936],
" -v- aureispina.....
" -f. calva.....[Bo-2], [Hi-4], [Iv 6-1957],
" -v- fuscatispina (Invalid name - See M. doelziana),
" -f. cristata.....[Af 13-4],
sericata (See Oreocereus sericata) [Ay], [Hi-4], [Ad-], [Ah-1981], [Aj 31-3-1980], [Aj 34-2-1983],
variicolor (See Arequipa variicolor, Matucana variicolor, Oreocereus variicolor, Oreocereus knizei),

MYRTGEROCACTUS

lindsayi (Hybrid - Myrtillocactus cochal X Bergerocactus emoryi).....[Hx*],

MYRTILLOCACTUS
chiotilla (See Escontria chiotilla),
cochal.....[CC], [KK], [OO], [Ay], [Az], [Bo-4], [Bv-1], [Iz], [Af 14-7], [Af 60-3], [Bm-1985],
eichlamii.....[Af 60-3],
geometrizans (See Cereus geometrizans).....[C], [H], [PP], [RR], [XX], [Ar], [Av], [Ay], [Bt], [Cx], [Cz], [Dq], [Gr], [Ha], [Hv], [Hx*], [Hy], [Il], [Jd], [Ad June, 1978], [Af 60-3], [Aw 2-1986], [Bi Vol. 4, #6],
" -v- geometrizans.....[Az],
" -v- grandiareolatus.....[JJ], [Az], [Bo-4], [Af 10-2], [Aw 1-1970], [Aw 1-1977], [Ba 2-1980],
" -v- pugionifera (See M. geometrizans -v- geometrizans), [Fw 1:10 Fig. 4, 1897],

MYRTILLOCACTUS.....(Cont'd),

geometrizans.....(cont'd)
" -f. cristata.....[Ay], [Hy], [Aj 24-7-1973], [Aj 27-5-1976], [Db 10-1980],
mixtecensis.....
multiareolatus.....[Af 59-1],
schenckii.....[RR], [Az], [Bo-4], [Bv-1], [Cx], [Dq], [Hx*], [Hy], [Il], [Af 60-3],
X Myrtillocactus lindsayi.....[Aw 3-1963],

NAVAJOA

durispina.....[Hy],
fickeisenii (fickeiseniae) (See Pediocactus peeblesianus -v-fickeiseniae),
[Q], [DD], [An], [Ar], [Bo-5], [Hx], [Hy], [Ba 1-1980], [Ba 7-1980], [Hl 1-1970], [Hs II-1-1981],
" -v- maianus (See N. maia),
maia.....[Ba 5-1978], [Cu 22:49, 1960], [Cu 1-1962], [Hl 1-1970],
peeblesiana (See Pediocactus peeblesianus -v- peeblesianus),
[M*], [Q], [DD], [Ar], [Bo-5], [Hx*], [Hy], [Jd], [Af 15-6], [Af 54-6], [Aj 31-7-1980], [Hl 1-1970], [Hl 5-1970],

NEOABBOTTIA

paniculata.....[Flower - DD], [RR], [Ay], [Bo-4], [Ct-1984], [Dq], [Hy], [Jd], [Af 27-2], [Af 56-3], [Aj 30-1-1979],
" -v- humbertii.....[Bo-4], [Af 27-2],

NEOBESSEYA

asperispina (See Coryphantha asperispina),
[Ar], [Il], [Ad April, 1979], [Aj 8-1978], [Hl 5-1972],
cubensis (See Coryphantha cubensis, Escobaria cubensis, Neolloydia cubensis),
[Hx*], [Aj 24-6-1973], [Ba 7-1977], [Hh July, - 1975],
macdougalii (See Escobaria macdougalii, Ortegocactus macdougalii).....[Jd],
missouriensis (See Coryphantha missouriensis, Escobaria missouriensis),
[F], [M*], [P], [LL*], [RR], [XX], [Ar], [Ay], [Bo-5], [Bt], [Bv-1], [Bz], [Dq], [Go], [Hn], [Hv], [Hx*], [Hy], [N-1980], [Af 53-3], [Am 10-1983], [Cu 6-1-1937],

NEOBESSEYA.....(Cont'd),

missouriensis.....(cont'd),
" -v- caespitosa (See Coryphantha missouriensis -v- caespitosa),
" -v- marstonii (See Coryphantha missouriensis -v- marstonii),
[An],
" -v- nuttallii (See Coryphantha missouriensis -v- missouriensis),
muehlbauerana (See Coryphantha muehlbauerana, Escobaria muehlbauerana),
notesteinii (See Coryphantha missouriensis -v- missouriensis),
odorata (See Neolloydia odorata),
roseiflora (See Coryphantha missouriensis -v- caespitosa).....[DD], [Bo-5], [Hx], [Hy], [Jd], [Cu 6-3-1938],
similis (See Coryphantha missouriensis -v- caespitosa, Coryphantha similis, Mammillaria similis), [Y], [GG], [NN], [RR], [Ar], [Bo-5], [Bt], [Hx], [Ba 3-1979],
" -v- robustior (See Coryphantha missouriensis -v- robustior),
wissmannii (See Coryphantha missouriensis -v- robustior, N. similis -v- robustior), [RR], [Bo-5], [Hx], [Hy], [Aj 16-6-1965], [Da 3-1978], [Hl 3-1967], [Hl 5-1972],
zilziana (See Coryphantha zilziana, Escobaria zilziana),

NEOBINGHAMIA

climaxantha (See Binghamia climaxantha, Peruvocereus albicephalus), [DD], [Bo-4], [Hx*], [Hy], [Jd],
" -v- armata.....[Bo-4],
" -v- lurinensis.....[Bo-4],
" -v- subfusciflora.....[Bo-4],
mirabilis.....[Ay], [Bo-4], [Bu], [Hy],
multiareolata (See Haageocereus albisetatus), [Bo-4], [Bo-6],
" -v- superba.....[Bo-4],
villigera (See Haageocereus albisetatus), [DD], [Ay], [Bo-4],

NEOBUXBAUMIA
----

euphorbioides (See Rooksbya euphorbioides), [Az], [Gr], [Jd], [Af 51-6],
" -v- euphorbioides.....[Aj 28-3-1977],
" -v- olfersii.....

NEOBUXBAUMIA.....(Cont'd),

macrocephala (See Cephalocereus ruficeps, Mitrocereus ruficeps),
[JJ], [Av], [Az], [Af 51-6], [Af 24:173, 1952], [Aj 16-6-1965], [Aw 1-1970], [Aw 1-1971], [Ba 2-1980], [Hl 2-1980],
mezcalaensis (See Cephalocereus mezcalaensis, Pilocereus mezcalensis, Pilosocereus mezcalaensis),
[JJ], [Az], [Bo-4], [Af 51-6], [Af 55-2], [Aw 1-1970],
" -v- multiareolata (See N. multiareolata), [Af 24:169-1952],
" -v- robusta (See N. mezcalaensis), [Bo-4], [Af 24-6],
" -v- typicus (See Cephalocereus mezcalaensis, N. mezcalaensis),
multiareolata (See Neobuxbaumia mezcalaensis -v- multiareolata, Cephalocereus multiareolatus),
[Az], [Aw 3-1973],
polylopha (See Cephalocereus polylophus, Pilosocereus polylophus),
[W], [DD], [UU], [Ar], [Ay], [Az], [Bd], [Bo-4] [Bt], [Cz], [Gr], [Hu], [Hv], [Hx*], [Hy], [Il], [Jd], [Aj 16-6-1965], [Aj 29-3-1978], [Aj 30-1-1979], [Ak-1981], [Aw 4-1970],
scoparia (See Cephalocereus scoparius, Pilosocereus scoparius),
[Az], [Bo-4], [Gr], [Af 24-6], [Aw 2-1964], [Aw 4-1969],
tetetzo (See Cephalocereus tetetzo, Pilosocereus tetetzo),
[I], [JJ], [Ar], [Ay], [Bo-4], [Af 36-5], [Af 51-6], [Ba 4-1980], [Ba 5-1980],
" -v- nuda.....[Az], [Bo-4], [Af 24-6],
" -v- tetetzo.....[Az],

Spec. (from Teotila).....[Bi 4-1986],
Spec. (from Santiago Nuyoo).....[Bi 4-1986],
Spec. (crested).....[Bi 4-1-1986],

NEOCARDENASIA

herzogiana (See Neoraimondia herzogiana)
[DD - Flower], [Ay], [Bo-2], [Bv-1], [Hx*], [Hy], [Af 41-3], [Af 57-4], [Cu 2-1980],
" -f. monstrosus.....[Cu 2-1980],

NEOCHILENIA

aerocarpa (See Neoporteria reichei -v- reichei -f. aerocarpa, Thelocephala aerocarpa),
[NN], [QQ], [Bo-3], [Bo-6], [Aj 24-3-1973],
" -v- fulva (Invalid name - See Neoporteria reichei -f. aerocarpa),
[X], [Z], [Bo-3], [Hx], [Aj 24-3-1973],
ambigua (See Weingartia ambigua),
andicola (See Horridocactus andicolus, Pyrrhocactus andicolus),
andreaeana (Invalid name - See Acanthocalycium andreaeanum, Neoporteria andreaeana, Pyrrhocactus andreaeana),
[DD], [QQ], [Bo-3], [Hx*], [Aj 31-9-1980],
aricensis (See Horridocactus aricensis, Neoporteria aricensis, Pyrrhocactus aricensis),
[Hx], [Ad-],
aspillagai (See Neoporteria aspillagai, Pyrrhocactus aspillagai).....[Jd], [Af 56-4], [Ah-1986],
atra (Invalid name - See Neoporteria atra, Thelocephala reichei, Neoportia fusca).....[DD], [Hb],
atroviridis (See Neoporteria tuberisulcata -v- atroviridis, Horridocactus atroviridis),
cachetiensis (cachytayensis) (See Horridocactus cachytayensis),
" -v- rubriflorus.....
calderana (See Neoporteria calderana, Pyrrhocactus calderanus).....[Ar],
carneoflora (Invalid name - See Neoporteria carneoflora, Neoporteria reichei -f. carneoflora),
[DD], [Il],
carrizalensis (See Horridocactus carrizalensis, Neoporteria carrizalensis, Neoporteria curvispina -v- carrizalensis, Pyrrhocactus carrizalensis),
chilensis (See Horridocactus chilensis, Neoporteria chilensis, Pyrrhocactus chilensis),
[H], [L], [S], [Bo-3], [Hx*], [Hy],
" -v- albidiflora (See Neoporteria chilensis -v- albidiflora, Pyrrhocactus chilenia -v- albidiflorus),
" -v- australis.....
" -v- borealis.....
" -v- chilensis (See Pyrrhocactus chilensis -v- chilensis),
" -v- confinis (See Horridocactus confinis, Pyrrhocactus confinis),
chorosensis (See Pyrrhocactus chorosensis, Neoporteria chorosensis.....[Bo-6],
confinis (See Horridocactus confinis, Pyrrhocactus confinis),

NEOCHILENIA.....(Cont'd),

crispa (See Horridocactus crispus, Neochilenia nigriscoparia, Neoporteria crispa, Pyrrhocactus crispus).....[Ae-1981],
deherdtiana (Invalid name - See N. occulta, Neoporteria occulta, Pyrrhocactus deherdtiana, Pyrrhocactus occultus), [DD], [Af 55-3],
dimorpha (See Pyrrhocactus dimorphus, Neoporteria dimorpha.....[DD], [Hx*],
duripulpa (See Thelocephala duripulpa, Neoporteria reichei -f. duripulpa).....[DD], [Hx], [Aj 22-10-1971],
ebenacantha (Backbg.) (See N. hankeana (Foerst.)), [NN], [Hy],
echina (See Horridocactus echinus, Pyrrhocactus echinus],
eriocephala (Invalid name - See Neochilenia floccosus, Pyrrhocactus floccosus) [Bo-3], [Hx*], [Aj 30-4-1979],
eriosyzoides (See Horridocactus eriosyzoides, Neoporteria eriosyzoides, Pyrrhocactus eriosyzoides).....[Bo-3], [Bo-6],
" -v- roseiflora.....[Hy],
esmeraldana (See Chilirebutia esmeraldana, Neoporteria esmeraldana, Thelocephala esmeraldana), [DD], [EE], [NN], [Bo-3], [Bo-6], [Hx], [Hy], [Ih], [Jd],
floccosa (See Horridocactus floccosus, Neoporteria eriocephala, Pyrrhocactus floccosus, Neochilenia eriocephala).....[Hx*], [Ae-1981], [Aj 29-4-1978], [Aj 31-8--1980],
fobeana (See Neoporteria jussieui -v-fobeana)..... [Bo-3], [Hx*],
fulva (See Chileorebutia fulva, N. aerocarpa -v- fulva, Thelocephala fulva),
" -v- esmeraldana (See N. esmeraldana),
fusca (Muehlpf.) (See Echinocactus ebenacanthus (Monv.), Neoporteria fusca (Muehlpf.)), [Bo-3], [Hx*], [Hy], [Hl 6-1978],
gerocephala (See Neoporteria gerocephala, Neoporteria nidus -v- gerocephala),
gigantea (See Horridocactus giganteus),
glabrescens (See Chileorebutia glabrescens, Thelocephala glabrescens).....[Av], [Bo-3], [Hu], [Hx*], [Aj 24-3-1973], [Db 7-1979],
glaucescens (See Pyrrhocactus glaucescens),
gracilis (See Pyrrhocactus gracilis),
hankeana (See Horridocactus hankeanus, Neoporteria hankeana).....[T], [Ar], [Bo-3], [Hx], [Af 56-2], [Ba 4-1978 Flower], [Hl 3-1971],

NEOCHILENIA.....(Cont'd),

hankeana.....(cont'd),
" -v- minor.....
" -v- roseiflora.....
" -v- taltalensis (Invalid name).....
(See Horridocactus taltalensis, Neoporteria taltalensis, Pyrrhocactus taltalensis),
[Bo-3],
heinrichiana (See Horridocactus heinrichianus),
huascensis (See Neoporteria huascensis, Pyrrhocactus huascensis),
[Ae-1981], [Af 53-6],
" -v- gigantea.....
hypogaea (Invalid name - See Copiapoa hypogaea),
[Bo-3],
imitans (Invalid name - See Neochilenia lembckei, N. neoreichei, Reicheocactus neoreichei, Thelocephala lembckei),
[Bo-3], [Hx],
intermedia (See Neoporteria intermedia, Pyrrhocactus intermedius),
iquiquensis (See Neoporteria iquiquensis, Pyrrhocactus floribundus, Reicheocactus floribundus),
[Hy], [Ad-], [Aj 21-9-1970],
jussieui (See Neoporteria jussieui, Pyrrhocactus jussieui),
[J*], [N], [NN], [Bo-3], [Hx], [Hy],
" -v- spinosior.....
" -f. cristata.....(Hl 4-1975],
krausii (krausei, kraussii) (Invalid name - See Chileorebutia krausii, Thelocephala krausii),
[QQ], [Bo-3], [Bo-6], [Hx], [Jd], [Ae-1981],
kunzei (See Pyrrhocactus kunzei).....[Hx], [Hy],
" -v- illapel.....
lembckei (Invalid name - See N. imitans, N. neoreichei, Thelocephala lembckei).....[Bo-3], [Hu],
lissocarpa (See Pyrrhocactus lissocarpus),
" -v- gracilis (See Horridocactus lissocarpus -v- gracilis),
" -v- luteospina (See Horridocactus lissocarpus -v-luteospina),
malleolata (See Chileorebutia malleolata, Neoporteria malleolata, Thelocephala malleolata),
[NN], [Hx], [Ae-1981], [Af 56-3],
" -v- solitaria (See Chileorebutia malleolata -v- solitaria, Thelocephala malleolata -v- solitaria),
mebbesii (See Echinocactus odieri -v- mebbesii),
[Bo-3], [Jd], [Af 53-6],

mebbesii.....(cont'd),
" -v- centrispina (See Echinocactus odieri -v- mebbesii, N. mebbesii)...
minima (See N. simulans -v- minima),
mitis (See Neoporteria mitis, Thelocephala mitis), [DD], [Ay], [Bo-6], [Hx], [Cu 1-1973],
" -v- glabrescens (See N. glabrescens, Thelocephala glabrescens),
monte-amargensis (Invalid name - natural hybrid - See Neoporteria monte-amargensis), [DD], [Bo-3:9, 1963],
napina (See Chileorebutia napina, Neoporteria napina, Thelocephala napina), [F], [T], [DD], [XX], [Ar], [Bo-3], [Bt], [Hb], [Hx*], [Hy], [Jd], [Af 54-6], [Af 56-3], [Af 56-6], [Aj 24-3-1973], [Cu 1-1968],
" -v- atroroseiflora.....[Hy],
" -v- falkenbergii (See Neoporteria napina -v- falkenbergii),
" -v- glaucescens (See Pyrrhocactus glaucescens), [Hy],
" -v- lanigera (See Neoporteria lanigera),
" -v- mitis (See N. mitis, Neoporteria mitis), [Am 8-1979], [Am 2-1983],
" -v- spinosior (See Neoporteria napina -v- spinosior, N. napina), [NN], [Bo-3], [Bo-6], [Hy],
" -v- " -f. mebbesii (See N. mebbesii),
neofusca (Invalid name).....[Bo-6],
" -v- trapichensis.....
neoreichei (Invalid name - See N. imitans, N. lembckei, Reicheocactus neoreichei, Thelocephala lembckei).....[Bo-3], [Hy],
nigricans (See Horridocactus nigricans, Neoporteria nigricans),
nigriscoparia (Invalid name - See N. crispa, Neoporteria crispa, Horridocactus crispus, Pyrrhocactus crispus)...[Bo-6:3784], [Hx*], [Aj 26-9-1975],
occulta (See N. deherdtiana, Neoporteria occulta, Pyrrhocactus deherdtianus, Pyrrhocactus occultus).....[R], [NN], [RR], [Bo-3], [Cy], [Hy], [Af 55-1], [Aj 24-3-1973], [Hl 6-1978],
" -v- applanata.....[Am 8-1979],
" -v- .....[Hx],
odierii (LEM.) Backbg. (See Chileorebutia odierii, Neoporteria odierii, N. pseudoreichei, Neoporteria reichei -v- reichei -f. pseudoreichei, Thelocephala odierii).....[Ar], [Bo-3],
" -v- mebbesii (See Echinocactus odierii -v- mebbesii, N. mebbesii).....[Bo-3],

odierii.....(cont'd),
" -v- pseudoreichei (See N. pseudoreichei),
[Ae-1982],
odoriflora (See Pyrrhocactus odoriflorus),
[Ae-1982],
paucicostata (See Neoporteria paucicostata, Horridocactus paucicostatus, Pyrrhocactus paucicostatus),.....[NN], [Ar], [Bo-3], [Bo-6], [Hb], [Hx], [Hy], [Ih], [Am 2-1983],
" -v- viridis (See Horridocactus paucicostatus -v- viridis, Pyrrhocactus paucicostatus -v- viridis),
[Ar], [Bo-6:3780], [Hu], [Hy], [Il], [Aj 22-10-1971],
pilispina (See Neoporteria pilispina, Pyrrhocactus pilispinus),
[DD], [Hx], [Jd],
" -f. pygmaea (See Horridocactus pygmaeus, N. pygmaeea, Neoporteria pilispina -f. pygmaea, Pyrrhocactus pygmaeus),
[Am 8-1979],
pseudoreichei (Invalid name - See Chileorebutia odieri, N. odieri -v- pseudoreichei, Neoporteria odieri, Thelocephala odieri),
[NN], [Bo-3], [Aj 15-1-1964],
pulchella (Invalid name - See Pyrrhocactus pulchellus),
[Bo-6], [Hx],
pygmaea (See Horridocactus pygmaeus, Neochilenia pilispina -v- pygmaea, Neoporteria pilispina -v- pygmaea, Pyrrhocactus pygmaeus),
[Av], [Bo-6], [Hb],
recondita (See Pyrrhocactus recondita, Thelocephala recondita),
[Bo-6], [Hx],
reichei (See Malacocaarpus reichei, Neoporteria reichei),
[Bo-3], [Cr], [Hx*], [Aj 24-3-1973], [Aj 26-2-1975], [Am 2-1983],
" -v- aerocarpa (See Chileorebutia aerocarpa, N. aerocarpa, Neoporteria reichei -v- reichei -f. aerocarpa, Thelocephala aerocarpa),
" -v- carneoflora (See N. carneoflora, Neoporteria reichei -v- reichei -f. carneoflora),
" -v- duripulpa (See N. duripulpa, Thelocephala duripulpa),
" -v- malleolata (See Chileorebutia malleolata, N. malleolata, Thelocephala malleolata),

reichei -v- malleolata.....(cont'd)
" -v- " -f. krausii (See Chileorebutia krausii, N. krausii, Thelocephala krausii),
" -v- " -f. solitaria (See Chileorebutia malleolata -v- solitaria, Thelocephala malleolata -v- solitaria),
" -v- odierii (See Chileorebutia odierii, N. odierii, Neoporteria odierii, Thelocephala odierii),
residua (See Horridocactus residuus, Pyrrhocactus residuus).....[DD], [Hx],
robusta (See Neoporteria coimasensis -v- robusta),
" -v- vegasana (See Neoporteria tuberisulcata -v- vegasana),
roseiflora (See Neochilenia napina, Thelocephala roseiflora).....[Hy],
rostrata (See Neoporteria subgibbosa),
rubriflora (See N. taltalensis -v- rubriflora),
rupicola (See Horridocactus rupicolus, N. taltalensis, Neoporteria rupicola, Neoporteria taltalensis (Hutch), Pyrrhocactus taltalensis).....[H],
saxifraga (See Chileorebutia saxifraga, Pyrrhocactus saxifragus).....[Hx*],
scoparia (See Pyrrhocactus scoparius),
setosiflora (Ritt.) (See Pyrrhocactus setosiflorus (Ritt.)),.....[Ae-1981],
" -v- intermedia (See Pyrrhocactus setosiflorus -v- intermedius),
simulans (See Horridocactus simulans, Pyrrhocactus simulans).....[Hy], [Af 50-6],
" -v- minima (See N. minima),.....
" -v- trapiche.....
subikii (subicii).....[Ar],
taltalensis (Hutch.) (See Horridocactus rupicolus, N. rupicola, Neoporteria rupicola, Neoporteria taltalensis, Pyrrhocactus rupicolus, Pyrrhocactus taltalensis).....[Bo-3], [Bo-6], [Hy],
" -v- flaviflora (See Neoporteria taltalensis -v- flaviflora, Pyrrhocactus neohankeanus -v- flaviflorus), [DD], [NN], [Hy],
" -v- rubriflora (See N. rubriflora).
tenebrica (See Thelocephala tenebrica),
totoralensis (See Pyrrhocactus totoralensis).. [Af 56-6],
transitensis (See Horridocactus transitensis, Pyrrhocactus transitensis),

NEOCHILENIA.....(Cont'd),

trapichensis (See Pyrrhocactus trapichensis),
vallenarensis (See Horridocactus vallenarensis),
[Ae-1981],
vanbaellii.....[Af 50-6], [Af 53-6],
vexata (See Pyrrhocactus vexatus),
[Aj 31-4-1980],
" -v- horridus (See Horridocactus vexatus -v- horridus, Pyrrhocactus vexatus -v- horridus),
violaciflora.....
wagenknechtii (See Neoporteria wagenknechtii, Pyrrhocactus wagenknechtii),
[Bo-6:3783],
woutersiana (See Delaetia woutersiana, Neoporteria woutersiana),

Spec. FR-803.....[NN],

NEODAWSONIA

apicicephalium (See Cephalocereus apicicephalium),
[Ay], [Bo-4], [Hy], [Jd], [Af 35-4], [Af 39-1], [Af 42-6], [Af 54-6], [Bm-1987],
" -f. cristata.....[Hy],
guiengolensis (See Cephalocereus apicicephalium, Neodawsonia apicicephalium),
nana (See N. apicicephalium),
nizandensis (See Cephalocereus nizandensis).....[Bo-4],
totolapensis (See Cephalocereus totolapensis),
[DD], [Bo-4], [Hx*], [Af 35-4],

NEOEVANSIA

diguetii (See Cereus striatus, Neoevansia striata, Peniocereus diguetii, Wilcoxia striata),
[P], [Q], [Dq], [Af 56-2],
haackeana (See Peniocereus haackeanus),
lazarocardenasii.....[Aw 3-1980],
striata (See N. diguetii, Cereus striatus, Peniocereus diguetii, Wilcoxia striata),
[P], [Az], [Aw 1-1973],
zapilotensis (See Wilcoxia zapilotensis),
[Az], [Aw 1-1973],

NEOGOMESIA

agavoides (See Ariocarpus agavioides),
[F], [DD], [Ar], [Ay], [Bo-5], [Dq], [Hx*], [Hy], [Jd], [Af 13-6], [Ba 4-1980], [Cu 9:29, 1947], [Dd 9-1980],

EOLLOYDIA

aguirreana (See Escobaria aguirreana, Gymnocactus aguirreanus).....[Iz],
beguinii (Invalid name - See Echinomastus erectocentrus, Gymnocactus beguinii, Neolloydia erectocentra, Neolloydia smithii -v- beguinii),
[Cm], [Cx], [Af 1-10],
" -v- senilis (See Gymnocactus beguinii -v- senilis, N. smithii -v- senilis),
[Dq],
ceratites (See N. conoidea -v- ceratites),
[DD], [RR], [Bo-5], [Hx*], [Hy],
clavata (See Coryphantha clavata).....[RR],
conoidea (See Echinocactus conoideus),
[F], [H], [M*], [Y], [GG], [QQ], [RR], [Av], [Ax], [Ay], [Bt], [Ct-1986], [Cx], [Df], [Dq], [Ha], [Hx*], [Hy], [Il], [Iu], [Iz], [Jd], [Aj 27-8-1976], [Aw 2-1963], [Hl 2-1972], [Af 60-1-1988],
" -v- ceratites (See N. ceratites)....[Af 56-4],
" -v- grandiflora (See N, grandiflora),
[Hv], [Iz],
" -v- matehualensis (See N. matehualensis),
" -v- texensis (See N. texensis).....[CC],
" -f. cristata.....[Hy],
cubensis (See Coryphantha cubensis, Escobaria cubensis, Neobesseya cubensis),
[Bo-5], [Aj 23-7-1972], [Hl 5-1970],
emskoetterana (See Escobaria emskoetterana, Escobaria muehlbaueriana),
erectocentra (See Echinocactus erectocentrus, Echinomastus erectocentrus, Gymnocactus beguinii, N. beguinii, N. smithii -v- beguinii),
[An], [Ax], [Au 37-1],
" -v- acunensis.....[Q], [Ax],
" -v- erectpcentra.....[An], [Ax],
" -v- pallidus (See Echinmastus pallidus),
gautii.....
gielsdorfiana (See Gymnocactus gielsdorfianus, Thelocactus gielsdorfianus).....[Ct-1986], [Iz],
grandiflora (See N. conoidea -v- grandiflora),
[F], [DD], [Ar], [Av], [Bo-5], [Cx], [Hx], [Il], [Ae-1982], [Aj 33-6-1982], [Hl 2-1972] [Hl 1-1978], [Hl 2-1981],
" -v- robusta.....[Aj 6-1981],
horripila (See Gymnocactus horripilus, Thelocactus goldii, Thelocactus horripilus),
[RR 4:16], [Ct-1986], [Iz], [Im], [Iu], [Af 6-7]
intertexta (See Echinocactus intertextus, Echinomastus intertextus),
[M*], [U], [An], [Df], [Ib], [Am 6-1983],

NEOLLOYDIA.....(Cont'd),

intertexta.....(cont'd),
" -v- dasyacantha (See Echinocactus intertextus -v- dasyacantha, Echinomastus intertextus -v- dasyacanthus),
[U], [Ax], [Df],
" -v- intertesta (See Echinocactus intertextus -v- intertextus, Echinomastus intertextus -v- intertextus),
[D], [U], [Y], [Ax], [Df],
johnsonii (See Echinomastus johnsonii, Sclerocactus johnsonii),
[M*], [P], [Q], [An], [Ax], [Ca],
" -v- lutescens (See Echinomastus johnsonii -v- lutescens),
[P], [FF], [An],
knuthiana (See Echinocactus knuthianus, Gymnocactus knuthianus, Thelocactus knuthianus),
[NN], [Cr], [Ct-1986], [Ig*], [Iu], [Iz],
[Af 2-6], [Aj 16-9-1965],
laui (See N. pilispina, Turbinicarpus laui),
[Ct-1986],
lophophoroides (See Toumeya lophophoroides, Turbinicarpus lophophoroides),
[Ct-1986],
macdowellii (See Echinomastus macdowellii, Thelocactus Thelocactus conothelos -v- macdowellii, Thelocactus macdowellii).....[Hv],
mandragora (See Gymnocactus mandragora, Rapicactus mandragora, Thelocactus mandragora),
[Ct-1986], [Jd],
mariposensis (See Echinocactus mariposensis, Echinomastus mariposensis, Sclerocactus mariposensis).....[Ax], [Bh], [Df],
matehualensis (See N. conoidea -v- matehualensis),
[Hl 2-1972],
megalantha (See N. grandiflora).....[Hy],
" -f. cristata.....
odorata (See Coryphantha odorata, Cumarinia odorata),
[Ar], [Bo-5], [Il], [Jd],
orcuttii (See Escobaria orcuttii),
pilispina (See N. laui, Turbinicarpus laui),
[RR], [Hy*],
pseudomacrochele (See Strombocactus pseudomacrochele, Toumeya pseudomacrochele, Turbinicarpus pseudomacrochele).....[Ct-1986],
" -f. krainziana (See Strombocactus pseudomacrochele -v- krainziana, Toumeya pseudomacrochele -v- krainziana, Turbinicarpus pseudomacrochele -v- krainziana).....[Ct-1986],

pseudopectinata (See Normanbokea pseudopectinata, Pelecyphora pseudopectinata, Turbinicarpus pseudopectinata), [Ct-1986], [Aw 3-1987],
pulleineana (See Coryphantha pulleineana), [Z], [Ar], [Bo-5], [Aj 20-11-1969],
raphidacantha (See Coryphantha clavata), [Af 14-12],
roseana (See Escobaria roseana).....[Iu], [Iz],
saueri (See Gymnocactus saueri, Thelocactus saueri), [Ct-1986], [Aj 16-7-1965],
schmiedickeana (See Strombocactus schmiedickeanus, Turbinicarpus schmiedickeanus), [Ct-1986],
" -v- dickisoniae (See Turbinicarpus schmiedickeanus -v- dickisoniae).....[Ct-1986],
" -v- flaviflora (See Strombocactus schmiedickeanus -v- flaviflorus, Turbinicarpus flaviflorus, Turbinicarpus schmiedickeanus -v- flaviflorus)..... [Ct-1986],
" -v- gracilis (See Strombocactus schmiedickeanus -v- gracilis, Turbinicarpus gracilis, Turbinicarpus schmiedickeanus -v- gracilis),
" -v- klinkeriana (See Strombocactus schmiedickeanus -v- klinkerianus, Toumeya klinkerianus, Turbinicarpus klinkerianus, Turbinicarpus -v- klinkerianus),
" -v- macrochele (See Strombocactus macrochele, Turbinicarpus macrochele, Turbinicarpus schmiedickeanus -v- macrochele).....[Ct-1986],
" -v- schmiedickeanus (See Turbinicarpus schmiedickeanus -v- schmiedickeanus), [Ct-1986],
" -v- schwarzii (See Strombocactus schwarzii, Turbinicarpus schmiedickeanus -v- schwarzii, Turbinicarpus schwarzii).....[Ct-1986],
smithii (See Gymnocactus beguinii -v- smithii), [Ct-1986], [Jd],
" -v- beguinii (See Gymnocactus beguinii, N. beguinii, Thelocactus beguinii), [Iz],
" -v- senilis (See Gymnocactus beguinii -v- senilis).....[Ct-1986],
stuetzlei (See N. conoidea),

NEOLLOYDIA.....(Cont'd),

subterranea (See Gymnocactus subterraneus, Rapicactus subterraneus).....[Ct-1986], [Hv], [Jd],
" -v- subterranea (See Gymnocactus subterraneus -v- subterraneus),
" -v- zaragosae (See Gymnocactus subterraneus -v- zaragosae).....[Iz],
texensis (See N. conoidea -v- texensis).....[LL], [RR],
unguispina (See Echinomastus unguispinus),
valdeziana (See Echinocactus valdezianus, Gymnocactus valdezianus, Normanbokea valdeziana, Pelecyphora valdeziana, Thelocactus valdezianus, Turbinicarpus valdezianus).....[Ct-1986],
viereckii (See Gymnocactus viereckii, Thelocactus viereckii).....[Ct-1986], [Hv], [Iz],
" -v- major (See Gymnocactus viereckii -v- major, Thelocactus viereckii -v- major),
" -v- viereckii (See Gymnocactus viereckii -v- viereckii, Thelocactus viereckii - -v- viereckii),
warnockii (See Echinocactus warnockii, Echinomastus warnockii).....[Af 41-4], [Aj 32-10-1981],
ysabelae (See Gymnocactus ysabelae, Thelocactus ysabelae),
" -v- brevispina (See Gymnocactus ysabelae -v- brevispina, Thelocactus ysabelae -v- brevispina),

NEOLOBIVIA (See Lobivia)

divaricata (See Lobivia divaricata).....[Hi-4],
echinata (See Lobivia echinata).....
hertrichiana (See Lobivia hertrichiana),
incaica (See Lobivia incaica).....[Hi-4],
minuta (See Lobivia minuta).....[Hi-4],
prolifera (See Lobivia prolifera, Lobivia hertrichiana).....[Hi-4],
vilcabambae (See Lobivia vilcabambae).....Hi-4],
winteriana (See Lobivia winteriana).....[Hi-4],
wrightiana (See Lobivia wrightiana).....[Hi-4],

NEOMAMMILLARIA (See Mammillaria)

NEOPORTERIA
acutissima (See N. subgibbosa).....[L],
aerocarpa (See Chileorebutia aerocarpa, Neochilenia aerocarpa, N. reichei -v- reichei -f. aerocarpa, Thelocephala aerocarpa),
" -v- fulva (See N. reichei -f. aerocarpa),
ambigua (See Weingartia ambigua),
andicola (See Horridocactus andicolus

NEOPORTERIA.....(Cont'd),

andreaeana (See Acanthocalycium andreaeanum, Neochilenia andreaeana),
aricensis (See Neochilenia aricensis, Horridocactus aricensis, Pyrrhocactus aricensis)....[Hx*],
armata (See Horridocactus armatus, N. tuberisulcata -v- armata, Pyrrhocactus armatus).....[S],
aspillagai (See Neochilenia aspillagai, Horridocactus aspillagai, Pyrrhocactus aspillagai), [E], [Af 58-2], [Jd],
atra (See Malacocarpus reichei, Neochilenia atra, Neochilenia reichei, N. fusca, N. reichei, Thelocephala reichei),
atrispinosa (See N. cephalophora, N. polyrhaphis, N. villosa), [Ar], [Ay], [Bo-3], [Bt], [Hx*], [Hy], [Af 28-5],
atrispina.....[Bv-1], [Af 34-1], [Hs II-2-1981],
backebergii (See Malacocarpus catamarcensis, Pyrrhocactus atrospinosus, Pyrrhocactus catamarcensis, Pyrrhocactus setiflorus),
bicolor (See Islaya bicolor).
bulbocalyx (See Pyrrhocactus bulbocalyx).....[Hv],
calderana (See Neochilenia calderana, Pyrrhocactus calderanus),
" -f. gracilis (See Pyrrhocactus gracilis),
carneoflora (See Neochilenia carneoflora, Neoporteria reichei -f. carneoflora),
carrizalensis (See N. curvispina -v- carrizalensis, Pyrrhocactus carrizalensis).....[Hv],
castanea (See Neoporteria subgibbosa -v- subgibbosa -f. castanea), [Hi-3], [Hy], [Bm-1988], [Hl 1-1982],
" -v- tunensis (See N. subgibbosa -v- subgibbosa -f. tunensis),
castaneoides (See N. subgibbosa -v- subgibbosa -f. castaneoides).....[NN], [Bo-3], [Hx], [Hy],
" -f. spirala.....[Hy],
catamarcensis (See Malacocarpus catamarcensis, Pyrrhocactus atrospinosus, Pyrrhocactus catamarcensis, Pyrrhocactus setiflorus),
cephalophora (See N. atrispinosa, N. polyrhaphis, N. villosa, N. villosa -v- cephalophora), [E], [Bo-3], [Aj 10-10-1959], [Hs II-2-1981],
chaniarensis (See Pyrrhocactus chaniarensis),
chilensis (See Horridocactus chilensis, Neochilenia chilensis, Pyrrhocactus chilensis), [RR], [Hx], [Af 34-1],
" -v- albidiflora (See Pyrrhocactus chilensis -v- albidiflorus).....[Iz],

NEOPORTERIA.....(Cont'd),

chilensis.....(cont'd),
" -v- chilensis (See Pyrrhocactus chilensis -v- chilensis).
" -v- " -f. confinis (See Horridocactus confinis, Pyrrhocactus confinis),
choapensis (See Horridocactus choapensis, Pyrrhocactus choapensis),
chorosensis (See Neochilenia chorosensis, Pyrrhocactus chorosensis),
clavata.....[DD], [Ar], [Hi-3], [Hx], [Hy], [Af 52-5], [Hl 6-1965],
" -v- nana.....
" -v- parviflora.....[Hi-3],
" -v- procera.....[Hi-3], [Hx], [Hy],
" -v- serenana.....
" -f. FK 27-77.....[Af 52-5],
" -f. FK 184-79.....[Af 52-5],
coimasensis.....[Bo-6], [Hi-3], [Hy], [Hl 6-1965],
" -v- robusta (See N. robusta).....[Hi-3],
coliguayensis (See Pyrrhocactus coliguayensis),
confinis (See Horridocactus confinis, Pyrrhocactus confinis),
coquimbana (See N. nigrihorrida -v- coquimbana)...... [Bo-3],
crassispina (See N. nigrihorrida -v- crassispina),
crispa (See Horridocactus crispus, Neochilenia crispa, Neochilenia nigriscoparia, Pyrrhocactus crispus), [Iz],
curvispina (See Horridocactus curvispinus, Pyrrhocactus curvispinus),
" -v- aconcaguensis (See Pyrrhocactus horridus -v- aconcaguensis),
" -v- " -f. orientalis (See Pyrrhocactus horridus -v- orientalis)
" -v- andicola (See Horridocactus andicolus, Pyrrhocactus andicolus),
" -v- " -f. descendens (See Horridocactus andicolus -v- descendens, Pyrrhocactus andicolus -v- descendens),
" -v- " -f. mollensis (See Pyrrhocactus andicolus -v- mollensis),
" -v- " -f. robusta (See Horridocactus robustus, Pyrrhocactus horridus -v- robustus),
" -v- carrizalensis (See Pyrrhocactus carrizalensis),
" -v- echinus (See Horridocactus echinus, Pyrrhocactus echinus),

NEOPORTERIA.....(Cont'd),

curvispina.....(cont'd),
" -v- garaventai (See Horridocactus garaventai, Pyrrhocactus garaventai),
" -v- geissei (See Horridocactus geissei, Neoporteria geissei),
" -v- " -f. albicans.....
" -v- grandiflora (See Pyrrhocactus grandiflorus),
" -v- heinrichianus (See Horridocactus heinrichianus, Neoporteria heinrichiana),
" -v- kesselringiana (See Horridocactus kesselringianus, Neoporteria kesselringiana),
" -v- " -f. subaequalis (See Horridocactus kesselringianus -v- subaequalis)
" -v- lissocarpa (See Pyrrhocactus lissocarpus,
" -v- " -f. gracilis (See Pyrrhocactus lissocarpa -v- gracilis, Pyrrhocactus gracilis).
" -v- transitensis (See Horridocactus transitensis, Pyrrhocactus transitensis),
" -v- vallenarensis (See Horridocactus vallenarensis, Neochilenia vallenarensis, N. vallenarensis, Pyrrhocactus vallenarensis),

deherdtiana (See Neochilenia deherdtiana, Neochilenia occulta, Neoporteria occulta, Pyrrhocactus deherdtianus, Pyrrhocactus occultus),

densispina (See N. nidus -f. densispina), [K], [Hy],

dimorpha (See Neochilenia dimorpha, Pyrrhocactus dimorphus),

divaricatiflora (See Islaya divaricatiflora)

dubia (See Pyrrhocactus dubius, Pyrrhocactus catamarcensis),

duripulpa (See Neochilenia duripulpa, Thelocephala duripulpa),

ebenacantha (See Echinocactus ebenacanthus (Monv.), [If], [Aj 1937:183 Fig.],

engleri (See Horridocactus engleri, Pyrrhocactus engleri),
" -v- krausii (See Horridocactus engleri -v- krausii, Neochilenia krausii, Pyrrhocactus krausii),

eriocephala (See Neochilenia eriocephala, Pyrrhocactus floccosus),
" -v- glaucescens (See Neochilenia glaucescens, Pyrrhocactus glaucescens),

eriosyzoides (See Horridocactus eriosyzoides, Neochilenia eriosyzoides, Pyrrhocactus eriosyzoides),

eriosyzoides -v- domeykoensis (See Pyrrhocactus eriosyzoides -v- domeykoensis),
esmeraldana (See Chileorebutia esmeraldana, Neochilenia esmeraldana, Thelocephala esmeraldana), [Hv], [Jd], [Hh Vol. 3 #3&4, 1978],
fankhauseri (See Thelocephala fankhauseri),
flavida (See Islaya flavida),
floccosa (See Horridocactus floccosus, Neochilenia floccosa, Pyrrhocactus floccosus, N. eriocephala),
" -v- minor (See Pyrrhocactus floccosus -v- minor),
floribunda (See Neochilenia iquiquensis, Neoporteria iquiquensis, Reicheocactus floribundus, Pyrrhocactus floribundus),
fobeana (See Neochilenia fobeana, N. jussieui -v- fobeana),
fulva (See Chileorebutia fulva, Neochilenia aerocarpa -v- fulva, Thelocephala fulva),
fusca (Muehl.) (See Echinocactus ebenacanthus (Monv.), Echinocactus fuscus (Muehl.), Neochilenia fusca (Muehl.)).....[I], [RR], [Dq],
geissei (See Horridocactus geissei, Neoporteria curvispina -v- geissei),
gerocephala (See N. nidus -v- gerocephala, N. senilis), [AA], [DD], [NN], [Ar], [Ay], [Bo-3], [Gr], Hb], [Hu], [Hx], [Hy], [Il], [Jd], [Af 54-1], [Ag-1985], [Ah-1985],
" -f. schwarzstachelige.....[NN],
" -f. weisstachelige.....[NN],
glabrescens (See Neochilenia glabrescens, N. napina -v-mitis -f. glabrescens, Thelocephala glabrescens),.....[Ij], [Hh Vol. 3, #3&4, 1978],
glaucescens (See N. eriocephala -v- glaucescens, Pyrrhocactus glaucescens),
gracilis (See N. calderana -v- gracilis, Pyrrhocactus gracilis),
grandiflora (See N. curvispina -v- grandiflora, Pyrrhocactus grandiflorus).....[UU], [Av],
grandis (See N. islayensis -v- grandis),
hankeana (See Echinocactus hankeanus, Neochilenia hankeana, Horridocactus hankeanus)....[T], [Iz],
" -v- minor (See Echinocactus ebenacanthus (Monv.)) [Cu 28:56 Aug. 1966],
" -v- taltalensis (See Neochilenia hankeana -v- taltalensis)....[Cu 28:56 Aug. 1966],
heinrichianus (See Horridocactus heinrichianus, N. curvispina -v- heinrichiana),
heteracantha (See N. subgibbosa -v- subgibbosa -f. heteracantha)....[Ar], [Bo-3], [Bv-1], [Cz], [Hy], [Af 34-1], [Aj 15-4-1964],

NEOPORTERIA.....(Cont'd),

horrida (See Horridocactus horridus, Horridocactus tuberisulcatus, Pyrrhocactus horridus, Pyrrhocactus tuberisulcatus),
" -v- aconcaguensis (See Pyrrhocactus aconcaguensis, Pyrrhocactus horridus -v- aconcaguensis),
" -v- mutabilis (See Pyrrhocactus horridus -v- mutabilis),
" -v- orientalis (See Pyrrhocactus horridus -v- orientalis),
" -v- robusta (See Pyrrhocactus horridus -v- robustus),
huascensis (See Neochilenia huascensis, Pyrrhocactus huascensis), [Cu 28:56-Aug. 1966],
imitans (See Neochilenia imitans, Neochilenia lembckei, Neochilenia neoreichei, N. reichei -v- reichei f. imitans, Reicheocactus neoreichei, Thelocephala lembckei, Copiapoa hypogaea).
intermedia (See Neochilenia intermedia, Pyrrhocactus intermedius).....[Cu 28:56 Aug. 1966],
iquiquensis (See Neochilenia iquiquensis, Pyrrhocactus floribundus, Pyrrhocactus iquiquensis, Reicheocactus floribundus).....[Cu 28:56 Aug. 1966],
islayensis (See Islaya islayensis, Islaya mollendensis) [Iz], [Cu 28:56 Aug. 1966],
" -f. brevicylindrica (See Islaya brevicylindrica).....[Cu 28:56, Aug. 1966],
" -f. copiapoides (See Islaya copiapoides), [Cu 28:56 Aug. 1966],
" -f. " -f. chalaensis.....
" -f. " -f. pseudomollendensis.....
" -f. divaricatiflora (See Islaya divaricatiflora), [Cu 28:56 Aug. 1966],
" -f. grandiflorens (See Islaya grandiflorens, Islaya maritima, N. maritima), [Cu 28:56 Aug. 1966],
" -f. grandis (See Islaya grandis), [Cu 28:56 Aug. 1966],
" -f. " -f. brevispina (See Islaya grandis -v- brevispina, Islaya grandis -v- neglecta)....[Cu 28:56 Aug. 1966],
" -f. minor (See Islaya islayensis -v- minor, Islaya minor), [Cu 28:56 Aug. 1966],
" -f. mollendensis (See Islaya islayensis, Islaya mollendensis), [Cu 28:56 Aug. 1966],

NEOPORTERIA.....(Cont'd),

islayensis.....(cont'd),
" -f. spinosior (See Islaya grandiflorens -v- spinosior, Islaya maritima, N. islayensis -f. grandiflorens, N. maritima), [Cu 28:56 Aug. 1966],
jussieui (See Neochilenia jussieui, Pyrrhocactus jussieui), [L], [RR], [Cr], [Dh-I*],
" -v- australis.....
" -v- fobeana (See Neochilenia fobeana), [Cu 28:56 - 1966],
" -v- spinosior.....
kesselringiana (See Horridocactus kesselringiana, Pyrrhocactus kesselringiana),
krainziana (See Islaya krainziana), [Cu 28:57 Aug. 1966],
krausii (See Chileorebutia krausii, Neochilenia krausii, Neoporteria reichei -v- malleolata -f. krausii, Pyrrhocactus krausii).....[Jd],
kunzei (See Neochilenia kunzei, Pyrrhocactus kunzei),
laniceps (planiceps) (See N. planiceps), [Hi-3], [Ae-1981], [Af 59-2], [Hs II-2-1981],
lembckei (See Neochilenia imitans, Neochilenia lembckei, Neochilenia neoreichei, Reicheocactus neoreichei, Thelocephala lembckei),
limarensis (See N. nigrihorrida f. limarensis),
lindleyi (See Islaya copiapoides -v- chalaensis), [Cu 28:57 Aug. 1966],
" -f. curvispina (See Islaya paucispina -v- curvispina)...[Cu 28:57 Aug. 1966],
" -f. lindleyi (See Islaya paucispina, Islaya paucispinosa).
lissocarpa (See N. curvispina -v- lissocarpa, Neoporteria lissocarpus),
litoralis (See N. gibbosa -v- gibbosa -f. litoralis), [Bo-3], [Bo-6], [Hi-3], [Hy], [Af 52-5], [Aj 27-10-1976], [Am 2-1983], [Hl 6-1965],
" -v- intermedia (See N. intermedius, Pyrrhocactus intermedius)..[Hi-3], [Hl 1-1982],
" -v- microsperma (See N. microsperma),
" -v- " -f. serenana (See M. microsperma -v- serenana).....
" -v- nigrihorrida (See N. nigrihorrida),
" -v- " -f. atra (See Neochilenia atra),
" -v- " -f. coquimbana (See Neoporteria coquimbana),
" -v- robustispina.....[Hy],
malleolata (See Chileorebutia malleolata, Neochilenia malleolata, N. reichei -v- malleolata, Thelocephala malleolata),

NEOPORTERIA......(Cont'd
malleolata.....(cont'd),
" -v- solitaria (See Chileorebutia malleolata -v- solitaria, Neoporteria reichei -v- malleolata -f. solitaria, Thelocephala malleolata -v- solitaria),
mamillarioides (See N. subgibbosa -v- mamillarioides), [W], [Bo-3], [Hy], [Aj 23-2-1972], [Au 35-2],
maritima (See Islaya maritima, Islaya grandiflorens), [Iz],
marksiana (See Horridocactus marksianus, Pyrrhocactus marksianus).....[Cu 28:57 Aug. 1966],
" -v- tunensis (See Horridocactus marksianus -v- tunensis, Pyrrhocactus tunensis), [Cu 28:57 Aug. 1966],
mebbesii (See Echinocactus odierii -v- mebbesii, Neochilenia mebbesii),
melanacantha (See Pyrrhocactus melanacanthus), [Cu 28:67-Aug. 1966],
microsperma (See N. subgibbosa -v- microsperma), [Hb], [Hi-3], [Hy], [Ae-1983], [Af 52-5], [Ba 1963:6], [Hl 1-1982],
" -v- graciana.....[Hi-3],
" -v- procerea (See N. procera),
" -v- serenana.....[Hi-3], [Hy], [Hl 1-1982],
minima.....
minuscula (See Islaya minuscula),
mitis (See Neochilenia mitis, N. napina -v- mitis, Thelocephala mitis).....[Ay],
monte-amargensis (See Neochilenia monte-amargensis), [Iz - Hybrid],
multicolor.....[F], [NN], [Ar], [Bo-3], [Hi-3], [Hx], [Hy], [Ae-1981], [Am 2-1983], [Hs I-4-1980],
" -v- litoralis (See N. litoralis),
" -v- nigrispina.....
napina (See Chileorebutia napina, Neochilenia napina, Thelocephala napina), [Bt], [Cg], [Cr], [Hv], [Ig], [Jd], [Af 34-1], [Aj 10--10-1959], [Aj 16-9-1965], [Hw-1989],
" -v- falkenbergii.........
" -v- lanigera.............
" -v- mitis (See Neochilenia mitis, Thelocephala mitis).....[Iz],
" -v- " -f. glabrescens (See Chileorebutia glabrescens, N. glabrescens, Neochilenia glabrescens, Thelocephala glabrescens),
" -v- spinosior (See Neochilenia napina -v- spinosior).....[Iz],
" -v- " -f. mebbesii (See Neochilenia mebbesii, Echinocactus odierii -v- mebbesii)
neofusca (See Neochilenia neofusca, Neoporteria fusca),

nidus.....[L], [N], [S], [NN], [RR], [Ar], [Av], [Bo-3], [Bu], [Dq], [Hb], [Hx], [Hy], [Iu], [Iz], [Jd], [Af 15-12], [Af 34-1], [Hl 4-1976],
" -f. densispina (See N. densispina, N. nidus),
" -v- gerocephala (See N. gerocephala, N. senilis, N. nidus -f. senilis)...[Hi-3], [Hv],
" -f. matancillana.....
" -f. nidus (See N. multicolor).....[Hi-3],
" -f. senilis (See N. nidus -v- gerocephala, N. senilis),
[F], [H], [Ab-1985], [Au Vol. 2, 1977 - Zone 3 Compendium], [Bu],
nigricans (See Horridocactus nigricans, N. tuberisulcata -v- nigricans, Pyrrhocactus nigricans), [RR], [Dq],
-v- coquimbana (See N. coquimbana),
nigrihorrida (See N. subgibbosa -v- nigrihorrida), [PP], [Bo-3], [Bu], [Bv-1], [Hi-3], [Hx*] [Hy], [Af 21-2], [Af 34-1], [Af 52-5], [Hl 6-1979],
" -v- atra (See Neochilenia atra, Thelocephala reichei),
" -v- coquimbana (See N. coquimbana, N. nigrihorrida)..[Hb], [Hy], [Am 1-1985],
" -v- crassispina (See N. nigrihorrida),
" -v- flaviflora.....
" -f. limariensis.....[Af 52-5],
" -v- major (See N. nigrihorrida)...[Bo-3], [Ha],
" -v- minor (See N. nigrihorrida), [Af 23:118, 1951],
" -v- parviflora (See N. clavata -v- parviflora),
" -v- procera (See N. clavata -v- procera),
" X Neoporteria litoralis -v- intermedia....[Hi-3],
nuda (See Thelocephala nuda),
occulta (See Neochilenia occulta, Neochilenia deherdtiana, Pyrrhocactus deherdtiana, Pyrrhocactus occultus).....[NN], [RR], [Hv], [Hh Vol. 3, #3&4, 1978], [Hw-1981],
odierii (See Chileorebutia odierii, Neochilenia odierii, Neochilenia pseudoreichei, Thelocephala odierii), [Iz], [Hh Vol. 3, #3 & 4, 1978],
" -v- mebbesii (See Echinocactus odierii -v- mebbesii).....[Jd],
" -v- araneifera (See Echinocactus araneifer).
odoriflora (See Neochilenia odoriflora, Pyrrhocactus odoriflorus),
pamaensis (See Pyrrhocactus pamaensis),
paucicostata (See Horridocactus paucicostatus, Neochilenia paucicostata, Pyrrhocactus paucicostatus).....[Hv], [Jd],

paucicostata.....(cont'd),
" -v- viridis (See Horridocactus paucicostatus -v- viridis, Neochilenia paucicostata -v- viridis, Pyrrhocactus paucicostatus -v- viridis),
pilispina (See Neochilenia pilispina, Pyrrhocactus pilispinus).....[Jd], [Aj 14-7-1963],
" -f. pygmaea (See Horridocactus pygmaeus, Neochilenia pygmaea, Neochilenia pilispina -v- pygmaea, Pyrrhocactus pygmaeus),
[Iz], [Am 2-1983],
planiceps (Invalid name - See N. laniceps).....[NN],
platyacantha (See Pyrrhocactus platyacanthus),
polyrhaphis (See N. atrispinosa, N. cephalophora, N. villosa -v- polyrhaphis, N. villosa),
[Bo-3], [Hy],
procera (See N. clavata -v- procera),
" -v- serenana (See N. clavata -v- serenana),
pseudoreichei (See Neochilenia pseudoreichei, N. reichei -v- reichei -f. pseudoreichei),
Neochilenia odierii, Neoporteria odierii, Thelocephala odierii).....[Gr],
pulchella (See Neochilenia pulchella, N. pilispina, Pyrrhocactus pulchellus),
pygmaea (See Neochilenia pygmaea),
rapifera (See N. wagenknechtii -v- napina),
[Bo-6], [Hx], [Hy], [Aj 23-2-1972], [Aj 27-12-1976], [Flower - Aj 29-10-1978],
recondita (See Neochilenia recondita, Pyrrhocactus reconditus, Thelocephala recondita),
reichei (See Malacocarpus reichei, Neochilenia reichei, Neochilenia atra, Thelocephala reichei),
[Cr], [Ig], [Aj 10-10-1959], [Iv 6-1946],
" -v- malleolata (See Chileorebutia malleolata, Neochilenia malleolata, N. malleolata, Thelocephala malleolata),
" -v- " -f. krausii (See Chileorebutia krausii, Neochilenia krausii, Thelocephala krausii),
" -v- " -f. solitaria (See Chileorebutia malleolata -v- solitaria, Thelocephala malleolata -v- solitaria),
" -v- reichei -f. aerocarpa (See Chileorebutia aerocarpa, Neochilenia aerocarpa, Thelocephala aerocarpa).....[Hv],
" -v- " -f. carneoflora (See Neochilenia carneoflora)..[Hv], [Ad Feb. 1979],

reichei -v- reichei.....(cont'd),
" -v- " -f. duripulpa (Neochilenia duripulpa, Thelocephala duripulpa)...
" -v- " -f. floribunda (See Neochilenia iquiquensis, Pyrrhocactus floribundus, Reicheocactus floribundus)
" -v- " -f. imitans (See Neochilenia imitans, Neochilenia lembckei, Neochilenia neoreichei, Reicheocactus neoreichei, Thelocephala lembckei),
" -v- " -f. lembckei (See Neochilenia lembckei, Neochilenia imitans, Neochilenia neoreichei, Reicheocactus neoreichei, Thelocephala lembckei),
" -v- " -f. neoreichei (See Neochilenia neoreichei, Reicheocactus neoreichei, Thelocephala lembckei),
" -v- " -f. pseudoreichei (See Neochilenia pseudoreichei, Reicheocactus pseudoreicheanus, Thelocephala odieri),
residua (See Horridocactus residuus, Neochilenia residua, Pyrrhocactus residuus),
ritteri (See Neochilenia wagenknechtii, Neoporteria wagenknechtii, N. wagenknechtii, Pyrrhocactus wagenknechtii)..... [Iz],
robusta (RITT.) (See N. coimasensis -v- robusta, N. tuberisulcata -v- robusta),
rupicola (See Horridocactus rupicolus, Neochilenia rupicola, Neochilenia taltalensis (Hutch), N. taltalensis (Hutch), Pyrrhocactus taltalensis (Hutch.), Pyrrhocactus rupicolus).....[Iz], [Am 9-1979],
sanjuanensis (See Pyrrhocactus sanjuanensis),
saxifraga (See Neochilenia saxifraga, Pyrrhocactus saxifragus),
scoparia (See Pyrrhocactus scoparius),
senilis (See N. nidus -v- gerocephala, N. gerocephala), [NN], [Aj 8-11-1957], [Am 2-1983],
" -v- nidus (See N. nidus),
" -f. monstrosus.....[Aj 23-4-1972],
setiflora (Backbg.) (See Malacocarpus catamarcensis, Pyrrhocactus catamarcensis, Pyrrhocactus atrospinosus, Pyrrhocactus setiflorus),
setosiflora (Ritt.) (See Neochilenia setosiflora (Ritt.), Pyrrhocactus setosiflora (Ritt.)), [Ae-1981],
" -v- intermedia (See Pyrrhocactus setosiflorus (Ritt.) -v- intermedius),

NEOPORTERIA.....(Cont'd),
simulans (See Horridocactus simulans, Neochilenia simulans, Pyrrhocactus simulans),
sociabilis.....[Hi-3], [Ba 1963:3],
" -v- napina.....[Hi-3],
strausiana (See Malacocarpus strausianus, Pyrrhocactus strausianus),
subaiana (See Horridocactus garaventai, Pyrrhocactus garaventai, Pyrrhocactus subaianus),
subcylindrica (See Chilenia subcylindrica, N. subgibbosa -v- subgibbosa -f. subcylindrica), -v- subgibbosa -f. subcylindrica),
subgibbosa.....[F], [NN], [RR], [ZZ], [Ar], [Bo-3], [Bv-1], [Dq], [Gr], [Hi-3], [Hx*], [Hy], [Il], [Jd], [Af 26-3], [Af 58-2], [Aj 9-1-1958], [Aj 24-3-1973], [Hl 1-1982], [Hw-1982],
" -v- chionacantha.....[Hy],
" -v- intermedia (See Neochilenia intermedia, Pyrrhocactus intermedius)..[Bo-3],
" -v- mamillarioides (See N. mamillarioides),
" -v- microsperma (See N. microsperma),
" -v- " f. serenana(See N. microsperma-v-serenana)
" -v- nigrihorrida (See N. nigrihorrida),
" -v- " -f. major (See Chilenia nigrihorrida -v- major),
" -v- " -f. minor (See Chilenia nigrihorrida -v- minor),
" -v- " -f. nigrihorrida (See Chilenia nigrihorrida).
" -v- orientalis.....[Hi-3], [Hl 1-1982],
" -v- subgibbosa -f. castanea (See N. castanea),
" -v- " -f. castaneoides (See N. castaneoides),
" -v- " -f. heteracantha (See Chilenia heteracantha, N. heteracantha),
" -v- " -f. litoralis (See N. litoralis),
" -v- " -f. subcylindrica (See Chilenia subcylindrica, N. subcylindrica),
" -v- " -f. tunensis (See N. castanea -v- tunensis),
" -f. cristata.....[Hi-3], [Hy],
taltalensis (Hutch), (See Horridocactus rupicolus, Neochilenia rupicola, Neochilenia taltalensis, N. rupicola, Pyrrhocactus taltalensis (Hutch.)). [Af 27:181 Fig. 167, 1955], [Hl 6-1965],
" -v- flaviflora (See Neochilenia taltalensis -v- flaviflora, N. hankeana).....[Hv],
tenebrica (See Thelocephala tenebrica),
tenuis (See Pyrrhocactus tenuis),
thiebautiana (form of N. nidus),
totoralensis (See Neochilenia totoralensis, Pyrrhocactus totoralensis),
transiens (See Pyrrhocactus transiens),

NEOPORTERIA.....(Cont'd),

transitensis (Horridocactus transitensis, Pyrrhocactus transitensis, N. curvispina -v- transitensis),
trapichensis (See Pyrrhocactus trapichensis),
truncatipetala (See Pyrrhocactus truncatipetalus),
tuberisulcata (See Horridocactus horridus, Horridocactus tuberisulcatus, Pyrrhocactus horridus, Pyrrhocactus tuberisulcatus),
" -v- armata (See N. armata, Horridocactus armatus, Pyrrhocactus armatus),
" -v- atroviridis (See Horridocactus atroviridis, Pyrrhocactus atroviridis),
" -v- cupreata (See Copiapoa cupreata, Echinocactus cupreatus).
" -v- froehlichiana (See Horridocactus froehlichianus, Pyrrhocactus froehlichianus)
" -v- nigricans (See Horridocactus nigricans, N. nigricans),
" -v- robusta (See Pyrrhocactus robustus, Horridocactus robustus),
" -v- vegasana (See Horridocactus robustus, Pyrrhocactus robustus),
umadeave (See Pyrrhocactus umadeave).....[Iz],
" -v- marayesensis (See Pyrrhocactus umadeave -v- marayesensis),
unguispina (See Islaya unguispina),
vallenarensis (See Horridocactus vallenarensis, Neochilenia vallenarensis, Pyrrhocactus vallenarensis),
vexata (See Neochilenia vexata, Pyrrhocactus vexatus),
villosa (See N. atrispinosa, N. cephalophora, N. polyrhaphis)...[F], [I], [QQ], [Ar], [Av], [Bo-3], [Hi-3], [Hb], [Hx*], [Hy], [Il], [Iz], [Af 56-6], [Am 2-1983], [Hl 6-1978],
" -v- atrispinosa (See N. atrispinosa),
" -v- cephalophora (See N. cephalophora),
" -v- nigra.....
" -v- nigrispina.....
" -v- polyrhaphis (See N. polyrhaphis).....[UU],
" -v- villosa.....
volliana (See Pyrrhocactus vollianus),
" -v- breviaristata (See Pyrrhocactus vollianus -v- breviaristata),
wagenknechtii (See Neochilenia wagenknechtii, N. ritteri, Pyrrhocactus wagenknechtii), [H], [NN], [UU], [Av], [Hi-3], [Hu], [Hx*], Hy], [Jd], [Af 52-5], [Aj 15-4-1964], [Hl 5-1968], [Hs I-3-1980],
" -v- multiflora.....
" -v- napina (See N. napina)..[Hi-3], [Ba 1963:5],
" -v- rapifera (See N. rapifera),

NEOPORTERIA.....(Cont'd),

woutersiana (See Delaetia woutersiana),
[Hh July 1974],
Spec. FK 185-79.....[Af 52-5],
Spec. J. Moreau, Algarroba.....[Af 52-5],

NEORAIMONDIA
----

arequipensis (See Cereus peruvianus, N. macrostibas, N. peruviana).....[Ay], [Bo-2], [Bo-6], [Hx*], [Hy], [Jd],
" -v-aticensis (See N. aticensis, N. peruviana), [Ay], [Hl 5-1983],
" -v- gigantea (See N. gigantea).....[Ay], [Jd],
" -v- rhodantha.....[Bo-2],
" -v- riomajensis.....[Bo-2],
" -v- roseiflora (See N. roseiflora), [Ij], [Af 57-2],
" -v- saniensis (See N. saniensis, N. gigantea -v- saniensis),
aticensis (See N. arequipensis -v- aticensis), [DD], [Bo-2], [Hx*], [Il], [Hl 2-1982], [Hl 4-1982],
gigantea (See N. arequipensis -v- gigantea), [DD], [Ay], [Bo-2], [Bo-6], [Bv-1], [Hx*], [Hy], [Ad-], [Af 50-1], [Hl 5-1982],
" -v- saniensis (See N. saniensis).....[Bo-2],
herzogiana (See Neocardenasia herzogiana).....[Hi-2],
longispina (See N. roseiflora -v- longispina),
macrostibas (See Cereus peruvianus, N. arequipensis, N. peruviana),
[RR], [Ay], [Bv-1], [Af 18-7], [Cu 8-1-1939],
" -v- gigantea (See N. arequipensis -v- gigantea, N. gigantea).....[Af 22-5],
peruviana (See Cereus peruvianus, N. arequipensis, N. macrostibas).....[Hi-4],
" -f. gigantea (See N. gigantea),
" -f. roseiflora (See N. roseiflora),
" -f. sayanensis (See N. roseiflora -v- sayanensis),
rosada.....
roseiflora (See N. arequipensis -v- roseiflora), [Ar], [Ay], [Bo-2], [Bo-6], [Hx*], [Af 55-2], [Da 2-1980], [Hl 4-1966], [Bi Vol. 4, #6],
" -v- longispina.....
" -v- sayanensis (See N. peruviana -v- sayanensis).....[Bo-6],
saniensis (See N. arequipensis -v- saniensis, N. gigantea -v- saniensis),

NEOWERDERMANNIA

chilensis (See Weingartia chilensis)....[Ar], [Bo-3], [Hi-3], [Hx*], [Hh July, 1973], [Hh 3,#5,1979],
" -v- putrensis.....
peruviana.....[Hi-4], [Iz],
" -v- tacunensis.....
putrensis.....
vorwerkii (See Gymnocalycium vorwerkii, Weingartia vorwerkii),
[NN], [Ar], [Ay], [Bo-3], [Hi-2], [Dq], [Hr], [Hv], [Hx*], [Hy], [Il], [Iz], [Jd], [Aj II-12-1930], [Aj 26-10-1975], [Hl 1-1983],
" -v- erectispina.....[Bo-3],
" -v- gielsdorfiana.....[Bo-3],
westii (See Weingartia westii),

NOPALEA

angustifrons (See N. dejecta, Opuntia dejecta),
auberi (See Opuntia auberi),
[I], [RR], [Ar], [Az], [Cx], [Af 20-8],
brittonii (See Cylindropuntia brittonii, Opuntia brittonii, Opuntia leptocaulis -v- brittonii),
[Az], [Bv-1], [Af 1-5],
" -v- variegata.....[Af 1-5],
cochenillifera (See Opuntia cochenillifera),
[I], [M*], [RR], [Az], [Bv-1], [Cx], [Cz], [Hy*], [Iu], [Af 1-5],
dejecta (See Opuntia dejecta),
[I], [L], [RR], [Av], [Bv-1], [Hx],
" -v- guamacciana.....
escuintiensis (See Opuntia escuintiensis),
gaumeri (See Opuntia gaumeri).....[RR],
guatamalensis (See Opuntia guatamalensis).....[RR],
inaperta (See Opuntia inaperta).....[RR],
karwinskiana (See Opuntia karwinskiana),
[RR], [Az], [Cx],
lutea (See Opuntia lutea).....[RR],
moniliformis (Consolea moniliformis, Opuntia moniliformis),
nuda (Invalid name - See Opuntia nuda),
[DD], [Ay], [Bo-6],

NOPALXOCHIA

ackermannii (See Epiphyllum ackermannii),
[I], [Az], [Bo-2], [Bo-6], [Bu], [Bv-1], [Cz], [Hx*], [Jd], [Af 53-2], [Aw 3-1980],
" -v- conzattianum (See N. conzattianum),
conzattianum (See Pseudonopalxochia conzattianum),
[Az], [Af 19-2], [Af 23-5],

NOPALXOCHIA.....(Cont'd),

horichii (See Disocactus horichii),
[Ij], [Af 56-1],
macdougalii (See Lobiera macdougalii),
[Az], [Hh July, 1973]*,
phyllanthoides (See Phyllocactus phyllanthoides),
[Bo-6], [Dq], [Hx*], [Hy], [Il], [Iz],
[Jd], [Af 13-2], [Af 13-11], [Aj 28-3-1977], [Aw 3-1962],

(Hybrid).....[Ay], [Ag-1983],

NORMANBOKEA

pseudopectinata (See Neolloydia pseudopectinata, Pelecyphora pseudopectinata, Turbinicarpus pseudopectinatus),
[Av], [Hx], [Jd], [Ae-1981], [Af 42-6],
[Aj 32-10-1981], [Bi 6-2-1988],
" -v- rubriflora.....
valdeziana [See Echinocactus valdezianus, Gymnocactus valdezianus, Pelecyphora valdeziana, Thelocactus valdezianus, Turbinicarpus valdezianus)
[Bt], [Hx], [Jd], [Ae-1981], [Ba 11-1977],
" -v- albiflora.....
" -f. cristata.....[Hx],

NOTOCACTUS

acuatus (See Echinocactus erinaceus, Notocactus erinaceus, Wigginsia acuata, Wigginsia erinacea),
acutus (See Wigginsia acuta),
[Hx*], [Hi-1], [Am 9-1980], [Hl 5-1975],
" -v- cadeaensis.....
" -v- depressus.....
" -v- lourencoensis.....
agnetae.....[Hx*], [Ah-1985], [Hl 6-1982],
" -v- aureispinus.....
" -v- minor.....
alacriportanus (See Brasiliparodia alacriportana, N. megapotamicus -v- alacriportanus, Parodia alacriportana),
[Hx], [Aj 27-10-1976],
" -v- cambarensis.....
" -v- fortalezensis.....
" -v- parvisetus.....
allosiphon....[Gp], [Hx*], [Af 50-3], [Ah-1980],
[Aj 24-1-1973], [Ba 2-1978], [Hh May 1974]
[Hl 1-1979],
ampliocostatus (See Eriocactus ampliocostatus),

apricus.....[K], [M*], [V], [AA], [DD], [Ar], [Bo-3],
[Bu], [Bv-1], [Cg], [Cm], [Dq], [Gp], [Hx],
[Hy], [N-1980],
" -v- concinnus (See N. concinnus).....[Hy],
" -v- durispinus.....
" -v- multicostatus (See N. multicostatus),
" -v- muricatus (See N. muricatus).....[Hy],
" -v- nigrispinus (See N. nigrispinus),
[Hy],
" -v- tabularis (See N. tabularis).....[Hy],
arachnites.....[R], [Hi-1], [Hx*], [Hl 5-1970],
" -v- minor.....[Hi-1],
araneolarius (See N. ottonis -v- uraguayanus),
arbolitensis (See N. mammulosus -v- arbolitensis),
arechavaletai (See Malacocarpus arechavaletai, N.
neoarechavaletai, Wigginsia arechavaletai),
[Hv], [Hl 3-1968],
" -v- alacriportanus (See N. alacriportanus),
[Hi-1],
" -v- aureus.....[Hi-1],
" -v- buenekeri.....[Hi-1],
" -v- horstii.....[Hi-1],
" -v- kovaricii (See N. kovaricii, N. neoarech-
avaletai -v- kovaricii),
" -v- limiticola.....[Hi-1],
" -v- nanus.....[Hi-1],
" -v- rubescens.....[Hi-1],
" -f. (Arroya-Cunapiru).....
" -f. (Estancia-Platon).....
artigas.....[Hl 3-1968],
aurisetus (See Parodia auriseta),
bertinii (See Austrocactus bertinii),
bezrucii (See Malacocarpus bezrucii,
" -v- centrispinus.....
" -v- corniger.....
blaauwianus.....[Hx*],
" -v- enormis.....[Ba 6-1976],
boomeljei (See N. muricatus).....[Hl 6-1972],
brasiliensis (See N. mammulosus -v- brasiliensis,
Parodia brasiliensis).....[Hx],
brevihamatus (See Brasiliparodia brevihamata,
Parodia brevihamata),
[BB], [Gp*], [Db 11-1980],
" -v- molispinus.....
buenekeri (See Brasiliparodia buenekeri, Parodia
buenekeri),
[Gp], [Il], [Iz],
" -v- capeloensis (capeolensis).....
" -f. conjugens (See Brasiliparodia buenekeri
-v- conjugens).....[Hi-1],

buenekeri.....(cont'd),
" -v-intermedius (See Brasiliparodia buenekeri -v- intermedia),
" -v- pelotensis.....
" -v- senescens.....
buiningii.....[DD-493], [Ar], [Ay], [Gp], [Hi-1], [Hu], [Hx*], [Il], [Iz], [N-1980], [Af 56-1], [Ag-1982], [Am 9-1980], [Au Vol. 2, 1977, Zone 3 Compendium], [Ba 2-1978], [Ba 9-1978], [Bi 4-1-1986], [Hl 5-1983],
" -v- alamoensis.....
" -v- fusca.....
" -v- pulchellus.....
" -v- schlosseri.....
caespitosus (See N. minimus).....[Bo-3], [Gp*], [Hx*],
" -v- albispinus.....
" -f. tenuicylindricus (See N. tenuicylindricus),
" -v- .....[Hx],
campestrensis.....[Hi-1],
carambeiensis.....[DD-494], [Ah-1985], [Aj 24-1-1973],
catarinensis.....
chrysocomus (See Brasiliparodia chrysocoma, Parodia chrysocoma),
" -v- rubrispinus.....
claviceps (See Eriocactus claviceps), [Gp*], [N-1980], [Af 43-5],
cobrensis (See N. scopa -v- cobrensis).....[Ba 1-1976],
cocinnus.....[J*], [L], [T], [TT], [Bu], [Bv-1], [Cs], [Dp], [Gp*], [Hu], [Hx], [Af 51-6], [Ba 10-1977],
" -v- apricus (See N. apricus).....
" -v- fimbriatiflorus.....
" -v- joadii (See N. joadii).....[Bo-3],
" -v- luysianus.....
" -v- maldoneda.....
" -v- montevideoensis.....
" -v- mosquitoensis.....[Ba 9-1977],
" -v- nigrispinus (See N. nigrispinus),
" -v- parviflorus.....
" -v- piriapolisensis (See N. piriapolisensis),
" -v- quervosensis.....
" -v- wuthuvianus.....
corynodes (See Malacocarpus corynodes, Wigginsia corynodes).....[Gp*], [Il],
courantii (See Wigginsia courantii, Wigginsia tephracantha -v- courantii),
coxii (See Austrocactus coxii, Austrocactus dusenii),
crassigibbus (See Parodia crassigibba), [G], [Z], [Ar], [Gp*], [Hi-1], [Il], [Iz], [Ag-1988], [Bi 1-3], [Cu 2-1973],

NOTOCACTUS.....(Cont'd),
crassigibbus.....(cont'd),
" -v- gracilior.....
" -v- minor.....
cristatoides.....
cupreatus (See N. roseiluteus),
curtinensis (See N. mammulosus -v- curtinensis), [Bi 2-1],
curvispinus.....
dalvae.....
dusenii (See Austrocactus coxii, Austrocactus dusenii)
elachisanthus (See Brasilicactus elachisanthus, N. scopa -v- elachisanthus),
elegans (See N. ottonis -v- elegans),
engleri.....
erectocylindricus.....
eremiticus.....[Hi-1], [Hx*], [Hl 5-1970], [Hl 2-1982],
erinaceus (See N. acuatus, Wigginsia acuata, Wigginsia erinacea),
" -v- kovaricii (See N. kovaricii),
" -v- leucocarpus (See Malacocarpus leucocarpus),
" -v- macrocanthus (See Malacocarpus macrocanthus)
" -v- spinosior (See N. tephracanthus),
erizo.....
erubescens.....[Bo-3],
erythracanthus.....
eugeniae.....[Hx*], [Bi 2-1],
" -v- microfloridus.....
euvelenovskyi (See N. floricomus -v- velenovskyi, N. velenovskyi),
ferrugineus.....
floricomus (See N. mammulosus -v- floricomus), [NN], [Ar], [Bo-3], [Hr], [Hx], [Iz], [Ja], [Jd], [Af 33-2], [Hl 1-1968], [Bi Vol.4, #6]
" -v- flavispinus.....
" -v- rubrispinus.....
" -v- spinosissimus.....[Iv 5-1958],
" -v- velenovskyi..... [Bo-3], [Eg 1945-1946:225 Fig.],
" -v- (de Rivera).....
fricii.....(See Malacocarpus fricii, Wigginsia fricii),
fuscus.....[Ar], [Gp*], [Hi-1], [Hx*], [Bi 3-3], [Hl 5-1970],
" -v- fuscus.....
" -v- longispinus.....[Hi-1],
gibberulus.....
glaucinus.....[Hi-1],
" -v- alegretensis.....
" -v- depressus.....[Hi-1],
" -v- gracilis.....[Hi-1],
" -v- gracilispinus (See N. gracilispinus),
" -v- miniatispinus.....

NOTOCACTUS.....(Cont'd),

globularis.....[Hi-1],
gracilis.....
graessneri (See Brasilicactus graessneri, Eriocactus graessneri),
[F], [H], [K], [X], [Bv-1], [Hy], [Il], [Iz], [Jd], [Ag-1985]. [Aj 27-3-1976], [Ba 4-1981], [Db 1-1981],
" -v- albisetus (See Brasilicactus graessneri -v- albisetus),
[HH], [Gp], [Ah-1980],
" -v- aureiflorus (See Brasilicactus graessneri -v- aureiflorus),
" -v- flaviflorus (See Brasilicactus graessneri -v- flaviflorus),
[Ah-1985],
" -v- fortalezensis.....
" -v- jaquiranensis.....
" -v- stellatus (See Brasilicactus graessneri -v- stellatus),
gracilispinus (See N. glaucinus -v- gracilispinus),
grossei (See Eriocactus schumannianus -v- grossei, Eriocactus grossei),
[Dj], [Hr],
" -v- aureispinus (See Eriocactus grossei -v- aureispina),
" -v- purpureiflorus.....
harmonianus.....[Hi-1],
haselbergii (See Brasilicactus haselbergii, Eriocactus haselbergii),
[C], [M*], [AA], [XX], [Av], [Ay], [Bv-1], [Cr], [Cz], [Gp], [Gr], [Hb], [Hy], [Il], [Iz], [Jd], [N-1980], [Ag-1981], [Aj 27-2-1976], [Am 2-1983], [Ba 6-1981], [Bi 1-3],
" -v- albispina.....
" -v- aureivariegatus.....
" -v- gracilispinus.....
" -v- lanatus..(PR-273).....
" -v- pseudograessneri (PR-274).....
" -v- robustispinus.....
" -v- stellatus (See Brasilicactus haselbergii -v- stellatus),
" -v- vacariensis.....
" -v- ......[Hy],
" -f. cristata.....[Hy],
" -f. monstrosus.....[Hy],
herteri.....[S], [T], [Ar], [Bo-3], [Gp], [Hb], [Hi-1], [Hv], [Hx], [Il], [Iz], [Ae-1981], [Af 54-2] [Ag-1982], [Ah-1980], [Db 1-1980], [Hl 3-1970], [Hl 2-1972],
" -v- muegelianus.....[Af 60-2],

NOTOCACTUS.....(Cont'd),

hibernus (See Austrocactus hibernus),
horstii.....[EE], [Gp], [Gr], [Hb], [Hi-1], [Hu], [Hx], [Iz], [Aj 27-7-1976], [Am 6-1982], [Da 2-1979], [Hl 1-1968], [Hl 3-1970],
" -v- erythrinus.....
" -v- juvenaliformis (See N. neohorstii -v- juvenaliformis, Wigginsia horstii -v- juvenaliformis).....[Hl 5-1970],
" -v- purpureiflorus.....[Hi-1],
hypocrateriformis (See N. mammulosus),
intricatus (See Malacocarpus arechavaletai, Wigginsia arechavaletai)
janousekianus (See N. ottonis -v- janousekianus),
joadii (See N. concinnus -v- joadii).....[Hx*],
juvenaliformis (See N. horstii -v- juvenaliformis, N. neohorstii -v- juvenaliformis, Wigginsia horstii -v- juvenaliformis),
kovaricii (See N. neoarechavaletai -v- kovaricii, Wigginsia kovaricii).....[Hl 6-1980],
laetivirens.....[Hi-1],
langsdorfii (See Wigginsia langsdorfii, Wigginsia polyacantha),
leninghausii (See Eriocactus leninghausii), [C], [H], [L], [M*], [S], [V], [Z], [AA], [QQ], [Av], [Ay], [Bv-1], [Cm], [Cr], [Cz], [Gr], [Hv], [Ia], [Il], [Iz], [Jd], [N-1980], [Ag-1985], [Aj 31-7-1980], [Aj 32-7-1981], [Am 9-1980], [Am 1-1984], [Bi 1-3], [Bi 1-4],
" -f. apelii.....
" -v- longispinus.....
" -v- major.....
" -v- minor.....[Ba 6-1981],
" -f. cristata.....[H], [Hb],
leprosorum (See Malacocarpus leprosorum, Wigginsia leprosorum).....[Iz], [Hl 1-1982],
leucocarpus (See Malacocarpus leucocarpus, Wigginsia leucocarpa),
linkii (See Malacocarpus linkii, N. ottonis -v- linkii)..... [Ba 11-1978],
" -v- altosensis.....
" -v- berlinensis.....
" -v- buenekeri (See N. buenekeri, Parodia buenekeri)......[Hi-1],
" -v- flavispinus.....[Aj 1-1978],
" -v- guaibensis (quaibensis).....[Hi-1],
" -v- jaquiranensis.....
" -v- laetivirens.....
" -v- linkii.....[Hi-1], [Bi 1-4],
" -v- megapotamicus (See N. megapotamicus),

NOTOCACTUS......(Cont'd),

linkii.....(cont'd)
" -v- multiflorus (See N. linkii -v- multiflorus, N. multiflorus),
" -v- quaibensis.....
" -v- paranaensis.....
" -v- vermelhensis.....
longispinus (See Wigginsia longispina),
macrocanthus (See Wigginsia macrocantha),
macrogonus (See Malacocarpus macrogonus, Wigginsia macrogona),
magnificus (See Eriocactus magnificus),
[R], [X], [Z], [Ay], [Gp], [Il], [Iz], [Ab-1985], [Ad-1983], [Ag-1985], [Ah-1980], [Aj 31-7-1980], [Am 9-1980], [Am 9-1983], [Bi 2-1987], [Bm-1988],
maldonadensis (See Malacocarpus arechavaletai, Wigginsia arechavaletai),
mammulosus (See Malacocarpus mammulosus),
[C], [H], [M*], [NN], [QQ], [Ar], [Ay], [Bo-3], [Bv-1], [Bv-3], [Cr], [Cz], [Gp], [Gr], [Hi-1], [Hu], [Hx], [Hy], [Jd], [Am 7-1983], [Ba 2-1978], [Db 3-1979], [Hl 3-1979], [Hl 1-1980], [Hw-1982],
" -v- albispinus.....
" -v- arbolitensis (See N. arbolitensis),
" -v- brasiliensis (See N. brasiliensis), [Ba 1-1981],
" -v- curtinensis (See N. curtinensis),
" -v- floricomus (See N. floricomus),
" -v- gracilior.....[Bo-3],
" -v- gracilispinus.....
" -v- marmarajensis.....
" -v- masollerensis.....
" -v- multiflorus.....
" -v- pampeanus (See N. pampeanus, N. submammulosus -v- pampeanus)....[Bo-3], [Dq],
" -v- paucicostatus.....[Ba 1-1976],
" -v- rubrispinus.....[Hs I-1-1980],
" -v- submammulosus (See N. submammulosus)..[Ar],
" -f. cristata.....[Ba 10-1977],
medinaensis.....
megalanthus.....[Aj 32-7-1981],
megapotamicus (See N. linkii -v- megapotamicus), [PP], [Gp*], [Hx*],
" -v- alacriportanus (See N. alacriportanus), [Hi-1],
" -v- crucicentrus.....[Hi-1],
" -v- horstii.....[Hi-1],
" -v- vulgatus.....[Hi-1], [Bi 1-4],
" -f. (Chorro Frio).....

NOTOCACTUS.....(Cont'd),

melchersii.....
meldianus.....
melianse.....
meonacanthus (PR-318).....
militaris.....
minimus (See N. caespitosus),
[Ar], [Bo-3], [Hr], [Hx*], [Iz], [Jd], [Af 54-6], [Hl 4-1971],
" -v- albispinus.....
" -v- maximus.....
moelleri.....
muegelianus (See N. purpureus -v- muegelianus),
[Ba 4-1980],
mueller-melchersii.....[H], [Ar], [Bo-3], [Gp], [Hr],
[Aj 11-12-1960], [Bi 1-3],
[Bi 2-1], [Db 9-1980],
" -v- colorispinus.....
" -v- elegans.....
" -v- gracilispinus.....[Bo-3], [Hx*],
" -v- longispinus.....
" -v- multicolorispina.....
" -v- taquarembensis.....
mueller-moelleri.....[Gp*], [Aj 28-5-1979], [Bi 2-1],
[Db 3-1980], [Hl 1-1968],
multicostatus.....[Hx*], [Hl 4-1970],
multiflorus (See N. linkii -v- multiflorus, N. ottonis -v- multiflorus).....[Ds],
muricatus (See N. bommeljei),
[T], [Bo-3], [Hb], [Hi-1], [Hx], [Hy], [Ia],
[Ba 1-1982], [Hl 1-1968], [Hs IV-4-1983],
" -v- flavifuscus.....
myriacanthus (See Parodia myriacanthus),
" -v- longispinus.....
napinus.....
neoarechavaletai (See N. arechavaletai),
[Ba 6-1977], [Ba 8-1981], [Hl 6-1980],
" -v- kovaricii (See N. arechavaletai -v- kovaricii, N. kovaricii),
[Ba 8-1981],
" -v- maldonadensis (See N. maldonadensis),
neobuenekeri.....[Ar], [Gp*], [Hi-1], [Bi 3-3],
neofuscus.....
neohorstii.....[Gp*], [Hs II-4-1981],
" -v- juvenaliformis (See Wigginsia horstii -v- juvenaliformis),
[Hs II-4-1981],
" -v- nova.....[Hs II-4-1981],
neo-ottonianus (See N. ottonis).....[Hy],
" -v- schuldtii (See N. ottonis -v- schuldtii),
[Hy],

neo-ottonianus.....(cont'd),
" -v- stenogonus (See N. ottonis -v- stenogonus), [Hy],
" -v- tortuosus (See N. ottonis -v- tortuosus), [Hy],
" -v- tenuispina (See N. ottonis -v- tenuispina),
nigrispina (See Eriocactus nigrispinus, Eriocactus schumannianus -v- nigrispinus, Malacocarpus nigrispinus, N. schumannianus -v- nigrispinus), [Z], [Gp*], [Db 3-1979],
orthacanthus (See Echinocactus orthacanthus, Wigginsia orthacantha).....[Ba 2-1978],
ottonis (See Malacocarpus ottonis).....[C], [H], [M*], [NN], [Bo-6], [Bu], [Bv-1], [Cm], [Dq], [Gp], [Gr], [Hi-1], [Hx], [Il], [Iu], [Jd], [N-1980], [Au 37-1*], [Hw-1981], [Hw-1986], [Hw-1988],
" -v- acutangularis.....
" -v- acutus (See N. acutus),
" -v- albispinus (Invalid name).....[Bo-6:3756],
" -v- arechavaletai (See Malacocarpus arechavaletai, Wigginsia arechavaletai),
" -v- " -sv. nanus (See N. arechavaletai -v- nanus),
" -v- blossfeldianus.....
" -v- brasiliensis (See N. brasiliensis, N. mammulosus -v- brasiliensis),
" -v- buenekeri (See N. buenekeri),
" -v- cacapavensis (cacapavanus).....
" -v- campestrensis (See N. campestrensis),
" -f. campoensis.....
" -v- denudatus.....
" -v- elegans.....(See N. elegans).....[Bo-3],
" -v- janousekianus (See N. janousekianus), [Hl 6-1975],
" -v- linkii (See N. linkii).....[Ar],
" -v- megapotamicus (See M. megapotamicus),
" -v- minasensis.....
" -v- minusculus.....
" -v- multiflorus (See N. multiflorus), [Hr], [Hl 4-1968],
" -v- nugualensis.....
" -v- ottonis.....[Ar],
" -v- paraguayensis.....[Hi-1], [Aj 31-4-1980], [2 forms in Ik],
" -v- pluricentralis.....
" -v- rubrispinus (See N. rubrispinus),
" -v- rufispinus.....
" -v- schuldtii (See N. neo-ottonis -v- schuldtii) [Ar],
" -v- segredoensis.....

NOTOCACTUS.....(Cont'd),

ottonis.....(cont'd),
" -v- stenogonus (Invalid name).....[Bo-6],
" -v- tenebrosus.....
" -v- tenuispinus (See Echinocactus tenuispina), [Z], [Ar], [Bo-3], [Bv-1],
" -v- tortuosus (See N. ottonis -v- linkii, Parodia ottonis -v- tortuosa), [Bo-3], [Iz],
" -v- uruguayanus (uruguayus) (uruguayensis).....
" -v- vargasensis.....
" -v- vencluisianus..... [Hr], [Aj 24-1-1973], [Flower - Ba 7-1979], [Bi 1-3], [Hl 7-1965]
" -v- villa-velhensis.....[Bo-3], [Hl 1-1975], [Iv 5-1958],
" -f. cristata.....[Hy], [Ah-1986],
" -f. (Abra de Zabalata).....
" -f. (Cerrode los Cuentos).....
" -f. (Maldonado Loreley).....
" -f. (Traint-a-tres).....
oxycostatus (See N. acutus).....[Hx*],
pampeanus (See N. submammulosus -v- pampeanus), [Bv-1], [Hy], [Ia], [Hl 1-1968],
" -f. variegatus.....
parvus.....
patagonicus (See Austrocactus patagonicus), [M*], [Iu],
pauciareolatus (See Malacocarpus pauciareolatus, Wigginsia sessiliflora -v- pauciareolata),
paulus.....[Aj 31-4-1980],
permutatus.....[Hi-1],
piriapolisensis (See N. concinnus -v- piriapolisensis),
pleiocephalus (PR-336)......
polyacanthus (See Wigginsia langsdorfii, Wigginsia polyacantha),
prolifera (See Wigginsia prolifera).....[Hs II-4-1981],
" -v- longispina.....
pseudoherteri.....
pseudomammulosus.....
pseudopulvinatus.....
pseudorutilans.....
" -v- microfloridus.....
" -v- minimus.....
pulvinatus (See Wigginsia pulvinata), [Gp*], [Hs*], [Db 3-1979], [Hs III-2-1982],
purpureus.....[Gr], [Hi-1], [Hx*], [Af 58-2], [Aj 30-9-1979], [Bm-1985], [Hl 3-1970],
" -v- muegelianus (See N. muegelianus),
rauschii.....[Av], [Hi-1], [Hu], [Hx*], [Aj 30-8-1979], [Am 9-1980], [Bi 3-3], [Hl 5-1970], [Hl 3-1975],

rauschii.....(cont'd),
" -v- caveraensis (PR-230).....
rechensis (See Brasiliparodia rechensis),
[Hx*], [Aj 19-2-1968], [Ba 6-1981],
reichei (See Neochilenia reichei, Neoporteria reichei, Thelocephala reichei),
roseiflorus.....[Aj 29-12-1978], [Bi 2-1],
roseoluteus.....[Gp*], [Hx*], [Il], [Bm-1985], [Db 3-1979*], [Hl 4-1975],
" -v-.....[DD-495],
ruberima.....[Cu 7-1960],
rubricostatus.....(Wigginsia rubricostata),
rubriflorus.....[Bv-1], [Hr], [Hx*],
rubrispinus (See N. ottonis -v- rubrispinus),
rubropedatus.....[Hi-1],
rutilans.....[G], [T], [NN], [TT], [Ar], [Bo-3], [Bv-1], [Cr], [Cz], [Gp*], [Hu], [Hv], [Hx*], [Hy], [Il, [Jd], [Am 6-1986], [Ba 4-1977], [Bi 1-3], [Bi 2-1], [Hl 3-1969], [Hw-1988],
" -f. storianus.....[Hl 4-1969],
" -f. jasminoflorus.....
schlosseri.....[Gp*], [Db 3-1979], [Ba 1-1974],
schumannianus (See Eriocactus schumannianus),
[M*], [Ar], [Av], [Ay], [Bv-1], [Gp*], [Gr], [Hr], [Jd], [Am 9-1980],
" -v- albispinus.....
" -v- amplicostatus.....
" -v- nigrispinus.....(See Notocactus nigrispinus) [Ay],
scopa (See Malacocarpus scopa),
[E], [F], [H], [L], [M*], [T], [AA], [Tt], [Ar], [Ay], [Be*], [Bo-3], [Bt], [Bv-3], [Cm], [Cz], [Gp], [Hi-1], [Hx], [Hy], [Il], [Jd], [Ag-1985], [Aj 28-4-1977], [Flower - Ba 11-1977], [Bi 3-3], [Da 2-1979], [Hw-1982],
" -v- albilanata.....
" -v- albispina.....
" -v- brunispinus.....[Bi 3-3],
" -v- candicans.....
" -v- candida.....[Cj], [Ds], [Hl 1-1968],
" -v- " -f. cristata.....[Hy],
" -v- cobrensis (See N. cobrensis),
[Bi 3-3], [Db 5-1980],
" -v- daenikerianus.....[Ar], [Hu], [Hx],
" -v- elachisanthus (See N. elachisanthus),
" -f. glauseriana.....[Hu], [Af 43-5], [Bi 3-3],
" -v- intermedius.....
" -f. longispinus.....[Bi 3-3],
" -v- machadonensis (machodoensis).....
" -v- major.....[Hl 3-1976],

NOTOCACTUS.....(Cont'd),

scopa.....(cont'd),
" -f. murielii.....[Ah-1985], [Bi 3-3], [Hl 6-1974],
" -v- prolifera (See N. prolifera),
" -v- ramosus.....[KK], [Bo-3], [Iz], [Hl 5-1970], [Hl 1-1973],
" -v- ruberimus.....[H], [NN], [Ar], [Bv-1], [Cr], [Gr], [Hy], [Af 14-10], [Af 15-7],
" -f. cristata.....[I], [Bv-1], [Cz], [Dl*], [Hy],

securituberculatus.....[Hi-1],
" -v- miniatispinus.....[Hi-1], [Flower - Ba 6-1981],

sellowii (See Malacocarpus sellowii, Wigginsia sellowii).....[Gp*],
" -v- courantii (See N. courantii),
" -v- macrogonus.....
" -v- macrocanthus (See N. macrocanthus),
" -v- tetracanthus.....
" -v- turbinatus.....
" -v- vorwerkianus (See N. vorwerkianus),

sessiliflorus (See Malacocarpus sessiliflorus, Wigginsia sessiliflorus).....[Jd], [Hl 5-1979],
" -v- martinii.....
" -v- stegmannii (See Malacocarpus stegmannii),
" -v- tetracanthus.....

seticeps.....

soldtianus.....

spinibarbis.....[Hi-1],

stegmannii (See Malacocarpus stegmannii),

stockingeri.....

submamillatus.....

submammulosus (See N. mammulosus -v- submammulosus), [W], [NN], [Tt], [Ar], [Bo-3], [Bv-1], [Hw-1981], [Bi 4-1-1986],
" -v- charrua.....
" -v- pampeanus (See N. pampeanus), [Ar], [Gp*], [Hv], [Hx], [Aj 26-6-1-1975],

sucineus (succineus).....[Av], [Gp*], [Gr], [Hi-1], [Hx*], [Iz], [Bi 3-3], [Hl 5-1970],
" -v- alba.....
" -v- albispinus.....[Bi 3-3],

sucrensis (See Parodia sucrensis),
" -v- albispinus.....
" -v- " -f. caespitosa (See N. caespitosus),

tabularis.....[T], [V], [UU], [Ar], [Ay], [Bt], [Cm], [Gp], [Hu], [Hx], [Il], [Cu 4-1967], [Hl 1-1969],

tarijensis.....

tenuicylindricus.....[Ar], [Hi-1], [Hx*], [Aj 29-4-1978]
" -v- major.....[Ba 5-1981],

NOTOCACTUS.....(Cont'd),

tenuispinus (See N. ottonis -v- tenuispinus),
tephracanthus (See Malacocarpus tephracanthus, Wigginsia tephracantha),
" -v- courantii (See Wigginsia courantii),
" -v- macracanthus (See Wigginsia macracantha),
" -v- martinii (See Wigginsia martinii),
" -v- turbinatus (See Wigginsia turbinata),
turbinatus (See N. sellowii -v- turbinatus),
uebelmannianus.....[DD-496], [NN], [Ar], [Av], [Bt], [Gp*], [Hb], [Hu], [Hv], [Il], [Iz], [N-1980], [Af 58-2], [Ag-1983], [Ba 3-1979], [Ba 11-1981], [Ba 7-1983] [Bi 1-3], [Bi 2-4], [Hl 4-1972],
" -v- flaviflorus.....(yellow flower),
" -v- uebelmannianus.....(red flower),
uruguayanus (See N. ottonis -v- uraguayanus),
vanvlietii.....[Ar], [Gp*], [Hx*], [Aj 21-5-1970], [Hl 5-1975],
" -v- antonianus.....[Hl 5-1975],
" -v- gracilis (gracilior).....
" -v- rubrispinus.....
veenianus.....[Bi 2-1],
velenovskyi (velenowskii, euvelenovskyi) (See N. floricomus -v- velenovskyi), [NN], [Am 9-1980],
vila-velhensis (See N. ottonis -v- vila-velhensis),
vorwerkianus (See Malacocarpus vorwerkianus, Wigginsia vorwerkiana), [R], [Hl 6-1979],
warasii (See Eriocactus warasii), [Aj 31-7-1980], [Ba 12-1979],
werdermannianus.....[Ar], [Bo-3], [Gp], [Hb], [Hx], [Aj 24-1-1973], [Da 2-1979], [Hl 5-1970], [Hl 1-1974],
" -v- albispinus.....
" -v- brunispinus.....
" -v- caespitosus.....
" -v- grandiflorus.....
" -v- miniflorus.....
winkleri.....[Hx*], [Ag-1988], [Bi 2-1],

Spec. nova "Bage" (Rio Grande do Sul).....[Ba 7-1981],
Spec. .....[Ba 11-1978], [Ba 6-1980], [Ba 7-1980],

NYCTOCEREUS

castellanosii.....
chontalensis.....[Az], [Bo-4], [Af 22-5], [Aw 3-1981], [Iv 9-1953],
guatamalensis.....[RR], [Ay], [Az], [Bo-4],

NYCTOCEREUS.....(Cont'd),

hirschtianus.....[Bo-4], [Hy],
neumannii.....
oaxacensis.....[DD], [Az], [Bo-4], [Hx*], [Aj 16-3-1965], [Aj 33-12-1982],
serpentinus.....[C], [M*], [R], [Z], [LL*], [NN], [RR], [Jd], [Af 14-4], [Aw 2-1987], [Bb-1984], [Hl 4-1982],
" -v- ambiguus (See N. serpentinus),
" -v- pietatis.....[Aw 4-1972],
" -v- splendens.....(See N. serpentinus),
" -v- strictior.....
" -f. cristata.....[W], [Bv-1], [Af 2-8],

OBREGONIA

denegrii (See Strombocactus denegrii), [E], [X], [Z], [AA], [DD], [HH], [NN], [QQ], [Ar], [Av], [Ay], [Bt], [Bv-1], [Cs], [Cx], [Cz], [Dq], [Gr], [Hr], [Hy], [Il], [Iu], [Jd], [Ae-1981], [Af 16-12], [Af 51-6], [Af 54-1], [Af 56-1], [Af 57-4], [Af 58-5], [Af 59-4], [Aj 27-1-1976], [Am 2-1984], [Au 2-1977-Zone 3- Compendium], [Aw 3-1987], [Ba 11-1978], [Hm 1-1979],

OEHMEA

aylostera (See Mammillaria beneckei),
balsasoides (See Mammillaria balsasoides, M. beneckei,
beneckei (See Dolichothele beneckei, Mammillaria beneckei),
callus.....[Bi 2-3],
colonensis (See Mammillaria beneckei),
guiiengolensis (See Mammillaria beneckei),
nelsonii (See Dolichothele nelsonii, Mammillaria beneckei).....[Jd],

OPUNTIA

abjecta (See Opuntia triacantha),
abyssii (See Cylindropuntia abyssii, Opuntia whipplei -v- multigeniculata).....[Af 15-12],
acanthocarpa (See Cylindropuntia acanthocarpa)..[P], [FF], [LL*], [OO], [RR], [UU], [Ax], [Cb], [Gr],
" -v- acanthocarpa.....[Ax], [Az],
" -v- coloradensis.....(See Cylindropuntia coloradensis)....[Q], [Ax], [Az], [Ca],
" -v- ganderi (See Cylindropuntia acanthocarpa -v- ganderi).....[Ax], [Az],
" -v- major (See Cylindropuntia acanthocarpa -v- major).....[Q], [Ax], [Az], [Ca],

OPUNTIA.....(Cont'd),

acanthacarpa.....(cont'd),
" -v- ramosa (See Cylindropuntia acanthocarpa -v- ramosa, Opuntia acanthocarpa -v- major), [Af 9-3],
" -v- thornberi (See Cylindropuntia acanthocarpa -v- thornberi).....[P], [Q], [Ax],
acaulis (See Consolea acaulis),
aciculata (See O. lindheimeri -v- aciculata, O. lindheimeri -v- lindheimeri), [W], [LL*], [RR], [Bn], [Bv-1], [Co], [Df], [Ad -], [Dg 2-3-1980],
" -v- orbiculata (See Opuntia orbiculata), [EE], [Ay], [Bo-1],
aequatorialis.....[RR],
affinis.....(Cx],
agglomerata (See Corynopuntia agglomerata),
alamosensis (See Cylindropuntia alamosensis, Opuntia thurberi -v- alamosensis).....[Az], [Cx],
albata (See Opuntia microdasys -v- albata), [NN], [Hx*], [Iz], [Aj 21-1-1970], [Bi 5-1],
albiflora (K. Schum.) (See Austrocylindropuntia salmiana, Opuntia salmiana, Platyopuntia salmiana),
albiflora (Baxter) (See O. whitneyana -v- albiflora, O. basilaris),
albisaetacens (See Platyopuntia albisaetacens), [Bo-1],
" -v- robustior (Invalid name),
albiscoparia (See Cumulopuntia boliviana, Opuntia boliviana (S.D.), Tephrocactus chichensis, Tephrocactus albiscoparia),
alboareolata (See Cumulopuntia alboareolata, Tephrocactus alboareolata),
albomarginata (See Maihueniopsis albomarginata),
alcahes (See Cylindropuntia alcahes), [C], [RR], [Bl], [Bv-1], [Cu 7-1960],
alexanderi (See Tephrocactus alexanderi),
" -v- bruchii (See O. bruchii).....[Bx],
alko-tuna (See Platyopuntia alko-tuna), [Bo-1], [Fc 23:23 Fig. 3, 1950],
allairei (See Opuntia compressa -v- allairei, Opuntia humifusa -v- humifusa), [LL*], [RR], [Bo-1], [Df],
alpina (See Opuntia tuberosa, Pterocactus tuberosus),
alta (See O. engelmannii -v- alta, O. lindheimeri -v- lindheimeri), [Df], [Hz], [Eo 21:165 pl. 19 & 20, 1910],
ammophila (See O. humifusa -v- ammophila), [RR], [Ha],
amyclaea.....[Bo-1], [Aj 33-3-1982], [Hl 2-1983],

OPUNTIA.....(Cont'd),

anacantha.....[RR], [Bn], [Dr],
anahuacensis (See O. stricta -v- dillenii),
andicola (See Tephrocactus andicolus, Tephrocactus glomeratus -v- andicola).....[Bx], [Af 6-7],
angustata (See O. phaeacantha -v- laevis), [A], [RR], [Bo-1],
antillana (See O. cubensis).....[RR], [Ar], [Af 56-3],
aoracantha (See O. ovata (HORT. ANGL.), Tephrocactus aoracanthus, Tephrocactus articulatus -v- ovatus, Tephrocactus ovatus), [L], [RR], [Bx - pages 43 & 73], [Dr], [If], [Af 6-7],
apurimacensis (See Platyopuntia apurimacensis),
arborea (See O. brasiliensis),
arborescens (See O. imbricata -v- imbricata), [Cf], [Go*], [Hz], [Ie],
arbuscula (See Cylindropuntia arbuscula, Cylindropuntia vivipara).....[A], [P], [Q], [CC], [LL*], [RR], [Az], [Bu], [Af 1-7],
arcei.....[Af 28-4], [Af 40-6],
archiconoidea.....
arechavaletai.....
arenaria.....[U], [Y], [RR], [Ax], [Df], [Ib],
argentina.....[C], [RR], [Am 5-1986],
arizonica (See O. phaeacantha -v- major), [Bo-1], [Fg], [Ic], [Eo 20:93, 1919],
armata.....[Bi-1],
" -v- panellana (See O. panellana).....[Bo-1],
arrastradillo.....[Bo-1], [Iv 6-1953],
articulata (See O. glomerata, Tephrocactus articulatus, Tephrocactus articulatus -v- calvus, Tephrocactus glomeratus),
" -v- diademata (See O. diademata, Tephrocactus diadematus, Tephrocactus articulatus -v- syringacanthus)...[Ay],
" -v- oligacantha (See Tephrocactus articulatus -v- oligacanthus).....[Iz],
" -v- ovata (See Tephrocactus ovatus, Tephrocactus russelii),
" -f. papyracantha (See O. papyracantha, Tephrocactus articulatus -v- papyracanthus).....[Bt], [Hv], [Jd],
" -f. polyacantha (See Tephrocactus articulatus -v- polyacanthus),
" -f. syringacantha (See O. diademata, O. turpinii, Tephrocactus articulatus -f. syringacantha),..... [Aj 28-1-1977],
asplundii (See Cumulopuntia pentlandii, O. dactilifera, Tephrocactus dactiliferus),

OPUNTIA......(Cont'd),

assumtionis.....
atacamensis (See Maihueniopsis atacamensis, Tephrocactus atacamensis),
atrispina.....[Y], [RR], [Ax], [Bo-1], [Bz], [Df], [Hz],
atrocapensis (See O. stricta -v- dillenii),
atroglobosa (See Platyopuntia atroglobosa, Tephrocactus atroglobosus),
atropes.....[Cx], [Az],
atrovirens (See O. mieckleyi),
atroviridis (See Austrocylindropuntia floccosa, O. floccosa, Tephrocactus atriviridis, Tephrocactus floccosus).....[Ay]
" -v- longicylindricus.....
" -v- parviflorus.....
" -v- paucispinus.....
auberi (See Nopalea auberi).....[Bo-1], [Jd],
aulacothele.....
aurantiaca.....[RR]. [Bo-1], [Cu 3-1975],
aurea (See O. basilaris -v- aurea).....[Af 5-6],
aureispina.....[Hx],
austrina (See O. humifusa -v- austrina)...[LL*], [RR],
australis (See Pterocactus australis).....[RR],
" -v- arnoldianus.....[Bo-1],
azurea.....[EE], [NN], [RR], [Ay], [Az], [Bo-1], [Bt], [Cx], [Hv], [Af 60-4], [Aj 33-6-1982],
backebergii (See Tephrocactus minor ?).....[Bx],
bahamana (See O. stricta).....[RR], [Bo-6],
bahiensis (See Brasiliopuntia bahiensis).....[RR],
ballii (See O. macrorhiza -v- pottsii), [Y], [LL*], [OO], [Df],
barkleyana (See Micropuntia barkleyana, Micropuntia spectatissima, Opuntia pulchella),
basilaris.....[C], [I], [M*], [P], [GG], [RR], [An], [Ar], [Ay], [Az], [Bt], [Cz], [Do], [Gr], [Hx*], [Ij], [Iz], [Ad Oct. 1980], [Af 15-10], [Aj 31-5-1980], [Bi 4-1985],
" -v- albiflorus (See Opuntia whitneyana -v- albiflora),
" -v- aurea (See O. aurea).....[An], [As], [Ax], [Af 5-6], [Am 2-1984],
" -v- basilaris.....[Ax], [Ca],
" -v- brachyclada (See O. brachyclada), [Ax], [Az], [Ca],
" -f. cordata (See O. basilaris -v- basilaris), [E], [HH], [Hu], [Aj 33-6-1982],
" -v- heilii (Hybrid).....[Af 57-4],
" -v- humistrata (See O. basilaris -v- basilaris)
" -v- longiareolata.....[An],
" -v- ramosa (See O. basilaris)..... [P], [FF], [PP], [An], [Cb], [Hx*], [Hs I-2-1980],

OPUNTIA.....(Cont'd),

basilaris.....(cont'd),
" -v- sanguinea.....[Bo-1],
" -v- treleasei (See Opuntia treleasei),
[P], [An], [Ax], [Ca], [Ad-],
" -v- woodburyi.....[An],
" -f. monstrosus.....[Aj 15-4-1964],
beckeriana.....[Cy],
bella.....[RR], [Hx], [Hy*], [Af 55-1],
bensonii.....[Az], [Aw 2-1972],
bentonii (See O. stricta -v- stricta),
[Eo 22:25 Pl. 1 & 2, 1912], [Hz],
bergeriana.....[C], [H], [V], [EE], [RR], [Ay], [Be*],
[Bo-1], [Cr], [Hx], [Hy], [Ig*],
bernichiana (Hybrid - See O. compressa, O. stricta),
[DD], [Bo-1],
bicolor (See O. kuehnrichiana),
bigelovii (See Cylindropuntia bigelovii),
[M*], [Q], [R], [GG], [HH], [LL], [OO], [QQ],
[RR], [Ay], [Az], [Cb], [Cd], [Cm], [Cz],
[Do], [Gr], [Jd], [Af 3-9], [Aj 31-7-1980],
[Au 37-1], [Aw 3-1962],
" -v- bigelovii.....[Ax], [Ca],
" -v- ciribe (See Cylindropuntia ciribe, O. ciribe, O. bigelovii -v- bigelovii)....[B],
" -v- hoffmannii (See O. hoffmannii).....[Ax],
bisetosa.....
bispinosa (Invalid name - See Platyopuntia bispinosa).
[Bo-6], [Aj 33-6-1982],
blancii (See Tephrocactus blancii).....[Cu 13:5, 1958],
boldinghii.....[RR], [Bo-1], [Ba 9-1979],
boliviana (S.-D.) (See Cumulopuntia boliviana),
[Bx], [Cu 7-1960], [Cu 1-1973],
boliviensis (Backbg.) (See Platyopuntia soehrensii,
O. soehrensii),
bonaerensis (See O. chakensis).....[Dl*], [Dr],
bonplandii.....[RR],
borinquensis.....[RR],
brachyacantha (See Platyopuntia brachyacantha),
brachyarthra (See O. fragilis -v- bracharthra),
brachyclada (See O. basilaris -v- brachyclada),
[FF], [Af 6-2],
brachyrhopalica (See Corynopuntia pulchella, Micropuntia brachyrhopalica, O. pulchella),
bradleyi.....
bradtiana (See Grusonia bradtiana).....[Az], [Jd],
[Af 42-5],
brasiliensis (See Brasiliopuntia brasiliensis)...[CC],
[RR], [Ax], [Bk], [Cz], [Gr], [Jd],
" -v- minor.....
bravoana.....[Bo-1], [Af 5-3],

brevispina (See Cylindropuntia brevispina),
brittonii (See Nopalea brittonii),
brunneogemmia (See Platyopuntia brunneogemmia),
brunnescens.....[RR], [Bo-1], [Bo-6],
bruchii (See O. alexanderi -v- bruchii, Tephrocactus alexanderi -v- bruchii),
[Dr], [Cu 1-1-1932],
bulbispina (See Corynopuntia bulbispina),
[RR], [Ay], [Az], [Aj 29-1-1978], [Bi 4-1-1986],
burrageana (See Cylindropuntia burrageana),
[RR], [Az], [Af 4-5],
cacanapa (See O. engelmannii -v- cacanapa, O. lindheimeri -v- lindheimeri),
calicola (See O. humifusa -v- humifusa),
californica (See Cylindropuntia californica, O. parryi -v- serpentina, Opuntia serpentina),
calmalliana (See Cylindropuntia calmalliana, O. molesta)
calva (See O. articulata, O. glomerata (Br. & R.), Tephrocactus articulatus, Tephrocactus articulatus -v- calvus),
camachoi espinosa (See Maihueniopsis camachoi espinosa, Tephrocactus camachoi),
[Ev 37:136 Fig. 29 & 31, 1933],
camanchica (See O. phaeacantha -v- camanchica),
[Be*], [Cm], [Df], [Di*],
campestris (See Cumulopuntia berteri, Tephrocactus dimorphus, Tephrocactus sphaericus),
[RR],
canada (See O. laevis -v- canada, O. phaeacantha -v- major),
[RR], [Eo 20:90 Pl. 2, 6, 12, & 13, 1909],
candelabriformis (See O. spinulifera),
canina (See O. retrorsa, Platyopuntia retrorsa),
[RR],
cantabrigiensis (See O. lindheimeri -v- cuija),
[RR], [Av], [Az], [Bd], [Bo-1], [Cx],
[Hy*], [Bi 3-2],
canterai.....[Bo-1], [Aj I-12-1929],
caracasana.....[RR], [Bo-1],
cardiosperma (See Platyopuntia cardiosperma),
[RR], [Bo-1],
cardona (See O. streptacantha),
caribaea (See Cylindropuntia caribaea),
[RR], [Bv-1], [Af 56-3],
catacantha (See Tephrocactus catacanthus),
catingicola.....[Bo-6], [Gr],
" -v- fulviceps.....
cedergreniana (See O. soehrensii, Platyopuntia soehrensii).....[Bo-1],

cereiformis (See Grusonia bradtiana),
[Aj II-9-1930],
" -f. cristata.....[Aj II-9-1930],
chaffeyi.....[RR], [Az], [Bl], [Bo-6], [Af 53-2],
[Af 57-2], [Ah-1986], [Bi 4-1985],
[Bi 4-1-1986],
chakensis (See O. bonaerensis)...[Av], [Bo-1], [Dr],
chavena.....[Ax],
chichensis (See Cumulopuntia boliviana, O. boliviana,
Tephrocactus chichensis).....[Au 13:5, 1958],
" -v- colchanus (See O. pentlandii -v- colchana),
chihuahuaensis (See O. phaeacantha).....[Ie],
chisosensis (See O. lindheimeri -v- chisosensis),
[Af 58-3], [Af 60-1-1988],
chlorotica.....[B], [C], [P], [Q], [Y], [FF], [GG], [LL]
[RR], [An], [Ax], [Az], [Bn], [Bo-1], [Ca],
[Cb], [Cw], [Df], [Do], [Ie], [Ij], [Af 6-3],
" -v- gosseliniana (See O. gosseliniana, O. violacea -v- gosseliniana)....[Af 60-4],
" -v- santa-rita (See O. santa-rita, O. violacea
-v- santa-rita, O. gosseliniana -v-
santa-rita).....[Af 60-4],
cholla (See Cylindropuntia cholla)....[OO], [RR], [Az],
chrysacantha (See O. neochrysacantha),
chuquisacana (See Austrocylindropuntia chuquisacana,
Austrocylindropuntia vestita -v- chuquisacana),
cineracea (See Cylindropuntia tesajo -v- cineracea),
[Az],
ciriibe (See Cylindropuntia ciribe, O. bigelovii -v-
ciribe).....[RR], [Az],
clavarioides (See Austrocylindropuntia clavarioides,
Cylindropuntia clavarioides)...[K], [RR],
[Bk], [Cy], [Cz], [Hx], [Ig*], [Il], [Jd],
[Aj 28-8-1977], [Aj 34-1-1983], [Bi 5-1],
" -f. cristata.....[L], [Ay], [Be*], [Cr], [Ds],
" -f. monstrosus.....[Ay],
clavata (See Corynopuntia clavata),
[P], [Q], [Y], [RR], [Ax], [Df], [Go], [Ie],
[Am 10-1983], [Cu 3-1973],
clavellina (See Austrocylindropuntia molesta, Cylindropuntia molesta, O. molesta),
coccinea.....[Hy],
cochabambensis.....[Bo-6], [Ew No. 9:20 Fig. 1955],
cochenillifera (See Nopalea cochenillifera),
[Ax], [Bo-1], [Ia],
cognata (See Platyopuntia cognata, Platyopuntia
bispinosa),
colorea (See Maihueniopsis colorea, Tephrocactus
coloreus),
colubrina (See Austrocylindropuntia colubrina),

comonduensis (See O. tapona).....[RR], [Af 20-3],
compressa (See O. humifusa -v- humifusa),
[P], [Y], [LL*], [MM], [Ce], [Cm], [Ct-1987],
[Hx], [Ja], [Af 59-6],
" -v- allairei (See O. allairei, O. humifusa -v- humifusa).....[P], [Y], [Df],
" -v- ammophila (See O. ammophila, O. humifusa -v- ammophila),
" -v- austrina (See O. austrina, O. humifusa -v- austrina),
" -v- fuscoatra (See O. fuscoatra, O. humifusa -v- humifusa).....[Y], [Df],
" -v- grandiflora (See O. grandiflora, O. macrorhiza -v- macrorhiza),
[Y], [Df],
" -v- humifusa (See O. humifusa -v- humifusa),
[Y], [Df],
" -v- macrorhiza (See O. macrorhiza -v- macrorhiza).....[B], [E], [H], [P],
[Y], [LL], [Df],
" -v- microsperma (See O. humifusa -v- humifusa),
[P], [Y], [Df],
" -v- stenochila (See O. macrorhiza -v- macrorhiza, O. stenochila),
[Y], [Df],
congesta (See O. kleiniae -v- tetracantha),
[Af 4-11], [Ex 20:88, 1909],
conjugens (See Platyopuntia conjungens),
conoidea (See Maihueniopsis conoidea, Tephrocactus conoideus),
corallicola (See Consolea corallicola, Consolea spinosissima, O. spinosissima),
cordobensis (See Platyopuntia cordobensis),
[C], [RR], [Ck], [Dr],
corotillo (K.SCH.) (See Cumulopuntia berteri (COLLA) Ritt., O. campestris, O. sphaerica (FOERST.), Tephrocactus berteri (COLLA) Ritt., Tephrocactus sphaericus (FOERST)),
" -v- aurantiaciflora (See Cumulopuntia ignota, O. ignota, Tephrocactus ignotus),
corrugata (See O. eburnea, O. longispina -v- corrugata Platyopuntia corrugata, Tephrocactus corrugatus)
[Cu 7-1960], [Cu 28:30, 1966],
" -v- monvillei.....
covillei (See O. littoralis -v- vaseyi)....[RR], [Cb],
crassa.....[RR], [Bo-1],
crassicylindrica (See Cumulopuntia crassicylindrica, Tephrocactus crassicylindrica),
[Au 13:5, 1958],

cretochaeta (See O. hyptiacantha),
crinifera (See O. orbiculata, O. pailana).....[R],
crispicrinita (See Austrocylindropuntia floccosa, O. floccosa, Tephrocactus crispicrinitus, Tephrocactus floccosus),
[Jd], [Au 13:5, 1958],
crystalenia (crystallina).....
cubensis (See O. antillana).....[Ax],
cumulicola (See O. austrina, O. humifusa -v- austrina),.....[Ez 907-1933],
curassavica (See O. repens),
[RR], [Dq], [Hx*], [Ba 1-1979], [Ep 43:88, Pl. 2, 1916],
" -v- columbiana.....
curvospina (See O. phaeacantha -v- mojavensis)
[FF], [An],
cyanella (See O. lindheimeri).....[Hz],
cylindrarticulata (See Cumulopuntia pentlandii -v- dactylifera, O. dactylifera, Tephrocactus dactylifer, Tephrocactus cylindrarticulatus),
[Au 13:5, 1958],
cylindrica (See Austrocylindropuntia corynephora, Austrocylindropuntia cylindrica),
[D], [RR], [Bv-1], [Cm], [Cz], [Af 4-6],
" -f. cristata (See Cylindropuntia cylindrica -f. cristata).....
[D], [Be*], [Dl*], [Ds], [Af 1-9],
cylindrolanata (See Austrocylindropuntia floccosa, O. floccosa, Tephrocactus floccosus, Tephrocactus cylindrolanatus),
[Bx], [Au 13:5, 1958],
cymochila (See O. macrorhiza -v- macrorhiza),
[Q], [Y], [Df], [Af 59-6],
" -v- montana (See O. macrorhiza -v- macrorhiza),
dactylifera (See Cumulopuntia pentlandii -v- dactylifera, O. cylindrarticilata, Tephrocactus dactylifer, Tephrocactus cylindrarticulatus),
[Bo-1], [Bx],
darrahiana.....
darwinii (See Maihueniopsis darwinii, Tephrocactus darwinii),
[RR], [Bo-1], [Bx], [Cu 4-1965],
dasyacantha.....
" -v- albispina.....
" -v- rubra.....
davisii (See Cylindropuntia davisii, O. tunicata -v- davisii),
[Y], [CC], [LL*], [RR], [Df], [Go], [Af 3-12],
deamii.....[RR*], [Fb 13:309 Pl. 65, 1911],

decumbens.....[RR], [Az], [Bo-1], [Bu], [Cx],
" -v- silvicola.....[Aw 2-1964],
decummana (See O. maxima),
dejecta (See Nopalea dejecta),
delaetiana.....[RR*].
delicata (See O. macrorhiza -v- pottsii),
[A], [Q], [RR],
demissa (hybrid).....[Ax], [Hz],
densiaculeata (See Cylindropuntia densiaculeata),
depressa.....[Az], [Cx],
deserta (See Cylindropuntia echinocarpa, O. echinocarpa -v- echinocarpa).....[Az],
diademata (See O. glomerata (Br. & R.), Tephrocactus articulatus -v- diadematus, Tephrocactus articulatus -v- syringacanthus, Tephrocactus diadematus),
[K], [L], [Bx], [Cm], [Gr],
" -v- calva (See Tephrocactus articulatus -v- calvus).....[Bx],
" -v- chionacantha (See O. diademata),
" -v- inermis (See O. strobiliformis, Tephrocactus articulatus -v- inermis, Tephrocactus strobiliformis),
[Bx], [Gr],
" -v- molinensis (See O. molinensis, Tephrocactus molinensis),
" -v- oligacantha (See O. diademata).....[Bx],
" -v- papyracantha (See O. articulata -v- papyracantha, Tephrocactus articulatis -v- papyracanthus, Tephrocactus articulatus -f. syringacanthus).....[Gr],
" -v- polyacantha (See O. diademata, O. paediophila, Tephrocactus paediophilus),?
" -v- syringacantha (See O. articulata -f. syringacantha, Tephrocactus articulatus -f. syringacanthus)
dillenii (See O. stricta -v- dillenii),
[D], [LL*], [RR], [Bo-1], [Bv-1], [Bu], [Cx], [Hu], [Af 52-1], [Af 54-4], [Gu III-1953], [Hl 1-1972],
" -v- stricta (See O. stricta),
" -v- tehuantepecana (See O. tehuantepecana),
[Aw 3-1964],

dimorpha (Foerst.)(See Tephrocactus dimorphus (Foerst.))
[Bx], [Jd],

OPUNTIA.....(Cont'd),

discata (See O. megacarpa, O. phaeacantha -v- discata, O. procumbens),
[A], [RR], [Dx],
discolor (See Platyopuntia discolor).....[RR],
distans.....[RR*],
dobbieana.....[RR],
domeykoensis (See Maihueniopsis domeykoensis),
drummondii (See O. pusilla),
[Y], [LL*], [RR], [Bo-1], [Df], [Af 39-3],
dulcis (See O. engelmannii -v- dulcis, O. phaeacantha -v- phaeacantha).....[Ie],
dumetorum (See Corynopuntia dumetorum, Platyopuntia dumetorum).....[Ek 58, 1929],
durangensis.....[H], [RR], [Az], [Bo-1], [Cx],
duraznillo.....[Af 1-7],
eburnea (See O. corrugata, Platyopuntia corrugata, Tephrocactus corrugatus),
eburnispina (See O. humifusa -v- austrina),
[RR], [Af 5-2],
echinacea (See Cumulopuntia echinacea, Tephrocactus echinaceus),
echinocarpa (See Cylindropuntia echinocarpa),
[H], [P], [Q], [LL*], [RR], [As], [Ax],
[Ca], [Cb], [Do], [Af 6-5],
" -v- echinocarpa.....[Az],
" -v- major (See acanthocarpa -v- major).....[Q],
" -v- nuda.....
" -v- wolfii.....[Ax], [Az], [Ca],
echios (See O. galapageia -v- echios),
[Ch], [Aj 26-8-1975],
" -v- barringtonensis (See O. galapageia -v- echios -sv- barringtonensis),
[Ij], [Af 39-5],
" -v- echios (See O. galapageia -v- echios),
[Ch], [Af 34-4], [Af 39-5], [Aj 18-8-1967],
" -v- gigantea (See O. galapageia -v- gigantea),
[Ij], [Af 5-8], [Af 34-3], [Af 39-5]
[Af 56-6],
" -v- inermis (See O. galapageia -v- echios -sv- inermis),
[Af 34-4], [Af 39-5], [Aj 18-8-1967],
" -v- typica.....[Af 5-8],
" -v- zacana (See O. galapageia -v- zacana),
edelmannii.....[Hw-1982],
eichlamii.....[RR],
ekmannii.....[Fb 13:310 Pl., 1896],
elata.....[RR], [Bo-1], [Bu],
" -v- anacantha (See O. anacantha),
" -v- elongataa.....[Bu], [Bv-1], [Cz],

elata.....(cont'd),
" -v- oblongata.....[Af 1-11],
" -v- obovata.....[Af 1-11],
elatior.....[Bo-1], [Bv-1], [Dq], [Hx*], [Af 54-4], [Af 57-3],
ellisiana (hybrid - 'subarmata'),
[Bo-1], [Hz], [Eo 21:170 Pl.],
elongata (See O. lanceolata).....
emoryi (See O. stanlyi -v- stanlyi).....[Ie],
endura.....
engelmannii (See O. phaeacantha -v- discata),
[P], [X], [GG], [LL*], [RR], [Bv-1], [Df], [Ie], [Af 57-3], [Af 58-4], [Af 60-1-1988],
" -v- aciculata (See O. aciculata, O. lindheimeri -v- lindheimeri).....[Y], [Df],
" -v- alta (See Opuntia alta, O. lindheimeri -v- lindheimeri),
[Y], [Ad], [Df],
" -v- cacanapa (See O. lindheimeri -v- tricolor, O. tricolor),
[Y], [Df],
" -v- cyclodes (See O. lindheimeri -v- cyclodes, O. phaeacantha -v- major).....[Q],
" -v- dillei.....
" -v- discata (See O. discata, O. phaeacantha - -v- discata).....[Q],
" -v- dulcis (See O. dulcis, O. phaeacantha -v- phaeacantha),
[Q], [Y], [Df],
" -v- engelmannii (See O. phaeacantha -v- discata).....[Y], [Df],
" -v- flexispina (flexospina) (See O. strigil -v- flexospina),
[Y], [Df],
" -v- linguiformis (See O. lindheimeri -v- linguiformis, O. linguiformis),
[Y], [Df],
" -v- littoralis (See O. littoralis),
" -v- occidentalis (See O. occidentalis),
" -v- subarmata (Hybrid) (See O. subarmata, O. lindheimeri -v- subarmata),
[Y], [Df],
" -v- texana (See O. lindheimeri -v- lindheimeri)
[Y], [Ad], [Df],
" -v- wootonii (See O. wootonii, O. phaeacantha -v- wootonii),
erectoclada (See O. microdisca, Platyopuntia micro-disca),
[E], [DD - Flower], [Bo-1], [Bv-1], [Cz],

OPUNTIA.....(Cont'd),

erinacea (See O. hystricina -v- bensonii).....[B], [D], [P], [Q], [FF], [HH], [RR], [An], [Cb], [Dq], [Gr] [Hy], [N-1980], [Af 1-2], [Af 5-8], [Af 14-2], [Af 54-6], [Am 5-1985], [Au 37-2], [Cu 1-1973],

" -v- columbiana.....[Ax],

" -v- erinacea.....[As], [Ax], [Ca],

" -v- hystricina (See O. hystricina)....[B], [U], [LL*], [An], [Az], [Bn], [Df], [Aw 3-1988],

" -v- rhodantha (See O. erinacea -v- utahensis), [P], [Bn], [Af 23-1],

" -v- ursina (See O. ursina, O. hystricina -v- ursina).....[B], [M*], [LL], [An], [Ax], [Az], [Bv-1], [Ca], [Cz],

" -v- utahensis (See O. erinacea -v- rhodantha, O. rhodantha, O. xanthostemma)...[Q], [U], [An], [As], [Ax], [Az], [Ca], [Df],

exaltata (See Austrocylindropuntia exaltata), [P], [RR], [Ay], [Bu],

" -f. cristata.....[Ba 2-1974],

" -f. monstrosus (See Austrocylindropuntia exaltata -f. monstrosus),

excelsa.....[Az], [Aw 3-1972],

falcata (See Consolea falcata).....[Bv-1],

famatimensis (See Cumulopuntia famatimensis)

ferocior (See Cumulopuntia boliviana, O. boliviana, Tephrocactus ferocior).....[Au 13-5],

ficus-indica.....[M*], [LL], [RR], [Ax], [Ay], [Az], [Bt], [Bv-1], [Ca], [Cb], [Cj], [Cx], [Cz], [Gr], [Hx*], [Hy],

" -v- amyclaea (See O. amyclaea),

" -v- demissa (hybrid).....

" -v- occidentalis (hybrid).....

" -v- reticulata.....[Ay], [Bo-1],

flavescens (See O. phaeacantha -v- major), [B], [P], [Q], [Bn], [Bo-1], [Hy], [Af 9-5],

flexospina (See O. engelmannii -v- flexispina, O. strigil -v- flexospina),

flexuosa (See Cumulopuntia pentlandii -v- dactylifera, O. dactylifera, Tephrocactus dactylifer),

floccosa (See Austrocylindropuntia floccosa, Tephrocactus floccosus).....[CC], [RR], [Ay], [Bt], [Bx], [Dq], [Jd], [Af 41-6], [Cu 1-1-1932], [Cu 4-1962],

" -v- cardenasii (See O. malyana, Tephrocactus floccosus -v- cardenasii),

" -v- denudata (See Teph. floccosus -v- denudatus)

" -v- rauhii (See O. rauhii, Tephrocactus rauhii),

formidabilis (See O. aoracantha, Tephrocactus aoracanthus, Tephrocactus articulatus -v- ovatus, Tephrocactus ovatus).....[Bx],

OPUNTIA.....(Cont'd),

fragilis.....[B], [P], [Q], [Y], [DD], [LL*], [MM],
[RR], [Af 55-6], [Af 56-1], [Am 3-1984],
[Au 37-2], [Ba 5-1979],
" -v- brachyarthra.....[An], [Ax], [Bo-1],
[Af 56-1],
" -v- denudata (See O. fragilis -v- fragilis),
[Q], [Bo-1], [Af 56-1], [Af 55-6],
[Am 9-1983],
" -v- fragilis.....[Q], [Ax], [Am 9-1983],
" -v- parviconspicua (See O. fragilis -v- fragilis).....[Q], [Bo-1],
frigida (See Cumulopuntia frigida),
frutescens (See Cylindropuntia leptocaulis, O. leptocaulis -v- leptocaulis),
fulgida (See Cylindropuntia fulgida),
[M*], [P], [Q], [CC], [GG], [LL], [RR], [Ax],
[Ay], [Az], [Cm], [Cw], [Cx], [Ha], [Af 1-7],
[Af 7-7],
" -v- fulgida.....[Q], [Ax],
" -v- mamillata (See O. mamillata),
[A], [P], [Q], [GG], [Ax], [By], [Gr],
" -v- " -f. cristata.....[Af 32-1], [Af 34-5],
[Am 7-1979], [Am 8-1979],
" -f. monstrosus.....[Cz],
fuliginosa.....[Az], [Cx],
fulvicoma (See Tephrocactus fulvicoma).....[Bx],
furiosa (See O. tunicata),
fuscoatra (See O. compressa -v- fuscoatra, O. humifusa -v- humifusa).....[RR], [Df],
fusicaulis (See O. robusta -v- guerrana),
fusciformis (See O. macrorhiza -v- macrorhiza),
[Hy],
galapageia.....[RR], [Ay], [Bt], [Af 34-4],
" -v- brossetii (Invalid name).....[D],
" -v- echios (See O. echios),
[DD], [Ay], [Bo-1],
" -v- " -sv- barringtonensis (See O. echios -v- barringtonensis)
" -v- " -sv- inermis (See O. echios -v- inermis),
" -v- " -sv- prolifera.....
" -v- galapageia.....[Ar], [Af 37-5],
" -v- gigantea (See O. echios -v- gigantea),
[I], [Ay], [Bo-1],
" -v- hamiltonii.....
" -v- helleri.....[Bo-1],
" -v- insularis.....
" -v- macrocarpa.....[Af 37-5], [Af 39-5],
" -v- myriacantha.....
" -v- " -sv- orientalis.....[DD],

galapageia.....(cont'd),
  " -v- profusa.....
  " -v- saxicola.....[Bo-1],
  " -v- zacana (See O. echios -v- zacana).....[Bo-1],
galerasensis (See Cumulopuntia galerasensis),
gaumeri (See Nopalea gaumeri),
geometrica (See Tephrocactus geometricus),
  [Ea 1934:172 Fig.],
gilvescens (See O. canada, O. phaeacantha -v- major),
  [Bo-1],
gilvoalba (See O. stricta -v- dillenii).....[Hz],
glaucescens.....[Az], [Bo-1],
glomerata (See Tephrocactus articulatus, Tephrocactus glomeratus),
  [C], [RR], [Bx], [Dr], [Af 5-1], [Cu 4-1975],
  " -v- albispina (See O. glomerata),
  " -v- atratospinus (See Tephrocactus glomeratus - -v- atratospinus).....[Bx],
  " -v- flavispina (See O. glomerata),
  " -v- fulvispina (See Tephrocactus glomeratus -v- fulvispinus),
  " -v- minor (See O. glomerata),
  " -v- oligacantha (See O. diademata -v- polyacantha).....[Au 13-5],
  " -v- papyracantha (See O. diademata -v- papyracantha, Tephrocactus glomeratus)
  " -v- stenocantha......
gomei (See O. lindheimeri -v- lehmannii).....[Hz],
gorda (See O. robusta),
gosseliniana (See O. chlorotica -v- gosseliniana, O. violacea -v- gosseliniana),
  [B], [Y], [Bo-1], [Bv-1], [Gr], [Af 2-10],
  " -v - santa-rita (See O. chlorotica -v- santa-rita, O. santa-rita, O. violacea -v- santa-rita),
  [Y], [Bo-1], [Hx], [Bm- 1985],
gracilicylindrica (See Micropuntia gracilicylindrica),
grahamii (See Corynopuntia grahamii, O. schottii -v- grahamii),
  [E], [P], [U], [Y], [LL*], [Bk], [Bz], [Df], [Hn], [Af 16-9], [Af 60-1-1988],
grandiflora.....(hybrid).....[F], [CC], [LL*], [RR], [Df],
grandis.....[Az], [Bo-1],
grata (See Maihueniopsis -v- ovata -f. senilis, Tephracactus gratus).....[Bk], [Bx],
greenei (See O. macrorhiza -v- macrorhiza, O. tortispina),
gregoriana (See O. phaeacantha -v- major).....[Hz],

griffithsiana (See O. lindheimeri -v- lindheimeri),
grosseiana.....
guatamalensis (See Nopalea guatamalensis),
guerkei (See O. molinensis, Tephrocactus molinensis),
guerrana (See O. robusta -v- guerrana).....[Bo-1],
guilanchi.....[Bo-1],
haematocarpa (See O. lindheimeri -v- lindheimeri),
[Bo-1],
halophila (See O. alexanderi, Tephrocactus alexanderi),
[Dr],
hamiltonii (See Grusonia hamiltonii, Cylindropuntia
hamiltonii).....[Az],
hanburyana.....[RR], [Af 4-6],
heacockii.....
hegenbartiana (See Cumulopuntia pentlandii, Tephrocactus hegenbartianus),
heliae (See O. puberula).....[Aw 1:23 Fig. 14, 1955],
helleri (See O. galapageia -v- helleri),
[Af 37-5], [Af 39-5], [Aj 18-8-1967], [Aj 26-8-1975],
hempeliana (See O. floccosa),
[Aj 9-1937],
hernandezii (See O. tomentosa -v- hernandezii),
herrfeldtii (See O. rufida),
[L], [Cm], [Af 5-4],
heteromorpha (See O. miquelii -v- heteromorpha, Austrocylindropuntia miquelii -v- heteromorpha, Tephrocactus heteromorphus, O. weberi),
hickenii (Br.& R.) (See Tephrocactus hickenii - (Br. & R.) Speg.),
[RR], [Bx], [Hh July 1975],
hildmannii.....[Hr],
hirschii (See Tephrocactus hirschii).....[Au 13-5],
hitchcockii.....
hoffmannii (See O. bigelovii -v- hoffmannii),
[Cx],
hondurensis.....[Ay],
horstii (Invalid name).....[DD],
howeyi (See O. camanchica).....[L], [Bo-1],
huajuapensis.....Az],
hualpaensis (See Cylindropuntia hualpaensis, O. whipplei -v- whipplei).....[Af 15-12],
humahuacana (See Austrocylindropuntia humahuacana, Austrocylindropuntia jujuyense, Cylindropuntia humahuacana, Opuntia shaferi),
humifusa (See O. compressa -v- humifusa, O. humifusa -v- humifusa), O. allairei, O. fuscoatra, O. mesacantha, O. rafinesquei)...[Ar], [Ax], [Bo-1], [Gr], [Hv], [Hx], [Hy], [Af 55-3], [Aj 32-4-1981]

OPUNTIA.....(Cont'd),

humifusa.....(cont'd),
" -v- ammophila (See O. ammophila, O. impedata), [Ax],
" -v- austrina (See O. austrina, O. eburnispina, O. pollardii).....[Ax],
" -v- humifusa (See O. compressa -v- humifusa), [Ax], [Df], [Flower - Af 56-1],
humistrata (See O. basilaris -v- basilaris), [FF], [Af 5-8],
hypogaea (See Maihueniopsis hypogaea), [Aj IV- 3-1932],
hypsophila.....
hyptiacantha.....[RR], [Az], [Cx],
hystriciana (See O. erinacea -v- hystriciana), [I], [Y], [RR], [Ar], [Bo-1], [Df], [Go], [Hn], [Hv], [Hy], [Il], [Jd], [Af 39-2], [Bm-1986], [Aj 30-10-1979],
" -v- bensonii.....(Invalid name - See O. erinacea -v- erinacea).....[Bo-1], [Hv],
" -v- nicholii (See O. nicholii),
" -v- ursina (See O. erinacea -v- ursina), [Av], [Ay], [Hl 2-1970],
hystrix (See Cumulopuntia hystrix),
ianthinantha (See Platyopuntia ianthinantha),
ignescens (See Tephrocactus ignescens).....[RR],
" -v- steiniana (See Tephrocactus -v- steinianus)
ignota (See Cumulopuntia ignota, Tephrocactus corotilla (Backbg.), Tephrocactus ignotus).....[RR],
imbricata (See Cylindropuntia imbricata), [I], [M*], [P], [Q], [U], [GG], [LL*], [RR], [Ax], [Ay], [Az], [Bn], [Bz], [Ca], [Cx], [Ib], [Jd], [Af 1-7], [Af 14-9], [Af 58-2], [Aj 29-3-1978],
" -v- arborescens (See O. imbricata -v- imbricata) [Q], [Y], [Df],
" -v- argentea (See Cylindropuntia imbricata -v- argentea),
" -v- cardenche (See O. kleiniae),
" -v- imbricata.....[Ax], [Df],
" -v- lloydii (See O. lloydii),
" -v- vexans (See O. imbricta -v- imbricata), [Y], [Df],
" -v- viridiflora (See Cylindropuntia viridiflora, O. viridiflora, O. whipplei -v- viridiflora).....[Y], [Df],
" -f. cristata.....[Af 34-5],
impedata (See O. humifusa -v- ammophila), [RR], [Bo-1], [Af 4-8],
inaequilateralis (See Platyopuntia inaequilateralis), [RR*], [Hy*],

OPUNTIA.....(Cont'd),

inaequilateralis.....(cont'd),
" -v- angustior.....
inamoena (See Platyopuntia inamoena).....[RR], [Bo-1], [Hh March 1975], [Hh Sept. 1975],
" -v- flaviflora.....[Bo-1],
" -f. spinigera.....
inarmata (See Austrocylindropuntia inarmata, Cylindropuntia inarmata).....[Jd],
inaperta (See Nopalea inaperta),
incarnadilla.....[Hz],
inermis (See O. stricta -v- stricta),
infesta (See Platyopuntia infesta),
insularris (See O. galapageia -v- insularis), [Af 5-8], [Af 37-5],
interjecta (See Platyopuntia interjecta),
intermedia (See O. humifusa -v- humifusa),
" -v- prostrata (See O. humifusa -v- humifusa),
invicta (See Corynopuntia invicta, Cylindropuntia invicta).....[E], [NN], [RR], [Ay], [Az], [Iz], [Ah-1985], [Aw 4-1963], [Hh July, 1976], [Bi 6-2-1988],
ipatiana (See Austrocylindropuntia ipatiana, Austrocylindropuntia salmiana, O. salmiana),
jaliscana.....[Az], [Aw 4-1972],
jamaicensis.....[RR], [Bo-1],
joconostle (joconoxtle).....
johnstonii (See O. macbridei).....[Ay], [Bo-6],
juniperina (See O. polyacantha -v- juniperina), [P], [RR], [Bo-1], [Hx*], [Aj 24-10-1973],
karwinskiana (See Nopalea karwinskiana),
kelvinensis (O. fulgida X O. spinosior),
keyensis (See O. stricta -v- stricta).....[RR],
kiska-loro (See Platyopuntia kiska-loro).....[RR],
kleiniae (See Cylindropuntia kleiniae), [F], [P], [Q], [Y], [CC], [LL*], [RR], [Az], [Cx], [Df], [Af 60-1-1988],
" -v- kleiniae.....[Ax],
" -v- tetracantha (See Cylindropuntia kleiniae -v- tetracantha), [B], [Q], [Ax],
kuehnrichiana (See Cumulopuntia kuehnrichiana, Tephrocactus kuehnrichianus), [E], [Ay], [Bx],
" -v- applanata (See Tephrocactus kuehnrichianus -v- applanata).....[Bx],
kunzei (See O. stanlyi -v- kunzei).....[Bj],
laevis (See O. phaeacantha -v- laevis), [A], [B], [P], [Q], [LL*], [RR],
" -v- canada (See O. phaeacantha -v- major), [P], [Q],

OPUNTIA.....(Cont'd),

laetevirens.....(Invalid name).....[Cu 1-1973],
lagopus (See Austrocylindropuntia lagopus, Tephrocactus lagopus),
[RR], [Bi 3-1987], [Cu 4-1965],
" -f. rauhii (See Opuntia rauhii, Tephrocactus rauhii),
lagunae.....[Ar], [Bo-1],
lanceolata (See O. elongata).....[RR], [Bo-1], [Hy],
larreyi (See O. robusta -v- larreyi),
lasiacantha (See O. megacantha -v- lasiacantha),
[Az], [Bo-1], [Cx], [Hy*],
lata (See O. humifusa -v- ammophila).....[Bo-1],
lauliacoana (See Austrocylindropuntia lauliacoana),
lecoriensis (See Tephrocactus lecoriensis).....
leoncito (Werd.) (See Tephrocactus leoncito),
leonina (See Tephrocactus leoninus).....[Di*],
leptocarpa (See O. macrorhiza -v- macrorhiza)......[E],
[H], [Y], [RR], [Bo-1], [Df], [Bi 5-4-1987],
leptocaulis (See Cylindropuntia leptocaulis),
[D], [M*], [Q], [U], [Y], [LL*], [Ay],
[Az], [Bz], [Cx], [Cz], [Df], [Gr],
[N-1980], [Ad Nov. 1979],
" -v- badia.....
" -v- brevispina.....
" -v- brittonii (See Nopalea brittonii, Cylindropuntia brittonii),
" -v- frutescens (See O. leptocaulis),
" -v- longispina (See Cylindropuntia mortolensis, O. mortolensis, O. leptocaulis),
" -v- pluriseta.....
" -v- robustior.....
" -v- stipata (See O. leptocaulis),
" -v- tenuispina.....
" -v- vaginata (See Cylindropuntia leptocaulis -v- vaginata),
leptoclada (See Maihueniopsis leptoclada),
leucacantha (See Consolea leucacantha),
" -v- lavior (See O. ferox),
leucophaea (See Cumulopuntia berteri),
leucotricha.....[C], [RR], [Ar], [Ax], [Az], [Bo-1],
[Bv-1], [Cm], [Cx], [Cz], [Gr], [Hx*],
[Hy], [Aw 3-1970], [Aw 4-1971],
limitata (See Platyopuntia limitata),
lindheimeri.....[LL*], [Ax], [Az], [Bo-1], [Bz],
[Af 56-3], [Aw 1-1962], [Aw 4-1977],
[Aw 1980-1983 Index],
" -v- aciculata (See O. engelmannii -v- aciculata, O. lindheimeri -v- lindheimeri),
" -v- chisosensis (See O. chisosensis, O. lindheimeri),

OPUNTIA.....(Cont'd),

lindheimeri.....(cont'd),
" -v- cuija (See O. cantabrigiensis),
" -v- cyclodes (See O. engelmannii -v- cyclodes, O. phaeacantha -v- major),
" -v- dulcis (See O. phaeacantha -v- phaeacantha)
" -v- lehmannii (See O. gomei, O. cyanella), [Ax], [Df],
" -v- lindheimeri.....[Ax], [Af 60-1-1988],
" -v- linguiformis (See O. engelmannii -v- linguiformis, O. linguiformis), [Ax], [Df], [Af 57-4], [Af 60-1-1988],
" -v- tricolor (See O. engelmannii -v- cacanapa), [Ax], [Df],
" -v- subarmata (See O. subarmata), [Aw 32-1-1987],
linguiformis (See O. engelmannii -v- linguiformis, O. lindheimeri -v- linguiformis), [D], [LL*], [RR], [Ay], [Bj], [Bo-1], [Bz], [Hy],
" -v- maverick.....[X], [Bu], [Cz],
littoralis.....[FF], [LL*], [RR], [Ax], [Az], [Bu], [Bv-1], [Ca], [Cb], [Do], [Cu 1-1962],
" -v- austrocalifornica.....[Ax], [Az], [Ca],
" -v- covillei (See O. littoralis -v- vaseyi),
" -v- littoralis.....[Ax],
" -v- martiniana (See O. phaeacantha -v- charlestonensis), [An], [Ax], [Az], [Ca],
" -v- piercei (See O. piercei, O. occidentalis --v- piercei),
" -v- vaseyi (See O. covillei, O. vaseyi), [An], [Ax], [Az], [Ca], [Ij],
lloydii (See Cylindropuntia lloydii).....[RR], [Gr],
longiareolata (See O. basilaris -v- longiareolata),
longiarticulata.....
longispina (Haw.).....[Bo-1], [Hx*], [Hy],
" -v- agglomerata.....
" -v- brevispina.....
" -v- corrugata (See O. corrugata)..... [Bo-1], [Hy*],
" -v- flavidispina.....[Bo-1],
" -v- intermedia.....
loomisii (See O. macrorhiza -v- macrorhiza), [Af 9-7], [Af 14-1], [Af 59-6],
lubrica.....[Bo-1], [Hy*], [Hz],
" -v- aurea (See O. basilaris -v- aurea),
lurida .....[Af 2-8],
lutea (See Nopalea lutea),
" -v- carni-straminea.....[Af 56-1],

macateei (See O. pusilla (Haw.), O. compressa -v- fusca-atra),
macbridei (See O. quitensis, Platyopuntia quitensis), [D], [RR], [Ay], [Bo-1], [Af 5-2], [Iv 9-1957],
" -v- orbicularis (See O. quitensis, Platyopuntia quitensis).....[Bo-1],
macbridgei (See O. johnsonii),
macdougaliana (See O. velutina -v- macdougaliana), [RR], [Cx], [Hy], [Aw 1-1972], [Ba 4-1980],
machacana (See Austrocylindropuntia machacana),
machensenii (See O. macrorhiza -v- macrorhiza),
macracantha (See Consolea macracantha),
macrarthra (See O. stricta -v- stricta), [LL*], [RR], [Bv-1], [Af 32-4],
macrocalyx.....[RR], [Bo-1], [Hy],
" -f. cristata.....[Af 2-8],
macrocentra (See O. violacea)...[U], [Y], [CC], [LL*], [RR], [An], [Bj], [Bt], [Bz], [Df], [Hx*], [Hy*], [Ie], [Jd], [N-1980], [Ad 3-1983], [Am 6-1983],
" -v- aureispina.....[Af 60-1-1988],
" -v- macrocentra (See O. violacea -v- castetteri, O. violacea -v- violacea, O. violacea -v- macrocentra).....[Af 60-1-1988], [Af 60-4],
" -v- martiniana (See O. littoralis -v- martiniana, O. phaeacantha -v- charlestonensis)
" -v- minor (See O. atrispina).....[Af 60-1-1988] [Af 60-4],
macrorhiza (See O. compressa -v- macrorhiza), [M*], [Q], [RR], [An], [Ax], [Az], [Ca], [Df], [Hy], [Af 60-1-1988],
" -v- macrorhiza (See O. compressa -v- macrorhiza).....[Ax], [Df], [Af 58-2],
" -v- pottsii (See O. ballii, O. pottsii, O. setispina, O. tenuispina), [Q], [An], [Ax], [Df],
maculacantha (See O. pampeana, O. sulphurea, O. vulpina).....[Bk],
magnifica (See O. stricta -v- stricta).....[Bo-1],
maldonadensis.....(Hybrid ?)
malyana (See Tephrocactus malyanus).....[Ay],
mamillata (See O. fulgida -v- mamillata).....[Q],
" -f. cristata.....[N-1980],
mandragora (See Maihueniopsis mandragora, Maihueniopsis minuta, Tephrocactus mandragora),
marenae (See Marenopuntia maranae), [Ay], [Az], [Jd], [Af 49-6],
maritima (See O. dillenii, O. stricta -v- dillenii),
marneriana.....[NN], [Bo-1], [Bo-6], [Aj 31-2-1980],
martiniana (See O. littoralis -v- martiniana),

matudae.....
maxima.....[RR], [Ar], [Bo-1], [Hy],
maxonii (See O. heliae, O. puberula),
megacantha (See O. ficus-indica (spiny form)),
[FF], [RR], [Ax], [Az], [Ca], [Cb], [Cx],
[Hy], [Aw 1-1967],
" -v- lasiacantha (See O. lasiacantha),
megacarpa (See O. engelmannii -v- discata, O. phaeacantha -v- discata, O. procumbens),
[Cb], [Do],
megalantha.....
megapotamica.....[Bo-1],
megarhiza (Possibly a form of O. macrorhiza),
megasperma (See O. galapageia -v- myriacantha),
[Af 56-6],
" -v- megasperma.....[Af 39-5], [Aj 18-8-1967],
" -v- mesophytica.....
" -v- orientalis (See O. galapageia -v- myriacantha -sv- orientalis),
[Af 34-4],
meickleyi.....[Aj 23-12-1972],
melanacantha (See Cumulopuntia pentlandii -v- dactylifera, O. dactylifera, Tephrocactus dactyliferus, Tephrocactus melanacanthus),
melanosperma.....
mesacantha (See O. compressa, O. humifusa -v- humifusa),
" -v- cymochila (See O. cymochila, O. macrorhiza -v- macrorhiza, O. tortispina),
" -v- greenei (See O. macrorhiza -v- macrorhiza, O. tortispina),
" -v- macrorhiza (See O. macrorhiza -v- macrorhiza, O. compressa -v- macrorhiza),
" -v- oplocarpa (See O. macrorhiza -v- macrorhiza)
" -v- stenochila (See O. macrorhiza -v- macrorhiza),
metuenda (See Cylindropuntia metuenda),
microclada.....
microdasys.....[D], [R], [AA], [RR], [Ar], [Av], [Ax],
[Ay], [Az], [Bo-1], [Bt], [Cg], [Cs],
[Cx], [Cy], [Cz], [Hx*], [Hy], [Il],
[Jd], [Ad April, 1979], [Aw 1-1967],
" -v- alba.....[C], [H], [Ck],
" -f. albata (See O. albata),
[Bu], [Bv-1], [Af 32-4],
" -v- albispina.....[I], [EE], [QQ], [XX], [Bo-1],
[Bv-1], [Cz], [Dp], [Gr],
[Hv], [Hx], [Hy], [Il],
[Db 6-1980],

microdasys.....(cont'd),
" -v- aurantiaca (See O. aurantiaca).....[Bo-1],
" -v- laevior.....
" -v- lutea.....[Bv-1],
" -v- minima (See O. microdasys -v- minor, O. rufida -v- minima),
" -v- minor.....
" -v- pallida (See O. pallida), [Bo-1], [Ja],
" -v- rufida (See O. rufida), [I], [K], [NN], [Bo-1], [Co], [Il], [Aj 29-8-1978],
" -v- " -f. albiflora (invalid name),
" -v- " -f. minima.....[Bo-1],
" -v- " -f. minor.....
" -v- " -f. monstruosa.....[Aj 15-4-1964],
" -v- undulata (See O. undulata), [Bo-1], [Hu],
" -f. cristata.....[G], [Be*], [Hy], [Ag-1985],
microdisca (See Platyopuntia microdisca, O. erecto-clada).....[RR], [Bo-1],
microsphaerica (See Tephrocactus microsphaericus),
mieckleyi.....[Aj 23-12-1972*],
militaris (See O. triacantha).....[RR],
millspaughii (See Consolea millspaughii).....[Rr],
minor (See Cumulopuntia pentlandii, Opuntia pentlandii, Tephrocactus minor),
minuscula (See Cumulopuntia pentlandii, Tephrocactus minusculus),
" -v- silvestris.....
minuta (See Tephrocactus minutus),
miquelii (See Austrocylindropuntia miquelii, Cylindro-puntia miquelii, Miqueliopuntia miquelii), [E], [RR], [Jd], [Ad-],
" -v- gessei (See Austrocylindropuntia miquelii -v- geesei),
" -v- heteromorpha (See Austrocylindropuntia miquelii -v- heteromorpha),
mira (See Cumulopuntia kuehnrichiana, O. kuehnrichiana, Tephrocactus mirus).....[Au 13-6-1958],
missouriensis (See O. polyacantha -v- polyacantha),
" -v- erythrostemma.....
" -v- trichophora (See O. polyacantha -v- tricho-phora, O. trichophora),
mistiensis (See Tephrocactus mistiensis),
moelleri (See Corynopuntia moelleri), [L], [Az], [Af 4-5], [Af 57-5], [Ag-1988],
mojavensis (See O. curvospina, O. phaeacantha -v- mojavensis) [MM], [Af 3-6],

OPUNTIA.....(Cont'd),

molesta (See Austrocylindropuntia molesta, Cylindropuntia molesta),
[G], [P], [FF], [RR], [Ay], [Gr], [Af 51-3],
molfinoi (See Maihueniopsis molfinoi),
molinensis (See Maihueniopsis molinensis, Tephrocactus molinensis),
[F], [H], [Dr], [Iz], [Bi 5-1],
monacantha (See O. vulgaris)....[Co], [Hy], [Af 25-1],
" -f. albovariegatus.....[Hy],
" -f. areovariegatus.....
moniliformis (See Consolea moniliformis),
[RR], [Ct-1984], [Jd], [Af 56-3],
montevidensis.....
monticola.....
monvillii (See O. corrugata -v- monvillii),
mortolensis (See Cylindropuntia mortolensis, Opuntia leptocaulis -v- longispina).....[RR],
muelleriana ......[Au 13-6-1958],
multiareolata (Invalid name - See Cumulopuntia multiareolata).....[Bo-6:3614),
multigeniculata (See Cylindropuntia multigeniculata, O. whipplei -v- multigeniculata),
[As], [Fk 7:69 Tab. 4, Fig. a, Tab. 5, Fig. e, 1943],
munzei.....[Ax], [Az], [Ca],
myriacantha (See O. galapageia -v- myriacantha, O. megasperma),
nana.....(See O. opuntia, - Aw Vol, 27 -Page 57),
nashii (See Consolea nashii).....[RR],
nejapensis.....
nemoralis (See O. humifusa -v- humifusa),
neochrysacantha.....[Az], [Aw 1-1974],
neuquensis (See Maihueniopsis neuquensis, Tephrocactus neuquensis),
nicholii (See O. hystricina -v- nicholii),
[B], [P], [Q], [An], [Ax],
nigrispina (See Platyopuntia nigrispina, Tephrocactus nigrispinus).....[RR], [Bx],
nigrita.....[Hz],
nitens (See O. stricta -v- dillenii),
noodtiae (See Cumulopuntia pentlandii -v- dactylifera, O. dactylifera, Tephrocactus dactylifer, Tephrocactus noodtiae),
nuda (See Nopalea nuda),
[See - Rep. Plant. Succ. 22:11, 1971],
obliqua (Invalid name - Bo-6:3614),
occidentalis (See O. ficus-indica -v- occidentalis),
[Q], [FF], [RR], [Cb], [Cm], [Af 2-7],
[Hl 3-1967],

occidentalis.....(cont'd),
" -v- piercei (See O. phaeacantha -v- piercei, O. piercei).....[Do],
ochrocentra (See O. antillana, O. cubensis),
opuntia (See O. humifusa -v- humifusa, O. nana), [RR],
orbiculata (See O. crinifera, O. pailana), [RR], [Bo-1], [Hy],
oricola.....[GG], [Ax], [Az], [Ca], [Cb], [Aw 2-1988],
orurensis (See Platyopuntia orurensis).....[Af 28-4],
otaviana (See Tephrocactus otavianus),
ovallei (See Maihueniopeis ovallei, Tephrocactus ovallei),
ovata (Pfeiff.) (See Maihueniopsis ovata, O. russelii, Tephrocactus russelii), [E], [RR], [Gr],
" -f. calva.....
" -f. sterilis.....
pachona (See O. streptacantha), [Hz], [Eo 21:168 Pl. 22, 1910],
pachyclada.....
" -v- rosea.....[Aj 4-1982],
pachypus (See Austrocylindropuntia pachypus, Cylindropuntia pachypus), [C], [RR], [Bv-1], [Af 3-8], [Af 20-8], [Af 56-6]
" -v- cristata.....[H], [Ay], [Am 8-1984],
paediophila (See Tephrocactus articulatus -v- polyacanthus, Tephrocactus hossei, Tephrocactus paediophilus), [E], [H], [AA], [Af 51-4], [Bi 5-1],
pailana (See O. crinifera, O. orbiculata), [Af 6-7], [Af 9-1], [Aj I-9-1929],
pallida (See O. microdasys -v- pallida)....[RR], [Cx],
palmadora.....[RR], [Bo-1],
palmeri (See O. chlorotica),
pampeana (See O. maculacantha, O. sulphurea, O. vulpina).....[Af 40-6],
panellana (See O. armata -v- panellana) [Bo-6 - Description],
papyracantha (See O. articulata -f. syringacantha, O. diademata, Tephrocactus articulatus -v- papyracanthus, Tephrocactus articulatus -f. syringacanthus, Tephrocactus diadematus)....
paraguayensis (See O. bonaerensis, O. chakensis), [Bo-1],
parishii (See Corynopuntia parishii, O. stanlyi -v- parishii), [Cb], [Af 6-1],
parmentieri (See O. retrospinosa, Tephrocactus retrospinosus),

parryi (See Cylindropuntia parryi),
[LL], [Ax], [Az], [Ca], [Cb], [Do], [Af 3-7],
" -v- parryi.....[Ax],
" -v- serpentina.....[Ax],
parviseta (See Tephrocactus parvisetus),
pascoensis (See Platyopuntia nana).....[RR], [Bo-1],
peckii.....(In cultivation),.....
penicilligera......
pennellii.....[RR],
pentlandii (Backbg.) (See T. neopentlandii),
pentlandii (S/D) (See Cumulopuntia pentlandii, Tephrocactus pentlandii).....[RR], [Bx], [Dq], [Gr],
" -v- adpressa.....(Invalid name),
" -v- colchana.....(See Cumulopuntia pentlandii -v- colchana),
" -v- dactylifera.....(See Cumulopuntia pentlandii -v- dactylifera),
" -v- fauxiana (See Tephrocactus pentlandii -v- fauxianus),
" -v- rossiana (See Tephrocactus pentlandii -v- rossianus),
perrita.....[Hz],
pes-corvi (See O. pusilla),
pestifer (See O. pubescens, Platyopuntia nana),
[RR], [Bo-6],
phaeacantha.....[L], [P], [Q], [U], [LL*], [RR], [Ay], [Bn], [Bo-1], [Cm], [Co], [Go*], [Hv], [Hx], [Hy], [Jd], [Am 2-1982], [Cu 4-1964],
" -v- albispina.....[Hx*], [Aj 23-11-1972], [Aj 26-1-1975],
" -v- angustata (See O. angustata, O. phaeacantha -v- major).....[P],
" -v- brunnea (See O. phaeacantha -v- major), [P], [Y], [An], [Df], [Af 58-4], [Af 60-1-1988],
" -v- camanchica (See O. camanchica), [Ax], [Bo-1], [Df], [Gr], [Hu],
" -v- carneostemma.....
" -v- charlestonensis (See O. littoralis -v- martiniana).....[Af 56-1],
" -v- chihuahuensis.....[Az],
" -v- cuija (See O. catabrigiensis),
" -v- discata (See O. engelmannii -v- discata, O. discata, O. megacarpa, O. procumbens),.....[Q], [An], [As*], [Ax], [Az], [Ca], [Df], [Ib], [Ij], [Am 2-1982], [Au 37-2],
" -v- flavescens (See O. flavescens, O. phaeacantha -v- major).....[An],
" -v- flavispina.....[An], [Ax],

phaeacantha.....(cont'd)
" -v- gigantea.....
" -v- laevis (See O. laevis).....
[Q], [An], [Ax], [Az],
" -v- linguiformis (See O. engelmannii -v- linguiformis, O. lindheimeri -v- linguiformis, O. linguiformis),
" -v- longispina (See O. longispina),
[Aj 26-1-1975],
" -v- major.....[Q], [U], [Y], [An], [As*], [Ax],
[Df], [Ib], [Af 57-3],
" -v- mojavensis (See O. curvospina, O. mojavensis).....[Q], [Bo-1],
" -v- nigricans (See O. phaeacantha -v- phaeacantha).....[Y], [An], [Df],
" -v- pallida (See O. pallida),
" -v- phaeacantha.....[An], [Ax], [Az],
" -v- piercei (See O. littoralis -v- piercei, O. occidentalis -v- piercei, O. piercei),
" -v- rubra.....
" -v- salmonea.....
" -v- spinosibacca (See O. spinosibacca)....[Df].
" -v- superbospina.....[Ax],
" -v- tenuispina (See O. tenuispina),
" -v- tortispina (See O. macrorhiza -v- macrorhiza, O. tortispina),
" -v- trichophora (See O. polyacantha -v- trichophora, O. trichophora),
[Ib],
" -v- watsonii.....
" -v- wootonii (See O. wootonii).....[Ax], [Df],
phyllacantha (See O. sphaerica),
picardoi.....[Bo-6], [Gr], [Hx], [Iz], [Bi 5-1], [Iv 3-1960],
picnantha.....[Hu],
piercei (See O. littoralis -v- piercei, O. occidentalis -v- piercei, O. phaeacantha -v- piercei),
[Cb],
pilifera.....[RR], [Ar], [Az], [Bo-1], [Cx], [Hx*],
" -v- aurantisaeta.....[Bo-1],
pintacaillan.....
piriai.....[Hr],
pisciformis (See O. humifusa -v- ammophila), (Hybrid),
[RR], [Bo-1], [Af 4-9],
pittieri.....[RR], [Bo-1], [Ij],
pituitosa (See Platyopuntia pituitosa),
planibulbispina (See Corynopuntia planibulbispina, O. schottii -v- grahamii),

platyacantha (S/D) (See Tephrocactus platyacanthus),
" -v- angustispinus (See Tephrocactus platyacanthus -v- angustispinus)...[Bo-1],
" -v- deflexispina (See Tephrocactus platyacanthus -v- deflexispinus)...[Bo-1],
" -v- gracilior (See O. tuberosa -v- spinosior),
" -v- monvillei (See Tephrocactus platyacanthus -v- monvillei),
" -v- neoplatyacantha (See Tephrocactus platyacanthus -v- neoplatyocanthus), [Bo-1], [Bx],
plumbea (See O. macrorhiza -v- macrorhiza), [A], [Y], [RR], [Df],
poecilacantha (Invalid name).....[Bo-6 Description],
poeppigii (See Maihuenia -v- poeppigii),
pollardii (See O. humifusa -v- austrina), [LL*], [RR], [Af 40-1], [Af 53-4],
polyacantha.....[B], [P], [Y], [MM], [RR], [An], [Cg], [Df], [Go*], [Ie], [Af 32-5], [Af 51-6], [Af 56-1], [Af 58-4], [Am 3-1984],
" -v- albispina (See O. polyacantha -v- polyacantha).....[Go],
" -v- erythrostemma (See O. missouriensis -v- erythrostemma),
" -v- hystricina (See O. erinacea -v- hystricina),
" -v- juniperina (See O. sphaerocarpa), [An], [Ax], [Df],
" -v- polyacantha.....[Ax],
" -v- rufispina (See O. rutila). [Q], [An], [Ax], [Az], [Ca],
" -v- salmonea (See O. polyacantha -v- polyacantha),
" -v- schweriniana (See O. fragilis -v- fragilis) [Bo-1],
" -v- trichophora (See O. polyacantha, O. trichophora). [P], [Q], [U], [LL*], [An], [Ax],
polycarpa (See O. humifusa -v- austrina),
posnanskyana (See Austrocylindropuntia verschaffeltii, Cylindropuntia verschaffeltii, O. verschaffeltii).....[Bo-1],
pottsii (See O. ballii, O. macrorhiza -v- pottsii, O. tenuispina), [Y], [RR], [Cf], [Df], [Af 58-2], [Af 59-6],
prasina.....
procumbens (See O. phaeacantha -v- discata)..[A], [RR],
prolifera (See Cylindropuntia prolifera), [FF], [LL], [RR], [Ax], [Az], [Ca], [Cb], [Do], [Ij], [Af 2-1], [Af 5-12], [Af 60-3],

OPUNTIA.....(Cont'd),

pseudorauppiana (See Tephrocactus dimorphus -v- pseudo-rauppiana).....[Bx],
pseudo-udonis (See Austrocylindropuntia floccosa, O. floccosa, Tephrocactus pseudo-udonis), [Au 13:6, 1958],
puberula (See O. heliae).....[Az],
pubescens (See Platyopuntia nana), [RR], [Az], [Bd], [Bo-1], [Cx], [Hy*],
puelchana.....
pulchella (See Corynopuntia pulchella, Micropuntia brachyrhopalica), [P], [Q], [Z], [LL*], [RR], [Ax], [Ca], [Af 29-1],
pumila.....[RR], [Az], [Cx],
punta-caillan (See Tephrocactus punta-caillan),
purpurea.....[Au 13:6, 1958],
pusilla (Haw.) (See O. drummondii), [RR], [Ax], [Bo-1], [Df], [Af 53-4],
pycnantha.....[RR], [Av], [Az], [Hv], [Af 2-4],
" -v- margaritana.....[EE], [Ay], [Bo-1], [Hx], [Af 58-5], [Jd],
" -f. cristata.....[HH],
pygmaea (See Micropuntia pygmaea).....[Au 13:5, 1958],
pyriformis.....[RR], [Aw 1-1972],
pyrrhacantha (See Cumulopuntia pyrrhacantha, Tephrocactus pyrrhacanthus).....[Bo-1],
" -v- leucoluteus (See Cumulopuntia pentlandii -v- dactylifera, O. dactylifera, Tephrocactus pyrrhacanthus -v- leucoluteus).....
pyrrhantha (See Platyopuntia pyrrhantha),
quimilo (See Platyopuntia quimilo), [L], [RR], [Bo-1], [Bv-1], [Gr], [Hx], [Hy*], [Ij], [Af 32-4],
quipa.....[DD], [Bo-1], [Bo-6],
quitensis (See O. macbridei, Platyopuntia quitensis), [RR],
rafinesquei (See O. humifusa -v- humifusa), [E], [An], [Bv-1], [Go], [Im], [Af 32-4], [Cu 7-1-1938], [Hl 4-1981],
" -v- arkansana (See O. humifusa -v- humifusa),
" -v- macrorhiza.....[Be*],
" -v- microsperma (See O. humifusa -v- humifusa, O. compressa -v- microsperma),
rahmeri (See Maihueniopsis -v- rahmeri, O. tarapacana),
ramosissima (See Cylindropuntia ramosissima), [B], [M*], [P], [Q], [FF], [LL*], [Ax], [Az], [Bu], [Ca], [Cb], [Do], [Jd], [Af 2-7],
" -f. cristata.....[Af 1-9], [Af 34-5],
ramulifera (See O. leptocaulis),

rarissima (See Cumulopuntia pentlandii, O. pentlandii, Tephrocactus pentlandii),
rastrera.....[Az],
rauhii (See Tephrocactus rauhii).....[Ay], [Jd],
rauppiana (See Cumulopuntia rauppiana, Tephrocactus sphaericus -v- rauppianus).....[D], [RR],
recondita (See Cylindropuntia recondita), [Gr], [Bi Vol. 4, #3],
" -v- perrita (See Cylindropuntia recondita -v- perrita),
reflexispina (See Corynopuntia reflexispina),
reicheana (See Maihueniopsis leoncito, Tephrocactus reicheanus),
repens.....[LL*], [RR], [Bo-1], [Ad Jan., 1979],
retrorsa (See O. canina, Platyopuntia retrorsa), [RR], [Bo-1], [Hx], [Hh 2-4-1976],
retrospina.....[Bo-1],
retrospinosa (See Tephrocactus retrospinosus)....[Bo-1]
rhodantha (See O. erinacea -v- utahensis).....[G], [P], [RR], [Bo-1], [Df], [Go], [Hx], [Jd], [Ad Sept. 1978], [Af 32-4], [Aj 33-4-1982],
" -v- brevispina (See O. erinacea -v- utahensis),
" -v- flavispina (See O. erinacea -v- utahensis),
" -v- pisciformis (See O. erinacea -v- utahensis)
" -v- rhodantha (See O. erinacea -v- utahensis), [Y], [Df],
" -v- schumanniana (See O. erinacea -v- utahensis)
" -v- spinosior (See O. erinacea -v- utahensis), [Df],
" -v- xanthostemma (See O. erinacea -v- utahensis)
rileyi.....[Az],
riojana (See O. alexanderi, Tephrocactus riojanus),
ritteri.....[Bo-1],
riviereana (Invalid name).....[Bo-6], [Aj 23-5-1972],
robinsonii.....
roborensis.....[Af 42-1],
robusta.....[RR], [Ar], [Bd], [Bo-1], [Bu], [Cx], [Hb], [Hx*, [Im], [Af 30-1], [Au 37-1],
" -v- guerrana (See O. guerrana),
" -v- larreyi.....[Az],
" -v- longiglochidiata.....[Bo-1],
" -v- maxima (See O. maxima),
" -v- robusta.....[Az],
" -v- viridior.....
rosarica (See Cylindropuntia rosarica, Grusonia rosarica).....[Az], [Af 14-4],
rosea (See Cylindropuntia rosea, O. pallida), [Az], [Bd], [Di*], [Aj 28-11-1977], [
" -v- atrorosea (See Cylindropuntia rosea -v- atrorosea),

OPUNTIA.....(Cont'd),

roseiflora (See Austrocylindropuntia miquelii, Cylindropuntia miquelii, Miqueliopuntia miquelii, Opuntia miquelii),
rossiana (See Cumulopuntia rossiana, Tephrocactus pentlandii -v- rossiana),
rubescens (See Consolea rubescens),
[RR], [Bj], [N-1980], [Af 38-4], [Af 54-4],
rubiflora (Davidson) (See O. littoralis -v- vaseyi),
rubrifolia (See O. erinacea -v- ursina),
rubrogemmia (See Platyopuntia rubrogemmia),
rufida (See O. microdasys -v- rufida),
[M*], [Y], [AA], [QQ], [RR], [Ax], [Ay], [Az], [Bo-1], [Bz], [Cz], [Df], [Gr], [Hn], [Hx*], [Il], [Af 60-1-1988],
" -v- (dwarf form).....[Ab-1985],
" -v- minima.....
" -v- minor.....
" -v- tortiflora.....
" -f. monstrosus.....[Ag-1985],
russellii (See O. ovata (Pfeiff.), Maihueniopsis ovata, Tephrocactus russellii),
[C], [RR], [Bo-1], [Bx],
rutila (See O. polyacantha -v- rufispina),
[Q], [Go], [Af 56-1], [Aj 26-1-1975],
rzedowskii.....
salagria.....[Ay], [Bo-1],
salmiana (See Austrocylindropuntia salmiana, O. spegazzinii, Platyopuntia salmiana),
[C], [QQ], [RR], [Bv-1], [Cn], [Gr], [Aj II-10-1930],
" -v- albiflora.....[Au 13:25, 1958],
" -v- ipatiana (See Austrocylindropuntia ipatiana, Austrocylindropuntia salmiana -v- ipatiana),
" -v- spegazzinii.....[Au 13:25, 1958),
salvadorensis.....
sanguinea.....
santamaria (See Grusonia santamaria),
santa-rita (See O. chlorotica -v- santa-rita, O. gosseliana -v- santa-rita, O. violacea -v- santa-rita)...[D], [Q], [GG], [RR], [Bj], [Df], [Af 5-2], [Au Vol. 2 - 1977, Zone 3 Compendium),
saxatilis (See Platyopuntia saxatilis),
saxicola (See O. galapageia -v- saxicola),
[Af 5-8], [Af 56-6],
scheerii [C], [RR], [Az*], [Bo-1], [Cx], [Hy], [If], [Jd], [Bi 2-3],
schickendantzii.....[DD], [RR], [Bo-1], [Bu], [Cz],
" -v- similis.....

OPUNTIA.....(Cont'd),

schottii (See Corynopuntia schottii),
[Y], [GG], [RR], [Ax], [Df], [Ad Nov. 1979],
[Af 58-4],
" -v- grahamii (See O. grahamii),
[U], [GG], [Ax], [Df], [Ib],
" -v- schottii.....[Ax],
schulzii (See Brasiliopuntia schulzii),
schumannii (Speg.) (See Maihueniopsis molinensis,
Opuntia molinensis, Tephrocactus
molinensis),
schumannii (Web.).....[RR], [Aj 2-1937],
schweriniana (See O. fragilis -v- fragilis, O. polyacantha -v- schweriniana),
[Go],
securigera.....
senilis (See O. crinifera, O. orbiculata),
serpentina (See O. parryi -v- serpentina),
[FF], [RR], [Az], [Cb], [Cq], [Do],
[Af 1-11], [Af 3-12],
setigera (See O. weberi -v- setigera, Tephrocactus
weberi -v- setiger),
[Af 17:114, 1945],
setispina (See O. macrorhiza -v- pottsii).....[RR],
shaferi (See Austrocylindropuntia shaferi, Austrocylindropuntia vestita -v- shaferi, Cylindropuntia humahuacana),
[Cu 4-1980],
" -v- humahuacana (See Austrocylindropuntia humahuacana, Cylindropuntia humahuacana),
silvestris (See Cumulopuntia pentlandii, O. pentlandii, Tephrocactus pentlandii, Tephrocactus
minor, Tephrocactus silvestris),
sinclairii (See O. Lindheimeri -v- lindheimeri)...[Hz],
skottsbergii (See Pterocactus skottsbergii).....[RR],
soederstromiana.....[RR],
soehrensii (See O. boliviensis, Platyopuntia soehrensii)
[RR], [Bo-1], [Bo-6], [Hx*], [Af 59-1],
" -v- grandiflora.....
" -v- tilcarensis.....
" -v- transiens.....
spectatissima (See Micropuntia spectatissima, Micropuntia barkleyana),
spegazzinii (See Austrocylindropuntia salmiana -v-
spegazzinii, O. salmiana, Platyopuntia
salmiana).....[K], [V],
sphaerica (See Cumulopuntia berteri, Tephrocactus
dimorphus (sensu BACKBG.), Tephrocactus pseudorauppianus, Tephrocactus sphaericus (Foerst.),
[E], [Z], [RR], [Ay], [Bx], [Af 57-2],

OPUNTIA......(Cont'd),

sphaerica.....(cont'd),
-v- rauppiana (See Cumulopuntia rauppiana, O. rauppiana, Tephrocactus sphaericus -v- rauppianus).
-v- unguispina (See O. unguispina, Tephrocactus sphaericus -v- unguispinus),
sphaerocarpa (See O. polyacantha -v- juniperina),
[Y], [Df],
" -v- utahensis (See O. erinacea -v- utahensis),
spinibarbis (See Platyopuntia spinibarbis),
" -v- grandiflora.....
spinosibacca (See O. phaeacantha spinosibacca),
[Y], [Df], [Af 60-1-1988],
spinosissima (See Consolea corallicola, Consolea spinosissima, O. corallicola),
[TT], [Ax], [Bj],
spinosior (See Cylindropuntia spinosior),
[B], [M*], [P], [Q], [Y], [LL*], [RR], [Ax],
[Bn], [Df], [Af 7-7-1936], [Cu 1-1974],
spinulifera.....[RR], [Az], [Bo-1], [Hy], [Hl 6-1977],
spranguei (See O. tomentosa -v- spranguei).....[Cx],
stanlyi (See Corynopuntia stanlyi, Cylindropuntia stanlyi),
[B], [E], [P], [Q], [Y], [RR], [Ax], [Df],
" -v- kunzei (See Corynopuntia stanlyi -v- kunzei)
[P], [Q], [Ax],
" -v- parishii (See Corynopuntia stanlyi -v- parishii).....[LL*], [Ax], [Ca],
" -v- peeblesiana (See Corynopuntia stanlyi -v- peeblesiana).....[Q], [Ax],
" -v- stanlyi.....[LL*], [Ax],
" -v- wrightiana (See Corynopuntia stanlyi -v- wrightiana, Grusonia wrightiana, O. wrightiana),
[P], [Ax],
steiniana (See Austrocylindropuntia steiniana),
stenarthra (See O. canina, O. retrorsa, Platyopuntia retrorsa),
stenochila (See O. compressa -v- stenochila, O. macro-rhiza -v- macrorhiza)
[A], [RR], [Cf], [Df], [Hy*],
stenopetala.....[RR], [Ar], [Az], [Bo-1], [Gr], [Iu],
[Jd], [Ad April, 1979], [Aw 4-1963],
" -v- inerme.....
" -v- stenopetala.....
streptacantha.....[RR], [Az], [Bo-1], [Bv-1], [Ds],
[Fg], [Im], [Aj 33-5-1982], [Aw 3-1970],
stricta (See O. keyensis, O. macrarthra, O. magnifica)
[Y], [LL*], [RR], [Df], [Ds], [Hl 2-1970],
" -v- dillenii (See O. dillenii).....[Ax], [Az],

stricta.....(cont'd),
" -v- inermis (See O. stricta -v- stricta), [Bv-1],
" -v- stricta.....[Ax], [Df], [Af 53-4],
strigil.....[Y], [RR], [Ax], [Bo-1], [Df], [Af 5-6],
" -v- flexospina (See O. engelmannii -v- flexispina).....[Ax], [Df],
" -v- strigil.....[Ax*],
strobiliformis (See O. diademata -v- inermis, Tephrocactus strobiliformis, Tephrocactus articulatus -v- inermis), [E], [Ar], [Bv-1], [Cz], [Iz], [N-1980], [Cu 4-1965],
subarmata (See O. engelmannii -v- subarmata), [Bo-1], [Df],
subinermis (See Tephrocactus subinermis), [D], [Bx], [Cu 4-1965],
subferox (See O. leucacantha -v- laevior),
subsphaerocarpa.....[RR], [Dr],
subterranea (See Tephrocactus subterraneus), [G], [RR], [Iz], [Jd], [Bi 5-1],
subulata (See Austrocylindropuntia subulata, Cylindropuntia subulata), [C], [K], [AA], [EE], [RR], [Ax], [Ay], [Bv-1], [Cz], [Dq], [Gr], [Hv], [Il], [Jd], [N-1980], [Af 4-6], [Cu 2-1980],
" -v- exaltata (See Austrocylindropuntia exaltata),
" -f. cristata.....[Bv-1],
" -f. monstrosus (See Austrocylindropuntia subulata -f. monstrosus),
sulphurea (See O. maculacantha, O. pampeana, O. vulpina, Platyopuntia sulphurea), [RR], [Bo-1], [Gr], [Hx*], [Hy], [N-1980], [Af 51-4],
" -v- albisetosa.....[Hy],
" -v- hildmannii.....[Bo-1], [Bo-6],
" -v- pampana.....[Bo-1],
syringacantha (See O. diademata (Lem.), O. glomerata (sensu Br. & R.), Tephrocactus articulatus -v- syringacantha, Tephrocactus glomeratus (sensu SPEG.)),
tapona.....[RR],
tarapacana (See Tephrocactus tarapacana),
tardospina (See O. lindheimeri -v- lindheimeri), [T], [RR], [Bo-1], [Df], [Hz],
tayapayensis (See O. pascoensis, Platyopuntia nana), [Bo-1],
taylori.....[RR], [Bo-1], [Ct-1984], [Hx*],
tehuantepecana.....[Az],

OPUNTIA.....(Cont'd),

tenuiflora.....[Az],
tenuispina (See O. macrorhiza -v- pottsii),
[RR], [Bo-1], [Df], [Ie],
tephrocactoides (See Austrocylindropuntia tephrocactoides),
teres (See Austrocylindropuntia teres),
tesajo (See Cylindropuntia tesajo),
[Bv-1], [Iz], [Af 40-4], [Bi Vol. 4, #1, 1986],
" -v- cineracea (See Cylindropuntia tesajo -v- cineracea, O. cineracea),
tessellata (See Cylindropuntia ramosissima, O. ramosissima),
" -f. cristata.....[Ed 8:70 Pl. 1898],
tetracantha (See Cylindropuntia kleiniae -v- tetracantha, Cylindropuntia tetracantha, O. kleiniae -v- tetracantha),
[A], [RR], [Bv-1],
texana (See O. engelmannii -v- texana, O. lindheimeri -v- lindheimeri),
thurberi.....[RR], [Az],
" -v- alamosensis (See Cylindropuntia alamosensis, O. alamosensis),
" -v- thurberi.....
" -v- versicolor (See O. versicolor),
ticnamarensis (See Cumulopuntia ticnamarensis),
tilcarensis (See Platyopuntia soehrensii -v- tilcarensis).....[Bo-1],
" -v- rubellispina (Invalid name - See O. boliviensis, O. soehrensii, Platyopuntia soehrensii),
[Bo-6: 3617-1962 Description],
tomentella.....[RR*],
tomentosa.....[RR], [Ax], [Az], [Bc], [Bd], [Bo-1],
[Ca], [Cx], [Gr], [Hx*], [Cu 4-1960],
" -v- hernandezii (See O. hernandezii),
[Aw 2-1979],
" -v- rileyi (See Opuntia rileyi),
" -v- spranguei (See Opuntia spranguei),
" -v- typica.....
" -v- velutina (See O. velutina).....[Bv-1], [Cz]
tortispina (See O. macrorhiza -v- macrorhiza),
[A], [Q], [CC], [RR], [Bo-1],
" -v- cymochila (See O. cymochila, O. macrorhiza -v- macrorhiza),
tracyi (See O. pusilla, O. drummondii),
[RR], [Bv-1]
treleasei (See O. basilaris -v- treleasei),
[FF], [LL], [N-1980],
" -v- kernii (See O. basilaris -v- treleasei),
[Af 1-12],

triacantha (See O. militaris).....
[LL*], [RR], [Ax], [Bo-1], [Af 54-4],
[Da 1-1981],
trichophora (See O. polyacantha -v- trichophora),
[RR], [Go], [N-1980], [Af 5-9],
tricolor (See O. lindheimeri -v- tricolor),
[Df],
tuberculosirhopalica (See Micropuntia tuberculosirhopalica),
tuberosa (See Pterocactus tuberosus),
" -v- albispina (See Pterocactus tuberosus),
" -v- spinosior (See O. platyacantha (Pfeiff.) -v- gracilior (S.- D.)),
tubercularis (See Cumulopuntia berteri, Cumulopuntia tubercularis),
tumida (See Cumulopuntia tumida),
tuna.....[RR], [Bv-1], [Hb], [Hy], [Af 32-5], [Af 54-4],
[Aj 10-4-1959],
" -f. cristata.....[Bo-1], [Cm], [Af 32-5],
" -f. monstrosus.....[Aj 11-3-1960], [Ba 4-1976],
[Iv 5-1956],
tuna-blanca.....[Dr],
tunicata (See Cylindropuntia tunicata),
[C], [Y], [HH], [RR], [Ax], [Ay], [Az],
[Bo-1], [Bv-1], [Cj], [Cm], [Cx], [Df], [Fg],
[Gr], [Il], [Jd], [Aj 9-9-1958], [Am 9-1983],
" -v- aricensis (See Cylindropuntia tunicata -v- aricensis),
" -v- chilensis (See Cylindropuntia tunicata -v- chilensis),
" -v- davisii (See O. davisii).....[Ax], [Df],
" -v- tunicata.....[Ax],
turbinata.....
turgida (See O. humifusa -v- ammophila),
turpinii (See O. articulata -f. syringacantha, Tephrocactus articulatus -f. syringacanthus, Tephrocactus turpinii).....[E], [Bv-1], [Cz],
udonis (See Austrocylindropuntia floccosa, O. floccosa, Tephrocactus floccosus, Tephrocactus udonis).....[Ea 1933:71 Fig.],
undulata (See O. imbricata).....[Bo-1], [Cx], [Hz],
unguispina (See Cumulopuntia unguispina, Tephrocactus sphaericus -v- unguispinus),
[Fo 1937:7 gen. 10, Sp. 7, Tab.],
urbaniana (See Consolea urbaniana),
ursina (See O. erinacea -v- ursina, O. hystricina -v- ursina).....[NN], [Cb],
utahensis (See O. macrorhiza -v- macroriza),
utkilio.....[RR], [Dr],
variacantha.....[Iv 7-1956],

OPUNTIA.....(Cont'd),

vaseyi (See O. littoralis -v- vaseyi).....[P], [FF], [LL], [RR], [Bo-1], [Cb], [Hy*], [Af 2-8],
velutina.....[RR], [Az], [Bo-1], [Cx], [Hy],
" -v- affinis (See O. affinis),
" -v- macdougaliana (See O. macdougaliana).[Az],
" -v- velutina.....[Az],
verschaffeltii (See Austrocylindropuntia verschaffeltii, Cylindropuntia verschaffeltii)..[C], [H], [L], [RR], [Bv-1], [Cz], [Gr], [Jd],
" -v- digitalis (See Austrocylindropuntia verschaffeltii -v- digitalis)...[DD],
" -v- hypsophila (See Austrocylindropuntia verschaffeltii -v- hypsophila),
" -v- longispina (See Austrocylindropuntia verschaffeltii -v- longispina),
versicolor (See Cylindropuntia versicolor), [B], [M*], [P], [Q], [LL*], [RR], [Ax], [Az], [Bn], [Cx], [Gr], [Ad Oct. 1980], [Af 2-12], [Af 28-4],
verticosa (See Austrocylindropuntia floccosa, Opuntia floccosa, Tephrocactus verticosus, Tephrocactus floccosus),
vestita (See Austrocylindropuntia vestita, Cylindropuntia vestita).....[C], [H], [L], [R], [NN], [RR], [Cm], [Cz], [Af 4-6], [Af 14-4], [Af 39-1], [Aj 29-6-1978 - Flower], [Au 35-2],
" -v- chuquisacana (See Austrocylindropuntia vestita -v- chuquisacana),
" -f. cristata (See Austrocylindropuntia vestita -f. cristata)..[Bv-1], [Aj 23-4-1972],
vexans (See O. imbricata -v- imbricata).....[Hz],
vilis (See Corynopuntia vilis), [RR], [Az], [Bv-1], [Co], [Cz],
" -v- bernhardinii (See Corynopuntia vilis -v- bernhardinii),
violacea (See O. macrocentra).....[M], [Q], [AA], [Az], [Aw 1-1981],
" -v- castetteri (See O. macrocentra -v- macrocentra).....[Ib], [Af 57-3], [Af 58-4],
" -v- gosseliniana (See O. chlorotica -v- gosseliniana, O. gosseliniana)..... [Ax], [Az], [Ij],
" -v- macrocentra (See O. macrocentra -v- macrocentra).....[Q], [U], [An], [Ax], [Az], [Df], [Ib], [Ij],
" -v- santa-rita (See O. chlorotica -v- santa-rita, O. gosseliniana -v-santa-rita, O. santa-rita).....[Q], [U], [An], [Ax], [Az], [Df], [Ib], [Ij], [Iz],

violacea.....(cont'd),
" -v- violacea (See O. macrocentra -v- macrocentra.....[Q], [An], [Ax], [Az], [Ij],
virgulta (virgulata) (See Tephrocactus virgultus),
viridiflora (See O. imbricata -v- viridiflora, O. whipplei -v- viridiflora).....[RR],
viridirubra (See Platyopuntia viridirubra),
vitelliniflora (See Platyopuntia vitelliniflora),
vivipara (See Cylindropuntia arbuscula, Cylindropuntia vivipara, O. arbuscula).....[A], [Q], [RR],
vulgaris (See O. monacantha, Platyopuntia vulgaris), [UU], [Ar], [Ax], [Bo-1], [Bv-1], [Cz], [Gr], [Hx*], [Af 1-9],
" -v- lemaireana (See O. lemaireana).....[Bo-1],
" -v- variegata.....[Ay], [Bo-1], [Bv-1], [Cz],
vulpina (See O. maculacantha, O. pampeana, O. sulphurea).....[Dr],
wagenknechtii.....
weberi (See Tephrocactus weberi).....[EE], [RR],
" -v- dispar (See Tephrocactus weberi -v- dispar), [Fp 1:5 Fig.],
" -v- setiger (See O. setigera, Tephrocactus setiger, Tephrocactus weberi -v- setiger), [Au 13:25, 1958],
weingartiana (See Austrocylindropuntia weingartiana), [Aj 3-1937], [Cu 4-1980],
wentiana.....[Af 54-4], [Af 57-3],
wetmoorei (See Tephrocactus glomeratus -v- andicola), [RR], [Bo-1], [Af 4-7],
whipplei (See Cylindropuntia whipplei).....[B], [P], [Q], [Y], [LL*], [RR], [Cf], [Df], [Ie], [Af 60-5],
" -v- multigeniculata (See Cylindropuntia multigeniculata, Cylindropuntia whipplei -v- multigeniculata, O. multigeniculata).....[Ax],
" -v- viridiflora (See Cylindropuntia viridiflora, O. imbricata -v- viridiflora, O. viridiflora).....[Df],
" -v- whipplei (See Cylindropuntia whipplei, Cylindropuntia hualpaensis, O. hualpaensis).....[Ax],
whitneyana (See O. basilaris -v- basilaris),
" -v- albiflora.....[FF],
wiegandii (See Micropuntia wiegandii),
wigginsii (See Cylindropuntia wigginsii), [Ax], [Az], [Ca],
wilcoxii.....[RR], [Az], [Af 16-4],
wilkeana (See Tephrocactus wilkeanus)...[Au 13:6, 1958]

OPUNTIA.....(Cont'd),

winteriana (See O. lindheimeri -v- lindheimeri),
[Bo-1],
woodsii (Syn. O. phaeacantha -v- major),
wootonii (See O. phaeacantha -v- wootonii).....[Hz],
wrightiana (See Corynopuntia stanlyi -v- wrightiana,
Grusonia wrightiana, O. stanlyi -v- wright-
iana).....[Cb], [N-1980], [Af 35-3],
yanganucensis (See Tephrocactus yanganucensis),
[Au 13:6, 1958],
xanthoglochia (See O. humifusa -v- humifusa).....[Hz],
xanthostemma (See O. erinacea -v- utahensis, O. rhod-
antha).....[Aj 33-4-1982],
xerocarpa (Hybrid) (See O. erinacea -v- erinacea),
zacana (See O. galapageia -v- zacana),
[Af 5-9], [Af 37-5], [Af 39-5], [Aj 18-8-1967],
zacuapanensis.....[RR],
zebrina (See O. stricta -v- dillenii),
[RR], [Af 32-4],
zehnderi (See Cumulopuntia zehnderi, Tephrocactus
zehnderi).....[F],
Sp. .....[Ba 5-1976],
Sp. O. aff. horstii.....[Af 60-4],

OREOCEREUS

bruennowii (See O. celsianus -v- bruennowii, O. neo-
celsianus -v- bruennowii).....[Hy],
celsianus (See Borzicactus celsianus, O. neocelsianus)
[D], [H], [T], [Z], [AA], [RR], [Bv-1], [Cj],
[Cr], [Cz], [Dq], [Gr], [Hi-2], [Hu], [Hx],
[Il], [Jd], [N-1980], [Af 15-7], [Af 29-3],
[Am 2-1979], [Hl 2-1969], [Hw-1988],
" -v- bruennowii (See O. bruennowii),
[L], [T], [CC], [Iu],
" -v- celsianus.....[Hv],
" -v- fossulatus (See O. fossulatus)...[L], [Gr],
" -v- gracilior.....
" -v- hendriksenianus (See O. hendriksenianus),
[Gr],
" -v- lanuginosior (See Cereus celsianus -v- lan-
uginosior),
" -v- maximus (See O. celsianus, O. maximus),
[Am 8-1979],
" -v- ritteri (See O. ritteri).....[Hv],
" -v- ruficeps.....[F],
" -v- trollii (See O. trollii),
" -v- variicolor (See O. variicolor),
" -v- villosissimus.....[F],
crassiniveus (See O. trollii -v- crassiniveus),
[Eh 21:9, 1949],

OREOCEREUS.....(Cont'd),

churinensis.....
culpinensis.....
densilanatus (See Borzicactus hendriksenianus -v- densilanatus, O. hendricksenianus -v- densilanatus),
doelzianus (See Morawetzia doelzianus),
[C], [PP], [Bv-1], [Gr] [Ha], [Hv],
[Ad April, 1978], [Af 15-7], [Af 29-3],
" -v- calvus.....
fossulatus (See O. celsianus -v- fossulatus),
[D], [CC], [Bl], [Bo-2], [Bv-1], [Dq], [Hv],
[Hx*], [Hy], [Il], [Af 5-3], [Flower - Aj - 30-6-1979],
" -v- brunispinus.....
" -v- calvus.....
" -v- crassispinus.....
" -v- curvispinus.....
" -v- flavispinus.....[Hu],
" -v- gracilior (Invalid name).....
[Bo-2], [Bt], [Hx*], [Hy],
" -v- lanuginosior (See Cereus celsianus -v- lanuginosior),
" -v- pilosior (See O. fossulatus),
" -v- pireacantha.....
" -v- potosinus.....
" -v- robustior.....
" -v- rubrispinus.....[Bo-2], [Bv-1], [Hi-2],
[Hy], [Bi 3-3],
" -v- spinus-aureis.....
" -f. cristata.....[Ar].
hendriksenianus (See O. celsianus -v- hendriksenianus),
[T], [NN], [QQ], [Av], [Bf], [Bo-2],
[Bu], [Hv], [Hx], [Hy], [Jd], [Af 52-1], [Af 52-2],
" -v- brunispinus.....
" -v- densilanatus (See O. ritteri),
[X], [Ay], [Bo-2], [Bt], [Af 50-1]
[Af 52-1], [Af 52-3], [Af 55-2],
[Af 56-5], [Ah-1985],
" -v- gracilior.....
" -v- irigogenii (See O. irigogenii, O. trollii],
" -v- niger.....
" -v- spinosissimus (See O. ritteri).....[Bo-2],
" -f. cristata.....[Ba 2-1974],
horridospinus.....
irigogenii (See O. hendriksenianus -v- irigogenii, O. trollii).....[Hr],
knizei (See Morawetzia variicolor),
lanatus (See Espostoa lanata),
lecoriensis.....

OREOCEREUS.....(Cont'd),

leucotrichus (See Borzicactus hendriksenianus, O. celsianus -v- hendriksenianus, O. hendriksenianus).....[Hi Vol. 3], [Hw-1989],
luribayensis.....[Bi 3-3],
lurinensis.....
magnificus.....
maximus (See O. celsianus, O. celsianus -v- maximus), [V], [Bo-2], [Bv-1], [Hb],
" -v- albilanatus.....
" -v- crassiniveus.....
" -v- crassirubrus.....
" -v- crassispinus.....
neocelsianus (See O. celsianus), [T], [Ar], [Av], [Ay], [Bo-2], [Bt], [Hb], [Hx*], [Bi 3-2],
" -v- atrihorridispinus.....
" -v- bruennowii (See O. bruennowii, O. celsianus -v- bruennowii).....[T],
" -v- ruficeps.....[Hy],
" -v- villosissimus (See O. celsianus -v- villosissimus).....[Hy],
piscoensis (See Borzicactus piscoensis, Loxanthocereus piscoensis),
rettigii (See Arequipa rettigii).....[Jd],
ritteri (See O. hendriksenianus -v- densilanatus), [F], [H], [NN], [Bv-1], [Bv-2], [Hi-4], [Hx], [Aj 33-6-1982], [Au 37-1], [Au Vol. 2, 1977, Zone 3 Compendium),
sericatus (See Morawetzia sericata).....[Ad -],
spinosissimus (See O. hendriksenianus -v- spinosissimus)
tacnaensis (See O. variicolor -v- tacnaensis), [Hi Vol. 4],
trollii (See Borzicactus trollii).....[C], [H], [L], [T], [X], [AA], [DD], [Ar], [Av], [Bo-2], [Bt], [Cg], [Ck], [Cr], [Hb], [Hi-2], [Hx], [Hy], [Il], [Jd], [N-1980], [Af 51-5], [Af 59-1], [Ah-1985], [Hw-1982], [Hw-1989],
" -v- crassiniveus (See O. trollii).....[Bo-2],
" -v- major.....
" -v- tenuior (See O. trollii).....[Bo-2],
urmiriensis.....
variicolor (See Arequipa variicolor, O. celsianus -v- variicolor)...[Bo-6], [Hi-3], [Aj 22-12-1971],
" -v- tacnaensis (See O. tacnaensis), [Ay], [Bo-6], [Hl 4-1971],

OROYA

acollana (See Oroya peruviana),
" -v- fuscata.....
" -v- luteispina.....

OROYA.....(Cont'd),

baumannii (See O. neoperuviana -v- laxiareolata),(O. peruviana -v- pluricentralis - rejected name), [Ba 26:2, 49, 1971],
" -v- rubrispina (See O. neoperuviana -v- laxiareolata),
borchersii.....[NN], [Ar], [Ay], [Bo-3], [Hi-4], [Hv], [Hx], [Hy], [Aj 9-12-1958], [Aj 15-3-1964], [Am 8-1979], [Ba 4-1981],
" -v- ferruginea.....
" -v- fuscata.....[G], [Bo-3],
" -v- fuscispina (See O. borchersii),
" X Matucana yanganucensis (Hybrid).....[Hi-4],
" -f. cristata.....[Af 50-6],
citriflora (See O. gibbosa -v- citriflora, O. peruviana -v- depressa),
depressa (See O. peruviana -v- depressa), [Iv 7-1956],
gibbosa.....[Bo-3], [Hi-4], [Hx*], [Aj 9-12-1958],
" -v- citriflora (See O. peruviana -v- depressa),
gigantea (KZ-387) (See O. neoperuviana -v- laxiareolata),
laxiareolata (See O. peruviana), [G], [Ar], [Bo-3], [Ad -], [Hx*], [Aj - 9-12-1958], [Aj 10-6-1959], [Hl 4-1968],
" -v- orthorachis (See O. neoperuviana -v- laxiareolata),
" -v- pluricentralis (Invalid name - See O. peruviana -v- pluricentralis), [DD],
lutea (See O. borchersii).....
neoperuviana (See O. peruviana).....[Ay], [Bo-3], [Bv-1], [Gr], [Hu], [Hx*], [Hy], [Af 60-6], [Ah-1985], [Aj 9-12-1958], [Da 2-1980], [Hh May, 1975], [Hw-1982],
" -v- depressa (See O. peruviana).....[Bo-3],
" -v- ferruginea (See O. peruviana).....
" -v- laxiareolata (See O. laxiareolata, O. peruviana),
" -v- subocculta (See O. subocculta), [Hh May, 1975],
" -v- tarmensis.....
" -v- tenuispina (See O. peruviana),
peruviana (See O. laxiareolata, O. neoperuviana, O. subgibbosa), [G], [I], [NN], [QQ], [Ar], [Ay], [Bo-3], [Bv-1], [Cr], [Dq], [Hi-4], [Hv], [Hx*], [Hy], [Il], [Iu], [Iz], [Ad Sept. 1980], [Ae-1983], [Af 53-6], [Ag-1982], [Aj 9-12-1958], [Hh May, 1975], [Hl 4-1972], [Hw-1988],
" -v- conaikensis.....[Af 51-4], [Hh May, 1975],

OROYA.....(Cont'd),

peruviana.....(cont'd),
" -v- depressa (See O. peruviana).....[UU],
" -v- densiaurea.....
" -v- ferruginea (See O. peruviana),
" -v- laxiareolata (See O. laxiareolata, O. peruviana).....[Jd],
" -v- neoperuviana (See O. peruviana), [Ba 4-1981],
" -v- pluricentralis [See O. laxiareolata -v- pluricentralis), [Hi Vol. 4],
" -v- subocculta (See O. neoperuviana -v- subocculta, O. subocculta), O. peruviana),
" -v- tenuispina (See O. peruviana),
subgibbosa (See O. peruviana).....[Af 51-4],
" -v- citriflora (See O. peruviana -v- depressa),
subocculta (See O. peruviana), [Ar], [Bo-3], [Aj 9-12-1958], [Hl 2-1971], [Iv 3-1960],
" -v- albispina (See O. neoperuviana -v- subocculta, O. subocculta), [Bo-3],
" -v- fusca (See O. subocculta),
" -v- multicostata (See O. subocculta),

ORTEGOCACTUS

macdougallii (See Escobaria macdougallii, Neobesseya macdougallii), [Z], [DD], [Ar], [Av], [Ay], [Bo-6], [Bt], [Il], [Jd], [Ae-1983], [Af 33-2], [Af 54-4], [Af 58-6], [Ag-1987], [Aj 31-8-1980], [Aj - 34-2-1983], [Bi 4-1985], [Hl 4-1974], [Hm 1-1979],

PACHGEROCEREUS

orcuttii.....[Az], [Aw 3-1963],

PACHYCEREUS

calvus (See P. pringlei).....[Bo-4], [Af 2-4],
chrysomallus (See Mitrocereus chrysomallus, Mitrocereus fulviceps, Pseudomitrocereus chrysomallus).....[RR], [Az], [Bv-1], [Cx], [Jd], [Af 6-3], [Af 6-4],
columna-trajani sensu Br. & R. (See Haseltonia columnatrajani).....[RR], [Iu],

PACHYCEREUS

columna-trajani sensu Daws. (See Mitrocereus chrysomallus, Mitrocereus fulviceps, Pachycereus chrysomallus).....[Bj],
gaumeri (See Anisocereus gaumeri, Pterocereus gaumeri) [Dq], [Af 8-7], [Cu 2-1962],
gigas.....[Bo-4],
grandis (See Pachycereus weberi, Stenocereus weberi), [Ax*], [Ay], [Az], [Bl], [Bo-4], [Cx], [Ad Feb. 1978],
hollianus (See Lemaireocereus hollianus), [JJ], [Az], [Af 51-4], [Aj 31-9-1980], [Aw 1-1970], [Ba 5-1980],
" -f. cristata.....
lepidanthus (See Anisocereus lepidanthus, Escontria lepidantha).....[Af 37-3],
marginatus (See Cereus marginatus, Lemaireocereus marginatus, Marginatocereus marginatus, Stenocereus marginatus), [C], [W], [X], [RR], [Ay], [Bv-2], [Cx], [Il], [Jd], [Af 4-9], [Hw-1989],
" -v- gemmatus.....[Ij],
militaris (See Backebergia militaris, Cephalocereus chrysomallus, Mitrocereus militaris, Pilosocereus chrysomallus),
orcuttii (See Pachgerocereus orcuttii), [Bo-4], [Hy], [Af 22-4],
pectin-aboriginum.....[G], [RR], [UU], [Ar], [Ay], [Az], [Bo-4], [Bo-6, Flower], [Bv-1], [Cx], [Hx], [Hy], [Il], [Af 51-6], [Af 57-4], [Aj 27-8-1976], [Aw 2-1970], [Cu 3-4-1935], [Dg 4, 1-4, 1982],
pringlei (See Cereus pringlei), [G], [H], [CC], [DD], [HH], [RR], [UU], [Ay], [Az], [Bo-4], [Bv-1], [Bt], [Cx], [Cz], [Gr], [Ha], [Hx*], [Hy], [Il], [Im], [Iz], [Jd], [Ad Jan., 1978], [Af 3-6], [Af 6-11], [Af 61-1], [Aj 27-8-1976], [Aw 3-1962],
" -f. cristata.....[Ay], [Dp], [Hy], [Af 34-6], [Aw 1-1962],
" -f. monstrosus (spiral).....[Aw 4-1963],
queretaroensis (See Cereus queretaroensis, Lemaireocereus queretaroensis, Ritterocereus queretaroensis, Stenocereus queretaroensis)
ruficeps (See Mitrocereus ruficeps, Neobuxbaumia macrocephala).....[Im], [Iv 3-1961],
schottii (See Cereus schottii, Lophocereus schottii),
tehuantepecanus (See P. pectin-aboriginum)...[Af 40-5],
tetetzo (See Cephalocereus tetetzo, Cereus tetetzo, Neobuxbaumia tetetzo),

PACHYCEREUS.....(Cont'd),

thurberi (See Cereus thurberi, Lemaireocereus thurberi, Marshallocereus thurberi, Stenocereus thurberi).....[Aw 3-1962],
titan (See Pachycereus pringlei),
weberi (See Lemaireocereus weberi, Stenocereus weberi), [NN], [Ar], [Av], [Bo-4], [Bt], [Hx], [Hy], [Ij] [Aj 27-10-1976], [Aw 1-1980], [Ba 2-1980], [Bi 4-1986],

PARODIA

agasta (See P. roseoalba -v- australis).
aglaisma (See P. subtilihamata).....

alacriportana (See Brasiliparodia alacriportana, Notocactus alacriportanus, Notocactus megapotamicus -v- alacriportanus), Bo-3], [Am 8-1980],
albo-fuscata .....
amambayensis (See P. paraguayensis),
amblayensis (See P. hummeliana).....[Jf],
amblayoensis (See P. DH 101),
andreae (See P. subtilihamata), [DD-497], [Hx*], [Jf], [Aj 24-7-1973], [Hl 5-1973],
andreaeoides (See P. subtilihamata),
applanata (See P. schwebsiana),
argerichiana.....[Jf],
atroviridis (See P. dichroacantha).....[Hx*], [Aj 19-1-1968],
aureicentra.....[Bo-3], [Jf], [Af 27-4], [Af 51-5], [Af 55-1], [Aj 33-12-1982],
" -v- albifusca.....[Hi-2],
" -v- aureicentra.....[Hi-2],
" -v- lateritia.....
" -v- omniaurea (See P. omniaurea), [Hi-2],
" -v- rauschii (See P. aureicentra -v- omniaurea, P. rauschii),
aureispina.....[E], [H], [BB], [DD-499], [Bo-3], [Bv-1], [Bv-3], [Cr], [Cz], [Dq], [Hb], [Hu], [Hy], [Ig], [Ja], [N-1980], [Ad-], [Af 33-6], [Af 50-1],
" -v- australis (See P. microsperma -v- erythrantha).....[DD-498], [Gr], [Hx*], [Aj 20-9-1969],
" -v- elegans (See P. elegans).....[Aj 19-1-1968]
" -v- erythrantha (See P. microsperma -v- erythrantha), P. aureispina -v- australis),

PARODIA.....(Cont'd),

aureispina.....(cont'd),
" -v- erythrostaminea (See P. erythrostaminea),
" -v- grandiflora.....[Hy],
" -v- rubriflora (See P. rubriflora)......
[DD-498],
" -v- scopaoides (See P. scopaoides),
" -v- vulgaris.....[DD-499], [Aj 20-9-1969],
" -v- (Nr FR-922).....[Hi-2],
" -f. cristata.....[Hy], [Am 2-1983],
aurihamata (See P. mutabilis).....[Ar],
avilezii.....[Hl 5-1979],
axiosa.....[Hl 5-1979],
ayopayana.....[Ar], [Bo-3], [Hi-2], [Hx], [Jd], [Jf],
[N-1980], [Af 23-3], [Aj 10-2-1959],
[Hl 2-1980],
" -v- buxbaumiana.....
" -v- elata (Kultur).....[Aj 16-4-1965],
backebergiana ....[DD-500], [Bm-1986],
[Hl 3-1973], [Hl 3-1980],
belliata.....
bermejoensis.....
betaniana.....[Hi-2], [Jf],
bilbaoensis.....[Gr], [Hx*], [Af 38-4], [Aj 26-1-1975],
blanc.....[Hl 2-1979],
borealis.....[Hi-2], [Aj 16-4-1965], [Hl 2-1980],
brasiliensis (See Notocactus brasiliensis).....[Dr],
brevihamata (See Brasiliparodia brevihamata, Notocactus brevihamatus).....[NN], [Bo-3],

buenekeri (See Brasiliparodia buenekeri, Notocactus buenekeri),
[R], [DD], [Hu], [Ah-1986], [Aj 16-4-1965],
buiningiana (See P. fechseri),
buxbaumiana (See P. ayopayana -v- buxbaumiana),.....
cafayatensis (See P. microsperma -v- cafayatensis),
caineana.....
camarapana (comarapana).....[Gz No. 6, 24, 1951],

camargensis.....[R], [Hi-2], [Jf], [Aj 27-11-1976],
" -v- camblayana (See P. camblayana),
[Bo-6], [Bu], [Hi-2], [Iz],
" -v- castanea (See P. castanea,.........
[Ba 1962:21],
" -v- prolifera (See P. prolifera),
[DD], [Hi Vol. 2], [Jf],
" -v- rubra.....

PARODIA.....(Cont'd),

camblayana (See P. camargensis -v- camblayana),
[Av],
" -f. rubra.....
campestrae.....[Hx*], [Jf],
candidata.....
capillitaensis (See P. spegazziniana).....[Jf],
carapariana.....[Jf],
cardenasii .....[Hi-2],
" -v- maior.....
carminata (See P. tilcarensis),
[Bo-3], [Hx*], [Hl 2-1965],
carrerana (See P. rosealba),.....
[Gs #78-94, 1963],
castanea (See P. camargensis -v- castanea,
[Hi-2],
catamarcensis.....[H], [V], [DD], [Bo-3], [Hx*], [Jd],
[Jf], [Af 56-1], [Aj 24-5-1973], [Aj 29-6-1978], [Aj 32-5-1981], [Cu 2-1962],
" -v- rubriflorens (See P. lohaniana),
centrispina .....
" -v- variabilis.....
chaetocarpa .....[Hi-2],
challamarcana (See P. procera, P. pseudoprocea),
[DD-501], [Hx*],
chirimoyarana.....[Af 50-1],
chlorocarpa (See P. rubellihamata -v- chlorocarpa),
[Hi-2],
chrysacanthion.....[C], [H], [L], [S], [T], [EE], [NN],
[Ar], [Bo-3], [Bv-1], [Hi-2], [Hu],
[Hv], [Hx], [Il], [Iz], [Jd], [Jf],
[Af 50-2], [Ag-1985], [Ah-1981], [Au Vol.2, 1977 Zone 3 Compendium),
[Hw-1986],
chrysocoma (See Brasiliparodia chrysocoma),
cintiensis (See P. ritteri -v- cintiensis),
[Hi-2], [Jf], [Ba 1962:122],
columnaris (See P. legitima).....[Hi-2], [Jf],
[Af 23-3], [Aj 10-1-1959],
[Aj 11-1-1960], [Hl 2-1980],
" -v- legitima (See P. legitima),
" -v- ochraceiflora.....[Hi-2],
comarapana.....[Hi-2], [Hx], [Jd], [Jf], [Af 28-5],
[Aj 11-1-1960],
" -v- neglecta (See P. neglecta),
" -v- paucicostata (See P. neglectoides)..[Hi-2],
comata (See P. sanguiniflora -v- comata),
commutans .....[Hi-2], [Hx*], [Jf], [Hl 5-1979],
comosa ......[Bo-6], [Hi-2], [Hu], [Hx*], [Gt #75, 22, 1962],
compressa.....[Hi-2], [Jf], [Aj 15-6-1964],

PARODIA.....(Cont'd),

cotacajensis.....
cruci-nigricentra (P. faustiana -v- cruci-nigricentra), [Aj 16-6-1965],
culpinensis.....[DD-502], [NN], [Bo-6], [Hx*], [Jf], [Aj 16-6-1965], [Hl 5-1969],
dextrohamata (See P. rigida)..... [DD], [Aj 24-5-1973],
" -v- stenopetala......[DD],
dichroacantha.....[Hx*], [Jf], [Aj 32-9-1981],
echinopsoides (See P. tredecimcostata),
echinus.....[K], [DD], [Av], [Bo-6], [Hi-2], [Hx], [Jf], [Af 53-1],
elachista (See P. zecheri),.....
elata (See P. ayopayana -v- elata),.....
elegans (See P. aureispina -v- elegans).....[DD],
erythrantha (See P. microsperma -v- erythrantha), [I], [Bo-3], [Hx*], [Hy], [Aj 11-1-1960], [Aj 24-5-1973], [Ba 7-1980], [Cu 3-1961],
" -v- thionantha (See P. thionantha).....[DD],
erythrostaminea (See P. aureispina -v- erythrostaminea)
escayachensis (See P. maassii).....[Aj 21-4-1970],
exquisita.....
faustiana.....[K], [Bo-3], [Hi Vol. 2], [Hr], [Hx], [Jf], [Aj 11-1-1960],
" -v- cruci-nigricentra (See P. cruci-nigricentra)
" -v- tenuispina (See P. tenuispina).....[Bo-3],
fechseri (Invalid name - See P. buiningiana), [DD], [Hx*], [Jf], [Aj 21-7-1970], [Hl 3-1979],
firmissima.....
formosa (See P. pachysa).....[Hi-2], [Hx*], [Af 56-6], [Aj 16-6-1965],
" -v- prolifera.....[Ag-1985],
friciana (See P. gigantea, P. stuemeri -v- robustior), [DD 503], [Ar], [Hx*], [Hl 5-1973], [Hl 1-1980],
fulvispina .....[Jf], [Aj 16-6-1965], [Aj 21-4-1970],
" -v- brevihamata (See P. brevihamata)...... [Hi-2],
fuscato-viridis........[DD], [Aj 22-11-1971],
gibbulosa (See P. gibbulosoides).....[Hi-2], [Hy],
gibbulosoides....[DD-504], [Hx*], [Jf], [Af 56-4], [Hl 5-1973],
gigantea (See Microspermia gigantea), [Aj 11-1-1960], [Aj 16-7-1965], [Hl 1-1980],

glenosa.....[Hl 5-1979],

glischocarpa (See P. uebelmanniana), [DD], [Hi Vol. 2], [Aj 24-6-1973], [Aj 24-8-1973], [Hl 3-1979],

PARODIA.....(Cont'd),
gokrauseana (See P. tilcarensis),
gracilis.....[Ar], [Hi-2], [Hx], [Jf], [Aj 15-3-1964], [Hl 5-1970],
gummifera (See Uebelmannia gummifera).....[Bo-3],
gutekunstiana (See P. stuemeri), [Gv iii:1604, 1959],
haageana.....
haemantha (See P. sanguiniflora)..... [Hy],
" -f. cristata.....[Hy],
hamata.....[Hl 4-1980],
hausteiniana.....[Hx*], [Af 57-2], [Aj 21-3-1970],
herzogii ......[Jf], [Aj 2-1981],
heteracantha.....[Jf],
higueritas (See P. weberiana -v- higueritas), [Hl 3-1979],
horrida.....[Jf], [Db 10-1979],
hummeliana ..... [Jf], [Aj 10-1978],
idiosa (See P. prestoensis)..... [Jf], [Aj 6-1982],
ignorata.....[DD-505], [Hx*], [Hl 5-1973],
intermedia (See P. maassii -v- intermedia),
jujuyana (See P. faustiana).....
kilianana ....[Jf], [DD-506],
" -v- rigida (See P. rigida),
knizei.....
koehresiana.....[DD-507], [Hx*], [Hl 5-1973],
krahni (See P. taratensis).....[Jf],
krasuckana .....[DD-508], [Jf], [Aj 23-7-1972],
lamprospina (See P. maassii -v- albescens),
lauii (See P. hausteniana).....[DD-509], [Jf], [Ae-1981], [Aj-24-11-1973], [K & L Nursery Catalog -1988],
legitima (See P. columnaris -v- legitima).....[Hx*],
lohaniana.....[Jf], [Aj 6-1979],
" -v- rubriflorens.....
lychnosa (See P. gracilis).....[Hx*],
maassii (See Echinocactus maassii, Malacocarpus maassii, Wigginsia maassii), [C], [M*], [XX], [Ar], [Ay], [Bo-3], [Bv-1], [Cz], [Gr], [Hb], [Hu], [Hx*], [Hy], [Iz], [Jd], [Af 16-11], [Af 51-5], [Aj 11-1-1960], [Hl 5-1971], [Hi 5-1975],
" -v- albescens (See P. lamprospina), [Hi-2], [Aj 21-4-1970], [Hl 6-1982],

" -v- atroviridis.....[Hl 4-1979],
" -v- camargensis.....(See P. camargensis),
" -f. castanea (See P. castanea),
" -v- carminatiflora.....[Hi-2], [Ba 1963:179],

maassii .....(cont'd),
" -v- camargensis (See P. camargensis),
" -v- " -f. castanea (See P. castanea),
" -v- carminatiflora.....[Hi-2], [Ba 1963:179],
" -v- commutans (See P. commutans),
" -v- " -f. maxima (See P. maxima),
" -v- distortihamata (See P. maassii -v- shaferi -f. distortihamata), [Hl 4-1979],
" -v- intermedia (See P. intermedia), [Hi-2], [Jf], [Ba 1963:179],
" -v- maasii.....[Hi-2], [Hl 4-1979],
" -v- rectispina.....[Af 23:84, 1951],
" -v- rubida (See P. rubida),
" -v- rubrispina.....[Hy],
" -v- shaferi.....[Hi-2],
" -v- " -f. distortihamata (See P. maassii -v- distortihamata), [Hi-2],
" -v- subterranea (See P. subterranea).....[Hv],
" -v- suprema (See P. suprema),

macednosa (See P. gigantea),
macrancistra (See P. microsperma -v- macrancistra), [Hr], [Jf], [Hw-1981],

mairanana.....[NN], [Ar], [Bt], [Hi-2], [Hx], [Hy], [Il], [Jd], [Jf], [Ad-], [Aj 11-1-1960], [Aj 16-7-1965],
" -v- atra......[DD], [Hv], [Aj 16-7-1965],

malyana.....[Hx], [Iz], [Jf], [Aj 32-5-1981], [Aj 33-3-1982],

" -v- rubriflora (See P. rubriflora),

matthesiana.....[Hx*],
maxima (See P. maassii -v- commutans -f. maxima), [NN], [Hi-2], [Hx*], [Hy], [Jf], [Ae-1981], [Af 52-5], [Aj 16-7-1965], [Hh Sept., 1973], [Hl 1-1972],

mendezana (See P. maassii -v- intermedia, and KK 835, KK 1004)
mercedesiana.....[Jf],

PARODIA.....(Cont'd),

mesembrina.....(See P. fechseri),
microsperma (See Hickenia microsperma, Microspermia microsperma),
[F], [Ay], [Be*], [Bv-1], [Cr], [Dr], [Hi-2], [Hx], [Hy], [Af 16-11], [Af 56-2], [Af 56-3], [Flower - Ba 10-1980],
" -v- aurantiaca.....

" -v- cafayatensis (Invalid name - See P. cafayatensis, P. tuberculosa-costata -v- cafayatensis).....[DD],
" -v- erythrantha (See P. aureispina -v- australis, P. erythrantha),
[Dr],
" -v- gigantea (See P. gigantea),
" -v- macrancistra (See P. macrancistra),
[Bo-3], [Hl 2-1979],
" -v- microthele (See P. microthele),
" -v- opulenta.....
" -v- rigidissima (See P. rigidissima),
" -v- rubriflora (See P. sanguiniflora).....
" -v- thionantha (See P. thionantha).....[Hy],
" -f. cristata.....[Bl],
microthele (See P. microsperma -v- microthele),
[QQ], [Bo-3], [Hx], [Aj 11-1-1960], [Db 12-1980], [Hl 2-1979],

miguillensis.....[Jf], [Af 33-4], [Hl 2-1980],
minima (See P. minuscula),
minuscula (See P. minima).....[Jf],,
minuta.....

miranda.....
muhrii.... [Jf], [Aj 33-12-1982],
multicostata .....[Jf], [Hi-2], [Hy],
mutabilis .....[H], [M*], [T], [EE], [UU], [Ar], [Bo-3], [Bt], [Bv-1], [Cs], [Cy], [Hu], [Hv], [Hx], [Hy], [Il], [Jd], [Jf], [Af 33-6], [Ba 10-1977], [Ba 12-1977], [Hl 2-1979],
" -v- carneospina.....[Bo-3], [Cr], [Ig*],
" -v- elegans (See P. elegans),
" -v- ferruginea.....[Bv-1],

neglecta (See P. comarapana -v- neglecta),
[DD-510], [Jf], [Aj 24-3-1973], [Hl 5-1973],

neglectoides (See P. comarapana -v- paucicostata),
[DD-511],
negresca.....

PARODIA.....(Cont'd),

nivosa (See Microspermia nivosa),
[J], [Ar], [Bo-3], [Ck], [Hx], [Hy], [Il], [Iz],
[Jd], [Jf], [Aj 11-1-1960], [Ba 7-1980],
" -v- striatipetala.....[Hy],

obtusa.....[Gr], [Hi-2], [Aj 21-4-1970], [Aj 21-11-1970]
[Hl 5-1979],
ocampoi.....[Hi-2], [Jf], [Aj 6-1-1955], [Aj 11-1-1960],
[Aj 21-4-1970], [Hl 3-1980],
" -v- compressa (See P. compressa),

occulta (See P. suprema)..........[Hi-2],
omniaurea (See P. aureicentra -v- omniaurea),

otaviana .....[Jf], [Gs #78, 95, 1963],
otuyensis.....[DD], [Hi-2], [Hu], [Hx*], [Jf],
[Aj 16-1-1965], [Aj 21-10-1970],
[Hl 2-1965], [Hl 3-1980],
" -v- flaviflora.....

pachysa (See P. formosa),.....
papagayana.....[Hx*],
paraguayensis (SPEG.) (See P. amambayensis).....
parvula (See P. pusilla).....
penicillata.....[H], [DD], [NN], [PP], [Av], [Bo-6],
[Gr], [Hu], [Hx], [Il], [Iz], [Jf],
[N-1980], [Af 59-1], [Ah-1985],
" -v- fulviceps.....[Jf],
" -v- nivosa (See P. nivosa).....[Jf],

perplexa (See P. andreae).....
" -v- cupreo-aurea.....
peruviana.....[Aj 16-8-1965], [Aj 22-12-1971],
pilcomayoensis.....[Hl 5-1979],
piltziorum.....[Af 56-3], [Aj 31-7-1980],
pluricentralis.....[DD-512], [Hx*], [Jf],
" -v- erythroflora.....
" -v- xanthoflora.....
" -v- H 2 Brandt.....
prestoensis.....[Jf], [Aj 27-9-1976],
procera (See P. pseudoprocera),
[Hi-2], [Il], [Jf], [Aj 21-4-1970],
[Aj 21-7-1970], [Aj 22-1-1971],

prolifera (See P. camargensis -v- prolifera),
pseudoayopayana (similar to P. ayopayana).....[Af 42-4]
pseudoprocera (See P. procera),
[DD-513], [Hx*], [Aj 21-7-1970],
pseudostuemeri .....[DD], [Jf], [Aj 16-8-1965],
[Hl 1-1966],

PARODIA.....(Cont'd),

pseudosubterranea
punae.....[DD-514], [Jf], [Af 54-2],
" -v- exquisita.....[Jf],
purpureo-aurea.....[Hi-2], [Hu],
pusilla (See P. parvula),
quechua (See P. idiosa, P. stereospina)....
[Aj 33-6-1982],
rauschii......[DD], [NN], [Hx], [Jf], [Af 55-1], [Aj 29-4-1978], [Aj 32-9-1981], [Hl 6-1979],
rigida (See P. dextrohamata)......[DD], [Jf],
[Af 51-5],
rigidispina (See Microspermia rigidispina),
[H], [Hx*], [Aj 16-8-1965],
" -v- major.....[Aj 16-8-1965],
rigidissima (See P. microsperma -v- rigidissima.....
[Aj 11-1-1960],
riograndensis.....
riojensis.....[Jf],
ritteri.....[NN], [Ar], [Bo-6], [Hi-2], [Hy], [Iz],
[Ba 17-1959], [Hl 3-1975],
" -v- cintiensis (See P. cintiensis),
robustihamata.....
roseo-alba (See P. agasta).....[Hi-2], [Cu 3-1975],
" -v- australis.....[Hi-2],
rostrum-sperma (See P. camargensis).....[DD-515],
[Hx*], [Hl 5-1973],
rubellihamata.....[DD], [Hx*], [Jf],
" -v- aureiflora.....
" -v- chlorocarpa (See P. chlorocarpa),
" -v- paucicostata (See P. betaniana).....
[Hl 3-1979],
rubida (See P. maassii -v- rubida),
[NN], [Bo-6], [Hi-2], [Jf], [Aj 16-8-1965],
[Hl 5-1979],
rubristaminea.....[Hi-2], [Hl 3-1979],
rubricentra (See P. steumeri),
[Bo-3], [Hy], [Af 56-3], [Aj 11-1-1960],
rubriflora (See P. aureispina -v- rubriflora),
[D], [R], [Ay], [Gr], [Hb], [Hy], [Aj 5-1981], [Hl 2-1979],
rubrifuscata.....[D],
rubrispina ..... [as P.rubricentra in
[Aj 11-1-1960],
rubristaminea.....
rufidihamata (See P. rubellihamata, P. uebelmanniana),
saint-pieana.....[EE], [NN], [Bo-3], [Hu], [Hx], [Jf],
[Ad -], [Ag-1985], [Ba 7-1980], [Bf-1984],
salitrensis.....
salmonea (See P. schwebsiana -v- salmonea),
[DD-516], [Hx*], [Aj 24-5-1973],

salmonea.....(cont'd),
" -v- carminata (See P. carminata, P. tilcarensis) [Aj 24-5-1973],
" -v- lau-multicostata.....
sanagasta (See Microspermia sanagasta)....[Hi-2], [Hx*], [Aj 11-1-1960], [Gy 1936. No. 10, p.8],
" -v- minimiseminea.....[Bo-6],
" -v- saltensis (See P. glischrocarpa)......
" -v- viridior (Invalid name).....[DD],

sanguiniflora (See Microspermia sanguiniflora), [E], [J*], [M*], [T], [EE], [NN], [Ar], [Bt], [Bv-3], [Cr], [Cs], [Cy], [Dq], [Gr], [Hi-2], [Hu], [Hx], [Il], [Iz], [Jd], [Af 33-6], [Ag-1982], [Aj 24-5-1973], [Flower - Ba 3-1981],
" -v- comata (See P. comata).....[Hi-2],
" -v- violacea.....[Bo-3], [Cz], [Aj 11-1-1960],

schuetziana.....[Av], [Bo-3], [Hi-2], [Hx*], [Jd], [Jf], [Aj 11-1-1960], [Aj 15-12-1964], [Aj 27-3-1976], [Hl 3-1968],
schwebsiana (See Echinocactus schwebsianus, P. minima), [T], [NN], [Ar], [Bo-3], [Hv], [Hx], [Hy], [Ia], [Iz], [Jd], [Jf], [Aj 11-1-1960], [Hl 2-1965],
" -v- applanata.......[DD], [Bo-3], [Hi-2], [Hl 2-1980]
" -v- salmonea (See P. salmonea)..... [Af 23:84, 1951],
" -v- taratensis (See P. taratensis),
scopaoides (See P. aureispina -v- scopaoides), [H], [Bo-3], [Iv 9-1953],
scoparia (See P. friciana).....[Hi-2],
separata (See P. tredecimcostata -v- minor).....[Jf],
setifera .....[E], [H], [Bo-3], [Dq], [Hx*], [Jf], [Aj 11-1-1960],
" -v- longihamata.....
setispina.....[Hi-2], [Aj 24-6-1973],
setosa ......[DD], [Jf], [Aj 16-8-1965], [Hl 1-1966], [Hl 1-1980],
sotomayorensis (See P. ignorata).....[Hi-2], [Jf], [Hl 5-1980],
spanisa..... [Jf], [Aj 32-8-1981],
speciosa (Hybrid P. speciosa and P. subtilihamata) [Db -26th Anniv. Edition],
spegazziniana.....[DD-517], [Jf], [Aj 15-6-1964], [Aj 27-3-1976],
" -v- aurea.....
splendens.....[Af 33-4], [Aj 21-4-1970], [Bm-1983],

PARODIA.....(Cont'd),
steinmannii.....[Bo-6],
stereospina (See P. idiosa), ...
[Aj 33-6-1982], [Am 8-1980],
stuemeri ......
[E], [DD], [Bo-3], [Hi-2], [Hu], [Hx], [Hy],
[Jf], [Aj 11-1-1960], [Aj 16-8-1965],
[Hl 1-1966], [Hl 1-1980],
" -v- robustior (See P. friciana, P. gigantea),
[DD], [Aj 16-8-1965],
" -v- tilcarensis (See P. tilcarensis),
[Hl 1-1980],
subterranea (See P. maassii -v- subterranea),
[DD], [Av], [Hb], [Hi-2], [Hx], [Hy], [Iz],
[Jf], [Ae-1981], [Am 8-1980], [Hl 5-1969],
" -f. .....[Hl 5-1969],

subtilihamata (See P. andreae)..... [Hi-2], [Hx*],
sucrensis .....[Jf],
superba (See P. dextrohamata),.....[DD-518],
suprema (See P. escayachensis).....[NN], [Hi-2], [Hu],
[Hx*], [Hy], [Jf], [Aj 16-4-1965], [Hl 5-1969],
[Hl 5-1979], [Bi Vol. 4, #6],
tafiensis (See P. weskampiana),........
[DD], [Jf],
talaensis.....[Jf],
tarabucina.....[Af 33-4], [Aj 33-6-1982], [Hl 3-1980],
taratensis (See P. schwebsiana -v- taratensis),
[Af 53-3], [Aj 21-4-1970], [Hl 2-1980],
tenuispina (See P. faustiana -v- tenuispina),
thieleana (See P. sotomayorensis),
thionantha (See P. erythrantha -v- thionantha, P. microsperma -v- thionantha),
[DD-519], [Jf], [Aj 20-8-1969],
tilcarensis.....[Av], [Bo-3], [Hx], [Jf], [Aj 11-1-1960], [Aj 16-8-1965], [Am 8-1980],
[Aw 1-1965], [Hl 1-1980],
" -v- gigantea (See P. gigantea, P. friciana),
[Bo-3], [Hi-2], [Aj 16-8-1965],
tojoensis (See P. brevihamata).....
tredecimcostata (See P. echinopsoides).....[Hi-2],
" -v- aurea (See P. riograndensis).....
" -v- minor (See P. separata).....[Hi-2],
tuberculata.....[Ar], [Hi-2], [Jf], [Af 23-3], [Aj 10-2-1959], [Aj 16-8-1965], [Hl 2-1965],
tuberculosi-costata.....[I], [Jf], [Aj 27-3-1976],
" -v- amblayana.....
" -v- cafayatensis (See P. microsperma -v- cafayatensis).....[Jf],
uebelmanniana (See P. rubellihamata)....[Hi-2], [Hx*],
uligiana (See P. aureicentra -v- albifusca).....[Z],
[DD], [Hx], [Aj 24-10-1973], [Hl 1-1966],

PARODIA.....(Cont'd),
uhligiana.....(cont'd),
" -v- robustior.....[Hl 2-1965],
" -v- stuemerioides.....
variabilis (KK589).....
variicolor.....[Hi-2], [Jf], [Aj 33-12-1982],
" -v- robustispina.....[Jf],
wagneriana (See P. spegazziniana).....[Jf],
weberiana.....[DD-520], [Hx*], [Jf], [Aj 20-11-1969],
[Hl 5-1973], [Hl 3-1979],
" -v- higueritas (See P. higueritas),
weberioides.....(Hybrid)
weskampiana.....[Hx*], [Jf], [Af 42-2], [Aj 2-1978],
yamparaezi (See P. backebergiana).....[Hb], [Hx*], [Af
52-6], [Aj 24-5-1973], [Aj 33-9-1982],
" -v- culpinensis (See P. culpinensis).[Af 52-6],
zaletaewana.....[DD-521], [Hl 5-1973],
zecheri.....[Jf], [Aj 29-1-1978],
Sp. ..(Flower).....[Aj 28-3-1977],
Sp. ....[Bo-6], [Aj 32-2-1981],
Sp. L-458.....[Aj 27-5-1976],
Sp. L-576.....[Aj 27-5-1976],
Sp. L-913.....[Aj 28-2-1977],
Sp. P-27.....[Aj 32-5-1981],
Sp. P-37.....[Aj 32-5-1981],
Sp. P-51.....[Aj 32-8-1981],
Sp. P-74.....[Aj 32-5-1981],
Sp. P-145.....[Aj 32-8-1981],
Sp. P-146.....[Aj 32-8-1981],
Sp. P-147.....[Aj 32-8-1981],
Sp. P-151.....[Aj 32-9-1981],
Sp. P-153.....[Aj 32-9-1981],
Sp. P-157.....[Aj 32-8-1981],

Continued

Sp. DH 129.....[Jf],
Sp. DH 172.....[Jf],
Sp. F 9........[Jf],
Sp. L 381......[Jf],
Sp. L 570......[Jf],
Sp. P 235......[Jf],
Sp. P 239......[Jf],

PEDIOCACTUS
acunensis (See Echinomastus acunensis),
bradyi (See Pilocanthus bradyi, Toumeya bradyi),
[P], [Q], [An], [Ar], [Ax], [Ay], [Bh], [Hx*],
[Iz], [Af 51-2], [Af 53-1], [Aj 31-7-1980],
" -v- knowltonii (Invalid name - See P. knowltonii)
[DD], [Ay], [Hx], [Il], [Jd],
caespiticus (See P. simpsonii -v- caespiticus, P. simpsonii -v- simpsonii),
despainii.....[Af 53-1],
erectocentrus (See Echinomastus erectocentrus),
glaucus (See Sclerocactus glaucus),
intertextus (See Echinomastus intertextus),
johnsonii (See Echinomastus johnsonii),
" -v- lutescens (See Echinomastus johnsonii -v-
lutescens),
" -v- multicolor (See Echinomastus johnsonii -v-
multicolor),

PEDIOCACTUS.....(Cont'd),

knowltoni (See P. bradyi -v- knowltoni).....[F], [P], [Y], [Z], [GG], [PP], [An, [Ay], [Bh], [Df], [Hv], [Hy], [Iz], [Jd], [Af 53-1],
mariposensis (See Echinomastus mariposensis),
mesae-verdae (See Coloradoa mesae-verdae, Echinocactus mesae-verdae, Sclerocactus mesae-verdae), [Ar], [Ay],
pallidus (See Echinomastus pallidus),
papyracanthus (See Toumeya papyracantha).....[M*], [Q], [Y], [Z], [An], [Ar], [Ax], [Ay], [Bo-5], [Df], [Hv], [Il], [Jd], [Af 53-1]
paradinei (See Pilocanthus paradinei), [P], [Z], [UU], [An], [Av], [Ax], [Ay], [Il], [Iz], [Jd], [Af 29-5], [Af 39-2], [Af 53-1], [Dg 7-1],
parviflorus (See Echinomastus parviflorus),
" -v- intermedius (See Echinomastus parviflorus -v- intermedius),
peeblesianus (See Navajoa peeblesiana, Toumeya peeblesiana).....[M*], [Q], [An], [Ar], [Iz], [Jd], [Af 53-1],
" -v- fickeiseniae (See Navajoa fickeisenii), [Q], [An], [Ax], [Hv], [Af 39-2], [Af 53-1], [Aj 33-6-1982], [Dg 22-1980], [Hl 4-1971],
" -v- maianus (See P. peeblesianus -v- peeblesianus),
" -v- peeblesianus.....[Ax], [Bh], [Af 53-1], [Af 54-6],
polyancistrus (See Sclerocactus polyancistrus, Thelocactus polyancistrus),
robustior (See P. simpsonii -v- robustior),
sileri (See Utahia sileri), [M*], [Q], [An], [Ax], [Ay], [Bh], [Af 39-2], [Af 53-1], [Dg 3-1-1981], [Hl 6-1973],
simpsonii (See Mammillaria purpusii), [F], [K], [M*], [P], [DD], [GG], [LL*], [RR], [An], [Ar], [Ay], [Bo-5], [Go], [Hv], [Hx*], [Hy], [Iz], [Af 16-4], [Cu 2-1972], [Hl 1-1974],
" -v- caespiticus (Invalid name - See P. simpsonii -v- simpsonii), [Bo-5], [Af 56-1],
" -v- hermannii (See P. simpsonii -v- simpsonii) [Aj 8-12-1957],
" -v- minor.....[Ar], [Ax], [Go], [Hy], [Af 53-1],
" -v- nigrispina (See P. simpsonii -v- robustior)
" -v- robustior (See P. simpsonii -v- nigrispina, P. robustior),
" -v- simpsonii.....[Y], [Ax], [Df], [Af 53-1],

PEDIOCACTUS......(Cont'd).

unguispinus (See Echinomastus -v- unguispinus),
" -v- laui (See Echinomastus unguispinus -v- laui),
whipplei (See Echinocactus whipplei, Sclerocactus whipplei),
winkleri.....[Iz], [Af 51-1], [Af 53-1],
wrightiae (See Sclerocactus wrightiae).....[Hv],

PEIRESKIA (See Pereskia)

PEIRESKIOPSIS (See Pereskiopsis

PELECYPHORA

asselliformis.....[D], [I], [Z], [NN], [QQ], [RR], [XX], [Ar], [Av], [Ay], [Bo-5], [Bt], [Bv-1], [Cm], [Cx], [Cz], [Dq], [Gr], [Hr], [Hv], [Hx], [Hy], [Il], [Iz], [Jd], [Ae-1981], [Af 13-9], [Af 57-3], [Aj 29-8-1978], [Aj 32-10-1981], [Ah-1985], [Ba 1-1976], [Ba 11-1977], [Ba 4-1979], [Bi 1-3], [Bi 3-1], [Bm-1987], [Hm 2-1979], [Hw-1981], [Hw-1988],
" -v- concolor.....[Cn],
" -f. cristata.....[Ar], [Hy],
plumosa (See Echinocactus valdezianus, Gymnocactus valdezianus, Neolloydia valdeziana, Normanbokea valdeziana, P. valdeziana, Thelocactus valdezianus, Turbinicarpus valdezianus), [Aj II-6-1930],
pseudopectinata (See Neolloydia pseudopectinata, Normanbokea pseudopectinata, Turbinicarpus pseudopectinatus), [G], [X], [DD], [Ar], [Ay], [Hy], [Il], [Jd], [Aj 29-7-1978], [Aj 32-10-1981], [Aj 32-11-1981], [Aw 1-1973], [Da 3-1978], [Da 3-1979],
" -v- juisache.....
" -v- rubriflora.....
strobiliformis (See Encephalocarpus strobiliformis), [Z], [Iz], [Jd], [Hh May, 1975], [Hw-1986],
valdeziana (See Echinocactus valdezianus, Gymnocactus valdezianus, Neolloydia valdeziana, Normanbokea valdeziana, Thelocactus valdezianus, Turbinicarpus valdezianus), [G], [H], [NN], [Bv-1], [Il], [Aj 32-10-1981], [Ba 11-1977], [Da 3-1978],
" -v- albiflora.....[Fq 129:1960], [Hl 2-1971],

PENIOCEREUS

castellae.....[Az], [Aw 1-1974],
cuixmalensis.....[Az], [Aw 4-1973],
diguetii (See Cereus striatus, Neoevansia diguetii, Neoevansia striata, Wilcoxia striata),
[Q], [Bo-4], [Af 23:119, 1951], [Aj 34-1-1983], [Da 4-1980],
fosterianus.....[Af 18-2], [Af 19-11], [Aw 2-1962],
" -v- fosterianus.....[Az],
" -v- multipetalum,.....[Flower - Az*],
" -v- nizandensis.....[Flower - Az*],
greggii (See Cereus greggii).....[M*], [P], [U], [Y], [GG], [LL*], [RR], [Ay], [Az], [Bo-4], [Bo-6], [Df], [Dq], [Hx*], [Hy], [Iz], [Jd], [Af 58-2], [Af 60-1-1988], [Aj 29-5-1978],
" -v- cismontanus (See Cereus greggii -v- greggii)
" -v- roseiflorus (See Cereus greggii -v- transmontanus),
" -v- transmontanus (See Cereus greggii -v- transmontanus).....[Q],
haackeanus (Invalid name - See Neoevansia haackeana), [DD],
johnstonii.....[RR], [Ay], [Az], [Bo-4], [Cx], [Dq], [Hx*], [Hy], [Aw 4-1968],
macdougallii (See Cereus macdougallii).... [Bo-4], [Af 19-6],
" -v- centrispinus (Invalid name), [Bo-6:3843, 1962, Description]
" -v- macdougallii.....[Az],
" -v- centrispinus (fertile).....(Bo-6],
maculatus (See Acanthocereus maculatus, Cereus maculatus),
[Az], [Bo-4], [Af 23-3], [Au 37-4], [Da 1-1979],
marianus (See Wilcoxia mariana),
[Az], [Af 57-4], [Aw 4-1962],
marnieranus (Invalid name).....[DD], [Ay],
multipetalum (See P. fosterianus -v- multipetalum),
nizandensis (See P. fosterianus -v- nizandensis),
occidentalis (See Acanthocereus occidentalis), [Az], [Aw 4-1963],
rosei.....[Az], [Bo-4], [Cx], [Hx*],
tepalcatepecanus.....[Az], [Aw 1-1974],
viperinus (See Cullmannia viperina, Wilcoxia viperina -v- viperina).....[Jd],

PERESKIA

aculeata (See Pereskia pereskia).....[E], [H], [I], [K], [M*], [AA], [DD], [Ar], [Ax], [Ay], [Az], [Cx], [Gr], [Hi-1], [Hv], [Hy], [Il], [Iz], [Jd], [Je], [Aj 29-3-1978], [Aw 4-1979], [Gu III-1953],

aculeata.....(cont'd),
" -v- lanceolata.....(See P. aculeata),
" -v- rotundifolia.....(See P. aculeata),
" -f. rubescens (See P. rubescens, P. aculeata),
" -v- godseffiana.....(See P. aculeata)........
[L], [X], [NN], [Bv-1], [Cm],
[Hx*], [Aw 3-1983], [Bi 1-2],
amapola (See P. nemorosa).....[Af 1-7], [Af 3-5],
" -v- argentina.....(See P. nemorosa, P. amapola)
antoniana (See P. weberiana, Rhodocactus antonianus)..
[Ay],
aureiflora.....[Hi Vol. 1], [Je],
autumnalis (See P. lychnidiflora, Pereskiopsis opuntiiflora, P. pititache, Pereskiopsis pititache, P. nicoyana, Rhodocactus autumnalis, Rhodocactus nicoyanus, P. conzattii, Rhodocactus conzattii).....[RR], [Fg],
bahiensis....[RR], [Ar], [Bl], [Hi-1], [Hv], [Jd], [Je], [Bi 1-2],
bleo (See P. corrugata, P. panamensis, Rhodocactus bleo, Rhodocactus corrugatus).....[RR], [Bl], [Bv-1], [Je*],
colombiana (See P. guamacho, Rhodocactus colombianus, Rhodocactus guamacho).....[RR],
conzattii (See P. lychnidiflora, P. autumnalis, P. nicoyana, P. pititache, Pereskiopsis opuntiiflora, Pereskiopsis pititache, Pereskiopsis autumnalis, Rhodocactus conzattii, Rhodocactus lychnidiflorus, Rhodocactus nicoyanus, Rhodocactus autumnalis)
corrugata (See P. bleo, P. panamensis, Rhodocactus bleo, Rhodocactus corrugatus)...........
[Bu], [Af 23-6],
cubensis (See P. zinniiflora, Rhodocactus cubensis, Rhodocactus zinniiflorus).....[RR],
diaz-romeroana.....[Ay], [Hi-2], [Af 56-5], [Bi 1-2], [Je*],
foetens (See P. aculeata -v- aculeata),
fragrans (See P. aculeata -v- aculeata),
godseffiana (See P. aculeata -v- godseffiana, P. aculeata).....[Bv-1], [Hy], [Af 1-2],
grandiflora (Misspelling - See P. grandifolia, P. tampicana, Rhodocactus grandifolius, Rhodocactus tampicanus).....[Cr], [Cu 3-1964],
grandifolia (See P. tampicana, Rhodocactus grandifolius, Rhodocactus tampicanus, Pereskia grandiflora, Rhodocactus grandiflorus).....
[EE], [RR], [UU], [Ay], [Bt], [Bv-1], [Cr], [Cz], [Dq], [Ig*], [Ij], [It], [Hh Jan., 1976],

grandifolia.....(cont'd)
" -v- grandifolia.....[Je],
" -v- violacea.....[Je],
guamacho (See P. colombiana, Rhodocactus colombianus, Rhodocactus guamacho)...[M*], [RR], [Dq], [Je*],
higuerana (Name doubtful)
horrida (See P. humboldtii).....[Hi-4],
humboldtii (See P. horrida).....[Ay], [Bo-1],
" -v- humboldtii (See P. horrida, P. vargasii, Rhodocactus horridus).....[Je*]
" -v- rauhii (See P. vargasii -v- longispina, P. vargasii -v- rauhii).....[Je*],
lychnidiflora (See P. autumnalis, P. pititache, P. nicoyana, P. conzattii, Rhodocactus lychnidiflora, Pereskiopsis opuntiiflora, Pereskiopsis pititache, Rhodocactus nicoyanus, Rhodocactus autumnalis, Rhodocactus conzattii)...[RR], [Az], [Cx], [Je], [Aw 3-1980], [Aw 2-1983],
moorei (See P. sacharosa, P. saipenensis, P. sparsiflora, Rhodocactus sacharosa, Rhodocactus saipinensis).....[RR], [Bv-1], [Af 1-2],
nemorosa (See P. amapola, P. argentina, P. amapola -v- argentina).....[Je*],
nicoyana (See P. lychnidiflora, Rhodocactus lychnidiflorus, P. pititache, Pereskiopsis pititache, Pereskiopsis opuntiiflora, Rhodocactus nicoyanus, P. autumnalis, Rhodocactus autumnalis, P. conzattii, Rhodocactus conzattii).....[RR],
panamensis (See P. bleo, Rhodocactus bleo, P. corrugata, Rhodocactus corrugatus).....
pereskia (See P. aculeata), [LL*], [RR], [Af 1-2], [Af 11-3],
pflanzii (See Quiabentia pflanzii),
pititache (See P. lychnidiflora, Rhodocactus lychnidiflora, Pereskiopsis pitatache, Pereskiopsis opuntiiflora, P. nicoyana, Rhodocactus nicoyanus, P. autumnalis, Rhodocactus autumnalis, P. conzattii, Rhodocactus conzattii)...[Z], [Af 30-2], [Af 35-1], [Af 50-2],
portulacifolia (See Rhodocactus portulacifolius), [RR], [Je], [Ct 2-1984],
quisqueyana.....[Je*], [Af 56-3],
rotundifolia (See Pereskiopsis chapistle, Pereskiopsis rotundifolia, Pereskiopsis cyclophylla),
rubescens (See P. aculeata -f. rubescens).....[Af 1-2],
sacharosa (See Rhodocactus sacharosa, P. moorei, P. saipinensis, Rhodocactus saipinensis, P. sparsiflora).....[Ij], [Je*], [Db 6-1980],

saipinensis (See P. sacharosa, Rhodocactus sacharosa, P. moorei, P. sparsiflora, Rhodocactus saipinensis)....[Af 41-3],
sparsiflora (See P. sacharosa, P. moorei, P. saipinensis).....[Bo-6], [Hi-2],
spathulata (Name doubtful see Pereskiopsis spathulata),
stenantha.....[Hi-1], [Je*],
tampicana (See P. grandifolia -v- grandifolia, Rhodocactus grandifolius, Rhodocactus tampicanus).....[Cc], [Cx], [Aw 3-1964],
vargasii (See P. humboldtii).....[Ay], [Bo-1], [Jd],
" -v- longispina (See P. humboldtii -v- rauhii) [Bo-1],
" -v- rauhii (See P. humboldtii -v- rauhii)..... [Bo-1],
verticillata (See Quiabentia verticillata, Quiabentia chacoensis),
weberiana (See P. antoniana, Rhodocactus antonianus).. [Je*], [Af 39-1],
zehntneri (See Quiabentia zehntneri).....[RR],
zinniiflora (zinniaeflora) (See Rhodocactus zinniiflorus, P. cubensis, Rhodocactus cubensis), [RR], [Az], [Cx], [Di*], [Je*],

PERESKIOPSIS

aquosa.....[RR], [Fg], [Ic], [Im],
autumnalis (See Pereskia autumnalis, Rhodocactus autumnalis, P. lychnidiflora, Rhodocactus lychnidiflorus, P. opuntiiflora, Pereskia pititache, P. pititache, P. nicoyana, Rhodocactus nicoyanus, Pereskia conzattii, Rhodocactus conzattii),
blakeana.....[Az], [Cx],
chapistle (See P. rotundifolia, P. cyclophylla), [RR], [Bj], [Bo-1], [Cx], [Bi 1-2],
cyclophylla (See P. chapistle, P. rotundifolia)..... [Hy],
diguetii (See P. velutina), [RR], [Az], [N-1980], [Af 1-3],
gatesii.....[Bo-1], [Af 3-8],
kellermannii.....[RR],
opuntiiflora (opuntiaeflora) (See Pereskia lychnidiflora, Rhodocactus lychnidiflorus, Pereskia pititache, P. pititache, Pereskia nicoyana, Rhodocactus nicoyanus, Pereskia autumnalis, P. autumnalis, Rhodocactus autumnalis, P. conzattii, Rhodocactus conzattii).....[RR], [Bo-1], [Cx], [Hy*],

PERESKIOPSIS.....(Cont'd)

pititache (See Pereskia pititache, Pereskia lychnidiflora, Rhodocactus lychnidiflorus, P. opuntiiflora, Pereskia nicoyana, Rhodocactus nicoyanus, P. autumnalis, Pereskia autumnalis, Rhodocactus autumnalis, Pereskia conzattii, Rhodocactus conzattii).....[Bo-1],
porteri.....[G], [DD], [RR], [Az], [Bo-1], [Cx], [Dq], [Hx*], [Ij], [Bi 1-2],
rotundifolia (See P. chapistle, P. cyclophylla)..... [RR], [Az], [Bk], [Bo-1], [Cx], [Aw 3-1983],
scandens.....[Az],
spathulata (See Pereskia spathulata).....[K], [Hx*], [Af 1-3], [Cu 3-1964], [Cu 2-1980],
velutina (See P. diguetii).....[G], [QQ], [RR], [Ay], [Bu], [Hu], [Hv], [Hy], [Jd], [Af 54-6], [Bi 1-2],
" -v- variegata.....

PERUVOCEREUS

albicephalus (See Neobinghamia climaxantha)..... [Af 19-10],
albisetatus (See Haageocereus albisetatus)..... (Espostoa haagei X Haageocereus multangularis).....[Af 20-12],
" -v- robustus (See Haageocereus albisetatus -v- robustus).....[Af 20-12],
albispinus (See Haageocereus albispinus, Haageocereus chosicensis -v- albispinus).....[Af 20-10],
" -v- floribundus.....[Af 20-10],
chosicensis (See Haageocereus chosicensis, Haageocereus multangularis),
" -v- albispinus (See P. albispinus),
" -v- rubrospinus (See P. rubrospinus),
chrysacanthus (See Haageocereus chrysacanthus, Haageocereus multangularis -v- pseudomelanostele),
clavatus (See Haageocereus multangularis -v- pseudomelanostele, Haageocereus pseudomelanostele),
multangularis sensu AKERS 1950, (See Haageocereus multangularis -v- pseudomelanostele, Haageocereus pseudomelanostele), [Af 22-6],
rubrospinus (See Haageocereus chosicensis -v- rubrospinus, Haageocereus rubrospinus), [Af 19-8],

PERUVOCEREUS.....(Cont'd)

salmonoideus (See Haageocereus chosicensis -v- albispinus, Peruvocereus albispinus, Peruvocereus albicephalus, Peruvocereus albisetatus).....[Af 19-5],
setosus (See Haageocereus setosus, Haageocereus pseudomelanostele -v- setosus),
[Af 19-5],
" -v- longicoma (See Haageocereus pseudomelanostele -v- longicoma).....[Af 19-6],
" -f. cristata.....[Af 19-5],
viridiflorus (See Haageocereus viridiflorus, Haageocereus multangularis, Haageocereus chosicensis, Haageocereus acranthus),
[Af 19-9],

PFEIFFERA

cereiformis (See P. ianthothele),
erecta.....[NN], [Hi-2], [Aj 27-11-1976],
gibberosperma (See P. mataralensis -v- floccosa),
gracilis (See P. stricta).....[Hi-2],
ianththele (See Rhipsalis ianthothele),
[C], [DD], [RR], [Ay], [Bo-2], [Bv-3], [Cg], [Dq*], [Hx], [Hy], [Jd], [Ba 6-1981],
" -v- boliviana.....[Hi-2],
" -v- tarijensis.....

mataralensis.....
" -v- floccosa (See P. gibberosperma)...[Hi-2],
miyagawae (See Lepismium miyagawae).....[Af 59-2],
monacantha (See Acanthorhipsalis monacantha)...[Jd],
multigona.....
stricta (See P. gracilis),

PHELLOSPERMA

longiflora (See Mammillaria longiflora),
[Aj 16-11-1965],
tetrancistra (See Mammillaria tetrancistra)....[BB],
[DD], [FF], [GG], [An], [Bo-5], [Cb], [Hx],
[Hy], [Jd], [Ae-1981], [Af 5-4],

PHILIPPICEREUS

castaneus (See Eulychnia castanea),
[DD], [Bo-2], [Hx*], [Hy],
spinibarbis (See Eulychnia spinibarbis),

PHYLLOCACTUS

ackermannii (See Epiphyllum ackermannii, Nopalxochia
ackermannii).....[Cm], [Be*], [Hy],
anguliger (See Epiphyllum anguliger, Epiphyllum
darrahii, P. darrahii, P. serratus),
biformis (See Epiphyllum biformis, Disocactus biformis)
[Be*],
cartagensis (See Epiphyllum cartagense),
chiapensis (See Chiapasia nelsonii, Disocactus nel-
sonii, Epiphyllum nelsonii, P. nelsonii),
crenatus (See Epiphyllum crenatum).....[If],
darrahii (See Epiphyllum anguliger, Epiphyllum
darrahii).....[Hy],
elegans.....[Ds],
gaertneri (See Epiphyllum gaertneri, Schlumbergera
gaertneri).....[Ds],
grandilobus (See Epiphyllum grandilobum),
hookeri (See Epiphyllum hookeri, Epiphyllum phyllanthus
-v- hookeri, Epiphyllum stenopetalum).....[Hy],
nelsonii (See Chiapasia nelsonii, Disocactus nelsonii)
oxypetalum (See Epiphyllum oxypetalum).....[Hy],
phyllanthoides (See Epiphyllum phyllanthus, Nopal-
xochia phyllanthoides),
[Be*], [Bk], [Aj 9-4-1958],
serratus (See Epiphyllum anguliger, Epiphyllum
darrahii),
stenopetalus (See Epiphyllum stenopetalum, Epiphyllum
hookeri, Epiphyllum strictum, P. hookeri),

strictus (See Epiphyllum strictum, Epiphyllum hookeri,
Epiphyllum stenopetalum, P. hookeri),
[Bk],

Sp. cristata.....[Cy],
Sp. mexicanus (Hybrid).....[Ds],

PILOCANTHUS

bradyi (See Pediocactus bradyi, Toumeya bradyi),
paradinei (See Pediocactus paradinei),
[DD], [Av], [Bo-5], [Hx], [Hy], [Jd], [Aj 8-12-1957], [Aj 26-5-1975],

PILOCEREUS

albispinus (See Cephalocereus albispinus),
" -v- weberi (Invalid name - See Subpilocereus repandus -v- weberi),
alensis (See Cephalocereus alensis, Pilosocereus alensis).....[Hy],
arrabidae (See Cephalocereus arrabidae, Pilosocereus arrabidae, Pilosocereus exerens),
[Hy], [It],
aurisetus (See Pilosocereus aurisetus, Pilosocereus coerulescens),
[It],
atroviridis (See Cephalocereus atroviridis, Subpilocereus atroviridis),
backebergii (See Cereus backebergii, Pilosocereus backebergii).....[Hy],
bahamensis (See Cephalocereus bahamenis, Pilosocereus bahamensis),
barbadensis (See Cephalocereus barbadensis, Pilosocereus barbadensis),
bradei (See Pilosocereus bradei, Pseudopilocereus bradei),
brasiliensis (See Cephalocereus brasiliensis, Pilosocereus sublanatus),
brooksianus (See Cephalocereus brooksianus, Pilosocereus brooksianus).....[Gu III-1953],
bruennowii (See Oreocereus celsianus -v- bruennowii),
catalinii (catalani) (See Pilosocereus catalinii),
catingicola (See Cephalocereus catingicola, Pilosocereus catingicola).....[Hy], [It],
celsianus (See Oreocereus celsianus),
chrysacanthus (See Cephalocereus chrysacanthus, Cereus chrysacanthus, Pilosocereus chrysacanthus),
chrysostele (See Pilosocereus chrysostele).....
[Hy], [It],

claroviridis (See Pilosocereus claroviridis),
coerulescens (See Pilocereus aurisetus, Pilocereus glaucescens, Pilosocereus coerulescens, Pilosocereus glaucescens),
collinsii (See Cephalocereus collinsii, Pilosocereus collinsii).....[Hy],

PILOCEREUS.....(Cont'd)

columbianus (See Cephalocereus columbianus, Pilosocereus columbianus).....[Hy],
cometes (See Cephalocereus cometes, Pilosocereus cometes),
cuyabensis (See Cephalocereus cuyabensis, Pilosocereus cuyabensis),
dautwitzii (See Espostoa lanata).....[Ir],

deeringii (See Cephalocereus deeringii, Cereus robinii -v- deeringii, Pilosocereus deeringii),
erythrocephalus (See Denmoza rhodacantha),

euphorbioides (See Cephalocereus euphorbiodes, Neobuxbaumia euphorbioides, Rooksbya euphorbioides, Pilosocereus euphorbioides),
fluminensis (See Cephalocereus fluminensis, Coleocephalocereus fluminensis).....[Hr],
fulvispinosus (See Cephalocereus royenii),

gaumeri (See Cephalocereus gaumeri, Pilosocereus gaumeri),
glaucescens (See Cephalocereus glaucescens, P. aurisetus, Pilosocereus aurisetus, Pilosocereus coerulescens, Pilosocereus glaucescens), [Hy], [It],
glaucochrous (See Cephalocereus glaucochrous, Pilosocereus glaucochrous, Pseudopilocereus glaucochrous).....[It],
gounellei (See Cephalocereus gounellei, Pilosocereus gounellei).....[Hy], [It],
haagei (See Espostoa haagei),

hapalacanthus (See Pilosocereus hapalacanthus), [Hy], [It],
hermentianus (See Cephalocereus hermentianus, Pilosocereus hermentianus),
horrispinus (See Cereus horrispinus, Pilosocereus horrispinus, Subpilocereus horrispinus),
houlletii (See Cephalocereus leucocephalus, Pilocereus leucocephalus, Pilosocereus houlletii, Pilosocereus leucocephalus).....[Hy],
keyensis (See Cephalocereus keyensis, Cephalocereus robinii -v- keyensis, Cereus keyensis, Cereus robinii -v- keyensis, Cereus robinii -v- robinii),
lanatus (See Espostoa lanata),
lanuginosus (See Cephalocereus lanuginosus, Pilosocereus lanuginosus),

PILOCEREUS.....(Cont'd),

lanuginosus.....(cont'd),
" -f. cristata.....[Ba 6-1979],
leucocephalus (See Cephalocereus leucocephalus, P. houlletii, Pilosocereus houlletii, Pilosocereus leucocephalus), [Dq],
" -f. cristata.....[Hy],
luetzelburgii (See Cephalocereus luetzelburgii, Pilosocereus luetzelburgii).....[Dq], [It],
macrocephalus (See Cephalocereus macrocephalus, Mitrocereus ruficeps, Neobuxbaumia macrocephalus, P. macrocephalus).....[Hy],
macrostibas (See Cereus peruvianus (Mill.), Neoraimondia arequipensis, Neoraimondia gigantea, Neoraimondia macrostibas, Neoraimondia peruviana),
maxonii (See Cephalocereus maxonii, Pilosocereus maxonii),
mezcalensis (See Cephalocereus mezcalensis, Neobuxbaumia mezcalensis),
millspaughii (See Cephalocereus millspaughii),
minensis (See Cipocereus minensis, Pilosocereus minensis, Pseudopilocereus minensis),
monoclonus (See Cephalocereus monoclonus),
moritzianus (See Cephalocereus moritzianus, Pilosocereus moritzianus),
nobilis (See Cephalocereus nobilis, Pilosocereus nobilis).....[Dq], [Iv 6-1953],
" -f. cristata.....[Hy],
oligolepis (See Cephalocereus oligolepis, Pseudopilocereus oligolepis),
palmeri (See Cephalocereus palmeri, Cereus victoriensis, Pilosocereus palmeri), [Hy], [Im],
" -f. cristata.....[Hy],
pentaedrophorus (See Cephalocereus pentaedrophorus, Pilosocereus pentaedrophorus).....[It],
perlucens (See Cereus perlucens, Pilosocereus perlucens),
phaeacanthus (See Brasilicereus phaeacanthus, Cephalocereus phaeacanthus),
piauhyensis (See Cephalocereus piauhyensis, Pilosocereus piauhyensis).....[[Hy], [It],
polygonus (See Cephalocereus polygonus, Pilosocereus polygonus).....[Af 24-2], [Bi Vol. 4, #6],
polylophus (See Cephalocereus polylophus, Neobuxbaumia polylopha, Pilosocereus polylophus),
purpusii (See Cephalocereus purpusii, Pilosocereus purpusii),

PILOCEREUS.....(Cont'd),

remolinensis (See Cereus remolinensis, Subpilocereus remolinensis),
robinii (See Cephalocereus robinii, Cereus robinii, Pilosocereus robinii),
royenii (See Cephalocereus royenii, Pilosocereus royenii),
rupicola (See Cephalocereus rupicolus, Pilosocereus rupicola, Pseudopilocereus rupicola),
russelianus (See Cephalocereus russelianus, Pilosocereus russelianus, Subpilocereus russelianus).....[Af 20-7],
salvadorensis (See Cephalocereus salvadorensis, Pilosocereus salvadorensis).....[It],
sartorianus (See Cephalocereus sartorianus, Cephalocereus palmeri -v- sartorianus, Pilosocereus sartorianus).....[Hy], [Ig*],
sergipensis (See Cephalocereus sergipensis, Pilosocereus sergipensis),
sublanatus (See Cephalocereus brasiliensis, Pilosocereus sublanatus),
swartzii (See Cephalocereus swartzii, Pilosocereus swartzii),
tehuacanus (See Cephalocereus tehuacanus, Pilosocereus tehuacanus),
tetetzo (See Cephalocereus tetetzo, Cereus tetetzo, Neobuxbaumia tetetzo, Pilosocereus tetetzo),
tuberculatus (See Pilosocereus tuberculatus),
tweedyanus (See Cephalocereus tweedyanus, Pilosocereus tweedyanus).....[Af 34-3],
ulei (See Pilosocereus ulei), [Hi-1],
urbanianus (See Cephalocereus urbanianus, Pilosocereus urbanianus),
vollii (Natural hybrid - P. arrabidae X Cephalocereus fluminensis),

PILOCOPIAPOA

solaris (See Copiapoa conglomerata, Copiapoa ferox, Copiapoa solaris), [DD], [Ar], [Ay], [Bo-6], [Hx*], [Hy], [Jd], [Aj 26-4-1976], [Hl 1-1968],

PILOSOCEREUS

albicephalus (See Austrocephalocereus albicephalus),
albispinus (See Cephalocereus albispinus),
alensis (See Cephalocereus alensis, Pilocereus alensis) [Bo-4], [Af 21-4], [Am 8-1979],
arenicola.....[Fu 109, 1933],

PILOSOCEREUS.....(Cont'd),

arrabidae (See Cephalocereus arrabidae, Cephalocereus exerens, Pilocereus arrabidae),
[Bo-4], [Hi-1],
atroflavispinus.....
aurisetus (See Pilocereus aurisetus, Pilosocereus coerulescens).....[Fu 103, Fig., 1933],
aureispinus (See Coleocephalocereus aureispinus),
aurilanatus (See Pseudopilocereus aurilanatus),
[Hi-1], [Aj 34-3-1983],
azureus (See Cephalocereus azureus, Cephalocereus cyaneus, Pseudopilocereus azureus),
backebergii (See Cephalocereus backebergii, Cephalocereus moritzianus -v- backebergii, Cereus backebergii, Pilocereus backebergii),
[Bo-4], [Dv 45:81, 82, Fig. March, 1930],
" -v- backebergii (See Cephalocereus moritzianus -v- backebergii),
bahamensis (See Cephalocereus bahamensis),
barbadensis (See Cephalocereus barbadensis),
bradyi (bradei) (See Pseudopilocereus bradyi),
[Bo-4],
" -v- longispinus.....
brasiliensis (See Cephalocereus brasiliensis, Pilosocereus sublanatus),
brooksianus (See Cephalocereus brooksianus, Pilocereus brooksianus),
[Bo-4], [Hs IV-2, 1983],
carolinensis (See P. flavipulvinatus),
" -v- robustispinus.....[Hi-1],
catalani - (See Cephalocereus catalani, Pilocereus catalani),
catingicola (See Cephalocereus catingicola, Pilocereus catingicola),
chrysacanthus (See Cephalocereus chrysacanthus),
[JJ], [Av], [Bo-4], [Ds], [Hv], [Hx],
[Af 21-2], [Aj 33-4-1982], [Aw 1-1970],
[Ba 2-1980], [Ba 5-1980],
" -f. cristata (See Cephalocereus chrysacanthus -f. cristata),
chrysomallus (See Backebergia militaris, Cephalocereus chrysomallus).....,[Aj II-9-1930], [Aj 1-1937],
chrysostele (See Pilocereus chrysostele),
[Bo-4],
clarosensis (See Pseudopilocereus clarosensis),
claroviridis (See Pilocereus claroviridis),
[Hl 5-1981],
coerulescens (See Pilocereus aurisetus, Pilosocereus aurisetus, Pilosocereus glaucescens),
[Hi-1],

PILOSOCEREUS.....(Cont'd),

collinsii (See Cephalocereus collinsii).....[Bo-4],
columbianus (See Cephalocereus columbianus, Pilocereus columbianus).....[Bo-4], [Af 54-6], [Af 55-1]
cometes (See Cephalocereus cometes),
[Bo-4], [Hx*],
crassipetalus (See P. densiareolatus),
cuyabensis (See Cephalocereus cuyabensis),
[Bo-4], [Fo 1935 1 gen. 98 sp. 4, Fig.],
cyaneus (See Cephalocereus cyaneus, Pilosocereus azureus, Pseudopilocereus azureus),
deeringii (See Cephalocereus deeringii, Cereus robinii -v- deeringii).....[Bo-4],
densiareolatus (See P. crassipetalus),
[Hi-1], [Hu],
densilanatus (See P. werdermannianus -v- densilanatus),
diamantinensis (See Pseudopilocereus saxatilis -v- diamantinensis, P. werdermannianus -v- diamantinensis),
diersianus (See Pseudopilocereus diersianus).....
euphorbioides (See Cephalocereus euphorbioides, Neobuxbaumia euphorbioides, Rooksbya euphorbioides).....[Af 18-1],
exerens (See Cephalocereus arrabidae, Pilocereus arrabidae).....[Bk],
flavipulvinatus (See Pseudopilocereus flavipulvinatus),
" -v- carolinensis (See P. carolinensis),
floccosus (See Cephalocereus floccosus, Pseudopilocereus floccosus).....[Bo-4],
fricii (See Cleistocactus strausii -v- fricii, P. strausii -v-fricii),
fulvilanatus (See Pseudopilocereus fulvilanatus),
[Fig. 12 in [Af 58-1] is P. fulvilanatus],
gaturianensis (See Cephalocereus gaturianensis),
gaumeri (See Cephalocereus gaumeri),
gironensis (See Cephalocereus gironensis).....[Bo-4],
glaucescens (See Cephalocereus glaucescens, Pilocereus glaucescens, Pilosocereus coerulescens),
[Bo-4], [Aj 1-1937],
glaucochrous (See Cephalocereus glaucochrous, Pilocereus glaucochrous, Pseudopilocereus glaucochrous).....[Jd],
[NN], [Bo-4], [Hx*], [Ad Feb., 1980],
[Af 15-11],
gounellei (See Cephalocereus gounellei, Pilocereus gounellei, Pseudopilocereus gounellei),
[Bl], [Bo-4], [Hi-1], [Il],
" -v- zehntneri.....[Bo-4],
gruberi (See Cephalocereus gruberi).....[Aj 33-1-1982],
guerreronis (See Cephalocereus guerreronis).....[Bo-4],
haenssleri.....[Bo-4],

hapalacanthus (See Cephalocereus hapalacanthus, Pilocereus hapalacanthus, Pseudopilocereus hapalacanthus),
[Bo-4], [Af 14-6], [Au 1-1937],
hermentianus (See Cephalocereus hermentianus),
horrispinus (See Cereus horrispinus, Subpilocereus horrispinus),
[Dv 45:82 March 1930 - Description],
houlletii (See Cephalocereus houlletii, Cephalocereus leucocephalus, Pilocereus houlletii, Pilocereus leucocephalus, (Pilosocereus leucocephalus).....[NN], [Bo-4], [Di*],
keyensis (See Cephalocereus keyensis, Cephalocereus robinii -v- keyensis, Cereus robinii -v- keyensis),
lanuginosus (See Cephalocereus lanuginosus),
[Bo-4], [Ba 6-1979],
" -f. cristata.....[Ba 6-1979],
leucocephalus (See Cephalocereus leucocephalus, Pilocereus houlletii, Pilocereus leucocephalus, Pilosocereus houlletii).....[Bo-4],
luetzelburgii (See Cephalocereus luetzelburgii, Pilocereus luetzelburgii, Coleocephalocereus luetzelburgii).....[Bo-4], [Jd], [Aj 1-1937], [Aj 22-12-1971], [Ct-1988],
machrisii (See Cephalocereus machrisii, Pseudopilocereus machrisii),
macrocephalus (See Cephalocereus macrocephalus, Neobuxbaumia macrocephalus, Pilocereus macrocephalus),
magnificus (See Pseudopilocereus magnificus),
[Hi-1], [Aw 2-1981],
maxonii (See Cephalocereus maxonii),
[AA], [Ay], [Bo-4], [Af 52-2], [Ah-1985],
medirisii (See Cephalocereus medirisii, Pseudopilocereus medirisii),
mezcalensis (See Cephalocereus mezcalensis, Neobuxbaumia mezcalensis),
millspaughii (See Cephalocereus millspaughii),
minensis (See Cereus minensis, Cipocereus minensis, Pseudopilocereus minensis),
[Bo-4], [Aj 31-4-1980],
monoclonus (See Cephalocereus monoclonus, Cereus monoclonus),
moritzianus (See Cephalocereus moritzianus, Cereus moritzianus),
[Bo-4], [Af 21-1, 2], [Hl 1-1981],
" -v- backebergii (See Pilocereus backebergii, Pilosocereus backebergii),
mortensenii (See Cephalocereus mortensenii).....[Bo-4],

PILOSOCEREUS.....(Cont'd),

mucosiflorus (See Cephalocereus mucosiflorus, Pseudo-
pilocereus mucosiflorus),
multicostatus (See Cephalocereus multicostatus),
[Hi-1], [Hu],
nobilis (See Cephalocereus nobilis, Pilocereus nobilis,
Pseudopilocereus nobilis).....[Dq], [Da 1-1980],
oligolepis (See Cephalocereus oligolepis, Pseudopilo-
cereus oligolepis),
oreus (See Cephalocereus oreus),
ottonis (See Cephalocereus ottonis, Subpilocereus
ottonis),
pachycladus (pachycladus) (See Cephalocereus pachy-
cladus, Pseudopilocereus pachycladus).....
[Bt], [Hi-1], [Af 55-4],
palmeri (See Cephalocereus palmeri, Cereus victori-
ensis, Pilocereus palmeri)
[C], [H], [I], [T], [PP], [QQ], [XX], [Ar], [Av],
[Ay], [Bo-4], [Hu], [Hv], [Hx], [Il], [Ae-1981],
[Db 4-1979], [Hh Jan., 1975], [Hl 1-1982],
" -v- palmeri (See Cephalocereus palmeri -v- pal-
meri),
" -v- sartorianus (See Cephalocereus palmeri -v-
sartorianus),
" -v- victoriensis (See Cephalocereus palmeri,
Pilosocereus palmeri, Cereus victoriensis)
paraguayensis (See Cephalocereus paraguayensis),
parvus (See Pseudopilocereus parvus).....[Ct-1988],
pentaedrophorus (See Cephalocereus pentaedrophorus,
Pseudopilocereus pentaedrophorus),
[R], [Bo-1], [Hi-1],
" X - Micranthocereus purpureus (Natural hybrid),
[Hi-1],
perlucens (See Cereus perlucens),
pernambucoensis (See Cephalocereus pernambucoensis),
" -v- caesius.....
" -v- montealtoi.....
piauhyensis (See Cephalocereus piauhyensis, Pilocereus
piauhyensis, Pseudopilocereus piauhyensis),
[Bo-4],
pleurocarpus (See Cipocereus pleurocarpus).....
polygonus (See Cephalocereus polygonus, Pilocereus
polygonus)...[Bo-4], [Af 56-3], [Aj 30-1-1979],
polylophus (See Cephalocereus polylophus, Neobuxbaumia
polylophus).....[Af 21-4],
pulvilanatus (should be fulvilanatus - See P. fulvila-
natus).....[Af 58-1],
purpusii (See Cephalocereus purpusii),
[Bo-4], [Bo-6], [Aw 2-1970],
pusillibaccatus.....[Af 58-6],
pusilliflorus (See Subgen. Floribunda pusilliflorus)...

PILOSOCEREUS.....(Cont'd),

quadricentralis (See Cephalocereus quadricentralis), [Bo-4],
quadricostatus (See Cephalocereus quadricostatus), [Hi-1],
remolinensis (See Cephalocereus remolinensis, Cereus remolinensis, Subpilocereus remolinensis),
robinii (See Cephalocereus robinii, Cereus robinii), [Hl 1-1972],
robustus (See Cephalocereus robustus).....[Hi-1],
" -v- ulei (See Cephalocereus ulei),
rosae (See Cephalocereus rosae),
royenii (See Cephalocereus royenii), [Hx*], [Af 54-6],
" -v- cortezii.....[Af 16-5],
rupicola (See Cephalocereus rupicolus, Pseudopilocereus rupicola),
ruschianus.....
russelianus (See Pilocereus russelianus, Subpilocereus russelianus).....[Aj 1-1937],
" ssp. margaritensis.....
salvadorensis (See Cephalocereus salvadorensis, Pilocereus salvadorensis), [Jd], [Aj 1-1937],
sartorianus (See Cephalocereus sartorianus, Pilocereus sartorianus).....[Bl], [Cr], [Jd],
saudadensis (See Cephalocereus saudadensis).....[Hi-1],
saxatilis (See Pilosocereus werdermannianus, Pseudopilocereus werdermannianus),
" -v- densilanatus (See P. werdermannianus -v- densilanatus),
schoebelii.....[Af 59-4],
scoparius (See Cephalocereus scoparius, Neobuxbaumia scoparius),
sergipensis (See Cephalocereus sergipensis, Pseudopilocereus sergipensis), [DD], [Bo-4],
smithianus (See Cephalocereus smithianus, Monvillea smithiana),
splendidus (See Cephalocereus splendidus), [Hi-1], [Hl 1-1979],
strausii (See Cleistocactus strausii)
" -f. cristata.....[Aj II 3, 4, 1930],
" -v- fricii.....[Aj II 3, 4, 1930],
sublanatus (See Cephalocereus brasiliensis), [Bo-4],
superbus.....[Hi-1],
" -v- gacapaensis.....[Hi-1],
" -v- lanosior.....
" -v- regius.....[Hi-1],

PILOSOCEREUS .....(Cont'd),

superfloccosus (See Cephalocereus superfloccosus, Pseudopilocereus superfloccosus),
supremus (See Pilosocereus magnificus, Pseudopilocereus magnificus, Pseudopilocereus supremus),
supthutianus (See Cephalocereus supthutianus),
swartzii (See Cephalocereus swartzii),
[Ay],
" -f. cristata (See Cephalocerus swartzii -f. cristata),
tehuacanus (See Cephalocereus tehuacanus).....[Bo-4],
tetetzo (See Cephalocereus tetetzo, Neobuxbaumia tetetzo),
tillianus (See Cephalocereus tillianus).....
[Aj 33-8-1982],
tuberculatus (See Cephalocereus tuberculatus, Pseudopilocereus tuberculatus),
[Fu 101 Fig., 1933],
tuberculosus (See Cephalocereus tuberculosus, Pseudopilocereus tuberculosus).....[Bo-4],
tweedyanus (See Cephalocereus tweedyanus, Pilocereus tweedyanus)..[Bo-4], [Af 34-3], [Hl 5-1982],
" -f. cristata.....[Bo-4],
ulei (See Pilocereus ulei).....[Bo-4],
urbanianus (See Cephalocereus urbanianus),
vilaboensis (See Pseudopilocereus vilaboensis).....
[Ct-1988],
vollii (See Pilocereus vollii),
werdermannianus (See Cephalocereus werdermannianus, Pseudopilocereus werdermannianus),
[Hi-1], [As Pilosocereus aurisetus in Af 58-1],
" -v- densilanatus (See Pilosocereus saxatilis -v- densilanatus).....[Hi-1],
" -v- diamantinensis (See Pseudopilocereus saxatilis -v- diamantinensis)....[Hi-1],
zehntneri (See Cephalocereus zehntneri, Pilosocereus gounellei -v- zehntneri),
Sp. nov. .....[Ct-1988],

PIPTANTHOCEREUS

aethiops (See Cereus aethiops, Cereus coerulescens),
[Hi-2],
alacriportanus (See Cereus alacriportanus).....[Hi-1],
bageanus (See Cereus bageanus),
cabralensis (See Cereus cabralensis).....[Hi-1],
calcirupicola (See Cereus calcirupicola),.....[Hi-1],
" -v- pluricostata.....
chalybaeus (See Cereus chalybaeus),
cipoensis (See Cereus cipoensis).....[Hi-1],
colosseus (See Cereus colosseus).....[Hi-2],,

PIPTANTHOCEREUS.....(Cont'd)

crassisepalus (See Cereus crassisepalus),
dayamii (See Cereus dayamii).....[Hi-2],
forbesii (See Cereus forbesii),
" -v- bolivianus (See Cereus cochabambensis), [Hi-2],
goiasensis (See Cereus goiasensis, Cereus jamacaru, Piptanthocereus jamacaru),
gonianthus (See Cereus gonianthus),
huilunchu (See Cereus huilunchu).....[Hi-2],
jamacaru (See Cereus jamacaru).....[Hi-1],
labouretianus (See Cereus labouretianus), [Fw 8:231, 1909, Description],
lamprospermus (See Cereus lamprospermus, Cereus colosseus, Piptanthocereus colosseus),
lanosus (See Cereus lanosus),
lindenzweigianus (See Cereus lindenzweigianus, Cereus spegazzinii -v- hassleri, Monvillea lindenzweigianus),
neonesioticus (See Brachycereus nesioticus, Cereus nesioticus, Cereus neonesioticus),
" -v- interior.....
obtusus (See Cereus fernambucensis, Cereus obtusus, Cereus pernambucensis),
pachyrhizus (See Cereus pachyrhizus),
sericifer (See Cereus sericifer).....[Hi-1],
stenogonus (See Cereus stenogonus),
uruguayanus (See Cereus uruguayanus, Cereus peruvianus (WERD.)),
validus (See Cereus validus),
xanthocarpus (See Cereus xanthocarpus),

PLATYOPUNTIA

albisaetacens (See O. albisaetacens).....[Hi-2],
" -v- robustior.....
alko-tuna (See Opuntia alko-tuna),
apurimacensis (See Opuntia apurimacensis).....[Hi-4],
atroglobosa (See Tephrocactus atroglobosus).....[Hi-2],
azurea (See Opuntia azurea),
bispinosa (See Opuntia bispinosa),
brachyacantha (See Opuntia brachyacantha),
brunneogemmia (See Opuntia brunneogemmia).....[Hi-1],
cardiosperma (See Opuntia cardiosperma),
chilensis (See Platyopuntia soehrensii),
cognata (See Opuntia cognata, Platyopuntia vitelliniflora),
conjugens (See Opuntia conjugens).....[Hi-2],
cordobensis (See Opuntia cordobensis).....[Hi-2],
corrugata (See Opuntia corrugata, Tephrocactus corrugatus).....[Hi-2],

PLATYOPUNTIA.....(Cont'd),

discolor (See Opuntia discolor).....[Hi-2],
dumetorum (See Corynopuntia dumetorum, Opuntia dumetorum),
ficus-indica (See Opuntia ficus-indica),
ianthinantha (See Opuntia ianthinantha).....[Hi-2],
inaequilateralis (See Opuntia inaequilateralis), [Hi-4],
" -v- angustior (See Opuntia inaequilateralis -v- angustior).....(Hi-4],
inamoena (See Opuntia inamoena).....[Hi-1],
" -f. spinigera.....[Hi-1],
infesta (See Opuntia infesta).....[Hi-4],
interjecta (See Opuntia interjecta).....[Hi-2],
kiska-loro (See Opuntia kiska-loro),
limitata (See Opuntia limitata),
microdisca (See Opuntia microdisca),
nana (See Opuntia nana, Opuntia pascoensis, Opuntia pubescens).....[Hi-2],
nigrispina (See Opuntia nigrispina, Tephrocactus nigrispinus).....[Hi-2],
orurensis (See Opuntia orurensis).....[Hi-2],
pituitosa (See Opuntia pituitosa),
pyrrhantha (See Opuntia pyrrhantha).....[Hi-2],
quimilo (See Opuntia quimilo).....[Hi-2],
quitensis (See Opuntia macbridei, Opuntia quitensis),
retrorsa (See Opuntia canina, Opuntia retrorsa), [Hi-2],
roborensis (See Platyopuntia vitelliniflora),
rufida (See Opuntia rufida),
rubrogemmia (See Opuntia rubrogemmia).....[Hi-1],
salmiana (See Austrocylindropuntia salmiana, Opuntia salmiana, Opuntia spegazzinii), [Hi-2],
saxatilis (See Opuntia saxatilis).....[Hi-1],
soehrensii (See O. boliviensis, Opuntia soehrensii), [Hi-3],
" -v- grandiflora.....[Hi-2],
" -v- tilcarensis (See Opuntia soehrensii -v- tilcarensis, Opuntia tilcarensis),
" -v- transiens (See Opuntia soehrensii -v- transiens),
spinibarbis (See Opuntia spinibarbis).....[Hi-2],
" -v- grandiflora (See Opuntia spinibarbis -v- grandiflora).....[Hi-2],
sulphurea (See Opuntia sulphurea).....[Hi-2],
viridirubra (See Opuntia viridirubra).....[Hi-1],
vitelliniflora (See Opuntia vitelliniflora), [Hi-2],
vulgaris (See Opuntia vulgaris).....[Hi-2],

POLASKIA

chichipe (See Cereus chichipe, Lemaireocereus chichipe) [D], [I], [V], [DD], [JJ], [Az], [Bo-4], [Hx*] [Il], [Jd], [Af 60-2], [Aw 1-1970],

PORFIRIA

coahuilensis (See Mammillaria coahuilensis, Porfiria schwartzii), [Bo-5], [Hy], [Af 56-4], [Dy 2:210 Fig. 1926]
" -v- albiflora (See Mammillaria albiarmata, Mammillaria coahuilensis -v- albiflora).....[Bo-5],
schwartzii (See Mammillaria schwartzii, Porfiria coahuilensis)..... [F], [K], [Hx], [Jd], [Hl 4-1970],
" -v- albiflora (See Mammillaria albiarmata),

PRAECEREUS

amazonicus (See Monvillea amazonica),
apoloensis (See Monvillea apoloensis),
campinensis (See Monvillea campinensis, Cereus campinensis),
diffusus (See Monvillea diffusa).....[Jd],
jaenensis (See Monvillea jaenensis),
" -v- paucispinus (See Monvillea jaenensis -v- paucispina),
maritimus (See Monvillea maritima).....[Jd],
smithianus (See Cephalocereus smithianus, Monvillea smithiana).....[Jd],

PSEUDOACANTHOCEREUS

brasiliensis (See Acanthocereus brasiliensis), [Hi-1],

PSEUDOESPOSTOA

dautwitzii (See Espostoa dautwitzii),
" -f. cristata (See Espostoa dautwitzii -f. cristata),
melanostele (See Espostoa haagei, Espostoa melanostele)....[T], [DD], [Ar], [Bo-4], [Bv-1], [Hx], [Jd], [Da 2-1980], [Hl 4-1969],
" -v- canetensis (See Espostoa melanostele -v- canetensis),
" -v- churinensis (See Espostoa melanostele -v- churinensis),
" -v- inermis (See Espostoa melanostele -v- inermis).....[Bo-4], [Bt],

PSEUDOESPOSTOA.....(Cont'd),

melanostele.....(cont'd)
" -v- longispina (See Espostoa melanostele -v- longispina),
" -v- rubrispina (See Espostoa melanostele -v- rubrispina).....[Bo-6],
" -f. cristata.....[Hx],
nana (See Espostoa nana).....[Ar],

PSEUDOLOBIVIA

acanthoplegma (See Cinnabarinea acanthoplegma, Lobivia acanthoplegma, Lobivia cinnabarina ssp. acanthoplegma).....[Bo-6],
ancistrophora (See Echinopsis ancistrophora), [NN], [Av], [Bo-2], [Hu - Flower], [Jd], [Aj 29-2-1978],
arachnacantha (See Echinopsis ancistrophora ssp. arachnacantha, Echinopsis arachnacantha, Lobivia arachnacantha),
aurea (See Echinopsis aurea, Lobivia aurea, Salpingolobivia aurea),
[C], [I], [J*], [T], [EE], [HH], [NN], [Ar], [Bo-2] [Bv-1], [Cs], [Cz], [Hb], [Jd], [Aj 15-7-1964],
" -v- albiflora.....
" -v- calochrysea (See Lobivia aurea -v- fallax),
" -v- cylindrica (See Lobivia aurea -v- shaferi),
" -v- dobeana (See Lobivia aurea -v- dobeana),
" -v- elegans (See Lobivia aurea).....[Bo-2],
" -v- elongata (See Lobivia elongata),
" -v- fallax (Invalid name - See Echinopsis aurea -v- fallax, Lobivia aurea -v- fallax), [Bo-2],
" -v- grandiflora (See Echinopsis aurea, Lobivia aurea, Pseudolobivia aurea, Sal- Salpingolobivia aurea -v- grandiflora),
" -v- lariojensis.....
" -v- leucomalla (See Echinopsis aurea -v- leucomalla, Lobivia aurea -v- leucomalla),
" -v- quinesensis (See Echinopsis aurea -v- quinesensis, Echinopsis quinesensis, Lobivia aurea -v- quinesensis),
" -v- robustior (See Lobivia aurea -v- fallax, Salpingolobivia aurea -v- robustior),
" -v- shaferi (See Echinopsis aurea -v- shaferi, Echinopsis shaferi, Lobivia aurea -v- shaferi, Lobivia shaferi),
" -v- spinosissima.....
backebergii (See Lobivia backebergii).....[F], [M],

PSEUDOLOBIVIA.....(Cont'd),

baldiana (See Echinopsis baldiana),
boyuibensis (See Echinopsis boyuibensis, Echinopsis obrepanda).....
bridgesii (See Echinopsis bridgesii, Trichocereus bridgesii),
cachensis (See Lobivia cachensis, Lobivia saltensis),
caespitosa (See Lobivia maximiliana -v- caespitosa), [E],
callichroma (See Echinopsis obrepanda -v- purpurea),

calorubra (See Echinopsis calorubra, Echinopsis obrepanda)....
candicans (See Trichocereus candicans),
cardenasiana (See Echinopsis ancistrophora ssp. cardenasiana),
carmineoflora (Invalid name - See Echinopsis carmineoflora, Echinopsis obrepanda -v- calorubra, Lobivia carmineoflora),
[DD], [Av], [Bo-2],
comarapana (See Echinopsis comarapana),
ducis-pauli (See Lobivia ferox -v- longispina, P. longispina),
" -v- rubriflora (See ferox -v- lomgispina),
eyriesii (See Echinopsis eyriesii),
" -v- grandiflora (See Echinopsis eyriesii -v- grandiflora),
" -v- inermis (See Echinopsis eyriesii -v- inermis),
falcata (See Echinopsis falcata),
ferox (See Echinopsis ferox, Lobivia ferox, Soehrensia ferox).....[RR], [Bo-2],
" -v- nigra (See Lobivia ferox -v- longispina),
fiebrigii (See Echinopsis fiebrigii, Echinopsis obrepanda)....
frankii (See Echinopsis obrepanda),
[Bo-6],
graulichii (See Echinopsis ancistrophora, Lobivia graulichii),
hamatacantha (See Echinopsis ancistrophora -v- hamatacantha, Echinopsis hamatacantha),
[Ar], [Ay], [Bo-2], [Bv-1], [Bu], [Aj 27-7-1976 - Flower], [Bi 1-2],
kermesina (See Echinopsis kermesina, Echinopsis mamillosa -v- kermesina, Lobivia kermesina),
[I], [J*], [EE], [PP], [QQ], [Av], [Ay], [Bo-2] [Cr], [Cs], [Hb], [Jd], [Af 27-6], [Af 32-1],
kratochviliana (See Echinopsis ancistrophora -v- kratochviliana, Echinopsis kratochviliana, Lobivia kratochviliana),
[C], [DD], [NN], [Bo-2], [Bi 1-1],

PSEUDOLOBIVIA.....(Cont'd),

lecoriensis (See Lobivia ferox -v- longispina, Echinopsis lecoriensis),
leucorhodantha (See Echinopsis ancistrophora, Echinopsis leucorhodantha)....[Flower - Hu],
longispina (See Echinopsis longispina, Lobivia ferox -v- longispina).....[H], [S], [RR], [Av], [Bo-2], [Hl 1-1972]
" -v- nigra (See Lobivia ferox -v- longispina), [Bo-2], [Hy],
luteiflora (Invalid name - See Echinopsis aurea, Lobivia aurea, Lobivia luteiflora, Pseudolobivia aurea).....[DD],
mamillosa (See Echinopsis mamillosa),
" -v- kermesina (See Echinopsis kermesina, Pseudolobivia kermesina),
obrepanda (See Echinopsis obrepanda, Echinopsis fiebrigii).....[Ar], [Bo-2], [Hx*], [Jd],
" -v- fiebrigii (See Echinopsis obrepanda, Echinopsis fiebrigii).....[Bo-2],
" -v- purpurea (See Echinopsis obrepanda -v- purpurea, Echinopsis purpurea),
orozasana (See Echinopsis mamillosa -v- orozasana, Echinopsis mamillosa).....[DD], [Hx*],
orurensis (See Lobivia formosa ssp. tarijensis -v- bertramiana, Helianthocereus orurensis),
pelecyrhachis (See Echinopsis ancistrophora), [Bo-2], [Cr], [Ig*],
" -v- lobivioides (See Echinopsis ancistrophora),
pictiflora (See Lobivia ferox -v- longispina, Lobivia pictiflora),
pojoensis (See Echinopsis ancistrophora ssp. pojoensis, Echinopsis pojoensis, Lobivia pojoensis),
polyancistra (See Echinopsis ancistrophora -v- polyancistra, Echinopsis polyancistra), [I], [DD], [Bo-2], [Bi 1-2],
potosina (See Echinopsis potosina, Lobivia ferox -v- potosina, Lobivia potosina).....[Bo-2], [Hy],
purpurea (See Echinopsis obrepanda -v- purpurea, Lobivia purpurea),
rojasii (See Echinopsis obrepanda, Echinopsis rojasii)
" -v- albiflora (See Echinopsis obrepanda)..[Hu],
tamboensis (See Echinopsis mamillosa, Echinopsis mamillosa -v- tamboensis, Echinopsis tamboensis),
tapecuana (See Echinopsis tapecuana),
toralapana (See Echinopsis obrepanda -v- purpurea),
torrecillasensis (See Echinopsis ancistrophora ssp. arachnacantha -v- torrecillasensis, Echinopsis torrecillasensis, Lobivia arachnacantha -v- torrecillasensis, Lobivia torecillasensis).....[F], [X], [Ar], [Bo-2],

PSEUDOLOBIVIA.....(Cont'd),

wilkeae (See Echinopsis longispina, Lobivia ferox -v- longispina, Lobivia longispina),
[Bo-6],
" -v- carminantha (Invalid name - See Lobivia ferox -v- longispina),

PSEUDOMAMMILLARIA

camptotricha (See Dolichothele camptotricha, Mammillaria camptotricha),
[NN], [Gr],
decipiens (See Dolichothele decipiens, Mammillaria decipiens),

PSEUDOMITROCEREUS

chrysomallus (See Mitrocereus fulviceps, Pseudomitrocereus fulviceps).....[Hx:131],
fulviceps (See Mitrocereus fulviceps, Mitrocereus columna-trajani sensu Daws., Pachycereus chrysomallus, Pseudomitrocereus chrysomallus),
[Jd], [Aj 15-12-1964],

PSEUDONOPALXOCHIA

conzattianum (See Nopalxochia conzattianum),

PSEUDOPILOCEREUS

arrabidae (See Cephalocereus arrabidae, Pilocereus arrabidae, Pilosocereus arrabidae),
aurilanatus (See Pilosocereus aurilanatus),
[Aj 31-2-1980],
aurisetus (See Pilocereus aurisetus, Pilosocereus aureisetus, Pilosocereus coerulescens, Pilosocereus glaucescens),
azureus (See Cereus azureus, Pilosocereus azureus),
[Jd], [Ah-1980], [Aj 26-11-1975], [Aw 2-1981],
[Ba 7-1981],
bradyi (See Pilosocereus bradyi),
[Aj 31-2-1980],
catingicola (See Cephalocereus catingicola, Pilosocereus catingicola),
chrysacanthus (See Cephalocereus chrysacanthus, Pilosocereus chrysacanthus),
chrysostele (See Cephalocereus chrysostele, Pilocereus chrysostele, Pilosocereus chrysostele)
clarosensis (See Pilosocereus clarosensis),
cuyabensis (See Cephalocereus cuyabensis, Pilosocereus cuyabensis),

PSEUDOPILOCEREUS.....(Cont'd),

diersianus (See Pilosocereus diersianus).....
[Aj 32-4-1981],
flavipulvilanatus (See Pilosocereus flavipulvilanatus)
[Ba 6-1979],
floccosus (See Cephalocereus floccosus, Pilosocereus
floccosus),
fulvilanatus (See Cephalocereus fulvilanatus, Pilosocereus fulvilanatus),
[DD-522], [Hx*], [Jd], [Ad July, 1978],
[Aj 24-7-1973], [Aj 33-12-1982],
glaucescens (See Pilocereus aurisetus, Pilocereus
glaucescens, Pilosocereus coerulescens,
Pilosocereus glaucescens),
" -v- multiflorus.....
glaucochrous (See Cephalocereus glaucochrous, Pilocereus glaucochrous, Pilosocereus glaucochrous).....[NN], [Jd],
gounellei (See Cephalocereus gounellei, Pilosocereus
gounellei),
" -v- zehntneri (See Pilosocereus gounellei -v-
zehntneri),
hapalacanthus (See Cephalocereus hapalacanthus, Pilocereus hapalacanthus, Pilosocereus
hapalacanthus),
houlletii (See Cephalocereus leucocephalus, Pilocereus
houlletii, Pilocereus leucocephalus, Pilosocereus houlletii, Pilosocereus leucocephalus),
juaruensis.....[Aj 29-7-1978],
luetzelburgii (See Cephalocereus luetzelburgii, Pilocereus luetzelburgii),
machrisii (See Cephalocereus machrisii),
[Ba 4-1977],
magnificus (See Cephalocereus magnificus, Pilosocereus
magnificus).....[Z], [DD-523], [Ag-1981],
[Hx*], [Af Vol. 44-1972],
medirisii (See Cephalocereus medirisii, Pilosocereus
medirisii),
minensis (See Cereus minensis, Cipocereus minensis,
Pilosocereus minensis).....[Aj 31-4-1980],
mucosiflorus (See Cephalocereus mucosiflorus, Pilosocereus mucosiflorus).....
[Aj 28-9-1977],
nobilis (See Cephalocereus nobilis, Pilocereus nobilis,
Pilosocereus nobilis),
oligolepis (See Cephalocereus oligolepis, Pilosocereus
oligolepis),
pachycladus (See Cephalocereus pachycladus, Pilosocereus pachycladus),
[Ar], [Av], [Ay], [Ah-1980],
" -v- chapensis.....

PSEUDOPILOCEREUS.....(Cont'd),

parvus (See Cephalocereus parvus, Pilosocereus parvus).....[Aj 33-5-1982],
pentaedrophorus (See Cephalocereus pentaedrophorus, Pilocereus pentaedrophorus, Pilosocereus pentaedrophorus),
piauhyensis (See Cephalocereus piauyensis, Pilosocereus piauyensis),
rosae (See Cephalocereus rosae, Pilosocereus rosae),
rupicola (See Cephalocereus rupicola, Pilosocereus rupicola),
ruschianus (See Cephalocereus ruschianus), [Aj 31-2-1980],
salvadorensis (See Austrocephalocereus salvadorensis, Pilosocereus salvadorensis).....[Jd],
saxatilis (See Pilosocereus werdermannianus, Pseudopilocereus werdermannianus),
" -v- densilanatus (See Pilosocereus densilanatus, Pilosocereus werdermannianus -v- densilanatus),
" -v- diamantinensis (See Pilosocereus werdermannianus -v- diamantinensis), [Aj 31-4-1980],
sergipensis (See Pilosocereus sergipensis),
superfloccosus (See Cephalocereus superfloccosus, Pilosocereus superfloccosus), [Ar], [Ay], [Ba 10-1980],
supremus (See Pilosocereus magnificus, P. magnificus)
tuberculatus (See Cephalocereus tuberculatus, Pilosocereus tuberculatus),
tuberculosus (See Pilosocereus tuberculosus),
ulei (See Cephalocereus ulei, Cephalocereus robustus -v- ulei, Pilocereus ulei, Pilosocereus ulei),
vilaboensis (See Cephalocereus vilaboensis, Pilosocereus vilaboensis).....
werdermannianus (See Pilocereus werdermannianus), [Aj 26-4-1975],
Sp. .....[Hv],

PSEUDORHIPSALIS

acuminata (See Disocactus acuminatus),
alata (See Disocactus alatus).....[RR], [Dq*], [Hx*],
harrisii (See P. alata),
himantoclada (See Disocactus himantocladus, Wittia costaricensis).....[RR], [Jd],
macrantha (See Disocactus macranthus).....[Al*], [Ar], [Ay], [Bb], [Bo-2], [Bv-1], [Co], [Cz], [Hx*], [Ia], [Jd], [Af 14-2], [Aw 4-1976],

PSEUDOSELENICEREUS

St. Vincent.....[HH],

PSEUDOSOLICIA

valdeziana (See Gymnocactus valdezianus, Normanbokea valdeziana, Pelecyphora valdeziana, Thelocactus valdezianus, Turbinicarpus valdeziana).....[Hy],
" -v- albiflora.....[Hy],
" -f. cristata.....[Hy],

PSEUDOZYGOCACTUS

epiphylloides (See Hatiora epiphylloides, Rhipsalis epiphylloides),
[Bo-2], [Hx*], [Hy],
" -v- bradei (Invalid name),
[DD], [Ay], [Bo-2], [Dh-1]
Sp. .....[HH], [Bo-2],

PTEROCACTUS

araucanas.....[Af 59-1],
australis (See Opuntia australis).....
[Eh 22:17, 1950 - Description],
" -v- arnoldianus.....[Bo-1],
decipiens (See P. tuberosus -v- decipiens),
[Ay], [Bt], Hx*], [Aj 19-5-1968],
fischeri.....[G], [RR], [Bo-1], [Af 53-3], [Af 57-2],
[Aw 2-1968], [Bi Vol. 4, #1, 1986],
" -f. monstrosus.....[Aw 2-1968],
hickenii.....[RR*], [Bo-1], [Hx*], [Hy*],
kuntzei (See P. tuberosus),
[D], [Bk], [If], [Iz], [Bi 4-1-1986],
marenae (See Marenopuntia marenae, Opuntia marenae),
pumilus.....[Bo-1],
skottsbergii.....[Bo-1],
tuberosus (See P. kuntzei).....[H], [RR], [Ar], [Av],
[Bl], [Bo-1], [Hv], [Hx*], [Hy], [Jd],
[Af 1-4], [Af 6-7], [Af 50-1], [Af 56-3],
[Ah-1985], [Au 37-3], [Aw 2-1968],
" -v- decipiens (See P. decipiens),
" -f. monstrosus.....[Aw 2-1968],
valentinii (See P. australis),

PTEROCEREUS

foetidus (See Anisocereus foetidus),
[DD], [Az], [Bo-4], [Hx*], [Hy],
[Af 26-6], [Fw 4:135 Fig., 1954],

PTEROCEREUS.....(Cont'd),

gaumeri (See Anisocereus gaumeri, Pachycereus gaumeri), [Az],

PYGMAEOCEREUS

akersii (See Arthrocereus akersii, P. bylesianus, P. rowleyanus).....[Flower - Ba 2-1979],
bylesianus (See Arthrocereus bylesianus)....[DD], [Bo-2], [Bo-6], [Hi-4], [Hx*], [Hy], [Jd], [Ag-1983], 1988], [Hl 3-1968], [Hl 4-1982],
densiaculeatus (Invalid name - See Arthrocereus densi-aculeatus).....[DD], [Ay],
" -f. cristata.....[Ah-1980],
familiaris.....[Hi-4],
" -f. (von Atico).....[Hi-4:1581],
napinus.....[Aj 28-8-1977], [Hl 3-1968],
rowleyanus (Invalid name - See P. bylesianus), [DD], [Ay], [Bo-6],
vespertinus.....(See P. bylesianus),

PYGMAEOLOBIVIA

haagei (See Digitorebutia haagei, Mediolobivia haagei).....[NN],

PYRRHOCACTUS

aconcaguensis (See Horridocactus aconcaguensis, Neopor-teria curvispina -v- aconcaguensis, P. horridus -v-aconcaguensis),
" -v- orientalis (See P. horridus -v- orientalis)
andicolus (See Horridocactus andicolus, Neoporteria curvispina -v- andicola).....[Ba 10:131, 1959],
" -v- andicolus.....[Hi-3],
" -v- descensdens (See Horridocactus andicolus -v- descendens).....[Hi-3],
" -v- mollensis (See Horridocactus andicolus -v- mollensis).....[Ba 10:131, 1959],
" -v- robustus (See Horridocactus robustus), [Hi-3],
andreaeanus (See Acanthocalycium andreaeanum, Neochil-enia andreaeana, Neoporteria andreaeana)..... [Ba 10:131, 1959],
aricensis (See Horridocactus aricensis, Neochilenia aricensis, Neoporteria aricensis).....[Hi-3],
armatus (See Horridocactus armatus, Neoporteria armata).....[Hi-3],
" -f. cristata.....[Hi-3],
aspillagai (See Horridocactus aspillagai, Neochilenia aspillagai, Neoporteria aspillagai)....[Hi-3],

PYRRHOCACTUS.....(Cont'd),

atrospinosus (Invalid name - See Malacocarpus catamarcensis, P. catamarcensis, P. setiflorus), [DD],
atroviridis (See Horridocactus atroviridis, Neoporteria tuberisulcata -v- atroviridis), [Hi-3],
bulbocalyx (See Neoporteria bulbocalyx)...[DD], [PP], [Ay], [Bo-6], [Gr], [Hi-2], [Hx*], [Hy], [Il], [Jd], [Af 58-3], [Ah-1985], [Hl 4-1972],
calderanus (See Neochilenia calderana).....[Hi-3],
carrizalensis (See Neoporteria carrizalensis, Neoporteria curvispina -v- carrizalensis)....[Hi-3],
catamarcensis (See Malacocarpus catamarcensis, P. atrospinosus, P. setiflorus), [Bo-3], [Hi-2], [Hy], [Jd],
centeterius (See Echinocactus centeterius),
chaniarensis.....[Hi-3],
chilensis (See Horridocactus chilensis, Neochilenia chilensis, Neoporteria chilensis), [Hi-3], [Ba 10:131, 1959], [Hl 1-1982],
" -v- albidiflorus.....[Hi-3],
" -v- borealis.....
" -v- chilensis.....[Hi-3],
choapensis (See Horridocactus choapensis, Neoporteria choapensis).....[Hi-3],
chorosensis (See Neochilenia chorosensis, Neoporteria chorosensis).....[Hi-3],
coliguayensis.....[Hi-3],
confinis (See Horridocactus confinis).... [Hi-3], [Ba 4, 1961],
crispus (See Horridocactus crispus, Neochilenia crispa, Neochilenia nigriscoparia, Neoporteria crispa), [Hi-3],
curvispinus (See Horridocactus curvispinus, Neoporteria curvispina).....[Hi-3],
" -v- australis.....[Hi-3],
" -v- campanaensis [Hi-3], [Af 54-3],
" -v- combarbalensis (See Horridocactus curvispinus -v- combarbalensis),
" -v- curvispinus.....[Hi-3],
" -v- echinus (See Horridocactus echinus, P. echinus),
" -v- " -f. minor (See P. echinus -v- minor, P. floccosus -v- minor),
" -v- felipensis (See Horridocactus curvispinus -v- felipensis).....[Hi-3],
" -v- fusciflorus.....[Hi-3],
" -v- mostazalensis.....[Hi-3],
" -v- " -f. (von Coliguay).....[Hi-3],
" -v- nidularis.....

curvispinus.....(cont'd),
" -v- petorcensis.....
" -v- santiagensis (See Horridocactus curvispinus -v- santiagensis),
" -v- tilamensis.....
deherdtianus (See Neochilenia deherdtiana, Neochilenia occulta, Neoporteria occulta, P. occultus), [Af 55-3],
dimorphus (See Neochilenia dimorpha, Neoporteria dimorpha).....[Hi-3],
dubius (See P. catamarcensis).....[Bo-3],
echinus (See Horridocactus echinus, Neochilenia echina) [Hi-3],
" -v- minor (See P. floccosus -v- minor) [Aj 22-12-1971],
engleri (See Horridocactus engleri),.....[Hi-3],
eriosyzoides (See Horridocactus eriosyzoides, Neoporteria eriosyzoides)...[Hi-3], [Ba 49:1959],
" -v- domeykoensis.....[Hi-3],
floccosus (See Horridocactus floccosus, Neochilenia floccosa, Neoporteria eriocephala).....[Hi-3],
" -v- minor (See P. echinus -v- minor)...[Hi-3],
floribundus (See Neochilenia iquiquensis, Neoporteria iquiquensis, Reicheocactus floribundus), [Hi-3],
froehlichianus (See Horridocactus froehlichianus),
garaventai (See Horridocactus garaventai, P. subaianus).....[Hi-3],
glaucescens (See Neochilenia glaucescens, Neoporteria eriocephala -v- glaucescens).....[Hi-3],
gracilis (See Neochilenia gracilis), [Hi-3], [Ba 1962:129],
grandiflorus (See Horridocactus grandiflorus, Neoporteria curvispina -v- grandiflora, Neoporteria grandiflora).....[Hi-3],
griseus (See Acanthocalycium griseum).....[Bo-6],
heinrichianus (See Horridocactus heinrichianus),
horridus (See Horridocactus horridus, Horridocactus tuberisulcatus, P. tuberisulcatus), [Hi-3],
" -v- aconcaguensis (See P. aconcaguensis), [Hi-3],
" -v- minor.....
" -v- mutabilis.....[Hi-3],
" -v- orientalis (See P. aconcaguensis -v- orientalis),
" -v- robustus (See Horridocactus robustus, P. robustus).....[Hi-3],
huascensis (See Neochilenia huascensis, Neoporteria huascensis).....Hi-3],

PYRRHOCACTUS.....(Cont'd),

intermedius (See Neochilenia intermedia, Neoporteria intermedia).....
[Hi-3],
iquiquensis (See Neochilenia iquiquensis, Neoporteria iquiquensis, P. floribundus, Reicheocactus floribundus),
jussieui (See Neochilenia jussieui, Neoporteria jussieui).....[Hi-3],
" -v- australis.....[Hi-3],
" -v- spinosior.....[Hi-3],
kesselringianus (See Horridocactus kesselringianus),
krausii.....[Hi-3],
kunzei (See Neochilenia kunzei).....[Hi-3],
limariensis (See Neoporteria nigrihorrida -v- limariensis).....[Hi-3],
lissocarpus (See Horridocactus lissocarpus, Neochilenia lissocarpa).....[Hi-3],
" -v- gracilis (See Horridocactus lissocarpus -v- gracilis),
mamillarioides (See Neoporteria mamillarioides),
marksianus (See Horridocactus marksianus).....[Hi-3],
" -v- tunensis.....
megliolii.....[Hx*],
melanacanthus (Invalid name).....[DD],
neohankeanus (See Neochilenia hankeana -v- taltalensis)
[Hi-3],
" -v- densispinus .....[Hi Vol. 3],
" -v- elongatus.....[Hi Vol. 3],
" -v- flaviflorus .....[Hi Vol. 3],
" -f. woutersianus (See Delaetia woutersiana),
[Hi Vol. 3],
nigricans (See Horridocactus nigricans, Neoporteria nigricans),
occultus (See Neochilenia deherdtiana, Neochilenia occulta, Neoporteria occulta, P. deherdtianus).....[Hi Vol. 3], [Af 55-3],
odoriflorus (See Neochilenia odoriflora)...[Hi Vol. 3],
pachacoensis.....[Hx*], [Aj 26-4-1975],
pamaensis.....[Hi Vol. 3],
paucicostatus (See Horridocactus paucicostatus, Neochilenia paucicostata, Neoporteria paucicostata).....[Hi Vol. 3],
" -v- viridis (See Horridocactus paucicostatus -v- viridis, Neochilenia paucicostata -v- viridis).....
[Hi Vol. 3],
pilispinus (See Neochilenia pilispina, Neoporteria pilispina).....
[Hi Vol. 3],
(cont'd)

pilispinus.....(cont'd),
" -f. pygmaeus (See Horridocactus pygmaeus, Neochilenia pilispina -f. pygmaea, Neochilenia pygmaea, Neoporteria pilispina -f. pygmaea, P. pygmaeus)
platyacanthus.....
pulchellus (See Neochilenia pulchella), [Bo-6], [Hi Vol. 3],
pygmaeus (See Horridocactus pygmaeus, Neochilenia pygmaea, Neoporteria pilispina -f. pygmaea), [Bo-6], [Hi Vol. 3], [Hl 5-1982],
" -f. (3 forms).....[Hi Vol. 3],
reconditus (See Neochilenia recondita, Thelocephala recondita),.....[Hi Vol. 3],
residuus (See Horridocactus residuus, Neochilenia residua).....[Hi Vol.3],
robustus (See Horridocactus robustus, P. horridus -v- robustus),
" -v- vegasanus (See Neoporteria tuberculatus -v- vegasana, P. robustus),
rupicolus (See Horridocactus rupicolus, Neochilenia rupicola, Neoporteria rupicola, P. taltalensis),
sanjuanensis (See Pyrrhocactus catamarcensis), [Bo-3], [Hx*],
saxifragus (See Neochilenia saxifraga)....[Hi Vol. 3],
scoparius.....[Hi Vol. 3], [Succulenta 5:51-53, 1962],
setiflorus (Backbg.) (See Malacocarpus catamarcensis, Pyrrhocactus catamarcensis, Pyrrhocactus atrospinosus (Backbg.)).....[Bo-6], [Hx*],
setosiflorus (Backbg.) (See P. setiflorus, P. catamarcensis),
setosiflorus (Ritt), (See Neochilenia setosiflorus), [Hi Vol. 3], [Ba 1962:70],
" -v- intermedius (See Neochilenia setosiflora -v- intermedia).....[Hi Vol. 3],
simulans (See Horridocactus simulans, Neochilenia simulans).....[Hi Vol. 3],
strausianus (See Malacocarpus strausianus, Neoporteria strausiana)..[E], [Bo-3], [Hx*], [Hy], [Iu],
subaianus (Invalid name - See Horridocactus garaventai, P. garaventai).....[Bo-3],
taltalensis (Hutch.) (See Horridocactus rupicolus, Neochilenia rupicola, Neochilenia taltalensis, Neoporteria rupicola).....[Hi Vol. 3],
taltalensis (Ritt) (See Pyrrhocactus neohankeanus
" -v- flaviflorus (See P. hankeanus -v- flaviflorus, Neoporteria hankeana),
totoralensis (See Neochilenia totoralensis, Neoporteria totoralensis).....[Hi Vol. 3],

PYRRHOCACTUS.....(Cont'd),

transiens.....[Hi Vol. 3],
transitensis (See Neochilenia transitensis),
[Hi Vol. 3], [Hx*], [Hl 5-1972],
trapichensis (See Horridocactus trapichensis, Neochilenia trapichensis),
[Hi Vol. 3],
truncatipetalus.....[Hi Vol. 3],
tuberisulcatus (See Horridocactus horridus, Horridocactus tuberisulcatus, Pyrrhocactus horridus).....[E], [Bv-1], [Dq],
[Ad March, 1980], [Af 34-2],
umadeave (See Neoporteria umadeave).....
[NN], [Bo-3], [Cq], [Hx*], [Hy], [Ae-1981],
[Aj 29-5-1978], [Hl 2-1974], [Hl 6-1969],
" -v- marayesensis (Invalid name).....
[DD], [Il], [Ae-1981],
vallenarensis (See Horridocactus vallenarensis, Neochilenia vallenarensis, Neoporteria vallenarensis).....[Hi Vol. 3],
vexatus (See Neochilenia vexata).....[Hi Vol. 3],
" -v- horridus.....
" -f. (2 forms).....[Hi Vol 3],
villicumensis.....[Hx*],
vollianus (See Neoporteria volliana),
" -v- breviaristatus (See Neoporteria volliana -v- breviaristata),
wagenknechtii (See Neochilenia wagenknechtii, Neoporteria wagenknechtii, Neoporteria ritteri).....[Hi Vol. 3], [Hl 2-1981],
Sp. FK 19-77 C .....[Af 54-3],

QUIABENTIA

chacoensis (See Q. pflanzii, Q. verticillata),
[D], [Ay], [Bo-1], [Dq], [Hx*], [Bi 1-2],
[Da 1-1981], [Iv 9-1953],
" -v- jujuyensis.....[Bo-1],
pereziensis.....[Bo-1], [Hi Vol. 3],
pflanzii (See Q. chacoensis, Q. verticillata),
[Cu 2-1980], [Hh Sept., 1976],
verticillata (See Q. chacoensis, Q. pflanzii),
zehntneri (See Pereskia zehntneri),
[K], [DD], [Bl], [Bo-1], [Hi Vol. 1], [Hx*],
[Im], [Af 1-4],
zinniaeflorus (See Pereskia zinniaeflora, Rhodocactus zinniaeflorus),

Sp. FR-1200 (Filadelfia).....

RAPICACTUS

mandragora (See Gymnocactus mandragora, Neolloydia mandragora, Thelocactus mandragora)....[Bl],
subterraneus (See Gymnocactus subterraneus, Neolloydia subterranea, Thelocactus subterraneus), [Bl], [Hx*], [Jd], [Aj 11-9-1960],

RATHBUNIA

alamosensis (See R. neosonorensis)....[G], [M*], [HH], [PP], [RR], [Ax], [Ay], [Az], [Bv-1], [Cx], [Cz], [Gr], [Hx*], [Hy], [Jd], [Af 2-11], [Cu 7-1-1938],
" -f. cristata.....[G],
kerberi (See Stenocereus kerberi), [G], [Az], [Ad Feb. 1978], [Af 57-4],
neosonorensis (Invalid name - See R. alamosensis), [DD], [Bo-4],
pseudosonorensis (See R. sonorensis),
sonorensis.....[Dq], [Hy], [Af 16-5],

RAUHOCEREUS

riosaniensis (See Browningia riosaniensis)..[I], [DD], [Bo-2], [Bt], [Hx*], [Hy], [Jd], [Af 50-1], [Ah-1980], [Hl 1-1972], [Hl 6-1982],
" -v- jaenensis.....[Ay], [Hi Vol. 4],

REBULOBIVIA

einsteinii (See Mediolobivia einsteinii, Rebutia einsteinii),
haagei (See Rebutia haagei, Rebutia pygmaea, Lobivia haagei),

REBUTIA

albiareolata (See Aylostera albiareolata), [Hi Vol. 2], [Hh July, 1976],
albiflora (See Aylostera albiflora)....[Ay], [Bp], [Hi-Vol. 2], [Il], [Iz], [Af 53-4], [Ag-1986], [Ba 9-1977], [Hh July, 1975], [Hh 3, #5&6, 1979], [Bi Vol. 4, #6],
albipilosa (See Aylostera albipilosa), [Av], [Bp], [Hi Vol. 2], [Af 53-2], [Cu 3-1965], [Hh July, 1976], [Hw-1986],
albiseta (See R. auriflora -f. albiseta),
albopectinata (See Aylostera albopectinata, Digitorebutia albopectinata, Mediolobivia albopectinata, Rebutia densipectinacea)..[Hi Vol. 2], [Hv], [Hx*], [Aj 23-9-1972], [Aj 30-1-1979], [Hh Sept., 1973], [Hh Vol. 3 #5 & 6, 1979],

REBUTIA.....(Cont'd),
allegraiana (See R. chrysacantha -v- elegans),
almeyeri (Invalid name - See R. senilis -f. stuemeri),?
archibuiningiana (Ritt.) (See R. buiningiana),
[Hh Vol. 3 #1, 1978], [Hi Vol. 2],
arenacea (See Sulcorebutia arenacea),
[Ar], [Ay], [Bo-3], [Aj 5:85, 1954], [Aj 1 c, 9:149, 1958], [Hf 23:94, 1951],
areolata.....
auranitida (See Mediolobivia auranitida),
" -v- flaviflora (See Mediolobivia auranitida -v- flaviflora),
" -f. gracilis (See Mediolobivia auranitida -v- gracilis),
aurantiaca.....
" -v- rubelliflora (See Mediolobivia aureiflora -v- rubelliflora, R. aureiflora -f. rubelliflora).....[Bo-3],
aureicentra.....
aureiflora (See Mediolobivia aureiflora),
[Bp], [Iz], [Jd], [Af 15-6],
" -v- albilongiseta (See Mediolobivia aureiflora -v- albilongiseta),
" -f. albiseta (See Mediolobivia aureiflora -v- albiseta, R. aureiflora),
" -v- blossfeldii (See R. leucanthema -v- coccinifera).....[Hh 2-5-1976],
" -f. boedekeriana (See Mediolobivia aureiflora -v- boedekeriana, Mediolobivia boedekereiana),
" -v- densispina (See R. aureiflora),
" -f. duursmaiana (See Mediolobivia aureiflora - -v- duursmaiana),
" -v- elegans (See Mediolobivia elegans),
" -f. kesselringiana (See Mediolobivia kesselringiana, R. kesselringiana) [Hh March, 1976],
" -f. longiseta (See Mediolobivia aureiflora -v- longiseta, R. aureiflora), [Hh Sept., 1975],
" -v- nidulans.....
" -f. rubelliflora (See Mediolobivia aureiflora -v- rubelliflora, Mediolobivia rubelliflora)....[Jd], [Ad April, 1978], [Ba 7/8:100, 1963], [Hh 2-5-1976],
" -f. rubriflora (See Mediolobivia aureiflora -v- rubriflora, R. blossfeldii),
" -f. sarothroides (See Mediolobivia aureiflora -v- sarothroides, Mediolobivia sarothroides),
" -f. (from Quebrada del Toro, Salta, Arg.) [Hh 2-5-1976],

REBUTIA.....(Cont'd),

aureiflora.....(cont'd),
" -f. (from Tambo, Tarijam Bol.),
beryllioides (See R. wessneriana -v- beryllioides),
binnewaldiana.....
blossfeldii (See R. aureiflora -v- blossfeldii, R. melachlora, R. aureiflora -f. rubriflora), [Cu 2-1961],
boedekeriana (See Mediolobivia boedekeriana, Mediolobivia aureiflora -f. boedekeriana)...[Cr],
brachyantha (See Mediolobivia brachyantha, R. rauschii, R. steinmannii, Sulcorebutia breviflora), [Aj 16-4-1965],
brunescens (See Digitorebutia brunescens, Mediolobivia brunescens)....[DD-484], [Hx*], [Aj 23-9-1972]
brunneiradicata (See Mediolobivia brunneoradicata), [Hi Vol. 2],
buiningiana (Ritt.) (See R. archibuiningiana, R. spinosissima).....[Hh May, 1974],
buiningiana (Rausch) (See Aylostera buiningiana), [Hx*], [Aj 23-4-1972],
bulbispina (See R. brachyacantha),
caineana (See Sulcorebutia caineana).....[Af 38-4],
cajasensis (See Aylostera heliosa -v- cajasensis, R. heliosa -v- cajasensis).....[Hi Vol. 2],
calliantha (See R. wessneriana -v- calliantha), [T], [DD], [Bo-3], [Bu], [Hx*], [Hy], [Jd],
" -v- beryllioides (See R. wessneriana -v- beryllioides)...[F], [H], [Cu 2-1965],
" -v- bewerunge.....[Cu 4-1961],
" -v- breviseta (See R. wessneriana -v- calliantha),
" -v- calliantha (See R. wessneriana -v- calliantha),
" -v- densiseta.....
" -v- kariusiana (See R. minuscula -v- violaciflora -f. kariusiana).....[D],
" -v- krainziana (See R. krainziana), [H], [Bp], [Ba 12-1977],
" -v- " -f. breviseta.....[Cu 2-1965],
camargoensis (camarguensis) (See Aylostera camargoensis) [Ba 3-1976], [Hh Vol. 3, #3 & 4, 1978],
canacruzensis (Digitorebutia canacruzensis), [Aj 27-3-1976],
candiae (Sulcorebutia candiae).....[Af 33-4],
canigueralii (See Sulcorebutia canigueralii),..... [Af 36-1], [Hw-1986],
caracarensis (See Sulcorebutia caracarensis), [Af 42-1],
carmeniana (See Digitorebutia carmeniana),..... [Aj 29-5-1978],

REBUTIA.....(Cont'd),

carminea (See R. minuscula -v- violaciflora, R. violaciflora -v- carminea).....[Hx],
christinae (See Aylostera christinae, Digitorebutia christinae].....[Ah-1986], [Aj 26-7-1975],
chrysacantha (See R. senilis ssp. chrysacantha), [Ar], [Bo-3], [Bp], [Cr], [Hy], [Ig*],
" -v- elegans (See R. senilis -v- elegans), [Hx],
" -v- iseliniana (See R. senilis -v- iseliniana)
" -v- kesselringiana (See R. senilis -v- kesselringiana),
" -v- lutea (Hybrid),
cincinnata (See Digitorebutia cincinnata, R. steinmannii -v- cincinnata).....[Aj 27-1-1976],
cintiensis (See Aylostera cintiensis), [Hi Vol. 2], [Hh May, June, 1977],
citricarpa (See R. xanthocarpa -f. citricarpa)....[Ia],
" -v- salmonea (See R. xanthocarpa -f. salmonea),
colorea (See Mediolobivia colorea), [Hi Vol. 2], [Aj 28:78],
conoidea (See R. einsteinii -f. conoidea, Mediolobivia conoidea),
" -v- columnaris (See R. einsteinii -f. columnaris),
corroana (See Sulcorebutia corroana, Weingartia corroana).....[Hx*], [Af 43-6],
costata (See Mediolobivia costata)..... [Bo-3], [Bp], [Bt],
" -f. eucaliptana (See R. eucaliptana), [Gc 414], [Ee 7/8:102],
" -f. pilifera.....[Cu 27-40],
crispata (See Sulcorebutia crispata).....[Bu],
cylindrica (See Sulcorebutia cylindrica),
dasyphrissa (See R. xanthocarpa -f. dasyphrissa),
deminuta (See Aylostera deminuta).....[Bp], [Bu], [Bv-1], [Il], [Iu], [Cu 4-1961], [Hh July, 1976],
densipectinacea (See R. albopectinata),
densipectinata.....[Hh Vol. 3, #5 & 6, 1979],
diersiana (See Aylostera diersiana), [Hx*], [Aj 26-2-1975],
" -v- atrovirens (See Lobivia pygmaea -v- nigrescens),.....[Aj 26-2-1975],
" -v- minor.....[Ba 11-1979],
digitiformis (See Digitorebitia digitiformis, Mediolobivia digitiformis, R. pygmaea)...[Bu],
donaldiana.....[Hh July, 1974], [Iz], [Af 50-2], [Af 52-1], [Af 55-4], [Hw-1986],
durispina.....[Ba 12-1977],
dumcresmaniana.....[Bv-1],
duursmaiana (See R. aureiflora -f. duursmaiana).[Bv-1],

einsteinii (See Mediolobivia einsteinii, Mediolobivia schmiedcheniana -v- einsteinii),
[Hh May, 1974], [Hh May, 1976], [Hl 4-1977],
" -f. columnaris.....[Beitr. Sukk. u. pflege 4],
" -f. conoidea.....[Beitr. Sukk. u. pflege 4],
[Hh May, 1976],
" -v- gonjianii (See Rebutia gonjianii),
[Gd 15:132], [Iz], [Hl 5-1983],
" -f. karreri.....[Hh May, 1976],
" -f. rubriviridis.....[Hh May, 1976],
" -f. schmiedcheniana (See Mediolobivia schmiedcheniana),
[Bp], [Beitr. Sukk. u. pflege-37], [Hh May, 1976],
" -f. steineckei....[Ba 11-1976], [Hh May, 1976],
elegans (See Mediolobivia elegans, R. aureiflora -v- elegans),
" -v- gracilis (See R. aureiflora),
eos (See Mediolobivia eos),
[G], [Hv], [Hh July, 1974],
espinosae (See R. narvaecensis).....[Bt],
euanthema (See Acantholobivia euanthema, Mediolobivia euanthema).....[Bo-3], [Bp],
" -f. fricii (See Mediolobivia euanthema -v- fricii).....[DD], [Bo-3],
" -f. neopygmaea (See Mediolobivia neopygmaea),
" -f. oculata (See Digitorebutia euanthema -f. oculata, Mediolobivia euanthema -v- oculata),
eucaliptana (See Mediolobivia eucaliptana, R. costata -f. eucaliptana).....
[Bo-3:1517], [Gc-414], [Hi Vol. 2],
fabrisii.....[Hx*], [Af 53-2], [Af-1984], [Aj 28-3-1977]
" -v- aureiflora......[Aj 28-3-1977],
famatimensis (See Lobivia famatimensis),
fidaiana (See Weingartia fidaiana),
fiebigiana.....
fiebrigii (See Aylostera fiebrigii, Rebutia pulchella)
[E], [H], [L], [R], [RR], [Ar], [Bo-3], [Bp], [Bu], [Bv-3], [Hi Vol. 2], [Iu], [Af 54-1], [Ah-1985], [Aj 27-2-1976], [Cu 7-2-1938], [Hh Vol. 2, #9, 1977],
" -v- densiseta (See Aylostera fiebrigii -v- densiseta).....[Iz],
" -v- vulpes (See Mediolobivia ithyacantha, R. ithyacantha).....[Hi Vol. 2],
" -f. (orange flower).....[Hh Vol. 2, #9, 1977],
flavistyla (See Aylostera flavistylus),
[Bp], [Hi Vol. 2], [Iz], [Hh July, 1976], [Hh Vol. 3, #1, 1978],

REBUTIA.....(Cont'd),

friedrichiana (See Digitorebutia friedrichiana, Mediolobivia friedrichiana),
[Hi Vol. 2], [Ba 55:101],

froehlichiana (See Aylostera froehlichiana),.....
[Ba 54:225],

fuauxiana (See Mediolobivia fuauxiana, R. pygmaea -f. fuauxiana),

fulviseta (See Aylostera fulviseta),
[Aj 21-2-1970], [Aj 29-6-1978],

fusca (See Aylostera fusca).....[Bp], [Hi Vol. 2],
[Aj 28-4-1977], [Hh Vol. 3, #2, 1978],

glomeriseta (See Sulcorebutia glomeriseta),
[Hb], [Af 23-3], [Af 39-1],

glomerispina (See Sulcorebutia glomerispina),
[Af 36-2-1964],

gonjianii (See Rebutia einsteinii -v- gonjianii),
[Gd 15:132-135, 1973],

graciliflora (Invalid name - See R. mamillosa -v- australis, Aylostera graciliflora, R. graciliflora).....[DD], [Il],

" -v- borealis (See R. camargensis),

" -v- occidentalis (See R. mamillosa),

" -v- orientalis (See R. mamillosa -v- orientalis)

gracilispina (See Weingartia gracilispina),
[Aj 28-4-1977], [Hl 1-1979],

grandiflora (See minuscula -v- grandiflora).....[C],
[TT], [Bo-3], [Bu], [Hb], [Hu], [Hy], [Ia],

" -f. cristata.....[E],

grandoleata.....[ZZ],

haagei (See Digitorebutia haagei, Digitorebutia pygmaea, Lobivia haagei, Mediolobivia haagei, Mediolobivia pygmaea),
[H], [L], [T], [Bo-3], [Bp], [Bv-1], [Cp], [Cr],
[Gr], [Iz], [Ba 11-1976],

" -v- chamaeleon.....[Cu 1-2-1932], [Cu 8-4-1946],

haefneriana (See Mediolobivia haefneriana, R. pygmaea -f. haefneriana),

haseltonii (See Sulcorebutia breviflora, Sulcorebutia haseltonii).....[Af 38-4],

heliosa (See Aylostera heliosa),
[I], [S], [W], [Z], [AA], [Ar], [Bp], [Bt],
[Hv], [Il], [Iz], [N-1980], [Ad Jan., 1979],
[Ag-1985], [Aj 30-1-1979], [Ba 7-1978],
[Bm-1986], [Hh Sept. 1973],

" -v- cajasensis (See R. cajasensis)...[Af 52-4],
[Ag-1982], [Cu 4-1982], [Hh 3-5&6,1979]

" -v- condorensis (See Aylostera heliosa -v- condorensis, R. solisioides)...[Iz], [Ab-1985], [Af 52-4], [Ag-1981], [Ah-1985], [Hh Vol. 3, #5 & 6, 1979],

heliosa.....(cont'd),
" -v- heliosa.....[Af 52-4],
" -v- (hybrid - R. heliosa X R. albiflora), [Aj 29-8-1978],
" -f. cristata (See Aylostera heliosa -f. cristata).
hirsutissima (See Mediolobivia hirsutissima, Lobivia tiegeliana -v- pusilla).....[Af 43:244],
hoffmannii (See Aylostera hoffmannii), [Af 55-2], [Aj 28-5-1977], [Bm-1989], [Hh Vol. 2, #9, 1977],
huasiensis (See Aylostera huasiensis)..... [Aj 28-2-1977], [Hh Vol. 3, #3 & 4, 1978],
hyalacantha (Invalid name - See R. senilis -v- hyalacantha, R. wessneriana).....[Bo-3],
inflexiseta (See Sulcorebutia inflexiseta)....[Af 42-1]
iridescens.....[Bo-3], [Hi Vol. 2], [Af 28-4],
iscayachensis (See Digitorebutia iscayachensis, Mediolobivia iscayachensis)..... [Hi Vol. 2], [Ba 1-1977],
iseliniana (See R. senilis -v- iseliniana),
ithyacantha (See Aylostera ithyacantha, Mediolobivia ithyacantha, R. fiebrigii -v- vulpes), [Aj 23-12-1972], [Hh May, 1974],
jujuyana (See Aylostera jujuyana)..... [Aj 24-7-1973], [Hh July, 1976],
kariusiana (See Rebutia calliantha -v- kariusiana, R. minuscula -v- violaciflora -f. kariusiana),
kesselringiana (See Mediolobivia kesselringiana, Mediolobivia aureiflora -v- rubriflora, R. aureiflora -f. kesselringiana, R. aureiflora -f. rubriflora).....[Cp],
kieslingii.....
knuthiana (See R. minuscula -v- violaciflora -f. knuthiana, R. violaciflora -v- knuthiana),
krainziana (See R. calliantha -v- krainziana, R. wessneriana -v- krainziana), [D], [R], [T], [V], [AA], [EE], [QQ], [Ar], [Av], [Bo-3], [Bp], [Bv-3], [Hb], [Hu], [Hv], [Hx], [Hy], [Af 52-2], [Ah-1981], [Aj 24-8-1973], [Bi 3-1987], [Hh Jan., 1976],
" -f. breviseta.....[Cu 4-1975],
" -v- gelbe.....
" -v- longiseta.....
" -v- rubiginosa (See R. rubiginosa).....
kruegeri (See Aylostera kruegeri, Sulcorebutia kruegeri).....[Bo-6],
kupperiana (See Aylostera kupperiana).....[C], [M*], [NN], [TT], [Bp], [Bu], [Bv-1], [Cm], [Cz], [Dq], [Hi Vol. 2], [Hv], [Ia], [Il], [Iz],

kupperiana.....(cont'd).
" -v- robustispina (See R. robustispina),
" -v- spathulata.....
" -v- spiniflora.....[Aj 28-4-1977],
" -f. (Lau-402).....[Hh July, 1976],
lanosiflora.....[Hi Vol. 2], [Aj 28-4-1977],
leucanthema (See Aylostera leucanthema),.....
[Aj 26-6-1975],
" -v- coccinifera (See R. melachlora)...[Hi Vol. 2], [Ba 8-1977], [Hh Vol.3, #2, 1978],
mamillosa (See Aylostera mamillosa, R. graciliflora), [Hi 2], [Ba 51:69, 1972], [Hh Vol. 3,#2, 1978],
" -v- australis (See R. graciliflora), [Hi Vol. 2], [Aj 28-4-1977],
" -v- orientalis (See R. graciliflora -v- orientalis),..[Hi Vol. 2], [Aj 28-4-1977],
margarethae.....[DD-528], [Aj 23-1-1972], [Hh Sept.1974] [Hh Vol.3, #3&4-1978], [Hh Vol.3,#5&6, 1979],
markusii (See Sulcorebutia markusii).....[Ay], [Cv-2],
" -v- longispina (See Sulcorebutia markusii -v- longispina),
marsoneri.....[E], [G], [I], [T], [NN], [TT], [UU], [Ar] [Bp], [Cr], [Hx], [Hy], [Il], [Af 52-2], [Af 57-2], [Ba 7-1978], [Hh 2-4-1976], [Hw-1988],
" -v- brevispina.....[Hv], [Aj 9-10-1958],
" -v- grandiflora (See R. marsoneri),
" -v- marsoneri.....[Iz],
" -v- sieperdaiana (See R. senilis -v- sieperdaiana, R. sieperdaiana), [NN], [Aj 18-12-1967],
" -v- spathulata.....
" -cv "Spatula".....[Hh 2-4-1976],
" -v- vatteri.....[Hx],
maxima.....
maximiflora (See R. pseudodeminuta),
melachlora (See R. leucanthema -v- coccinifera), [Hh Vol. 3, #2, 1978],
menesesii (See Sulcorebutia menesesii).....[Af 33-4],
mentosa (See Sulcorebutia mentosa).....[Ay], [W-1982],
minuscula.....[C], [M*], [T], [V], [RR], [Ar], [Be*], [Bo-3], [Bu], [Bv-3], [Cj], [Cm], [Cr], [Dn], [Hv], [Hx*], [Hy], [Af 5-10], [Af 16-8], [Hh Nov., 1975],
" -v- grandiflora (See R. grandiflora),..... [H], [Aj 30-6-1979], [Am 5-1985],
" -v- violaciflora (See R. violaciflora, R. carminea), [H], [K], [Hv], [Hh Nov., 1975],
" -v- " -f. kariusiana (See R. calliantha -v- kariusiana, R. kariusiana)....[D], [Hv], [Iz], [Aj 14:149, 1963],

REBUTIA.....(Cont'd),
minuscula -v- violaciflora.....(cont'd),
" -v- " -f. knuthiana (See R. knuthiana), [Gc 277 & 416], [Gr],
" -v- (hybrid).....[Ba 8-1980],
" -f. cristata.....[Hy], [Aj 8-11-1957],
minutissima.....[Hi Vol. 2], [Aj 28-4-1977],
mixta.....[Hi Vol. 2], [Aj 28-4-1977],
mixticolor.....[Hi Vol. 2], [Af 53-4], [Ag-1985], (Aj 27 -8-1976],
mudanensis (See Digitorebutia mudanensis, Mediolobivia mudanensis).....[Aj 27-8-1976],
muscula (See Aylostera muscula).....[G], [X], [UU], [Ar], [Bp], [Hi-2], [Hu], [Hv], [Iz], [Ab-1985], [Af 60-2], [Ah-1985], [Am 1-1981], [Bi 6-2-1988], [Cu 4-1965], [Hh Vol. 2, #9, 1977],
napina.....
narvaecense (narvaecensis) (See Aylostera narvaecense, R. espinosae).....[Bf], [Bm -], [Hv], [Iz], [Af 43:245], [Ah-1980], [Hh Sept. 1973], [Hh Vol. 3, #5 & 6, 1979],
nazarenoensis (See Digitorebutia nazarenoensis),
neopygmaea (See R. euanthema -f. neopygmaea, Mediolobivia neopygmaea),
nigricans (See Mediolobivia nigricans, R. ritteri -v- nigricans),
nitida.....[Hh 3, 14 - 1978],
nivea (See R. spinosissima).....[Hh Vol. 2, #9, 1977],
nivosa (See Aylostera muscula),
nogalensis (nogalesensis).....[Aj 28-4-1977],
oculata (See R. euanthema -f. oculata).....[Aj 1-1938],
odontopetala (See Aylostera odontopetala, R. friedrichiana).....[Aj 28-4-1977],
orurensis (See Mediolobivia orurensis, Mediolobivia pectinata -v- orurensis)..[Bp], [Hi Vol. 2], [Il],
padcayensis (See Aylostera padcayensis)..[Aj 21:65, Apr. 1970], [Ba 56:233, 1977], [Hh July, 1974], [Hh Vol. 3, #3 & 4, 1978],
pallida (See Digitorebutia pallida).....[Ba 10, 1977],
patericalyx (See Mediolobivia patericalyx), [Hi Vol.2], [Aj 28-4-1977],
" -v- odontopetala (See R. odontopetala),
pauciareolata.....[Hi Vol. 2], [Aj 28-4-1977],
paucicostata.....[Hi Vol. 2], [Aj 28-4-1977],
pectinata (See Mediolobivia pectinata, Mediolobivia pygmaea, R. pygmaea),
" -v- atrovirens (See R. pygmaea -f. atrovirens),
" -v- digitiformis (See Mediolobivia digitiformis, R. digitiformis, R. pygmaea), [Bo-3],
" -v- neosteinmannii (See R. pygmaea -v- neosteinmannii)

REBUTIA.....(Cont'd),
pectinata.....(cont'd)
" -v- orurensis (See R. orurensis),
permutata (Invalid name - See R. wessneriana -v- permutata).....[DD],
" -f. gokrausei (Invalid name - Hybrid ?)..[DD],
perplexa.....[Hv], [Iz], [Af 52-4], [Af 53-4], [Af 54-2] [Ag-1982], [Ah-1985], [Aj 34-1-1983], [Hh Vol. 3, #5&6, 1979],
peterseimii (See Digitorebutia peterseimii, R. ritteri -v- peterseimii).....[TT], [Am 1-1984],
petersonii (See R. minuscula).....[Bp], [Ja],
pilifera.....
poecilantha.....[Hi Vol. 2], [Hh Vol. 3, #2, 1978],
polymorpha (See Sulcorebutia polymorpha, Sulcorebutia steinbachii -v- polymorpha, Weingartia polymorpha).....[Aj 16:115, 1965],
potosina.....[Hi Vol. 2], [Aj 28-4-1977],
pseudodeminuta (Backbg.) (See Aylostera pseudodeminuta), [H], [L], [M*], [NN], [QQ], [Av], [Ay], [Bt], [Bu], [Bv-3], [Cm], [Cu 4-1961],
" -f. albiseta (See Aylostera pseudodeminuta -f. albiseta).....[Bo-3], (Af 23:82, 1951]
" -f. grandiflora (See Aylostera pseudodeminuta -f. grandiflora)..[Bo-3], [Af 23:82, 1951],
" -f. rubrifilamentosa.....[Aj 16-5-1965], [Hh July, 1976], [Hl 1-1981]
" -f. schneideriana.....[Af 23:82, 1951],
" -f. schumanniana.....[Fy 2:7 Fig.
pseudominuscula (See Aylostera pseudominuscula, R. deminuta -f. pseudominuscula), [C], [M*], [RR], [Af 59-1], [Cu 4-1981],
pseudopygmaea (See R. iscayachensis),
pulchella (See Aylostera pulchella, R. fiebrigii), [H], [Aj 23-12-1972], [Hh Vol. 2, #9, 1977],
pulchra (pulchera) (See Sulcoreb. pulchra)....[Af 42-1],
pulvinosa (See Aylostera pulvinosa)...[G], [Gr], [Hi-Vol. 2], [Cu 4-1965], [Hh Vol.3, #5&6, 1979],
pygmaea (See Digitorebutia haagei, Digitorebutia pygmaea, Mediolobivia haagei, Mediolobivia pygmaea, Rebutia haagei), [NN], [RR], [Bp], [Hi Vol. 2], [Hv],
" -f. atrovirens (See Mediolobivia pectinata -v- atrovirens).....[Gc 242],
" -f. flavovirens (See Mediolobivia pygmaea -v- flavovirens).....[Gc 243],
" -f. fuauxiana (See Mediolobivia fuauxiana, R. fuauxiana),
" -f. haefneriana (See Mediolobivia haefneriana), [Ba 6:119, 1955],
" -f. neosteinmannii (See R. neosteinmannii), [Ah-1986], [Cu 27:40, 1965],

raulii (See Digitorebutia raulii, Mediolobivia raulii)
[Aj 31-6-1980],
rauschii (See Digitorebutia rauschii, Mediolobivia rauschii, Rebutia brachyantha, Rebutia steinmannii),
[Ay], [Cv-2], [Hx*], [Aj 28-4-1977],
reichii.....[Cu 4-1960],
residua.....
ritteri (See Digitorebutia ritteri, Mediolobivia ritteri, Lobivia ritteri),
[NN], [Bo-3], [Bv-1], [Hi Vol. 2], [Iz],
" -v- nigricans (See Mediolobivia nigricans),
[Dk],
" -v- " -f. hahniana.....[Cu 27:40, 1965],
[Hl 5-1981],
" -v- paucicostata (See R. paucicostata),
" -f. peterseimii (See R. peterseimii),
robustispina (See Aylostera robustispina),
[Hi Vol. 2], [Ba 8-1977],
" -v- minor.....[Hi Vol. 2], [Ba 8-1977],
rosalbiflora (See Mediolobivia rosalbiflora),
[Hi Vol. 2], [Aj 28-4-1977],
" -v- amblypetala.....[Hi Vol. 2], [Aj 28-4-1977],
rubelliflora (See Mediolobivia rubelliflora, Mediolobivia aureiflora -f. rubelliflora, R. aureiflora -f. rubelliflora),
rubiginosa (See Aylostera rubiginosa),
[Hi Vol. 2], [Af 56-2], [Hh July, 1976],
rubriflora (See R. aureiflora -f. rubriflora),
rutiliflora (See Rebutia diersiana).....
[Hi Vol. 2], [Aj 28:76],
salpingantha (See R. steinmannii).........
[Hi Vol.2], [Af 56-2],
sanguinea (See Aylostera sanguinea).....
[Hi Vol. 2], [Ba 8-1977], [Hh May, 1977],
[Hh Vol.3, #2, 1978],
" -v- minor.....[Hi Vol. 2],
sarothroides (See Mediolobivia aureiflora -v- sarothroides, Mediolobivia sarothroides, R. aureiflora -f. sarothroides),
scarlatea.....
" -v- brevispina.....
schatzliana (See Aylostera schatzliana).....
[Aj 26-11-1975], [Hh Vol. 3, #5 & 6, 1979],
schmiedcheniana (See Cylindrorebutia schmiedcheniana, Mediolobivia schmiedcheniana, R. einsteinii -f. schmiedcheniana),
" -v- einsteinii (See R. einsteinii).....[Bo-3],

schmiedcheniana.....(cont'd),
" -v- karreri (See R. einsteinii -f. karreri, R. einsteinii -f. rubriviridis),
" -v- rubriviridis (See R. einsteinii -f. rubriviridis),
" -v- steineckei (See R. einsteinii -f. steineckei),
senilis.....[E], [H], [L], [M*], [S], [AA], [NN], [Av], [Ay], [Bm -], [Bo-3], [Bv-1], [Cs], [Hx*], [Hy], [Aj 15-7-1964],
" -v- aurescens (See R. senilis -v- senilis), [TT],
" -v- breviseta (See R. krainziana), [Bo-3], [Hx], [Aj 24-8-1973],
" -v- cana (See R. senilis -v- senilis),
" -v- calva.....
" ssp. chrysacantha (See R. chrysacantha), [Hh Nov. 1975],
" " " -v- schieliana (See R. senilis -v- schieliana)..[Hh Nov. 1975],
" -v- elegans R. chrysacantha -v- elegans), [Bp], [Hy], [Af 23:83, 1951],
" -v- hoiferi.....
" -v- hyalacantha (See R. hyalacantha, R. wessneriana).....[Aj 24-8-1973],
" -v- iseliniana.....[E], [Hu], [Hx], [Hy],
" -v- kesselringiana (See R. kesselringiana), [E], [I], [T], [V], [EE], [Bo-3], [Hb], [Hu], [Hx*], [Hy], [Jd], [Aj 18-12-1967, [Bi -Vol. 2, #1],
" -v- lilacinorosea..[D], [H], [NN], [Bo-3], [Hb], [Iz], [Aj 8-2-1957], [Hh Nov. 1975],
" -v- " -f. cristata.....[EE], [Bv-1],
" -v- longispina.....
" ssp. senilis.....[Hh Nov. 1975],
" -v- sieperdaiana (See R. marsoneri -v- seiperdaiana).....[T], [Bo-3],
" -v- schieleana.....[Hx], [Bo-6], [Aj 8-7-1957],
" -v- semperflorens.....[Af 11-5],
" -v- stuemeri (See R. stuemeri).....[E], [Bo-3], [Bv-1], [Hy], [Ba 12-1977],
" (hybrid).....[N],
sieperdaiana (See R. marsoneri -f. sieperdaiana, R. senilis -v- sieperdaiana)....[Ba 11-1976],
singularis.....[Hi Vol. 2], [Hh Vol. 3, #1, 1978], [Hh Vol. 3, #3 & 4, 1978],
solicioides (See R. rutiliflora),
spegazziniana (See Aylostera spegazziniana), [C], [F], [I], [Bo-3], [Bm -], [Hh May - June, 1977],

REBUTIA.....(Cont'd),
spegazziniana.....(cont'd)
" -v- atroviridis..............[PP], [Af 23:82], [Hh May-June-1977],
" -v- waltheriana (See Aylostera waltheriana, R. spegazziniana),
sphaerica (See R. tarijensis),
spinosissima (See Aylostera spinosissima), [Ar], [Av], [Bo-3], [Bp], [Jd], [Aj 9-8-1958], [Hh Vol. 2, #9, 1977],
" -v- buiningiana.....
" -v- cardenas.....
" -v- torquata (See R. torquata),
spiralisepala.....
steinbachii (See Sulcorebutia steinbachii).... [F], [Ay], [Hw-1988],
steinmannii (See Aylostera steinmannii, Mediolobivia steinmannii, R. brachyantha), [RR], [Bu], [Ia], [Hh Vol. 3, #3 &4, 1978],
" -v- brachyantha.....
" -v- cincinnata (See R. cincinnata).[Hi Vol. 2],
stuemeri (See R. senilis -v- stuemeri).....[Bi 2-3],
supthutiana (See Aylostera supthutiana)..... [Aj 27-6-1976], [Hh Vol. 3, # 5&6, 1979],
tamboensis (See Aylostera tamboensis,..... [Hi Vol. 2], [Hh May-June, 1977],
taratensis (See Sulcorebutia taratensis),..... [Af 36:26, 1964],
tarijensis (See Aylostera tarijensis)....[Hi Vol. 2], [Aj 26-9, 1975], [Hh Vol. 3, # 3&4, 1978],
tarvitaensis (See Mediolobivia tarvitaensis), [Hi Vol. 2], [Aj 23-1-1972], [Hl 6-1978], [Hh Vol. 3, #2, 1978],
tiraquensis (See Sulcorebutia tiraquensis), [Af 40-6], [Iv 12-1957],
" -v- longiseta (See Sulcorebutia tiraquensis -v- longiseta).....[Af 42:188, 1970],
torquata.....[Hi Vol. 2], [Hv], [Ba 8-1977],
totorensis (See Sulcorebutia totorensis), [Af 40-6], [Gs 13:259, 1957], [Iv 12, 1957],
tropaeolipicta.....[Hi Vol. 2], [Aj 28-4-1977],
tuberculato-chrysacantha (See Sulcorebutia tuberculato-chrysacantha),
tuberosa (See Aylostera tuberosa).....[Hi Vol. 2],
tunariensis (See Sulcorebutia tunariensis).... [Af 36-2], [Af 38-2],
turbinata (Hort.) -f. (short-spined)...[Hh Jan. 1976],
vallegrandensis (See Aylostera vallegrandensis), [Af 42-1], [Hh Vol. 2, #9, 1977],
vasqueziana.....[Il],
vatteri.....
violacea.....[Hy],

REBUTIA.....(Cont'd),

violaciflora (See R. minuscula -v- violaciflora),
[D], [M*], [EE], [NN], [TT], [Bo-3],
[Bv-3], [Dq], [Hx], [Hy], [Ia], [Il],
[Jd], [Af 15-12], [Af 52-6], [Aj 8-2-1957], [Aj 33-3-1982], [Bm -],
" -v- carminea (See R. carminea, R. minuscula -v- violaciflora),
[Aj 8-2-1957], [Ba 11-1976],
" -v- knuthiana (See R. minuscula -v- violaciflora -f. knuthiana),
[T], [NN], [Bt], [Hx], [Hy],
violascens.....[Hi Vol. 2], [Aj 28-4:76, 1977],
vizcarrae (See Sulcorebutia vizcarrae),.....
[Hh July, 1973],
vulpina (See Rebutia fusca).....
[Hi Vol. 2], [Hu], [Af 53-2], [Ba 8, 1977],
wessneriana.....[Ar], [Bp], [Hx], [Hy], [Aj 24-8-1973],
" ssp. wessneriana.....
" " " -f. beryllioides (See R. calliantha -v- beryllioides).....
[Hh Jan. 1976],
" " " -f. calliantha (See R. calliantha),
[Hh Jan. 1976],
" " " -f. krainziana (See R. krainziana),
[Iz],
" " " -f. permutata (See R. permutata),
" -f. cristata.....[Hy],
" -cv "Ruby"......[Hh Jan.,1976],
xanthocarpa.....[NN], [Bo-3], [Cj], [Hx*], [Hy],
[Ba 7-1978], [Hh Nov., 1975],
" -f. aurescens.....
" -f. citicarpa (See R. citicarpa).....[Af 23:83]
" -f. coerulescens.....[Bo-3], [Aj 11-7-1960],
" -f. dasyphrissa.....[Hy], [Aj 11-7-1960],
" -v- elegans (See Rebutia senilis -v- elegans),
" -f. graciliflora (See R. graciliflora),
" -v- luteirosea (See R. xanthocarpa -f. xanthocarpa).....[EE], [Af 23:83],
" -f. salmonea.....[C], [H], [TT], [Bp],
" -v- sanguinea (See R. sanguinea),.....
[Hh May-June, 1977],
" -f. violaciflora.....[Bp], [Hx],
yuquinensis (See Digitorebutia yuquinensis).....
[Bt], [Aj 31-10-1980],
zecheri (See Aylostera zecheri).....
[Ba 2-1977], [Hh Vol. 3, #3 & 4, 1978],
Sp. Lau-346.....[Hh May, 1974],
Sp. Lau-405.....(Aylostera Sp.).....[Aj 31-3-1980],
Sp. Lau-409.....[Hh Jan., 1976],
Sp. Lau-416.....[Hh July, 1974],

REBUTIA.....(Cont'd),

Sp. Lau-520.....[Hh Sept., 1974],
Sp. Lau-533.....[Hh Sept., 1974],
Sp. Uebelmann Nr 924.....[Hu],

X grandililacea (R. grandiflora X R. violaciflora),
[N-1980], [Af 52-6],

REICHEOCACTUS

floribundus (Invalid name - See Neochilenia iquiquensis,
Pyrrhocactus floribundus),
[Bo-6], [Hx*], [Aj 31-6-1980],
neoreichei (Invalid name - See Neochilenia lembckei,
Neochilenia neoreichei, Neochilenia imitans,
Thelocephala lembckei).....[Bo-6],
pseudoreicheanus (See Echinocactus famatimensis, Lobivia famatimensis, Neochilenia pseudoreichii).....[DD], [NN], [Ar], [Av],
[Ay], [Hx], [Jd], [Ba 3-1978],
reichei (See Malacocarpus reichei, Neochilenia reichei,
Neoporteria reichei, Thelocephala reichei),

Sp. .....[DD],

RHIPSALIDOPSIS

epiphylloides (See Hatiora epiphylloides, Rhipsalis
epiphylloides, Pseudozygocactus epiphylloides),
gaertneri (See Epiphyllopsis gaertneri, Schlumbergera
gaertneri),
[X], [EE], [NN], [QQ], [Ay], [Bt], [Bv-3],
[Cz], [Hu], [Hv], [Ig*], [Il], [Ja], [Jd],
" -v- makoyana (See Schlumbergera gaertneri -v-
makoyana).....[D], [Bv-3],
[Aj 28-4-1977],
" -v- tiburtii.....
rosea (See Hatiora rosea, Rhipsalis rosea),
[D], [K], [L], [V], [AA], [PP], [TT], [UU], [Av],
[Ay], [Bb], [Bo-6], [Bt], [Bv-3], [Dq], [Gr],
[Hv], [Hx*], [Hy], [Il], [Jd], [Aj 26-6-1975],
[Ba 8-1979], [Bi 2-3],
" -v- remanens (Invalid name),
[Bo-6], [Af Vol. 47 Supplement],
serrata (See Epiphyllopsis gaertneri -v- serrata),
X graeseri (See Rhipsaphyllopsis graeseri),

RHIPSALIS

aculeata (See Lepismium aculeatum, R. lumbricoides),
[RR], [Ay],

alboareolata.....[Hi Vol. 1],
anceps (See Lepismium cruciforme -v- anceps),
angustissima (See Disocactus ramulosus -v- angust-
issimus, R. coriacea, R. leiophloea),
[Bo-2],
asperula (See Lymanbensonia micrantha).....[Af 56-3],
baccifera (See R. cassutha).....[I], [Ax], [Ay], [Az],
" ssp. erythrocarpa (See R. erythrocarpa),
" ssp. horrida (See R. horrida),
" ssp. mauritiana (See R. cassytha -v- mauritiana)
[Ct-1987],
" -v- rhodocarpa.....[Ay],
bambusoides (See Hatiora bambusoides),
bartlettii (See R. cassutha, R. baccifera),
[Ep 65:567 Fig. 6 & 7, 1938],
bermejensis.....[Bv-1],
biolleyi (See Weberocereus biolleyi),
boliviana (See Lepismium cruciforme -v- anceps).....
[Bv-1],
brevispina (See Acanthorhipsalis brevispinum, Lepismium
brevispinum).....[Af 56-3*],
burchellii (See R. cribrata).....[Bu], [Bv-1],
campos-portoana.....[Fz 2:35 Pl. 7, 1918],
capilliformis.....[RR], [XX], [Al*], [Ay], [Bo-2], [Bu],
[Bv-1], [Cz], [Gr], [Hx*], [Jd],
cassutha (See R. baccifera),
[W], [LL*], [RR], [YY], [Bl], [Bo-2], [Bu],
[Bv-1], [Cm], [Cn], [Cz], [Hx*], [Gu III, 1953]
" -v- hookeriana (See R. cassutha),
" -v- mociniana (See R. cassutha),
" -v- swartziana (See R. cassutha),
cassytha.....(See R. cassutha),
" -v- mauritiana (See R. baccifera ssp. mauritiana),
cassythoides (See R. cassutha),
cassuthopsis (Similar to cassutha),
cavernosa (See Lepismium cruciforme).....[Bk],
cereoides (See Lepismium cereoides).....
[Bo-2], [Bv-1], [Jd],
cereuscula.....[D], [M*], [R], [RR], [Ay], [Bl], [Bo-2],
[Bv-1], [Co], [Cz], [Dq*], [Jd], [Bb-1985],
" -v- rubrodisca.....
chloroptera (See R. eliptica -v- helicoidea),....[Hy*],
chrysanthum (See Lepismium chrysanthum),
chrysocarpa (See R. puniceo-discus),
clavata (See Hatiora clavata),
[Ay], [Bu], [Bv-1], [Hv], [Af-1985],
" -v- delicatula (See R. clavata)...[Dv 25:f. 11,
No. 16], [Fz 2:45 Tab. 17, 1918],
clavellina.....[Hi Vol. 1],
conferta (See R. teres).....[Ds],

RHIPSALIS.....(Cont'd),

coralloides (Invalid name - See R. saxicola)....[Bo-6],
coriacea (See R. angustissima, R. leiophloea), [RR], [Bv-1],
crenata (See Acanthorhipsalis crenata, Lepismium crenatum),
cribrata (See R. burchellii, R. penduliflora), [RR], [Hx],
crispata.....[D], [RR], [YY], [Bl], [Bo-2], [Bv-1], [Hy], [Af 34-5],
" -v- latior (See R. crispata),
crispimarginata.....[RR], [Bl], [Bv-1], [Hv], [Hx],
cruciformis (See Lepismium cruciforme).[Ij], [Iz], [Jd], (For varieties see Lepismium cruciforme),
cuneata.....[RR],
densiareolata (See R. lindbergiana). [Fz 2:41 Pl. 12, 1918],
dilimitia (See R. micrantha),?
dissimilis (See Lepismium dissimile, R. pacheco-leonii) [RR], [Bv-1], [Hx*],
" -v- setulosa (See R. dissimilis),
elliptica (See R. chloroptera).....[RR],
" -v- helicoides (See R. chloroptera),..... [Fz 2:44 Pl. 16, 1918],
epiphyllanthoides (See Lepismium epiphyllanthoides), [Hx*], [Jd],
epiphylloides (See Hatiora epiphylloides, Pseudo-zygocactus epiphylloides).....[I], [Ay],
erecta (See Pfeiffera erecta),
erythrocarpa (See R. baccifera ssp. erythrocarpa, R. densiareolata, R. lindbergiana),
fasciculata (See R. horrida, R. madagascarensis), [QQ], [Ah], [Ay], [Bo-6], [Hv], [Hx*], [Jd], [Hh March, 1974],
floccosa (See Lepismium floccosum).....[RR],
" -v- gibberula (See R. gibberula),
flosculosa.....[Hi Vol. 1],
funalis (See R. grandiflora),
gibberula (See Lepismium gibberulum).....[MM], [RR],
gobeliana.....[Bo-2], [Hx*], [Hy*],
gonocarpa.....[RR], [Bv-1],
grandiflora (See Lepismium grandiflorum), [D], [M*], [RR], [Bk], [Bv-1], [Hi Vol. 1], [Hx*], [Af 34-5], [Bb-1987],
hadrosoma (See R. grandiflora)..... [I], [Ay], [Bv-1], [Jd],
heptagona.....[Av], [Bo-2],
herminiae (See Hatiora herminiae).....[Bv-1], [Ij],
heteroclada.....[Z], [RR], [Bv-1], [Dq*], [Af 54-1],
himantoclada (See Pseudorhipsalis himantoclada),
hohenauensis.....[Hi Vol. 1],

RHIPSALIS.....(Cont'd),

horrida (See R. fasciculata, R. madagascarensis), [I], [Ay], [Ab-1985],
houlletiana (See Lepismium houlletianum)....[C], [M*], [NN], [RR], [Ay], [Bl], [Bo-2], [Bo-6], [Bv-1], [Cm], [Cz], [Hi Vol. 1], [Hx*], [Hy], [Il], [Iz], [Jd], [Af 16-2],
" -v- regnellii.....(See R. regnellii),
hylaea.....[Hi Vol. 4],
ianthothele (See Lepismium ianthothele, Pfeiffera ianthothele).....[Ij], [Iz],
incachacana.....[Af 57-6],
incahuasina (See Acanthorhipsalis incahuasina, R. monacantha),
jamaicensis.....[RR],
kirbergii.....[Ay],
latipomona (See R. platycarpa).....[Hy*],
leiophloea (See R. angustissima, R. coriacea), [Bl], [Bo-2],
leucorhaphis (See R. lumbricoides -v- leucorhaphis), [RR], [Ay],
lindbergiana (See R. densiareolata, R. erythrocarpa),[RR],
linearis.....[Bl],
loefgrenii (See R. leucorhaphis, R. lumbricoides -v- leucorhaphis).....[RR], [Bl],
lorentziana (See Lepismium lorentzianum),
lumbricoides (See R. aculeata, Lepismium lumbricoides), [RR], [Bk], [Bv-1], [Hi Vol. 1],
" -v- leucorhaphis (See R. leucorhaphis, R. loefgrenii),.....[Hi Vol. 1],
macrantha (See Pseudorhipsalis macrantha, Disocactus macranthus).....[Bv-3],
madagascarensis (See R. fasciculata, R. horrida), [Bo-6],
marnierana (See Lepismium cruciforme -v- marnieranum, Lepismium marnieranum).....[Ay],
megalantha (See Lepismium megalanthum)..... [RR], [Ay], [Jd],
mesembryanthemoides.....[EE], [RR], [Ay], [Bk], [Bl], [Bt], [Bu], [Bv-1], [Cz], [Hv], [Hx], [Jd], [N-1980], [Bb-1985],
micrantha (HBK) (See Acanthorhipsalis micrantha, Lymanbensonia micrantha).....[RR], [Ay], [Bo-2], [Bv-1], [Gr], [Hi Vol. 4], [Hx*],
minutiflora (See R. cassutha),
monacantha (See Acanthorhipsalis monacantha, Lepismium monacanthum).....[Bk], [Af 24-6],
" -v- samaipatana.....[Au 12:85, Fig., 1957],
monteazulensis.....[Hi Vol. 1],
myosurus (See Lepismium myosurus),
neocassutha (See R. cassutha (cassytha).....[Hy*],

neoshaferi (See R. shaferi),
neves-armondii (See Lepismium neves-armondii).....
[RR], [Ay], [Bu], [Af 60-2], [Bb-1987],
oblonga.....[RR], [Bv-1],
occidentalis.....[Ct-1987],
pacheco-leonii (See Lepismium pacheco-leonii), Lepismium dissimile, R. dissimilis)....[Hx*],
pachyptera (See R. robusta).....[M*], [RR], [YY],
[Be*], [Bv-1], [Cr], [Ds], [Gr], [Hi Vol. 1]
[Hx], [Ig*], [Ab-1985],
paradoxa (See Lepismium paradoxum),
[M*], [RR], [YY], [Ay], [Be*], [Bk], [Bv-1],
[Cm], [Cz], [Hx*], [Ij], [Bb-1985],
paranganiensis (See Acanthorhipsalis paranganiensis,
Lepismium paranganiense).....[Af 56-5*],
penduliflora (See R. cribata).....
[I], [Ay], [Bt], [Bv-1], [Af 2-9],
pentaptera.....[RR], [YY], [Bl], [Bv-1], [Hv], [Jd],
[N-1980], [Bb-1987],
pilocarpa (See Erythrorhipsalis pilocarpa),
[I], [Ay], [Ct-1987], [Hx], [Il], [Jd],
pilosa (See R. fasciculata, R. horrida, R. madagascarensis).....[DD],
pittieri (See Lepismium pittieri).....[RR], [Ay],
platycarpa (See R. latipomona).....
[I], [RR], [Bv-1], [Jd],
prismatica.....[V], [RR], [Bv-1], [Hx*],
pulchra.....[RR], [Bv-1],
pulvinigera (See Lepismium pulvinigerum).....[RR],
puniceo-discus (See Lepismium puniceo-discus),
[RR], [Af 27-1], [Ag-1984],
" -v- chrysocarpa.....
purpusii.....[Bo-2], [Hv], [Aj 23-2-1972], [Aw 1-1978],
quellebambensis (Invalid name).....
[Bu], [Bv-1], [Cz],
radicans (See Lepismium radicans),
ramulosa (See Disocactus ramulosus),
[RR], [Ay], [Hy*], [Jd],
rauhiorum.....[Ay], [Hx*], [Bb-1986],
regnellii (See R. houlletiana -v- regnellii),
[Be*], [Bk],
rhombea.....[R], [RR], [YY], [Ar], [Ay], [Bv-1], [Cz],
[Hy], [Ij], [Aj 31-9-1980],
" -v- crispata (See R. crispata),
rigida (See Lepismium rigidum, R. neves-armondii, ?)
robusta (See R. pachyptera),.....
[Ay], [Bo-2], [Bo-6], [Bv-1], [Hx*], [Jd],
rosea (See Hatiora rosea, Rhipsalidopsis rosea),
roseana.....[Bv-1], [Hb],
russelii.....[RR],

RHIPSALIS.....(Cont'd),

salicornioides (See Hatiora salicornioides),
saxatile (See Lepismium saxatile),
saxicola (See R. coralloides),
[Bv-1], [Am 6-1982],
shaferi (See R. neoshaferi),.....[RR], [Bv-1],
simmleri.....
suareziana (See R. prismatica),
sulcata.....[RR], [YY], [Bv-1],
teres (See R. conferta).....[RR], [Bv-1], [Cg], [Hx*],
tetragona (See R. prismatica),
tunduzii.....[RR], [Be*], [Bv-1], [Aj 3-5-1982],
triangularis.....
trigona (See Lepismium trigonum).....
[RR], [Ay], [Bv-1], [Bb-1986],
tucumanensis (See Lepismium tucumanensis).....[RR],
virgata.....[RR], [Bv-1],
warmingiana (See Lepismium warmingianum),
[M*], [V], [MM], [RR], [Ay], [Bl], [Bv-1],
[Hi Vol. 1], [Hx*], [If],
werklei (See R. tonduzii).....
zanzibarica (See R. cassutha).....[Bu], [Bv-1],

Sp. UE-153 (Columbia).....[Af 55-1],
Sp. .....[Cy], [Hv],
(-f. cristata).....[Bi 1-2],

RHIPSAPHYLLOPSIS

graeseri (Hybrid).....[Hv], [Hx*], [Il],

RHODOCACTUS

antonianus (Invalid name - See Pereskia antoniana,
Pereskia weberiana).....[DD], [Ay],
autumnalis (See Pereskia autumnalis, Pereskiopsis
autumnalis, Pereskia lychnidiflora,
Rhodocactus lychnidiflorus, Pereskiopsis
opuntiiflora, Pereskia pititache, Pereski-
opsis pititache, Pereskia nicoyana, Rhodo-
cactus nicoyanus, Pereskia conzattii, Rhodo-
cactus conzattii).....[Ay], [Bo-1],
bleo (See Pereskia bleo, Pereskia corrugata, Rhodo-
cactus corrugatus, Pereskia panamensis).....
[Ar],
colombianus (See Pereskia colombiana, Pereskia
guamacho, Rhodocactus guamacho).....

conzattii (See Pereskia lychnidiflora, Rhodocactus lychnidiflorus, Pereskiopsis opuntiiflora, Pereskia pititache, Pereskiopsis pititache, Pereskia nicoyana, Rhodocactus nicoyanus, Pereskia autumnalis, Pereskiopsis autumnalis, Rhodocactus autumnalis, Pereskia conzattii).. [Bo-1],
corrugatus (See Pereskia corrugata, Pereskia bleo, Rhodocactus bleo, Pereskia panamensis)..... [Bo-1],
cubensis (See Pereskia cubensis, Pereskia zinniiflora, Rhodocactus zinniiflorus).....[Bo-1],
guamacho (See Pereskia guamacho, Pereskia colombiana, Rhodocactus colombianus).....
grandiflorus (misspelling - See Pereskia grandifolia -v- grandifolia, Rhodocactus grandifolius, Pereskia tampicana, Rhodocactus tampicanus) [Bi 1-2].
grandifolius (See Pereskia grandifolia, P. grandiflora, Rhodocactus grandiflorus, Pereskia tampicana, Rhodocactus tampicanus) [Ba 12, 1978], [Da 1, 1981],
higueranus (Name doubtful - See Pereskia higuerana),
lychnidiflorus (See Pereskia lychnidiflora, Pereskiopsis opuntiiflora, Pereskia pititache, Pereskiopsis pititache, Pereskia nicoyana, Rhodocactus nicoyanus, Pereskia autumnalis, Pereskiopsis autumnalis, Rhodocactus autumnalis, Pereskia conzattii, Rhodocactus conzattii).....[Bo Vol. 1],
moorei (See Pereskia moorei, Pereskia sacharosa, Rhodocactus sacharosa, Pereskia saipinensis, Pereskia sparsiflora).....
nicoyanus (See Pereskia lychnidiflora, Rhodocactus lychnidiflorus, Pereskiopsis opuntiiflora, Pereskia pititache, Pereskiopsis pititache, Pereskia nicoyana, Pereskia autumnalis, Pereskiopsis autumnalis, Rhodocactus autumnalis, Pereskia conzattii, Rhodocactus conzattii).....
panamensis (See Pereskia panamensis, Pereskia bleo, Rhodocactus bleo, Pereskia corrugata, Rhodocactus corrugatus).....[Hy],
portulacifolius (See Pereskia portulacifolius),

sacharosa (See Pereskia sacharosa, Pereskia moorei, Pereskia saipinensis, Rhodocactus saipinensis, Pereskia sparsiflora)..... [DD], [Ay], [Hv], [Hx*],

RHODOCACTUS.....(Cont'd)

saipinensis (See Pereskia saipinensis, Pereskia moorei,
Rhodocactus moorei, Pereskia sacharosa,
Rhodocactus sacharosa, Pereskia sparsiflora)
tampicanus (See Pereskia tampicana, Pereskia grandi-
folia -v- grandifolia, Rhodocactus grandi-
folius, Pereskia grandiflora).....
zinniiflorus (See Pereskia cubensis, Pereskia zinn-
iiflora, Rhodocactus cubensis),
[Bo-1],

RITTEROCEREUS

chacalapensis (See Marshallocereus chacalapensis, Sten-
ocereus chacalapensis).....[Bo-4],
deficiens (See Cereus deficiens, Lemaireocereus
deficiens).....[RR], [Bo-4],
eburneus (See R. griseus),

eichlamii (See Lemaireocereus eichlamii, Stenocereus
eichlamii).....[RR], [Ay], [Bo-4],
fimbriatus.....[Aj 30-2-1979],

griseus (See Cereus griseus, Lemaireocereus griseus,
Stenocereus griseus),
[Av], [Az], [Bo-4], [Hx*], [Hy], [Ba 10-1978],
[Ba 1-1979],
" -f. cristata.....[Ba 1-1979],
hystrix (See Lemaireocereus hystrix, Stenocereus
hystrix).....[Bo-4], [Hs IV-2-1983],
laevigatus (See Cereus laevigatus, Lemaireocereus
longispinus, Stenocereus laevigatus),
[DD], [Az], [Bo-4],
montanus (See Lemaireocereus montanus, Stenocereus
montanus),
olivaceus (See Cereus hystrix, Lemaireocereus
hystrix, Stenocereus hystrix),
pruinosus (See Lemaireocereus pruinosus, Stenocereus
pruinosus),
[Bo-4], [Bt], [Hx*], [Hy],
queretaroensis (See Lemaireocereus queretaroenis,
queretaroensis).....[Bo-4], [Hy],
" -v- aureo-variegatus),
standleyi (See Lemaireocereus standleyi, Stenocereus
standleyi),.....[Az],
stellatus (See Cereus stellatus, Lemaireocereus
stellatus, Stenocereus stellatus),
[Ba 3-1981],
weberi (See Cereus weberi, Lemaireocereus weberi,
Pachycereus weberi, Stenocereus weberi),

RODENTIOPHILA

atacamensis (See Eriosyce rodentiophila),.....
[NN], [Ar], [Hy], [Hl 3-1980],
" -v- aureispina.....
lanata (See Eriosyce rodentiophila -v- lanata)
[Af 55-1],
megacarpa (See Eriosyce megacarpa),.....
[DD], [Hx*], [Hy], [Af 50-5],
" -v- lanata.....(See R. lanata),

ROOKSBYA

euphorbioides (See Neobuxbaumia euphorbioides, Stenocereus euphorbioides).....[DD], [Ar], [Av],
[Bo-4], [Hx*], [Hy], [Il], [Jd],
" -v- olfersii.....[Bo-4],

ROSEOCACTUS

fissuratus (See Ariocarpus fissuratus),
[T], [Z], [DD], [NN], [Ar], [Ay], [Bo-5],
[Cx], [Hx*], [Hy],
" -v- lloydii (See Ariocarpus fissuratus -v- lloydii, Ariocarpus lloydii, R. lloydii).....[Ay],
intermedius (See Ariocarpus fissuratus -v- lloydii, Ariocarpus lloydii, R. lloydii),
[Bo-5], [Aj 11-10-1960],
kotschoubeyanus (See Ariocarpus kotschoubeyanus),
[T], [Z], [Ar], [Ay], [Bo-5], [Hx],
[Hy], [Ae-1982], [Aw 1-1971], [Aw 2-1973]
" -v- albiflorus (See Ariocarpus kotschoubeyanus -v- albiflorus).....[Ar], [Bo-5],
" -v- elephantidens (See Ariocarpus kotschoubeyanus -v- elephantidens),
" -v- macdowellii (See Ariocarpus kotschoubeyanus -v- macdowellii)....[Ar], [Bo-5], [Hy],
lloydii (See Ariocarpus fissuratus -v- lloydii, Ariocarpus lloydii, R. fissuratus -v- lloydii),
[L], [Ar], [Bo-5], [Hx], [Hy], [Iu], [Ae-1982],
[Aj 11-10-1960],
" -f. .....[Ar],

ROSEOCEREUS

cinerispinus (See R. tephracanthus).....[Hy],
tephracanthus (See Cereus tephracanthus, Eriocereus tephracanthus, Harrisia tephracantha, Trichocereus tephracanthus).....
[DD], [Ay], [Bo-2], [Hx*],
" -v- ayopayanus.....

ROSEOCEREUS.....(Cont'd),

tephracanthus......(cont'd)
" -v- mizquensis.....

SALPINGOLOBIVIA

aurea (See Echinopsis aurea, Lobivia aurea, Pseudolobivia aurea).....[Hy],
" -v- grandiflora (See Echinopsis aurea, Lobivia aurea, Pseudolobivia aurea, S. aurea).....[Hy],
" -v- robustior (See Echinopsis aurea -v- fallax, Lobivia aurea -v- fallax, Pseudolobivia aurea -v- fallax).....[Hy],
cylindracea (See Echinopsis aurea -v- shaferi, Echinopsis shaferi, Lobivia aurea -v- shaferi, Lobivia shaferi).....[Hy],
cylindrica (See Echinopsis aurea, Lobivia aurea, Lobivia aurea -v- cylindrica, Lobivia cylindrica, Pseudolobivia aurea).....[Hy],
densispina (See Lobivia famatimensis -v- densispina, Reicheocactus pseudoreichianus).....[Hy],
leucomalla (See Lobivia famatimensis -v- leucomalla, Lobivia leucomalla),
shaferi (See Echinopsis aurea -v- shaferi, Echinopsis shaferi, Lobivia aurea -v- shaferi, Lobivia shaferi, Lobivia cylindracea, S. cylindracea),

SAMAIPATICEREUS

corroanus (See Corryocactus corroanus), [Bo-2], [Hi Vol. 2], [Hx*], [Hy], [Ij], [Jd], [Af 24-4], [Aj 26-5-1975],
inquisivensis (See Corryocactus inquisivensis, Haageocereus inquisivensis, Yungasocereus inquisivensis, Yungasocereus microcarpus), [Iv 12-1959],

SCHLUMBERGERA

bridgesii (hybrid) (See Epiphyllum bridgesii, Epiphyllum truncatus -v- bridgesii), (S. X buckleyi = S. truncata X S. russelliana)..... [C], [R], [X], [Bo-6], [Bu], [Cz],
gaertneri (See Epiphyllopsis gaertneri, Epiphyllum russelianum, Hatiora gaertneri, Rhipsalidopsis gaertneri), [H], [L], [M*], [T], [V], [X], [EE], [NN], [QQ], [TT], [UU], [Bv-1], [Af 16-3],
" -v- mackoyana (See Rhipsalidopsis gaertneri -v- mackoyana),......[D], [Bv-1],

SCHLUMBERGERA.....(Cont'd)

obtusangula (See Cereus obtusangulus, Epiphyllanthus microsphaerica, Epiphyllanthus obtusangulus),
opuntioides (See Epiphyllanthus obovatus, Zygocactus opuntioides).....[Ay], [Iz], [Jd], [Af Vol. 47 Supplement], [Aj 28-12-1977],
orssichiana.....[Av], [Ay], [Hx*], [Iz], [Jd], [Aj 28-12-1977],
russelliana.....[T], [Ay], [Ay], [Bo-2], [Bu], [Bv-1], [Cz], [Gr], [Hx*], [Af Vol. 47 Supplement], [Aj 28-12-1977],
truncata (See Epiphyllum truncatum, Zygocactus truncatus),
[I], [Z], [Av], [Ax], [Ay], [Il], [Af Vol. 47, Supplement], [Aj 28-12-1977], [Aj 29-8-1978], [Aj 29-12-1978],
" -v- alba.....
" -v- altensteinii.....
" -v- crenulatus.....
" -v- delicatus (See Zygocactus truncatus -v- delicatus),
[I], [Ay], [Hv], [Hh Nov. 1973],
X buckleyi.....[Aj 28-12-1977],
X exotica.....[Aj 28-12-1977],
(Hybrid).....[Hv],

SCLEROCACTUS

blainei (See S. spinosior var.),
contortus.....[Af 51-1],
erectocentrus.....(See Echinocactus erectocentrus, Echinomastus erectocentus, Neolloydia erectocentrus),
franklinii (See Pediocactus glaucus, S. glaucus), [Bo-5], [Af 11-5],
glaucus (See Pediocactus glaucus, S. franklinii), [Ar], [Ax], [Bh], [Aj 30-5-1979], [Hl 3-1972], [Hh Vol. 3, #5&6, 1979], [Hs I-3-1980],
havasupaiensis (See S. parviflorus -v- parviflorus), [Bo-5], [Af 24-3],
" -v- roseus (See S. parviflorus -v- parviflorus, S. whipplei -v- roseus),
heleii (See S. whipplei -v- heleii),
intermedius (See S. parviflorus -v- intermedius, S. whipplei -v- intermedius).....[Bo-5], [Jd], [Af 22-5], [Bi 1-4], [Hl 6-1971],
intertextus (See Echinocactus intertextus, Echinomastus intertextus, Neolloydia intertextus),
" -v- dasyacanthus (See Echinomastus intertextus -v- dasyacanthus),

SCLEROCACTUS.....(Cont'd),

johnsonii (See Echinomastus johnsonii, Neolloydia johnsonii),
" -v- lutescens (See Echinomastus johnsonii -v- lutescens),
mariposensis (See Echinocactus mariposensis, Echinomastus mariposensis, Neolloydia mariposensis),
mesae-verdae (See Coloradoa mesae-verdae, Echinocactus mesae-verdae, Pediocactus mesae-verdae), [An], [Ax], [Ay], [Bh], [Df], [Af 39-2], [Aj 30-5-1979], [Hl 6-1972], [Hs I-3-1980],
papyracanthus (See Pediocactus papyracanthus, Toumeya papyracantha),
parviflorus (See Pediocactus whipplei -v- roseus), [Ax], [Ad Oct., 1980],
" -v- blessingiae.....[An],
" -v- intermedius (See Echinocactus whipplei -v- intermedius, S. intermedius, S. whipplei -v- intermedius), [An], [Ax], [Df],
" -v- parviflorus.....(See S. havasupaiensis), [Ax],
" -v- roseus (See S. whipplei -v- roseus), [Hm 3-1979],
polyancistrus (See Pediocactus polyancistrus, Thelocactus polyancistrus).....[K], [M*], [P], [Z], [DD], [FF], [RR], [An], [Ar], [Ax], [Bo-5], [Ca], [Cb], [Cd], [Cm], [Dq], [Hx*], [Hy], [Jd], [Af 3-5], [Af 51-5], [Af 52-1], [Af 53-6], [Af 55-3], [Af 56-6], [Af 58-4], [Aj 30-5-1979], [Aj 31-5-1980], [Hh Sept., 1974], [Hl 2-1979],
pubispinus.....[Ax], [Af 38-3], [Aj 34-2-1983], [Hl 2-1972],
pygmaeus (See S. whipplei -v- pygmaeus),
scheeri (See Ancistrocactus scheeri, Echinocactus scheeri, Thelocactus scheeri),
schleseri (See S. spinosior var.).....[Af 60-5],
spinosior (See S. whipplei -v- spinosior), [An], [Ax], [Aj 34-2-1983],
" -v- blainei.....
" -v- schleseri (See S. schleseri)
terra-canyonae.....[Af 51-1],
uncinatus (See Ancistrocactus uncinatus, Ferocactus uncinatus, Glandulicactus uncinatus, Hamatocactus uncinatus),
" -v- crassihamatus (See Ancistrocactus uncinatus -v- crassihamatus, Glandulicactus crassihamatus, Hamatocactus crassihamatus, Thelocactus crassihamatus),

(cont'd)

SCLEROCACTUS.....(Cont'd),

uncinatus.....(cont'd),
" -v- wrightii (See Ancistrocactus uncinatus -v- wrightii, Echinocactus uncinatus -v- wrightii, Ferocactus uncinatus -v- wrightii, Glandulicactus uncinatus -v- wrightii, Hamatocactus uncinatus -v- wrightii),
unguispinus (See Echinomastus unguispinus, Neolloydia unguispina),
" -v- durangensis (See Echinomastus durangensis, Echinomastus unguispinus -v- durangensis),
warnockii (See Echinocactus warnockii, Echinomastus warnockii, Neolloydia warnockii),
whipplei (See Echinocactus whipplei, Pediocactus whipplei),
[E], [M*], [P], [Y], [LL*], [RR], [An], [Av], [Ax], [Bo-5], [Hx*], [Hy], [Aj 31-6-1980], [Bm-1988], [Hl 3-1972],
" -v- heleii (See Echinocactus whipplei -v- heleii) [An], [Df],
" -v- intermedius (See Echinocactus whipplei -v- intermedius, S. intermedius, S. parviflorus -v- intermedius), [M*], [Q], [Ay], [Bo-5], [Df], [Af 39-2], [Aj 30-5-1979], [Hs I-3-1980],
" -v- pygmaeus (See S. whipplei),
" -v- reevesii.....[An], [Df],
" -v- roseus (See S. parviflorus -v- parviflorus), [An], [Af 38-3], [Cu 3-1972], [Hl 3-1972],
" -v- spinosior (See S. spinosior), [Go], [Hy], [Aj 27-4-1976],
" -v- whipplei.....[Aj 30-5-1979], [Hs I-3-1980],
wrightiae (See Pediocactus wrightiae), [Ax], [Bh], [Hl 4-1972],

SCOPAREBUTIA

maresii.....[Hr],

SELENICEREUS

acutangulus (See Acanthocereus acutangulus, A. pentagonus),
anoplus (See S. inermis).....[Hy*],
atropilosus.....[Af 50-6], [Bb-1987],
barbadensis (See S. grandiflorus -v- barbadensis),
boeckmannii.....[RR], [Az], [Bv-1],

SELENICEREUS.....(Cont'd),

brevispinus.....[RR], [Bi-2],
coniflorus.....[V], [RR], [Az], [Bo-2],
donkelaari.....[RR], [Az], [Af 40-1],
grandiflorus (See Cereus grandiflorus, S. megalantha), [G], [H], [I], [L], [T], [V], [LL*], [QQ], [RR], [Ar], [Av], [Ay], [Be*], [Bt], [Bv-3], [Cm], [Cz], [Dq*], [Hb], [Hv], [Hx*], [Il], [Ad May, 1978], [Ba 4-1980], [Gu III-1953],
" -v- affinis.....[Bo-2],
" -v- armatus.....
" -v- barbadensis.....
" -v- irradians.....
" -v- ophites.....
" -v- telli.....[Hu],
" -v- uranos.....
grusonianus (See S. macdonaldiae -v- grusonianus),
hahnianus (See Harrisia hahniana, Mediocactus hahnianus),
hallensis.....[Bo-2],
hamatus.....[RR], [Ay], [Az], [Be*], [Bl], [Bo-2], [Bu], [Bv-1], [Cc], [Cx], [Cz], [Hx*], [Hy*], [Aw 1-1975],
hassleri (See Mediocactus hassleri),
hondurensis (See Acanthocereus hondurensis, Acanthcereus pentagonus -v- hondurensis), [RR], [Ay],
inermis (See S. anoplus).....[RR], [Jd],
innesii.....[Af 54-1],
jalapensis (See S. coniflorus).....[L],
knuthianus.....[RR], [Hy*],
leptocaulis (See S. donklaari),
lindmannii (See Mediocactus lindmannii),
macdonaldiae.....[D], [RR], [YY], [Ar], [Bo-2], [Bv-1], [Hx*], [Hy], [Af 8-4], [Af 58-1], [Flower - Bb-1985],
" -v- grusonianus (See S. grusonianus),
" -v- maxonii (See S. maxonii, Cereus urbanianus, Selenicereus urbanianus),
maxonii (See Cereus urbanianus, S. urbanianus),
megalantha (See S. grandiflora).....[Hy*],
mirandae.....[Az], [Aw 3-1967],
murrillii.....[RR-2], [Dq],
nelsonii.....[RR],
nycticallus (nycticalus) (See Cereus nycticallus, Cereus pteranthus, S. pteranthus),
" -v- armatus (See S. coniflorus),
pomifer (See Mediocactus pomifer),
pringlei (See S. coniflorus),..... [DD], [Bo-2], [Hl 3-1980],

SELENICEREUS.....(Cont'd),

pseudospinulosus (See S. spinulosus),
pteranthus (Cereus nycticallus, Cereus pteranthus),
[T], [RR], [Az], [Bu], [Bv-1], [Cr], [Dq],
[Hv], [Hx], [Hy], [Il], [Iz], [Af 8-4],
[Af 14-8], [Af 16-7], [Flowers - Gr,
Aj 27-6-1976, Bb-1986],
radicans....
rizzinii (See Hylocereus triangularis),
roseanus (See Cereus urbanianus, S. maxonii, S. urbanianus),
rostratus (See S. hamatus),
rothii.....
setaceus (See Cereus setaceus, Mediocactus coccineus, Mediocactus setaceus),
spinulosus (See Cereus spinulosus),
[RR], [Az], [Bc], [Bd], [Bo-2], [Hh May, 1974], [Hw-1986],
testudo (See Cereus testudo, Deamia testudo),
[Az], [Jd], [Aj 27-6-1976],
urbanianus (See Cereus urbanianus),
[G], [M*], [RR], [Av], [Bu], [Bv-1], [Cz],
[Af 56-3],
" -f. monstrosus.....[Af 26-6], [Af 59-3],
vagans.....[RR], [Az], [Bu], [Bv-1], [Cx], [Af 16-4],
vaupelii.....
werklei.....[RR], [Bv-1], [Cz], [Af 8-4], [Aj 27-5-1976],

(Hybrid).....[NN],

SETICEREUS

aurivillus (See Borzicactus aurivillus, Borzicactus icosagonus, Cereus aurivillus, Cereus icosagonus, S. icosagonus).....[NN],
chlorocarpus (See Gymnanthocereus chlorocarpus, Gymnocereus chlorocarpa),
[Bo-2], [Hx*], [Iv 9-1957],
chotaensis.....[Bo-2],
ferrugineus (See S. icosagonus -v- ferrugineus),
[Aj 11-1937],
humboldtii (See Borzicactus humboldtii, Cleistocactus neohumboldtii),
[NN], [Bo-2], [Hx*], [Hl 4-1971],
icosagonus (See Binghamia aurivillus, Borzicactus aurivillus, Borzicactus icosagonus, Cereus icosagonus, S. aurivillus),
[DD], [NN], [QQ], [Ah], [Ay], [Bo-2], [Bt],
[Cr], [Hu], [Hx*], [Hy], [Jd], [Af 50-1],
[Aj 11-1937], [Cu 3-1970], [Db 9-1980],

SETICEREUS.....(Cont'd),

icosagonus.....(cont'd),
" -v-auranticiflorus.....
" -v- ferrugineus (Invalid name - See S. ferrugineus),
" -v- ochmeanus (See S. ochmeanus).....[Bt],
microspermus (See Browningia microsperma, Gymnanthocereus microspermus, Gymnocereus microspermus),
neohumboldtii (See Binghamia humboldtii, Borzicactus humboldtii, S. humboldtii),
[Hy],
ochmeanus (See S. icosagonus -v- ochmeanus),
[Aj 11-1937],
roezlii (See Borzicactus roezlii, Borzicactus sepium -v- roezlii, Borzicactus neoroezlii),
[DD], [Ay], [Bo-2], [Hx], [Jd],
" -v- metachrous.....

SETICLEISTOCACTUS

dependens (See Cleistocactus dependens),
[DD], [Hx*], [Hy],
piraymirensis (See Cleistocactus piraymirensis),
[DD],

SETIECHINOPSIS

mirabilis (See Arthrocereus mirabilis, Echinopsis mirabilis),
[DD], [XX], [Ar], [Ay], [Bt], [Hv], [Hx*], [Il], [Jd], [Aj 27-9-1976], [Flower - Ba 2-1979],
" -v- gracilior (Invalid name).....[DD],

SOEHRENSIA

alagantha.....
bruchii (See Echinopsis bruchii, Lobivia bruchii, Lobivia formosa ssp. bruchii),
[H], [DD], [NN], [Ar], [Bo-3], [Bo-6], [Hy], [Jd], [Af 51-4], [Cu 1-1962],
" -v- aureorubriflora.....[Hy],
" -v- kieslingii.....(See Lobivia kieslingii),
" -v- longispina.....[Hy],
" -v- " -f. cristata.....[Hy],
" -v- nivalis (See Lobivia bruchii -v- nivalis, Lobivia formosa ssp. bruchii -v- nivalis, S. nivalis).....[Bo-3],
ferox (See Echinopsis ferox, Lobivia ferox, Lobivia potosina, Pseudolobivia ferox).....[X],

SOEHRENSIA.....(Cont'd),

formosa (See Lobivia formosa, Lobivia oreopepon, S. ingens, S. oreopepon),
[Ar], [Ay], [Bo-2], [Bu], [Bv-1], [Hx*], [Hy], [Hl 4-1972],
" -v- catamarca.....
" -v- maxima (Invalid name - See Lobivia formosa), [Bo-3],
" -v- polycephala (Invalid name - See Lobivia formosa).....[Bo-3],
" -v- velascoensis (See Lobivia formosa), [Bo-3],
grandis (See Lobivia formosa -ssp. grandis, Lobivia grandis).....[Bo-3], [Il],
" -v- pinchasensis.....
" -v- salta.....
ingens (See Lobivia formosa, S. formosa), [Bo-3], [Bu], [Bi 2-4],
korethroides (See Lobivia formosa -ssp. bruchii -v- nivalis).....[DD], [Bo-3], [Hy],
nivalis (See Lobivia formosa -ssp. bruchii -v- nivalis, S. bruchii -v- nivalis, S. korethroides).....
oreopepon (See Acanthocalycium oreopepon, Lobivia oreopepon, Lobivia formosa, S. formosa), [Bo-3], [Bu], [Gr], [Aj 30-3-1979],
schaeferi (See Trichocereus uebelmanianus, S. uebelmanniana, Lobivia formosa, S. formosa),
smrziana (See Trichocereus smrzianus),
tarijensis (See Lobivia formosa ssp. tarijensis, Lobivia tarijensis, Trichocereus poco),
uebelmanniana (See Trichocereus uebelmannianus, Lobivia formosa, S. formosa), [Bo-3],
walterii (See Lobivia walterii),

Sp. .....[Ar],

SOLICIA

pectinata (See Mammillaria pectinifera, Pelecyphora pectinata),
[G], [DD], [NN], [RR], [Ar], [Ay], [Bu], [Bv-1] [Cz], [Dq], [Hr], [Hx*], [Hy], [Jd], [Ae-1981], [Aj 31-2-1980], [Aj 32-10-1981], [Ba 4-1979],
" -f. cristata.....[Bl], [Dq], [Af 16-6], [Hl 4-1975],
pseudopectinata (See Neolloydia pseudopectinata, Normanbokea pseudopectinata, Pelecyphora pseudopectinata, Turbinicarpus pseudopectinata),

STENOCACTUS

acroacanthus (See Echinofossulocactus acroacanthus), [Dq],
albatus (See Echinofossulocactus albatus)..... [Am 2-1982],
anfractuosus (See Echinofossulocactus anfractuosus), [Cx],
arrigens (See Echinofossulocactus arrigens),
" -f. cristata.....
boedekerianus (See Echinofossulocactus boedekerianus),
brachycentrus (See Echinofossulocactus brachycentrus),
bravoae (See Echinofossulocactus bravoae),
bustamantei (See Echinofossulocactus bustamantei, Echinofossulocactus lexarzai, S. lexarzai), [Cx],
caespitosus (Backbg.) (See Echinofossulocactus caespitosus, Echinofossulocactus caespitosus -v- gracilispinus (Bravo))..... [Ia],
" -v- gracilispinus (Bravo) (See Echinofossulocactus caespitosus (Backbg.)),
confusus (See Echinofossulocactus confusus),
coptonogonus (See Echinofossulocactus coptonogonus), [Bu], [Cx],
crispatus (See Echinofossulocactus crispatus, E. violaciflorus).....[Iz],
debilispinus (See Echinofossulocactus debilispinus),
densispinus (See Echinofossulocactus densispinus),
dichroacanthus (See Echinofossulocactus dichroacanthus)
erectocentrus (See Echinofossulocactus erectocentrus),
esperanzaensis (See Echinofossulocactus esperanzaensis)
flavispinus (See Echinofossulocactus flavispinus),
flexispinus (See Echinofossulocactus flexispinus),
geminispinus (See Echinofossulocactus geminispinus),
gladiatus (See Echinofossulocactus gladiatus),
grandicornis (See Echinofossulocactus grandicornis),
guerraianus (See Echinofossulocactus guerraianus),
hastatus (See Echinofossulocactus hastatus),
heteracanthus (See Echinofossulocactus heteracanthus, Echinofossulocactus tetraxiphus),
hookeri (See Echinofossulocactus hookeri),
kelleriana (See Echinofossulocactus kellerianus), [Cu 4-1963],
lamellosus (See Echinofossulocactus lamellosus), [Cx], [Dq*],
lancifer (See Echinofossulocactus lancifer),
lexarzai (See Echinofossulocactus lexarzai, S. bustamantei).....
lloydii (See Echinofossulocactus lloydii).....[Ia],
longispina (See Echinofossulocactus longispina),
mocianus (See Echinofossulocactus mocianus),

STENOCACTUS.....(Cont'd),

multiareolatus (See Echinofossulocactus multiareolatus)
multicostatus (See Echinofossulocactus multicostatus), [Cx],
nerispinus (See Echinofossulocactus nerispinus),
obvallatus (See Echinofossulocactus obvallatus),
ocheterenaus (See Echinofossulocactus ocheterenaus), [Cx], [Am 9-1984], [Dv 48:397 Fig. 1933],
oligacanthus.....
parksianus (See Echinofossulocactus parksianus),
pentacanthus (See Echinofossulocactus pentacanthus),
phyllacanthus (See Echinofossulocactus phyllacanthus), [Cx], [Af 33-6], [Aw 2-1986],
rectospinus (See Echinofossulocactus rectispinus, Echinofossulocactus vaupelianus -v- rectospinus),
rentacantha (See Echinofossulocactus rentacantha),
rosasianus (See Echinofossulocactus rosasianus),
sphacelatus (See Echinofossulocactus sphacelatus, Echinofossulocactus ocheteranaus -v- sphacelatus),
sulphureus (See Echinofossulocactus sulphureus),
tegelbergii (See Echinofossulocactus phyllacanthus sensu Taylor),
tellii (See Echinofossulocactus tellii),
tetraxiphus (See Echinofossulocactus heteracanthus, Echinofossulocactus tetraxiphus),
tricuspidatus (See Echinofossulocactus phyllacanthus -v- tricuspidatus, Echinofosssulocactus tricuspidatus),
vaupelianus (See Echinofossulocactus vaupelianus), [Iz],
violaciflorus (See Echinofossulocactus violaciflorus, Echinofossulatus crispatus),
wippermannii (See Echinofossulocactus wippermannii),
xiphacanthus (See Echinofossulocactus xiphacanthus),
zacatecasensis (See Echinofossulocactus zacatecasensis) [N-1980],

STENOCEREUS

alamosensis (See Rathbunia alamosensis, Rathbunia neosonorensis, Rathbunia sonorensis)..... [Af 60-6],
aragonii (See Lemaireocereus aragonii, Marshallocereus aragonii).....[Aw 32-1],
beneckei (See Hertrichocereus beneckei, Lemaireocereus beneckei)...[NN], [Az], [Ij], [Iz], [Af 55-2],
chacalapensis (See Marshallocereus chacalapensis, Ritterocereus chacalapensis).....[Az],
chende (See Heliabravoa chende),

STENOCEREUS

chichipe (See Polaskia chichipe),
chrysocarpus.....[Az], [Aw 4-1972], [Aw 1-1980],
deficiens (See Lemaireocereus deficiens, Ritterocereus deficiens),
dumortieri (See Isolatocereus dumortieri, Lemaireocereus dumortieri).....[Az], [Bd], [Ag-1988], [Af 40-5],
eichlamii (See Lemaireocereus eichlamii, Ritterocereus eichlamii),
eruca (See Cereus eruca, Machaerocereus eruca), [Iz], [Af 57-4], [Hw-1986],
euphorbioides (See Neobuxbaumia euphorbioides, Rooksbya euphorbioides).....[Iz],
" -v- olfersii (See Rooksbya euphorbioides -v- olfersii),
fimbriatus (See Ritterocereus fimbriatus),
foetidus (See Anisocereus foetidus, Pterocereus foetidus),
fricii.....[Az], [Aw 4-1973], [Hl 6-1974],
gaumeri (See Anisocereus gaumeri, Pachycereus gaumeri, Pterocereus gaumeri),
griseus (See Cereus griseus, Lemaireocereus griseus, Ritterocereus griseus), [Az], [Gr], [Aw 1-1977], [Aw 4-1983],
gummosus (See Machaerocereus gummosus).....[Ij],
hollianus (See Lemaireocereus hollianus, Pachycereus hollianus),
hystrix (See Lemaireocereus hystrix, Ritterocereus hystrix).....[Gr], [Hv], [Af 60-1-1988],
kerberi (See Rathbunia kerberi).....[Af 60-6],
laevigatus (See Lemaireocereus longispinus, Ritterocereus laevigatus).....[Az],
lepidanthus (See Anisocereus lepidanthus, Pachycereus lepidanthus),
longispinus (See Lemaireocereus longispinus, Ritterocereus laevigatus, Stenocereus laevigatus),
marginatus (See Cereus marginatus, Lemaireocereus marginatus, Marginatocereus marginatus, Pachycereus marginatus),
" -v- gemmatus (See Marginatocereus marginatus -v- gemmatus, Pachycereus marginatus -v- gemmatus),
" -v- marginatus.....[Az],
martinezii (See Lemaireocereus martinezii), [Az], [Af 57-4],
montanus (See Cereus montanus, Lemaireocereus montanus) [Az],
pruinosus (See Lemaireocereus pruinosus, Ritterocereus pruinosus).....[Az], [Gr],
queretaroensis (See Lemaaireocereus queretaroensis, Ritterocereus queretaroensis).....[Az],

STENOCEREUS.....(Cont'd),

quevedonis.....[Az], [Aw 4-1979], [Aw 1-1980],
schumannii (See Lemaireocereus schumannii),
sonorensis (See Rathbunia alamosensis, Rathbunia sonorensis),
standleyi (See Ritterocereus standleyi),
[Az], [Af 60-6], [Aj 15-8-1964], [Aw 1-1980],
stellatus (See Cereus stellatus, Lemaireocereus stellatus, Ritterocereus stellatus),
[JJ], [Ar], [Ay], [Az], [Bo-4], [Hx*], [Jd],
[Af 51-4], [Af 61-1], [Aw 1-1970],
" -v- tonalianus.....
thurberi (See Cereus thurberi, Lemaireocereus thurberi, Marshallocereus thurberi, Pachycereus thurberi),
[Az], [Gr], [Iz], [Af 57-4],
" -v- littoralis (See Lemaireocereus thurberi -v- littoralis, Marshallocereus thurberi -v- littoralis)....[Az],
" -v- thurberi.....[Az],
treleasii (See Lemaireocereus treleasii)..........
[Az], [Bo-4], [Af 61-1],
weberi (See Cereus weberi, Lemaireocereus weberi, Pachycereus weberi),
[JJ], [Ay], [Az], [Bc], [Gr], [Af 51-4], [Aj 15-8-1964], [Aw 1-1970], [Aw 3-1973],

STEPHANOCEREUS

leucostele (See Cephalocereus leucostele),
[G], [R], [DD], [Ay], [Bo-4], [Dq], [Hi-1],
[Il], [Jd], [Au Vol. 2, 1977, Zone 3 - Compendium], [Aw 2-1980], [Ct-1988],

STETSONIA

coryne.....[D], [H], [K], [X], [RR], [Ay], [Bt], [Dq],
[Gr], [Hr], [Hx*], [Hy], [Il], [Iz], [Jd], [Af 57-2], [Ah-1985], [Aj 27-4-1976], [Hh Sept., 1976],
" -v- procera.....[Hi Vol. 2], [Hx*], [Af 42-3],
" -f. cristata.....[Hr],

STROMBOCACTUS

denegrii (See Obregonia denegrii).....[Bt], [Cr], [Iz],
disciformis.....[D], [T], [DD], [PP], [TT], [Ar], [Av],
[Ay], [Bo-5], [Bt], [Bz], [Cg], [Cm], [Cs],
[Cx], [Dq], [Gr], [Hr], [Hv], [Hx], [Hy],
[Il], [Iu], [Iz], [Jd], [N-1980], [Af 52-3],
[Af 55-1], [Af 56-2], [Aj 31-10-1980], [Ba 5-1978], [Ba 3-1981], [De 8-1980], [Hm 2-1979],

STROMBOCACTUS.....(Cont'd),

disciformis.....(cont'd),
" -f. cristata.....[Ar], [Hy], [Am 8-1984],
" (grafted).....[Af 55-3],
klinkerianus (See Neolloydia schmiedickeana -v- klinkeriana, Toumeya klinkeriana, Turbinicarpus schmiedickeanus -v- klinkerianus),
laui (See Neolloydia laui, Turbinicarpus laui),
[Iz],
lophophoroides (See Neolloydia lophophoroides, Toumeya lophophoroides, Turbinicarpus lophophoroides).....[Cr], [Iz], [N-1980],
macrochele (See Neolloydia schmiedickeana -v- macrochele, Turbinicarpus macrochele, Turbinicarpus schmiedickeanus -v- macrochele),
[Iz],
pseudomacrochele (See Neolloydia pseudomacrochele, Toumeya pseudomacrochele, Turbinicarpus pseudomacrochele).....[Iz],
" -v- krainzianus (See Neolloydia pseudomacrochele -f. krainziana, Toumeya krainziana, Turbinicarpus krainzianus, Turbinicarpus pseudomacrochele -v- krainzianus)
[Iz].
pseudopectinatus (See Neolloydia pseudopectinata, Normanbokea pseudopectinata, Pelecyphora pseudopectinata, Turbinicarpus pseudopectinatus).....[Iz],
schmiedickeanus (See Neolloydia schmiedickeana, Turbinicarpus schmiedickeanus),
[Dq], [Iz],
" -v- flaviflorus (See Neolloydia schmiedickeana -v- flaviflora, Turbinicarpus flaviflorus, Turbinicarpus schmiedickeanus -v- flaviflorus),
[Iz],
" -v- gracilis (See Turbinicarpus gracilis, Turbinicarpus schmiedickeanus -v- gracilis).....[Iz],
" -v- klinkerianus (See Toumeya klinkeriana, Turbinicarpus klinkerianus, Turbinicarpus schmiedickeanus -v- klinkerianus),
[Iz], [Hh May, 1975],
schwarzii (See Neolloydia schmiedickeana -v- schwarzii) Turbinicarpus schwarzii, Turbinicarpus schmiedickeanus -v- schwarzii).....[Ia],
turbiniformis.....
" -v- seidelii.....[Iz],

STROMBOCACTUS.....(Cont'd),

valdezianus (See Neolloydia valdeziana, Normanbokea valdeziana, Pelecyphora valdeziana, Thelocactus valdezianus, Turbinicarpus valdezianus).....[Iz],

STROPHOCACTUS

wittii.....[M*], [DD], [RR], [Ay], [Bl], [Bo-2], [Dq*], [Hx*], [Hy*], [Jd], [Af 52-4], [Cu 5-1-1936],

SUBMATUCANA

aurantiaca (See Arequipa aurantiaca, Borzicactus aurantiacus, Borzicactus calvescens, Matucana aurantiaca, S. calvescens), [NN], [Ar], [Bo-2], [Hx], [Hy], [Jd], [Hl 4-1965], [Hl 4-1969],
aureiflora (See Borzicactus aureiflorus, Matucana aureiflora).....[N-1980], [Af 57-4],
" -v- elata (See Matucana aureiflora -v- elata),
caespitosa (See S. paucicostata),
calliantha (See Borzicactus callianthus, Matucana calliantha),
calvescens (See Arequipa aurantiaca, Borzicactus aurantiacus, Borzicactus calvescens, Matucana aurantiaca, (S. aurantiaca).....[Ar],
" -v- semidenuda.....
celendinensis (See Borzicactus intertextus -v- celendinensis, Matucana celendinensis), [Hl 6-1971],
currundayensis (See Matucana currundayensis), [Bo-6], [Hx], [Hy],
formosa (See Borzicactus formosus, Matucana formosa),
" -v- minor (See Matucana formosa -v- minor),
fruticosa (See Matucana fruticosa),
glomerispina.....[NN],
huagalensis (See Borzicactus huagalensis), [Af 50-3], [Hs II-4-1981],
" -v- bruneispina.....
intertexta (See Borzicactus intertestus, Matucana intertesta), [Hx*], [Ae-1981], [Af 50-3], [Bi 5-4-1987],
krahnii (See Borzicactus krahni, Matucana krahnii),
madisoniorum (See Borzicactus madisoniorum, Matucana madisoniorum), [DD], [HH], [NN], [Ar], [Hx*], [Jd], [Af 50-1], [Af 56-5], [Am 3-1976], [Ba 5-1976],
" -v- asterium.....
" -v- horridispinum.....

SUBMATUCANA......(Cont'd),

madisoniorum.....(cont'd),
" -v- madisoniorum (See Borzicactus madisoniorum -v- madisoniorum, Matucana madisoniorum -v- madisoniorum), [Aj 4-1977],
" -v- paucicostata (See Borzicactus paucicostatus, S. paucicostata),
pujupatii (See Borzicactus madisoniorum -v- pujupatii, Matucana madisoniorum -v- puyupatii),
myriacantha (See Borzicactus myriacanthus, Matucana myriacantha),
oreodoxa (See Borzicactus oreodoxa, Eomatucana oreodoxa, Matucana oreodoxa),
paucicostata (Invalid name - See Borzicactus paucicostata), [Ar], [Bo-6], [Bt], [Hx], [Ae-1981], [Db 6-1980], [Hl 1-1974], [Hs I-3-1980],
" -v- pujupatii See Borzicactus madisoniorum -v- pujupatii),
ritteri (See Borzicactus ritteri, Matucana ritteri), [F], [NN], [Bo-6], [Hx*], [Hy],
samaipatanus (See Borzicactus samaipatanus, Bolivicereus samaipatanus),
tuberculata (See Borzicactus tuberculatus, Matucana tuberculata),
turbiniformis (See Matucana turbiniformis),
variicolor (See Matucana variicolor),
Sp. Lau-173.....[Ae-1981],
Sp. .....[Bo-2], [Bo-6],

SUBPILOCEREUS

albispinus (See Cereus repandus, Subpilocereus repandus),
atroviridis (See Cephalocereus atroviridis).....[Bo-4],
grenadensis (See Cereus grenadensis, Cephalocereus grenadensis).....[Bo-4], [Hx*],
horrispinus (See Cereus horrispinus, Pilosocereus horrispinus).....[Bo-4],
margaritensis (See Cereus margaritensis, S. russelianus -v- margaritensis),
neorusselianus (See S. russelianus),
ottonus (See Cephalocereus ottonis).....[Bo-4],
remolinensis (See Cereus remolinensis), [Bo-4], [Iv 9-1952],
repandus (See Cephalocereus repandus, Cereus repandus) [Bo-4], [Ba 10-1978],
" -v- weberi (See Cereus repandus -v- weberi), [Bo-4],
SUBPILOCEREUS.....(Cont'd),

russelianus (See Cephalocereus russelianus, Pilocereus russelianus, Pilosocereus russelianus), [Bo-4],
" -v- micranthus.....
" -v- margaritensis (See Pilocereus russelianus -v- margaritensis, S. margaritensis),
wagenaari (See S. ottonis),

SULCOREBUTIA

alba.....[G], [DD-529], [Bg], [Hx*], [Ah-1980], [Aj 9-1981], [Hc 25-1976], [Hm 1982 Yearbook],
albaoides (See Weingartia albaoides),
" ssp. subfusca (See Weingartia albaoides -ssp. subfusca),
albida (See S. flavissima -v- albida, S. albissima),
" -v- riograndensis.....
" -v- robustispina.....
albissima (See S. albida, S. flavissima -v- albida, Weingartia albissima), [Bg], [Aj 36:36, 1985],
arenacea (See Rebutia arenacea, Weingartia arenacea), [Av], [Bg], [Gr], [Hx*], [Iz], [Jd], [Ag-1988], [Ah-1981], [Am 7-1983], [Au 16:81, 1961], [Bi 6-2-1988], [Hm 1982 Yearbook],
ayopayana.....[Bg],
bicolor (See S. steinbachii).....[Bg],
breviflora (See Rebutia brachyacantha, S. haseltonii), [Bg], [Hx*], [Af 53-3], [Aj 34:257, 1983], [Hc 26, 1976], [He 3 (14):15, 1969],
brevispina (See S. verticillacantha -f. brevispina, Weingartia brevispina).....[Aj 31-11-1980],
caineana (See Rebutia caineana, S. breviflora), [Aj 34:257, 1983], [Hc 27, 1976],
callecallensis (See Weingartia callecallensis), [Hd 2 (3):1, 1981],
camachoi (See S. tiraquensis -v- bicolorispina), [Bg],
candiae (See R. candiae, S. xanthoantha), [DD], [Bg], [Hi-2], [Hx*], [Iz], [Af 53-3], [Af 54-6] [Ag-1985], [Ba 7/8:104, 1963], [Hc 27, 1976], [Hm 1982 - Yearbook],
canigueralii (See Rebutia canigueralii), [G], [Bg], [Bt], [Hv], [Hx*], [Iz], [Af 58-2], [Aj 15-5-1964], [Am 3-1980], [Bi 6-2-1988], [Cu 4-1981], [Hc 28-1976],
" -v- applanata.....
caracarensis (See Rebutia caracarensis), [Bg], [Hc 29-1976],
SULCOREBUTIA.....(Cont'd),

cardenasiana....[Bg], [Hx*], [Aj 26-3-1975], [Hc 29-1976]
clavata (See Weingartia clavata),
clizensis.....[Bg], [Hc 29-1976],
cochabambina (cochabambensis) (See S. clizensis, S. pojoniensis),.....
corroana (See Rebutia corroana, Weingartia chuquichuquiensis).....[Ba 52:192, 1973], [Hc 30, 1976],
crispata....[Z], [Bg], [Hx*], [Iz], [Jd], [Af 55-3], [Ag-1986], [Aj 21:102, 1970], [Aj 26-8-1975], [Aj 31-11-1980], [Hc 30: 1976],
" -f. (long-spined form).....[Bg],
" -f. (short-spined form).....[Bg],
croceareolata (See S. verticillacantha -v- cuprea)
'cupreata' (KK-1800) (See S. flavissima).....[Bg],
cylindrica.....[Bg], [Iz], [Jd], [Af 51-3], [Aj 32-3-1981], [Hc 30:1976], [Hh March, 1974], [Hm-1982-Yearbook],
flavida (Invalid name),
flavissima.....(See Weingartia flavissima), [DD-530], [Bg], [Hc 31:1976], [Hv], [Hx*], [Iz], [Ae-1981], [Af 54-3], [Ah-1985], [Aj 27-12-1976], [Aj 30-10-1979], [Aj 31-11-1980], [Hm 1982 Yearbook]
" -v- albida (See S. albissima)....[Aj 30-10-1979]
formosa (See S. markusii -v- longispina, Weingartia formosa),
frankiana [DD-532], [Bg], [Hx*], [Af 51-2], [Aj 21:102, 1970], [Ba 1-1974], [Bm-1988], [Hc 32:1976], [He 3 (14):15, 1969],
" -v- aureispina (See Weingartia aureispina), [Bg], [Ba 53:3, 1974], [Hc 32:1976],
" -v- frankiana (See Weingartia frankiana),
glanduliflora (See S. tunariensis).....[Hc 32:1976],
glomeriseta (See Rebutia glomeriseta), [Bg], [Hv], [Iz], [Af 51-4], [Af 57-2], [Ag 1986], [Aj 31-11-1980], [Au 16:81, 1963] [Hc 32:1976], [Hm 1982 Yearbook),
glomerispina (See Rebutia glomerispina), [Bg], [Ab-1985], [Af 61-1], [Aj 31-11-1980], [Cu 27:80, 1965], [Hc 33:1976],
haseltonii (Invalid name - See S. breviflora, S. menesesii -v- kamiensis), [Aj 34:257, 1983], [Hc 33:1976],
hoffmanniana (See Lobivia hoffmanniana)...[Bg], [Hb], [Hv], [Hx*], [Ad March, 1980], [Ba 4, 1977], [Bi 6-2-1988], [Hc 34:1976], [Hl 2-1974],
" -v- seinoiana.....[Bg],
" -f. (densely spined form).....[Bg],
SULCOREBUTIA.....(Cont'd),

inflexiseta (See Rebutia inflexiseta),
[Bg], [Hc 34:1976],
'jolandae' (See S. purpurea),
krahnii.....[Z], [DD-531], [Bg], [Hv], [Hx*], [Il],
[Af 54-3], [Ag-1981], [Aj 21:102, 1970],
[Aj 31-11-1980], [Hc 35:1976], [Hl 1-1978],
krugerii (kruegeri) (See Aylostera krugerii, Rebutia
krugeri),
[G], [K], [NN], [Bg], [Hu], [Hv], [Hx*], [Hy],
[Jd], [Ad April, 1978], [Af 57-2], [Af 61-1],
[Am 3-1980], [Au 16:79, 1961], [Bi 6-2-1988],
" -f. (white spined form).....[Bg],
" -v- hoffmanniana (See S. hoffmanniana),
" -v- mineata.....
langeri.....[Bg], [Iz], [Af 61-1], [Aj 32-2-1981],
[Bi 4-1986*], [Bi 6-2-1988],
latiflora (See S. purpurea).....[Hc 36:1976],
lepida (See S. totorensis -v- lepida),
[G], [DD], [Av], [Hi Vol. 2], [Hx*], [Jd], [Aj
31-11-1980], [Au 17:13, 1962], [Hc 36:1976],
[Hl 4-1967], [Hm 1982 Yearbook],
" -f. cristata.....[Hi Vol. 2],
lepida X S. tiraquensis.....[Hi Vol. 2],
losenickyana (See S. sucrensis, S. verticillacantha
-v- ritteri, S. verticillacantha -v-
losenickyana, S. verticillacantha -v-
albispina),
[Bg], [Hi Vol. 2], [Hx*],
'mairanana' (KK 1811 & KK 1905),.....[Bg],
markusii (See Rebutia markusii).....[Ay], [Bg], [Cv-2],
[Hx*], [Af 54-1], [Bi 6-2-1988],
" -v- longispina (See S. vertillacantha -v-
cuprea),
[Aj 21:102, 1970], [Hc 37:1976],
menesesii (See Rebutia menesesii),
[DD], [Bf], [Bg], [Hx*], [Af 51-4], [Af 53-2]
[Ag-1986], [Ba 7/8:96, 1963], [Hc 38:1976],
[Hm 1982 Yearbook],
" -v- 'kamiensis' (See S. breviflora).....[Bg],
mentosa (See Rebutia mentosa, Weingartia mentosa),
[DD], [Ay], [Bg], [Hb], [Hi Vol. 2], [Hv],
[Iz], [Aj 34-2-1983], [Ba 21:102, 1964],
[Bi 6-2-1988], [Hc 38:1976],
" -f. (from Tarata).....[Bg],
minima (See S.taratensis -v- minima, S. verticilla-
cantha -v- minima).....[Am 3-1976], [Am 3-1980],
mizquensis (See Weingartia mizquensis),
[Av], [Bg], [Hv], [Hx*], [Iz], [Jd],
[Aj 21:102, 1970], [Bi 6-2-1988], [Hc-
39:1976], [Hl 5-1972], [Hm 1982-Yearbook],
SULCOREBUTIA.....(Cont'd),

muschii (See Weingartia muschii).....[Bg], [Jd],
[Ah-1985], [Ba 3-1974], [Hc 39:1976],
neocorroana (See Rebutia corroana, Weingartia corroana,
Weingartia neocumingii -v- corroana),
nigro-fuscata (See S. tiraquensis -v- 'spinosior', S.
steinbachii -v- spinosior, Weingartia
nigro-fuscata),
[Cactus (Hoofstadt) 10:113, 1978, Belgium]
oenantha (See Weingartia oenantha).....[Bg], [Af 61-1],
[Ba 50:112, 1971], [Ba 50-102, 1970],
[Bi 6-2-1988], [Hc 39:1976], [Hl 5-1980],
" -v- 'Epizana'.....[Bg],
oligacantha (See S. tarijensis, Weingartia oligacantha)
pampagrandensis (See Weingartia pampagrandensis, S.
weingartioides),
[Bg], [Hx*], [Af 61-1], [Aj 25:97, 1974],
[Hc 40:1976],
pasopayana (See Weingartia pasopayana, S. pulchra -
-f. (long-spined form)),
perplexiflora (See Weingartia perplexiflora, S.
pulchra -f. (dark form)),
'pojoniensis' (See S. cochabambina),
[Bg], [Hc 40:1976],
polymorpha (See Rebutia polymorpha, S. steinbachii -v-
polymorpha, Weingartia polymorpha),
[Hx*], [Ah-1986], [Aj 16:115, 1965], [Hc 41-
1976],
pulchra (pulchera) (See Rebutia pulchera, Weingartia
pulchera).....[Bg], [Af43:39, 1971],
[Aj 30-9-1979], [Hc 41:1976],
" -f. (dark form).....(See S. perplexiflora),
" -f. (long-spined form).....(See S. pasopayana),
purpurea (See Cinnabarinea purpurea, Weingartia
purpurea)...[Bg], [Aj 32-11-1981], [Hh 1:55, 1974],
rauschii.....[G], [Z], [DD-534], [Ar], [Ay], [Bg], [Gr],
[Hv], [Hx], [Il], [Iz], [Jd], [Ab-1985],
[Ag-1985], [Aj 28-4-1977], [Am 8-1979],
[Am 8-1984*], [Ba 4-1978 - Flower], [Bi 6-
2-1988], [Hc 42:1976], [Hl 4-1980], [Hm
1982 - Yearbook].
" -f. .....[Iz],
" -f. (with prominent white areoles).....[Bg],
" -f. (green-bodied form).....[Bg],
" -f. (golden-spined form).....[Bg],
" -f. (purple body form).....[Bg],
" -f. (tall body form).....[Bg],
'ressiniana' (See S. verticillacantha -v- albispina),
rigidispina.....[Hy],
ritteri (See S. verticillacantha -v- albispina, S.
verticillacantha -v- ritteri, Weingartia ritteri)
SULCOREBUTIA.....(Cont'd),

'rojasii' (A Karel Knize catalog name),
rubro-aurea (See S. verticillacantha -v- aureiflora), [Hg 9 (2):34, 1984],
sanguineo-tarijensis (See S. tarijensis, Weingartia sanguineo-tarijensis),
santiaginensis.....[Bg], [Aj 30-10-1979],
'seinoiana' (See S. hoffmanniana).....[Hc 43:1976],
'senilis' (See S. tiraquensis -v- bicolorispina),
'singularis' (See Rebutia padcayensis),
steinbachii (See Rebutia steinbachii).....[X], [DD], [EE], [NN], [QQ], [Ar], [Ay], [Bg], [Bo-3], [Bo-6], [Bt], [Gr], [Hb], [Hv], [Hx*], [Hy], [Il], [Jd], [Af 61-1], [Am 3-1980], [Bi 6-2-1988], [Hg 46:1976], [Hh Vol. 3, #5 & 6, 1979],
" -v- australis.....
" -v- gracilior (Invalid name).....[Bg], [Hu], [Hx*], [Af 43-1], [Af 61-1], [De 12-1980], [Hw-1989],
" -v- horrida (See Weingartia backebergiana), [Bg], [Aj 24-9-1973],
" -v- polymorpha (See Rebutia polymorpha, S. polymorpha).....[Bg],
" -v- rosiflora (Invalid name - See S. steinbachii).....[Bg], [Hy],
" -v- violaciflora (Invalid name - See S. steinbachii).....[Bg], [Hy],
" -f. (black-spined form).....[Bg],
" -f. (brown-spined form).....[Bg],
'sucrensis' (See S. losenickyana, S. verticillacantha -v- albispina, S. verticillacantha -v- ritteri, S. ritteri)..[Hx*], [Af 43-1], [Aj 15-5-1964], [Hc 49:1976], [Hm 1982 -Yearbook],
swobodae (See S. mentosa).....[Bg], [Aj 35-6-1984],
tarabucoensis.....[Bg], [Hx], [Aj 15-5-1964], [Hc 50:1976], [Hm 1982 - Yearbook],
taratensis (See Rebutia taratensis).....[DD], [Bg], [Hv], [Cu 25:57, 1965], [Hc 50:1976],
" -v- minima (See S. minima, S. verticillacantha -v- minima).....[Hx*], [Af 42-4], [Aj 19-6-1968], [Hc 52:1976], [Hl 4-1976], [Hm 1982 - Yearbook],
tarijensis (See Weingartia tarijensis), [Bg], [Hi Vol. 2], [Aj 30-9-1979], [Hh 2-2-1975], [Hh 3 (1):13, 1978],
" -v- (from Villazon).....[Bg],
tiraquensis (See Rebutia tiraquensis, Weingartia tiraquensis)...[G], [DD], [Bg], [Bo-6], [Hi-Vol. 2], [Hx*], [Hy], [Af 56-2], [Af 61-1], [Au 16:79, 1961], [Hc 52:1976],
SULCOREBUTIA.....(Cont'd),

tiraquensis.....(cont'd),
" -v- aglaia (See S. tiraquensis -v- bicolorispina, Weingartia aglaia),
" -v- bicolorispina (See Weingartia aglaia) [Bg], [Iz], [Af 60-2], [Hc 53:1976],
" -v- electracantha.....[DD], [Bg], [Af 43-1], [Af 61-1], [Hc 53:1976],
" -v- 'horrida'.....[Bg],
" -v- longiseta (See Rebutia tiraquensis -v- longiseta).....[Af 42:188, 1970], [Af 43:39, 1971],
" -v- longispina.....
" -v- 'spinosior'.....(See Weingartia nigrofuscata).....[Bg], [Hc 54:1976],
" -v- tiraquensis (See Rebutia tiraquensis), [DD], [Bg], [Iz],
'tominense' (KK-1810).....[Bg],
torotorensis (See Weingartia torotorensis, Cinnabarinea torotorensis), [Bg], [Af 54-3], [Ah-1986], [Aj 32-11-1981], [Au 16:81, 1961], [De 12, 1981],
totoralensis (See Weingartia totoralensis).....
totorensis (See Rebutia totorensis, Weingartia totorensis).....[Au 16:79, 1961], [Hc 55:1976],
" -v- lepida (See S. lepida, Weingartia lepida), [Bg], [Au 17:13, 1962],
" -v- totorensis.....[Bg], [Bi 6-2-1988],
tuberculato-chrysantha (See S. steinbachii).....[Bg], [Hx*], [Ba 52 (11):193, 1973], [Hc 55:1976],
tunariensis (See Rebutia tunariensis, Weingartia tunariensis).....[Bg], [Hb], [Hx*], [Hy], [Af 53-3], [Ah-1983], [Aj 34:202, 1983], [Cu 27:80, 1965], [Hc 56:1976],
unguispina (See S. purpurea),
'vanbaelii'.....(See S. hoffmanniana, S. verticillacantha).....
vasqueziana (See S. verticillacantha)....[Bg], [Aj 21-6-1970], [Ba 52 (12):222-1973], [He 3 (14):15-1969],
" -v- albispina (See S. verticillacantha -v- albispina).....[Hc 56:1976],
verticillacantha.....[Bg], [Hi Vol. 2], [Hx*], [Af 43-1] [Au 17:13, 1962], [Cu 2-1980], [Hc 57:1976],
" -v- albispina (See S. verticillacantha -v- ritteri, Weingartia ritteri), [Bg], [Hc 56:1976],
" -v- applanata.....[Bg], [Cu 42 (2):37, 1980],
" -v- aurea.....[Af 57-5],
" -v- aureiflora...[DD-533], [Bg], [Iz], [Aj 23-5-1972], [Aj 31-11-1980], [Bi 6-2-1988],

SULCOREBUTIA.....(Cont'd),

verticillacantha.....(cont'd),
" -v- brevispina (See S. brevispina, Weingartia brevispina).....[Hc 58-1976],
" -v- chatajillensis (See S. alba)..... [Aj 35:216, 1984],
" -v- crispata (See S. crispata), [He 3 (14):15, 1969],
" -v- cuprea.....[Bg], [Iz], [Ag-1985], [Aj 23-5-1972], [Hc 58:1976],
" -v- losenickyana (See S. losenickyana), [Aj 35:223, 1984],
" -v- minima (See S. minima, S. taratensis -v- minima).....[Bg], [Hc 52,1976], [He 3 (14):14, 1969],
" -v- ressinii (See S. verticillacantha -v- albispina),
" -v- ritteri (See S. ritteri, S. verticillacantha -v- albispina)..... [Af 57-5], [Cu 42 (2):38, 1980],
" -v- vasqueziana (See S. vasqueziana),.... [He 3(14):15, 1969],
" -v- verticillacantha.....[Bg],
" -v- verticosior (See S. verticillacantha), [Au 17:13, 1962],

villicumensis.....
vizcarre (See Rebutia vizcarre, Weingartia vizcarre), [Bg], [Af 43:40, 1971], [Hc 59:1976],
" -v- minor,,,,,
weingartioides (See S. pampagrandensis), [Hb], [Hy], [Af 43:40, 1971], [Ba 50:1, 1971], [Ba 52:193, 1973], [Hc 59 & 60: 1976],
xanthoantha (Invalid name - See S. candiae), [Bg], [Hx*], [Hc 50:1976],
zavaletae (See Aylostera zavaletae),..... [Bg], [Hc 61:1976],

Sp. .....[DD], [Hv],
Sp. KK-808.....[Bg],
Sp. calle - calle.....[Hv],
Sp. (Mizque).....[Hh July, 1973],

TACINGA

atropurpurea (See T. zehntneri).....[Bo-1],
" -v- zehntnerioides.....[DD], [Bo-1],
funalis.....[E], [RR], [Ay], [Bo-1], [Hx*], [Hy], [Jd], [Cu 1-1963],

TEPHROCACTUS

albiscoparius (Backbg.) (Invalid name - See Cumulopuntia boliviana, Opuntia boliviana, T. albiscoparius, T. bolivianus, T. ferocior), [Bo-6],
alboareolatus (See Cumulopuntia alboareolata)...[DD],
alexanderi (See Opuntia alexanderi, Opuntia halophila), [Ar], [Bo-1], [Hi Vol. 2], [Il],
" -v- bruchii (See Opuntia bruchii, T. bruchii), [Bo-1], [Bx], [Af 54-6], [Hl 6-1973], [Hl 4-1978],
" -v- " -sv- brachyacanthus.....[Bo-1],
" -v- " -sv- macranthus.....[Bo-1],
" -v- subsphaericus.....[Bo-1],
andicola (See Opuntia andicola, T. glomeratus -v- andicolus, T. neoglomeratus -v- andicolus),
aoracanthus (See Opuntia aoracantha, T, articulatus -v- ovatus, T. ovatus).....[Af 59-1],
articulatus (See Opuntia articulata, Opuntia glomerata, Tephrocactus articulatus, T. articulatus -v- calvus, Tephrocactus glomeratus (Speg.)), [Ar], [Bo-1], [Hy], [Bi 2-4],
" -v- calvus (See T. articulatus)..... [Bo-1], [Bx], [Hx*],
" -v- diadematus (See Opuntia diademata, T. diadematus),.... [Ar], [Ay], [Bo-1], [Hx*], [Hy],
" -v- inermis (See Opuntia diademata -v- inermis, Opuntia strobiliformis, Tephrocactus strobiliformis),.....[Ar], [Bo-1], [Bx], [Hx*], [Hl 3-1978],
" -v- oligacanthus (See Opuntia diademata -v- oligacantha),..... [Bo-1], [Bx], [Hx*],
" -v- ovatus (See Opuntia aoracantha, T. aoracanthus, T. ovatus), [Bo-1], [Bx], [Hx*], [Hl 3-1978],
" -v- papyracanthus (See Opuntia papyracantha, Opuntia diademata, T. articulatus -f. syringacanthus, T. diadematus, T. articulatus -v- diadematus), [NN], [QQ], [Bo-1], [Hx], [Hl 3-1978],
" -v- polyacanthus (See Opuntia paediophila, T. hossei, T. paediophilus), [Bo-1], [Hx*], [Hy],
" -v- " -f. longispina.....
" -f. syringacanthus (See Opuntia diademata, Opuntia papyracantha, T. articulatus -v- diademata, T, diadematus, T. glomeratus (SPEG.)),

TEPHROCACTUS.....(Cont'd),

articulatus.....(cont'd),
" -v- syringacanthus.....
[Bo-1], [Bx], [Hy], [Af 56-1], [Aj 33-6-1982], [Bi-2-1987*], [Hl 1-1965],
" -f. monstrosus.....[Af 56-1],
asplundii (See Cumulopuntia pentlandii -v- dactylifera, Opuntia dactylifera, Tephrocactus dactylifera), [Bo-1],
atacamensis (See Maihueniopsis atacamensis, Opuntia atacamensis).....[Bo-1],
" -v- chilensis (See Maihueniopsis atacamensis), [Bo-1],
" -v- rubra.....
atroglobosus (Invalid name - See Platyopuntia atroglobosa).....[DD], [Bo-6:3905 - Description].
atroviridis (See Austrocylindropuntia floccosa, Opuntia atroviridis, Opuntia floccosa, Tephrocactus floccosus),
[I], [Ay], [Bo-1], [Bx], [Hx*], [Iv 7-1956],
" -v- longicylindricus.....
" -v- parviflorus......
" -v- paucispinus.....
berteri (See Cumulopuntia berteri, Opuntia sphaerica, Tephrocactus sphaericus, Tephrocactus dimorphus sensu Backbg.),
" -v- kuehnrichianus (See Cumulopuntia kuehnrichiana, Tephrocactus kuehnrichiana),
bicolor (See Cumulopuntia kuehnrichiana, Opuntia kuehnrichiana, Tephrocactus kuehnrichianus, T. fulvicomus -v- bicolor, Tephrocactus mirus),
blancii (See Opuntia blancii).....[Bo-1],
bolivianus (S/D) (See Cumulopuntia boliviana, Opuntia boliviana, Tephrocactus albiscoparius), [Bx], [Hy], [Hl 4-1978],
bolivianus (sensu Backebg.) (See Cumulopuntia pentlandii (S/D), Tephrocactus pentlandii (S/D), Opuntia pentlandii (S/D),),
bruchii (See Opuntia alexanderi -v- bruchii, Opuntia bruchii, T. alexanderi -v- bruchii),
camachoi (See Maihueniopsis camachoi, Opuntia camachoi) [Bo-6],
catacanthus (Invalid name - See Opuntia catacantha), [DD],
chichensis (See Cumulopuntia boliviana, Opuntia boliviana (S/D), Tephrocactus bolivianus sensu Backebg.).....[Bo-1], [Aj 6-3-1955],
" -v- colchanus (See Cumulopuntia pentlandii -v-colchanus)...[Bo-1], [Aj 6-3-1955],

chilensis (See Maihueniopsis atacamensis, Tephrocactus atacamensis, T. atacamensis -v- chilensis),
coloreus (See Maihueniopsis coloreus).....[DD],
conoideus (Backbg.) (See Maihueniopsis conoidea - (Backbg.)).....[Bo-1], [Bo-6],
conoideus (Ritt.) (See Maihueniopsis archiconoidea),
corotilla sensu Backbg. (See Cumulopuntia ignota, Opuntia ignota).....[Bo-1],
" -v- aurantiaciflorus (See Cumulopuntia ignota, Opuntia ignota).....[DD], [Bo-1],
corotilla (K. Sch.) (See Cumulopuntia berteri, Opuntia sphaerica, T. sphaericus),
corrugatus (See Opuntia corrugata, Opuntia longispina -v- corrugata, Platyopuntia corrugata), [Cu 28:30, 1966],
crassicylindricus (See Cumulopuntia crassicylindrica, Opuntia crassicylindrica), [Bo-1],
crispicrinitus (See Austrocylindropuntia floccosa, Opuntia floccosa, Opuntia crispicrinita, T. floccosus).....[Bo-1],
" -v- cylindraceus.....[Bo-1],
" -v- " -sv- flavicomus.....[Bo-1],
" -v- tortispinus.....[Bo-1],
curvispinus (Invalid name).....[DD],
cylindrarticulatus (See Cumulopuntia pentlandii -v- dactylifera, Opuntia cylindrarticulata, Opuntia dactylifera, T. dactyliferus).....[Bo-1],
cylindrolanatus (See Austrocylindropuntia floccosa, Opuntia floccosa, Opuntia cylindrolanata, Tephrocactus floccosus).....[Bo-1],
dactylifer (See Cumulopuntia pentlandii -v- dactylifera, Opuntia dactylifera, T. cylindrarticulatus), [Bo-1], [Bx], [Il],
darwinii (See Maihueniopsis darwinii, Opuntia darwinii) [Bo-1], [Bx],
" -v- hickenii (See Opuntia hickenii, T. hickenii).....
diadematus (See Opuntia diademata, Opuntia articulata -v- diademata, T. articulatus -f. syringacanthus, T. articulatus -v- syringacanthus, Opuntia glomerata, Tephrocactus glomeratus), [I], [X], [Bx], [Cr], [Ig*], [Im], [Hl 6-1973],
dimorphus sensu BACKBG. (See Cumulopuntia berteri, Opuntia sphaerica, T. sphaericus, T. pseudorauppianus).....[Bo-1], [Bx],
" -v- pseudorauppianus (See T. dimorphus, T. pseudorauppianus).....[Bo-1], [Hx],

TEPHROCACTUS.....(Cont'd),

duvalioides (See Cumulopuntia pentlandii -v- dactylifera), Opuntia dactylifera, T. dactyliferus),
echinaceus (See Cumulopuntia echinacea), [Taxon XIII:4, 145, 1964 (2)],
ferocior (See Cumulopuntia boliviana, Opuntia boliviana, T. bolivianus, T. albiscoparius), [Bo-1],
flaviscoparius (See Maihueniopsis ignescens, Opuntia ignescens, T. ignescens),
flexispinus (BACKBG.).....[DD],
flexuosus.....(See Cumulopuntia pentlandii -v- dactylifera, Tephrocactus dactylifer),
floccosus (See Austrocylindropuntia floccosa, Opuntia floccosa, Tephrocactus atroviridis), [X], [QQ], [Ay], [Bt], [Hx*], [Hy], [Ij], [Af 51-4], [Af 56-3], [Ah-1986], [Da 2-1980], [Hl 2-1970], [Hl 6-1973],
" -v- canispinus.....[Bo-1],
" -v- cardenasii .....[NN], [Bo-6],
" -v- crassior.....[Bo-1],
" -sv- aurescens.....
" -f. denudatus (See T. floccosus)..... [Bo-1], [Bx],
" -v- lapogus (See Austrocylindropuntia lapogus, Opuntia lapogus, T. lagopus, T. rauhii),
" -v- ovoides.....[Bo-1], [Bo-6],
" -f. rauhii (See Austrocylindropuntia lagopus -f. rauhii, T. rauhii),
" -f. cristata.....[Hx],
fulvicomus (See Opuntia fulvicoma, Cumulopuntia pentlandii (hybrid)),.....[Bo-1], [Bx], [Hy*],
" -v- bicolor (See Cumulopuntia kuehnrichiana, Opuntia kuehnrichiana, T. kuehnrichianus).....[Bo-1],
" -v- geometricus (See Opuntia geometrica), [Bo-1],
glomeratus (See Opuntia glomerata, T. glomeratus, T. hypogaea,), [DD], [Ar], [Bo-1], [Iu], [Hl 4-1978],
" -v- andicola (See T. andicola, T. neoglomeratus -v- andicolus).....[Bo-1],
" -v- atrospinus (atrotospinus) (Invalid name).... [DD],
" -v- elongatus.....
" -v- fulvispinus.....[Bo-1], [Bx],
" -v- gracilior (See Opuntia platyacantha -v- gracilior)....

TEPHROCACTUS.....(Cont'd),
gracilis.....
" -f. longispinus.....
" -v- longispinus (Invalid name),.....[DD],
" -v- major (See Opuntia andicola, T. andicolus),
" -v- oligacanthus (See Opuntia diademata -v- oligacanthus),
goldii .....[Aj 8-2-1957],
gratus (See Maihueniopsis ovata -f. sterilis),
halophilus (See Opuntia alexanderi, Opuntia halophilus, T. alexanderi),
hegenbartianus (See Cumulopuntia pentlandii).....[DD],
heteracanthus (See Maihueniopsis mandragora, T. mandragora),
heteromorphus (See Opuntia weberi, T. weberi),..... [Bo-1],
hickenii (See Opuntia hickenii),.... [Bo-1], [Bi Vol. 4, #1, 1986], [Hl 2, 1970],
hirschii (See Opuntia hirschii).....[Bo-1],
hossei (See Opuntia paediophila, T. articulatus -v- polyacanthus, T. paediophilus),..... [Bx], [Aj 3, 1953],
hypogaea (See Maihueniopsis hypogaea, Opuntia hypogaea),
" -v- rossianus.....
ignescens (See Cumulopuntia ignescens, Opuntia ignescens),.....[Bo-1], [Iv 4, 1957],
" -v- steinianus.....[Bo-1],
ignotus (See Cumulopuntia ignota, Opuntia ignota),
kuehnrichianus (See Cumulopuntia kuehnrichianus, Opuntia kuehnrichiana, T. fulvicomus -v- bicolor, Tephrocactus mirus), [Ay], [Bo-1], [Bx], [Hl 4, 1982],
" -v- applanatus (See Cumulopuntia kuehnrichiana) [Bo-1],
lagopus (See Austrocylindropuntia lagopus, Opuntia lagopus), [QQ], [Bo-1], [Bo-6], [Bx], [Hy], [Iv 7, 1956],
" -v- aureo-penicillatus.....[Bo-1],
" -v- aureus.....[Bo-1],
" -v- " -sv- brachycarpus.....(Bo-1],
" -v- leucolagopus.....[Bo-1],
" -v- pachycladus..... [Bo-1], [Bo-6],
" -v- rauhii.....
lecoriensis.....
leoncito (See Maihueniopsis leoncito, Opuntia leoncito)
leoninus (Invalid name - See Opuntia leoninus), [Bo-6],
longiarticulatus (Invalid name).....[DD],
malyanus (See Austrocylindropuntia malyana, Opuntia malyana),

TEPHROCACTUS.....(Cont'd),

mandragora (See Maihueniopsis mandragora, Opuntia mandragora),
[Bo-1], [Bo-6], [Af 59-1], [Aj 13-5-1962],
melanacanthus (Invalid name - See Cumulopuntia pentlandii -v- dactylifera, Opuntia dactylifera, T. dactylifer),.....[DD],
microcladus (Invalid name - See Cumulopuntia rossiana, T. pentlandii -v- rossianus),
[Bo-6:3599-3600],
microsphaericus (Invalid name).....[DD],
minor (See Cumulopuntia pentlandii, Opuntia pentlandii,
[Bo-1], [Bx],
minusculus (See Cumulopuntia pentlandii, Opuntia pentlandii).....[Bo-1],
minutus (See Opuntia minuta).....[Bo-1],
mirus (See Cumulopuntia kuehnrichiana, Opuntia kuehnrichianus, T. kuehnrichianus),.....[Bo-1], [Bo-6],
mistiensis.....[Bo-1],
molinensis (See Maihueniopsis molinensis, Opuntia molinensis),.....[Bo-1], [Bx], [Hy],
muellerianus (See Opuntia muelleriana),.....[Bo-1],
multiareolatus (See Cumulopuntia multiareolata, Opuntia multiareolata),
neoglomeratus (See T. glomeratus),
" -v- andicolus (See T. glomeratus -v- andicolus)
[Hy],
neopentlandii (See Opuntia pentlandii (Backbg.), T. pentlandii (Backbg.)),.....[Hy*],
neuquensis (See Maiqueniopsis neuquensis, Opuntia neuquensis),.....[DD],
nigrispinus (See Opuntia nigrispina, Platyopuntia nigrispina),..[Bo-1], [Bx], [Hy*], [N-1980],
noodtiae (See Cumulopuntia pentlandii -v- dactylifera, Opuntia dactylifera, T. dactylifer),
octavianus.....
" -v- giganteus.....
ovallei (See Maihueniopsis ovallei, Opuntia ovallei),
[Bo-1],
ovatus (See Opuntia aoracantha, T. articulatus -v- ovatus, T. aoracantha),
[Ar], [Bo-1], [Bx], [Hx*], [Hy], [Hl 6-1973],
paediophilus (See Opuntia diademata -v- polyacantha, Opuntia paediophila),.....[Bx], [Ah-1985],
pachypus (See Austrocylindropuntia pachypus, Opuntia pachypus),.....[Bv-1],
papyracanthus (See T. articulatus -v- papyracanthus),
parvisetus (Invalid name),.....[DD],
pentlandii (Backbg.) (See Cumulopuntia rossiana, T. pentlandii -v- rossianus, T. microclados, T. neopentlandii),

TEPHROCACTUS.....(Cont'd),

pentlandii (S/D) (See Cumulopuntia pentlandii, Opuntia pentlandii (S/D)),.....
[Bo-1], [Bx], [Hx], [Hl 6-1973],
" -v- adpressus (Invalid name),.....
" -v- fuauxianus.....[Bo-1],
" -v- rossianus (See Cumulopuntia rossianus, T. pentlandii (Backbg.), T. microclados),.....[Bo-1],
platyacanthus (See Opuntia platyacantha),
[Bo-1], [Bx], [Hx*],
" -v- angustispinus (See Opuntia platyacanthus -v- angustispina),.....[Bo-1],
" -v- deflexispina (See Opuntia platyacantha -v- deflexispina).....[Bo-1],
" -v- gracilior (See Opuntia platyacantha -v- gracilior),
" -v- monvillei (See Opuntia platyacantha -v- monvillei),.....[Bo-1],
" -v- neoplatyacanthus (See Opuntia platyacantha) [Bo-1], [Bx], [Hl 3-1978],
pseudorauppianus (See Cumulopuntia berteri, T. berteri, Opuntia sphaerica, T. dimorphus -v- pseudorauppianus, T. dimorphus (Backbg.),.....
pseudo-udonis (See Austrocylindropuntia floccosa, Opuntia floccosa, Opuntia pseudo-udonis, T. floccosus).....[Z],
punta-caillan (See Opuntia punta-caillan),.....[Bo-1],
pyrrhacanthus (See Cumulopuntia pyrrhacantha, Opuntia pyrrhacantha),.....[Bo-1],
" -v- leucoluteus (See Cumulopuntia pentlandii -v- dactylifera, O. dactylifera, T. dactilifer, T. melanacanthus),
[Bo-1],
rarissimus (See Cumulopuntia pentlandii (S/D), O. pentlandii (S/D], T. pentlandii (S/D),),
[Bo-1],
rauppianus (See Cumulopuntia rauppiana, Opuntia rauppiana, T. sphaericus -v- rauppianus),
reicheanus (See Maihueniopsis leoncito, Opuntia leoncito, Tephrocactus leoncito).....[Bo-1],
retrospinosus (See Opuntia retrospinosus),
[Bo-1:431-433],
riojanus (See Opuntia alexanderi, Opuntia halophila, T. alexanderi),
russelii (See Maihueniopsis ovata, Opuntia ovata, Opuntia russelii, T. ovatus),.....[Bo-1], [Bx],
schaeferi (See Maihueniopsis conoidea, T. conoideus),
setiger (See O. setigera, Opuntia weberi -v- setigera, Tephrocactus weberi -v- setiger),

silvestris (See Cumulopuntia pentlandii, Opuntia pentlandii (S/D), T. pentlandii (S/D),),
sphaericus (See Cumulopuntia berteri, Opuntia sphaerica, T. dimorphus sensu Backbg., T. pseudorauppianus).....[NN], [Ay], [Bo-1], [Bx], [Af 57-2], [Hl 4-1972],
" -v- glaucinus (See T. sphaericus).....
" -v- longispinus.....
" -v- rauppianus (See Cumulopuntia rauppiana, O. rauppiana, T. rauppianus), [Bo-1],
" -v- unguispinus (See Cumulopuntia unguispina, Opuntia unguispina),.....[Bo-1],
strobiliformis (See Opuntia strobiliformis, Opuntia diademata -v- inermis, Tephrocactus articulatus -v- inermis),..... [Ar], [Hi Vol. 2], [Hy],
subinermis (See Cumulopuntia pentlandii, Opuntia pentlandii (S/D), Opuntia subinermis, T. pentlandii (S/D),),.....[Bo-1], [Bx],
subterraneus (See Cumulopuntia subterranea, Opuntia subterranea, T. variiflorus),..... [DD], [Bo-1], [Hx*],
tarapacanus (See Maihueniopsis tarapacana, Opuntia tarapacana),
tortispinus (See Cumulopuntia tortispina),
turpinii (See Opuntia turpinii, O. diademata, O. glomerata sensu Br. & R., T. articulatus -f- syringacanthus, T. diadematus), [Bx], [Ij], [Af 9-7],
udonis (See Austrocylindropuntia floccosa, Opuntia floccosa, O. udonis, T. floccosus), [Bo-1], [Bo-6],
unguispinus (See Opuntia unguispina, T. sphaericus -v- unguispinus),
variiflorus (Invalid name - See Cumulopuntia subterranea, Opuntia subterranea, T. subterraneus) [Bo-6: 3594-5],
verticosus (See Austrocylindropuntia floccosa, Opuntia floccosa, T. floccosus, T. atroviridis), [Bo-1], [Bo-6],
virgulatus (virgultus) (Invalid name - See Opuntia virgulata).....[DD], [Hx*],
weberi (See Opuntia weberi, T. heteromorphus (Backbg)) [Bo-1], [Hi Vol. 2], [Hx*], [Hy], [Af 51-4],
" -v- dispar (See Opuntia weberi -v- dispar), [Bo-1],
" -v- setiger (See Opuntia setigera, Opuntia weberi -v- setiger).....[Bo-1],
wilkeanus (See Opuntia wilkeana),.....

TEPHROCACTUS.....(Cont'd),

yanganucensis (See Opuntia yanganucensis),.....[Bo-1],
zehnderi (See Cumulopuntia zehnderi),
[Bo-1], [Ij],

THELOCACTUS
aguirreanus (See Escobaria aguirreana, Gymnocactus aguirreanus),
beguinii sp. nov. (See Echinocactus erectocentrus, Echinomastus erectocentrus, Gymnocactus beguinii, Neolloydia beguinii, Neolloydia erectocentra),
bicolor (See Echinocactus bicolor)....[E], [G], [H], [I], [M*], [S], [LL], [NN], [PP], [RR], [Ar], [Av], [Ay], [Bt - Flower], [Bv-1], [Bz], [Cm], [Cx], [Cz], [Hb], [Hn], [Hx], [Iz], [Jd], [Bi 6-2-1988], [Dg 2-3-1980], [Hh Nov., 1974], [Hw-1989],
" -v- bicolor.....[Ct-1987],
" -v- bolansis (See T. bicolor -v- bicolor, T. bolansis), [E], [CC], [Ar], [Bo-5], [Hv], [Iu], [Iz], [Af 49-6], [Af 51-5], [Af 60-1, 1988], [Au 26-4-1971], [Hl 3-1966],
" -v- flavidispinus (See T. flavidispinus), [Df], [Af 55-6], [Hm 1982 Yearbook],
" -v- pottsii (See T. pottsii),.....[Bo-5], [Hx],
" -v- schottii (See Echinocactus bicolor -v- schottii), [Y], [Ax], [Df], [Hm 1982 Yearbook],
" -v- schwarzii (See T. schwarzii).....[Ct-1987],
" -v- tanquecillos.....[Di*], [Db 4-1979],
" -v- texensis (Invalid name) (See T. bicolor -v- schottii),.....[Bo-5],
" -v- tricolor.....[T], [CC], [NN], [Ar], [Ay], [Gr], [Hx], [Af 1-10], [Af 34-1],
" -v- " -f. rubrispina.....
" -v- vera.....[Ia],
" -v- wagnerianus.....[Di*],
bolansis (bolaensis) (See T. bicolor -v- bolansis), [CC], [Cu 1-1963],
buekii (bueckii) (See T. tulensis -v- buekii)..... [E], [RR], [Ar], [Bo-5], [Hy], [Il], [Af 49-5-1977], [Hh 1-8-1974], [Hl 1-1980], [Hm - 1982 Yearbook],
conothelos.....[CC], [Bu], [Gr], [Hv], [Hx], [Hy], [Af 36-5], [Hl 3-1982],
" -v- argenteus.....[Ae-1981], [Af 49-5], [Bi 2-3] [Hl 1-1975],
" -v- aurantiacus.....[HH], [Ar], [Av], [Ct-1987], [Iz], [Af 49-5], [Bi 2-3], [Hl 1-1975], [Hl 1-1981],

conothelos.....(cont'd),
" -v- conothelos (See T. saussieri),
[Ct-1987], [Af 49-5], [Af 50-4],
[Hm 1982 Yearbook],
" -v- " -f. cristata.....[Af 56-1],
" -v- crassihamatus (See Ancistrocactus crassihamatus, Glandulicactus crassihamatus, Hamatocactus crassihamatus, Thelocactus crassihamatus, Ferocactus crassihamatus)
" -v- macdowellii (See Neolloydia macdowellii, Echinomastus macdowellii, T. macdowellii),.....[Af 49-5], [Af 55-6],
crassihamatus (See Ancistrocactus crassihamatus, Ferocactus crassihamatus, Glandulicactus crassihamatus, Hamatocactus crassihamatus) [Dq], [Af 34-1],
ehrenbergii (See Echinocactus ehrenbergii, T. leucacanthus -v- ehrenbergii, L. leucacanthus), [Bo-5], [Gr], [Hx*],
erectocentrus (See Echinomastus erectocentrus)....[Iz],
flavidispinus (See T. bicolor -v- flavidispinus), [G], [Y], [DD], [Bo-5], [Df],
fossulatus (See T. hexadrophorus -v- fossulatus), [M*], [CC], [RR], [Bl], [Cx], [Ia],
gielsdorfianus (See Gymnocactus gielsdorfianus, Neolloydia gielsdorfiana).....[Hx*],
goldii (See Gymnocactus goldii, Gymnocactus horripilus, Neolloydia horripila, Thelocactus horripilus), [N-1980], [Af 27-5], [Af 34-1],
hastifer.....[Bo-5], [Ct-1987], [Dq], [Iu], [Af 54-3], [Hl 3-1966], [Hl 2-1983],
heterochromus (See Echinocactus heterochromus), [Ar], [Ay], [Bo-5], [Bu], [Hb], [Il], [Ae-1981], [Af 42-1], [Af 49-6], [Ag-1985], [Au 35-2], [Hl 2-1974],
hexaedrophorus....[G], [T], [NN], [QQ], [Ar], [Av], [Ay] [Ct-1987], [Hx], [Hy], [Jd], [N-1980], [Ad Jun., 1979], [Af 49-5], [Af 50-5], [Af 57-2], [Aj 15-3-1964], [Hm 1982 -Yearbook], [Hw-1981], [Hw-1982],
" -v- decipiens.....[Af 49-5], [Hl 2-1979], [Hl 5-1981], [Hm 1982 Yearbook],
" -v- droegeanus.....
" -v- fossulatus (See T. fossulatus), [Ar], [Bo-5], [Ad -], [Ae-1981], [Af 41-1], [Au 35-2], [Hm 1982 - Yearbook],
" -v- labourtianus.....
" -v- lloydii (See T. lloydii).....[Ct-1987], [Af 49-5], [Hm 1982 Yearbook],

THELOCACTUS.....(Cont'd),

hexaedrophorus.....(cont'd),
" -v- major.....[Hy], [Cu 3-1961],
" -v- roseus.....[Af 49-5)?
" -f. .....[Ar],
horripilus (See Gymnocactus goldii, Gymnocactus horripilus, Neolloydia horripila, T. goldii), [Hx*],
johnsonii (See Echinomastus johnsonii).....[Iz],
knuthianus (See Echinocactus knuthianus, Gymnocactus knuthianus, Neolloydia knuthiana, Turbinicarpus knuthianus).....[Hx*], [Jd],
krainzianus (See Neolloydia pseudomacrochele -f.- krainziana, Toumeya pseudomacrochele -v- krainzianus, Turbinicarpus pseudomacrochele -v- krainzianus),
[Bo-5], [Hx*], [Am 6-1979], [Hl 2-1966], [Hl 4-1983],
lausseri.....[Ct-1987], [Aj 37 (8):162, 164, 1986],
leucacanthus (See T. ehrenbergii),
[E], [M*], [CC], [RR], [XX], [Av], [Bo-5], [Bv-1], [Ct-1987], [Gr], [Hx], [Iz], [Jd], [Af 34-1], [Af 55-4], [Hl 2-1978], [Hm 1982 Yearbook),
" -v- ehrenbergii (See T. ehrenbergii, T. leucacanthus),
" -v- leucacanthus.....[Hl 6-1980],
" -v- porrectus.....[Bo-5],
" -v- sanchez-mejoradai (See T. sanchez-mejoradai)
" -v- schmollii (See T. schmollii),
[Bo-5], [Hb], [Af 43-1], [Hl 4-1978], [Hm 1982 Yearbook],
lloydii (See T. hexaedrophorus -v- lloydii).....[Hy],
lophothele (See T. rinconensis), ?
[V], [CC], [Ar], [Ay], [Bo-5], [Bt -flwr.] [Gr], [Hu], [Hx], [Hy], [Il], [Jd], [Ae-1981], [Af 34-1], [Am 3-1982], [Aw 2-1962], [Hm 1982 Yearbook],
" -v- longispina.....[Hy],
" -v- nidulans (See T. nidulans),
" -v- rubrispinus.....
" -f. cristata.....[Ar],
macdowellii (mcdowellii) (See Echinomastus macdowellii, Neolloydia macdowellii, T. conothelos -v- macdowellii),
[Ct-1987], [Aw 1-1969],
macrothele.....[Hl 3-1966],
mandragora (See Gymnocactus mandragora, Rapicactus mandragora),.....[Az],

THELOCACTUS.....(Cont'd),

mapimiensis (See Echinomastus durangensis -v- mapimiensis, Echinomastus mapimiensis, Echinomastus unguispinus, T. unguispinus)...[Iz],
mariposensis (See Echinocactus mariposensis, Echinomastus mariposensis, Neolloydia mariposensis),.....[Iz],
mathssonii (See Ancistrocactus crassihamatus),
matudae (See T. tulensis -v- matudae),
[Av], [Ae-1981], [Aj 31-6-1980], [Aw 3-1978],
[Hl 5-1979], [Hl 4-1983], [Hm 1982 Yearbook],
megalothelos.....[CC],
neglance.....[Ax],
nidulans (See Echinocactus nidulans, T. rinconensis -v- nidulans),
[M*], [Z], [CC], [NN], [RR], [UU], [Ar], [Ay],
[Bv-1], [Cx], [Gr], [Hb], [Hx], [Hy], [Au Vol. 2 - 1977, Zone 3 Compendium], [Hm 1982 Yearbook],
" -f. cristata.....[Hn],
paradensis.....
phymatothelos (phymatothele) (See T. rinconensis -v- phymatothele).....
[RR], [Ar], [Bo-5], [Hy], [Aj 27-1-1976],
[Hm 1982 Yearbook],
polyancistrus (See Sclerocactus polyancistrus),
[Iz],
porrectus (See T. leucacanthus -v- porrectus),
pottsii (See T. bicolor -v- pottsii),
[Bu], [Cx], [Hx*], [Hy],
rhodophalmus (See T. bicolor).....[Hy],
" -v- bolansis (See T. bicolor -v- bolansis),
[Hy],
" -v- tricolor (See T. bicolor -v- tricolor),
[Hy],
" -f. cristata.....[Hy],
rinconensis.....[Z], [Ct-1987], [Hx], [Iu], [Af 55-1],
[Hm 1982 Yearbook],
" -v- nidulans (See Echinocactus nidulans, T. nidulans, T. riconensis),
[Af 49-6],
" -v- phymatothelos (See T. phymatothelos, T. riconensis).....
[Iz], [Af 49-6],
" -v- rinconensis.....[Af 49-6],
roseanus (See Escobaria roseana, Gymnocactus roseanus)
[Af 37-4],
sanchez-mejoradai (See T. leucacanthus -v- sanchez-mejoradai).....
[Af 56-5],
saueri (See Gymnocactus saueri, Neolloydia saueri),
[Hv], [Hx*], [Aj 16-7-1965],

THELOCACTUS.....(Cont'd),

saussieri (See T. conothelos),
[NN], [Ar], [Bu], [Bv-1], [Cx], [Dq], [Hx*],
[Hy], [Iu], [Ad Oct. 1980], [Af 41-3], [Aw -
3-1975], [Ba 4-1981 - Flower],
" -v- longispina.....[Hy],
scheeri (See Ancistrocactus scheeri, Echinocactus
scheeri),.....[Iz],
schmollii (See T. leucacanthus -v- schmollii),
schwarzii (See T. bicolor -v- schwarzii),
[NN], [Ar], [Bo-5], [Hx], [Hy], [Ia], [Il],
[Ae-1981], [Ag-1988], [Aj 27-10-1976],
[Db 7-1979], [Hl 2-1978],
setispinus (See Echinocactus setispinus, Ferocactus
setispinus, Hamatocactus setispinus),
[Ct-1987], [Iz],
smithii (See Gymnocactus smithii, Neolloydia smithii)
subterraneus (See Gymnocactus subterraneus, Neolloydia
subterranea, Rapicactus subterraneus),
tulensis.....[CC], [RR], [Ar], [Bl], [Bo-5], [Dq*],
[Hx*], [Hy], [Il], [Iu], [Ae-1981],
[Af 34-1], [Af 49-6], [Hm-1982 Yearbook],
" -v- buekii (bueckii) (See T. buekii),....
[Ct-1987],
" -v- longispinus.....[Hy],
" -v- matudae (See T. matudae),.....[Ct-1987],
" -v- tulensis.....
uncinatus (See Ancistrocactus uncinatus, Ferocactus
uncinatus, Glandulicactus uncinatus, Hamato-
cactus uncinatus),
[Dq], [Iz], [Af 34-1], [Hl 6-1981],
unguispinus (See Echinomastus unguispinus, Neolloydia
unguispina),.....
-v- laui (See Echinomastus unguispinus -v- laui),
[Iz],
valdezianus (See Echinocactus valdezianus, Gymnocactus
valdezianus, Neolloydia valdeziana, Norman-
bokea valdeziana, Pelecyphora valdeziana,
Turbinicarpus valdezianus).....[Bv-1],
viereckii (See Gymnocactus viereckii, Neolloydia
viereckii),.....[Hx], [Ba 10-1976],
" -v- major (See Gymnocactus viereckii -v- major)
" -v- viereckii.....[Aj 32-6-1981],
wagnerianus (See T. bicolor -v- wagnerianus),
[Hb], [Hu], [Hy], [Iz], [Ae-1981],
[Hl 3-1971],
whipplei (See Sclerocactus whipplei).....[Iz],
ysabelae (See Gymnocactus ysabelae).....[Af 5-10],
" -v- brevispina.....(See Gymnocactus ysabelae -v-
brevispina).....
Sp. .....[Hv], [Aj 12-1979],

THELOCEPHALA

aerocarpa (See Chileorebutia aerocarpa, Neochilenia aerocarpa, Neoporteria aerocarpa, Neoporteria reichei -v- reichei -f. aerocarpa), [Hi Vol.3], [Aj 31-8-1980],
chorosensis (See Neochilenia chorosensis, Pyrrhocactus chorosensis),
duripulpa (See Chileorebutia duripulpa, Neochilenia duripulpa).....[Hi Vol. 3],
eriocephala (See Neochilenia eriocephala, Neoporteria eriocephala, Pyrrhocactus floccosus),
esmeraldana (See Chileorebutia esmeraldana, Neochilenia esmeraldana).....[Hi Vol. 3 -Grafted], [Aj 31-8-1980],
fankhauseri (See Neoporteria fankhauseri)..[Hi Vol. 3],
fulva (See Chileorebutia fulva, Neochilenia aerocarpa -v- fulva),.....[Hi Vol. 3], [Aj 31-8-1980],
glabrescens (See Chileorebutia glabrescens, Neochilenia glabrescens)..[Hi Vol. 3], [Aj 22-11-1971],
imitans (See T. lembckei),
krausii (See Chileorebutia krausii, Neochilenia krausii),.....[Hi Vol. 3],
lembckei (See Chileorebutia imitans, Chileorebutia lembckei, Neochilenia imitans, Neochilenia lembckei, Neochilenia neoreichei).....[Hi-3],
longirapa.....[Hi Vol. 3],
malleolata (See Chileorebutia malleolata, Neochilenia malleolata).....[Hi Vol. 3], [Hs III-2-1982],
" -v- solitaria (See Chileorebutia malleolata -v- solitaria),.....[Hi Vol. 3],
mitis (See Neochilenia mitis, Neoporteria mitis),
monte-amargoensis.....
napina (See Chileorebutia napina, Neochilenia napina, Neoporteria napina).....[Hi Vol. 3], [Aj 31-8-1980],
" -v- lanigera.....
" -v- spinosior.....
nuda.....[Hi Vol. 3],
" -v- albispina.....
occulta (See Neochilenia occulta, Neoporteria occulta, Pyrrhocactus occultus),
odierii (See Chileorebutia odierii, Neochilenia odierii, T. pseudoreichei),.....[Hi Vol. 3],
" X Pyrrhocactus confinis........[Hi Vol. 3],
pseudoreichei (See Neochilenia odierii, Neochilenia pseudoreichei, Neoporteria odierii, Reicheocactus pseudoreichei, T. odierii),
recondita (See Neochilenia recondita, Pyrrhocactus recondita),.....[Aj 31-8-1980],
reichei (See Neochilenia atra, Neochilenia reichei, Neoporteria reichei, Malacocarpus reichei),

THELOCEPHALA.....(Cont'd),

roseiflora (See Neochilenia roseiflora),
tenebrica (See Chileorebutia tenibrica, Neochilenia tenibrica).....[Hi Vol. 3],

sp. .....[Hi Vol. 3],

THRIXANTHOCEREUS

blossfeldiorum (See Cephalocerus blossfeldiorum, Espostoa blossfeldiorum, Facheiroa blossfeldiorum).....
[F], [I], [W], [DD], [NN], [QQ], [Ay], [Bo-4], [Hi Vol. 4], [Hx], [Hy], [Il], [Jd], [N-1980], [Af 50-1], [Ah-1981], [Ba 11-1979], [Hl 6-1982],
" -f. cristata.....[Hv],
cullmannianus.....[DD], [NN], [Hi Vol. 4], [Ag-1987],
longispinus.....[Hi Vol. 4],
mirabilis (See Espostoa mirabilis),
" -f. cristata (See Espostoa mirabilis -f.- cristata),.....[NN],
ritteri (See Espostoa ritteri),
senilis (See Espostoa senilis),
[F], [Z], [DD], [NN], [Ar], [Bo-4], [Hi Vol. 4] [Hv], [Hx], [Hy], [Il], [Af 50-3], [Au Vol. 3, Zone III Compendium),
" -v- jelinkyanus.....

TOUMEYA

bradyi (See Pediocactus bradyi),.....[P],
klinkeriana (See Strombocactus schmiedickeanus -v- klinkerianus, Turbinicarpus klinkerianus, Turbinicarpus schmiedickeanus -v- klinkerianus),
krainziana (See Neolloydia pseudomacrochele -f. krainziana, Toumeya pseudomacrochele -v- krainziana, Turbinicarpus krainzianus, Turbinicarpus pseudomacrochele -v- krainzianus),
[Af 33-2], [Aj 11-11-1960], [Aw 1-1962],
lophophoroides (See Neolloydia lophophoroides, Strombocactus lophophoroides, Turbinicarpus lophophoroides),
[F], [Iv 10-1960],
papyracantha (See Pediocactus papyracanthus),
[D], [P], [Z], [DD], [GG], [NN], [An], [Ar], [Ay], [Bv-1], [Cz], [Dq], [Hx], [Iz] [Jd], [Af 58-2], [Au 35-2], [Ba 5-1978], [Hl 1-1965], [Hl 1-1970], [Bi Vol. 4, #6],
" -f. .....[Hl 1-1970],

TOUMEYA.....(Cont'd),

peeblesiana (See Navajoa peeblesiana, Pediocactus peeblesianus).....[E], [P],
polaskii (See Neolloydia schmiedickeanus -v- schwarzii, Turbinicarpus polaskii, Turbinicarpus schmiedickeanus -v- schwarzii, Turbinicarpus schwarzii) [NN], [Bt],
pseudomacrochele (See Neolloydia pseudomacrochele, Strombocactus pseudomacrochele, Turbinicarpus pseudomacrochele).....[NN],
" -v- krainziana (See Neolloydia pseudomacrochele -f. krainziana, T. krainziana, Turbinicarpus krainzianus, Turbinicarpus pseudomacrochele -v- krainzianus),.....[NN],
schmiedickeana (See Neolloydia schmiedickeana, Strombocactus schmiedickeanus, Turbinicarpus schmiedickeanus),
" -v- klinkeriana (See Strombocactus klinkerianus Strombocactus schmiedickeanus -v- klinkerianus, Toumeya klinkeriana, Turbinicarpus klinkerianus, Turbinicarpus schmiedickeanus -v- klinkerianus),
Sp. DBG # 62-7089.....[P],
Sp. DBG # 56-5621.....[P],

TRICHOCEREUS

andalgalensis (See Helianthocereus andalgalensis, Helianthocereus huascha -v-rubriflorus),, [R], [Hi Vol. 2], [Af 59-1],
" -v- auricolor (See Helianthocereus huascha -v- auricolor, T. auricolor), [Hi Vol. 2],
angelesii.....
antezanae (See Helianthocereus antezanae, Helianthocereus orurensis, Lobivia formosa ssp. tarijensis -v- bertramiana, T. orurensis, T. tarijensis -v- orurensis),
atacamensis (See Helianthocereus atacamensis), [Hy], [Hi Vol. 3],
" -v- pasacana (See Helianthocereus pasacana, T. pasacana),.....[Hi Vol. 2],
auricolor (See Helianthocereus huascha -v- auricolor, T. andalgalensis -v- auricolor),..... [R], [Bv-1],
bertramianus (See Helianthocereus bertramianus, Lobivia formosa ssp. tarijensis -v- bertramianus), [Fo 1935 #6 gen. 51 sp.-2 Fig.],

TRICHOCEREUS.....(Cont'd),

boliviensis (See Lobivia boliviensis, Lobivia pentlandii -v- hardiana),
bridgesii (See Echinopsis bridgesii, Echinopsis lagenaeformis).....[Bo-2], [Bv-3], [Hi Vol. 2], [Hy], [Af 6-7], [Af 30-4],
" -v- brevispinus.....
" -v- lagenaeformis (See Echinopsis lagenaeformis, T. bridgesii),
" -v- longispina.....
" -f. monstrosus.....[Ay], [Bo-2],
bruchii (See Echinopsis bruchii, Lobivia bruchii, Lobivia formosa ssp. bruchii, Soehrensis bruchii)
cajasensis.....
callianthus (See Echinopsis calliantha, Lobivia callianthus, Helianthocereus huascha, Lobivia huascha -v- calliantha),.....[Aj 30-7-1979],
camarguensis (See Echinopsis camarguensis),.....[Hy],
campos-portoi (See Arthrocereus campos-portoi).....[It]
candicans (See Cereus candicans, Echinopsis candicans) [C], [J*], [L], [Av], [Bo-2], [Bt - Flower], [Hi Vol.2], [Hu], [Hv], [Hx], [Hy], [Ij], [Jd], [Af 6-7], [Af 56-1], [Au 37-1], [Bi 2-4],
" -v- courantii (See T. courantii),
" -v- gladiatus (See T. gladiatus),..... [C], [H], [Bo-2], [Hy],
" -v- longispina.....
" -v- nitens (See Cereus lamprochlorus (S/D), T. lamprochlorus, T. neolamprochlorus)
" -v- robustior.....
" -v- roseoflorus.....
" -v- rubriflorus (See Helianthocereus pseudocandicans, T. candicans),
" -v- tenuispina (See T. tenuispinus).....[Bo-2],
" -f. (hybrid).....[Hv],
catamarcensis (See Lobivia andalgalensis),
caulescens.....[Hi Vol. 2],
cephalomacrostibas (See Weberbauerocereus cephalomacrostibas, Haageocereus ferox)..... [Bo-2], [Hb], [Hy], [Af 20-9],
cephalopasacana.....[Dv 44:169 Fig., 1929],
chalaensis.....(See Echinopsis chalaensis).....[Bo-2],
chilensis (See Lobivia backebergii -v- chilensis, Lobivia backebergii ssp. wrightiana -v- chilensis, Lobivia chilensis), [H], [R], [QQ], [RR], [Ay], [Bo-2], [Gr], [Hy], [Jd], [Af 3-12], [Aj 27-10-1976],
" -v- australis.....[Hy],
" -v- " -f. cristata.....[Hi Vol. 3],
" -v- borealis.....

TRICHOCEREUS.....(Cont'd),
chilensis.....(cont'd),
" -v- chilensis.....[Hi Vol. 3],
" -v- conjungens.....[Hi Vol. 3],
" -v- eburneus.....[Bo-2], [Ij], [Aj 8-11-1959],
" -v- funkii.....
" -v- panhoplites.....[Hi Vol. 3], [Hy],
" -v- pycnacanthus.....
" -v- quisco.....[Aj 23-11-1972],
" -v- spinosissimus.....
" -v- zizkaanus.....
chuquisacanus.....[Hi Vol. 2],
clavatus.....[Hi Vol. 2],
coquimbanus (See Echinopsis coquimbana), [RR], [Bo-2], [Hi Vol. 3], [Af 17-11],
" -v- nigripilis.....[Bo-2],
courantii (See T. candicans -v- courantii).....[Bo-2],
crassiarboreus....(See Weberbauerocereus crassiarboreus, Weberbauerocereus cuzoensis),
crassicostatus.....[Hi Vol. 2],
culpinensis (See Lobivia culpinensis, Lobivia pugionacantha -v- culpinensis)
cuzcoensis (See Echinopsis cuzcoensis)........[Bo-2], [Hi Vol. 4], [Cu 8-1-1939], [Iv 4-1957], [Bi Vol. 4, #6],
" -v- knuthianus (See T. knuthianus, Echinopsis knuthianus, Echinopsis tarmaensis, T. tarmaensis, Azureocereus deflexispinus),
damazioi (Werd.).....[It],
deserticola (See Cereus deserticolus, Echinopsis deserticola).....[Hi Vol. 3], [Hy],
eburneus (See T. chilensis -v- eburneus),
enneacanthus (See Helianthocereus enneacanthus)..[NN],
eremophilus (See Echinopsis eremophila)....[Hi Vol. 2],
" -f. cristata.....[Hi Vol. 2],
escayachensis (See T. werdermannianus).....[Af 35-5],
fabrisii (See Echinopsis fabrisii, Echinopsis volcanensis, T. volcanensis),
fascicularis (See T. weberbaueri, Weberbauerocereus fascicularis, Weberbauerocereus weberbaueri).....[Aj 9-5-1958],
formosus (See Echinopsis formosa, Lobivia formosa, Soehrensia formosa),
friedrichii (See Echinopsis friedrichii, Echinopsis shaferi, T. shaferi),
fulvianus.....[Hu], [Hy],
fulvilanus (See Echinopsis fulvilana)........[Av], [Bo-6], [Hi Vol. 3], [Hv], [Il], [Aj 13-10-1962],
gigantea (See Echinopsis gigantea),
gladiatus (See T. candicans -v- gladiatus),

TRICHOCEREUS.....(Cont'd),

glaucus (See Echinopsis fulvilana)...[Hi Vol. 4], [Hy],
" -f. pendens.....[Hi Vol. 3],
glottsbergi.....[K], [Cp],
grandiflorus (See Lobivia grandiflora, Lobivia huascha -v- grandiflora),
[DD], [EE], [NN], [UU], [Gr], [Hv],
grandis (See Lobivia grandis, Soehrensia grandis),
herzogianus (See Helianthocereus herzogianus, Trichocereus formosa ssp. tarijensis -v- bertramiana, Trichocereus bertramianus),
[Gb 1 (5):19 Fig. 6, 1953],
" -v- totorensis (See Helianthocereus herzogianus -v- totorensis, Lobivia formosa ssp. tarijensis -v- totorensis, T. totorensis),
[Gb 1 (5):19 Fig. 7, 1953],
[Iv 1-1959],
hollianus.....
huascha (Web.) Ritt. (See Cereus huascha -v- flaviflorus),
huascha (Web.) Br. & R. (See Echinopsis huascha, Helianthocereus huascha),
[HH], [NN], [PP], [RR], [Cv-3], [Hh -],
" -v- grandiflorus (See Helianthocereus grandiflorus),
ingens (See Lobivia formosa, Soehrensia formosa, Soehrensia ingens),
korethroides (See Lobivia formosa ssp. bruchii -v- nivalis, Soehrensia korethroides),
knuthianus (See Azureocereus deflexispinus, Echinopsis knuthiana, T. cuzcoensis -v- knuthianus, T. tarmaensis),
lamprochlorus (See Cereus lamprochlorus, Echinopsis lamprochlora, T. neolamprochlorus),
[C], [DD], [RR], [Af 2-10], [Aj 22-9-1971],
litoralis (See Echinopsis litoralis)...[Bo-2], [Hi Vol. 3], [Hy], [Aj 9-1-1958], [Hl 1-1972],
lobivioides (See Lobivia grandiflora (Werd), Helianthocereus grandiflorus),
longicomus.....
macrogonus (See Echinopsis macrogona),
[C], [H], [Bo-2], [Bv-1], [Dn], [Hu], [Jd], [Af 12-8], [Af 15-4], [Af 24-5],
" -f. cristata.....[Hy],
manguinii (See Echinopsis manguinii),
[Bo-2], [Iv 3-1953],
maritimus.....[Aj 8-11-1957],
montanus.....
narvaecensis (narvaezensis) (See Lobivia huascha ssp. narvaecensis, T. tenuispina)

neolamprochlorus (Invalid name - See T. lamprochlorus), [Bo-2], [Hx*], [Hy],
neocoquimbanus (See T. coquimbanus).....[Hy],
nigripilis.....[Bo-2],
orurensis (See Helianthocereus orurensis, Lobivia formosa ssp. tarijensis -v- bertramiana, T. tarijensis -v- orurensis),
" -v- albiflorus (See Helianthocereus orurensis -v- albiflorus, Lobivia formosa ssp. tarijensis -v- bertramiana),
pachanoi (See Echinopsis pachanoi, Trichocereus peruvianus).....[RR], [Ay], [Bu], [Cz], [Hx*], [Hy], [Af 14-8], [Af 22-1], [Bo], [Hi Vol. 4],
" -f. peruvianus (See T. peruvianus, T. pachanoi)
pallarensis.....[Hi Vol.4],
pasacana (See Helianthocereus pasacana), [RR], [Ay], [Cz], [Dq], [Dr], [Gr], [Af 51-4], [Af 56-3], [Af 59-1], [Aj 29-4-1978],
" -v- albicephala.....
" -v- inermis.....
" -v- nigrus.....[Bv-2],
" -f. cristata.....[Aj 20-12-1969],
pecheretiana (See Helianthocereus pecheretianus, Lobivia huascha),
peruvianus (See T. pachanoi -f. peruvianus), [RR], [Ay], [Bo-2], [Bu], [Bv-2], [Hy],
" -v- tortuosus.....[Ay],
pircacantha.....
poco (See Helianthocereus poco, Lobivia formosa ssp. tarijensis, T. tarijensis -v- poco), [Hv], [N-1980], [Af 6-7],
" -v- albiflorus (See Helianthocereus poco -v- albiflorus, Lobivia formosa ssp. tarijensis)
" -v- fricianus (See Helianthocereus poco -v-fricianus, Lobivia formosa ssp. tarijensis)
" -v- sanguiniflorus (See Helianthocereus poco -v- sanguiniflorus, Lobivia formosa ssp. tarijensis),
pseudocandicans (Invalid name - See Helianthocereus pseudocandicans, T. candicans),
pseudothelegonus (Invalid name).....[Bo-2],
puquiensis (See Echinopsis puquiensis), [Bo-2], [Hi Vol. 4],
purpureopilosus (See Echinopsis purpureopilosus), [Bo-2], [Hy], [Flower - Aj 7-1979], [Flower - Gr],
quadratiumbonatus (See Echinopsis quadratiumbonatus), [Hi Vol. 2],
racaquiensis.....[Bu],

randallii (See Lobivia formosa ssp. tarijensis, Lobivia tarijensis),
[Cv-3], [Hi Vol. 2], [Af 35-5],
riomizquensis.....[Hi Vol. 2],
rivierei (See Leucostele rivierei).....[Aj 5-1981],
rowleyi (rowleyana) (See Echinopsis rowleyana, Lobivia huascha -v- grandiflora, Lobivia grandiflora),
rubinghianus (Invalid name -See Echinopsis rubinghianus) [DD],
santaensis (See Echinopsis santaensis),
[Bo-2], [Hi Vol. 4],
santiaguensis (See Echinopsis santiaguensis),
[Bo-2], [Af 15-10], [Hy],
" -v- rubrispinus.....
schickendantzii.....[R], [CC], [NN], [Ay], [Bo-2],
[Bv-3], [Dq], [Hu], [Hy], [Af 2-12],
[Af 4-7], [Af 12-8], [Aj 28-6-1977],
[Aj 31-4-1980], [Aj 31-12-1980],
schoenii (See Echinopsis schoenii).....[Bo-2],
scopulicola (See Echinopsis scopulicola).....
[Hi Vol. 2],
serenanus (See T. coquimbanus (Br. & R.),.....
[Hy], [Aj 16-11-1965],
shaferi (See Echinopsis friedrichii),
[F], [H], [Bo-2],
skottsbergii (See Echinopsis skottsbergii),
[Bo-2], [Hi Vol. 3], [Hy],
" -v- breviatus (See T. skottsbergii).....[Bo-2],
smrzianus (See Echinopsis smrzianus),.....
[D], [H], [Gr], [Hy],
spacianoides (See Echinopsis spacianoides, T. tenuispinus),
spacianus (See Echinopsis spacianus).....[C], [H], [V],
[LL], [RR], [UU], [Bv-1], [Cz], [Hu], [Hy],
[Il], [Jd], [Af 1-6], [Af 57-4],
" -f. cristata.....[Hy], [Af 16-12],
spinibarbis (See T. nigripilis).....[Hi Vol. 3],
strigosus (See Echinopsis strigosa),
[C], [RR], [Bo-2], [Bv-1], [Gr], [Au Vol. 3-1978 - Zone 3 Compendium],
" -v- flaviflorus (See Echinopsis strigosa -v- flaviflorus, Lobivia huascha),
" -v- intricatus.....
" -v- longispinus.....
" -v- variegatus.....
superbus (See T. randallii),
tacaquirensis (See Echinopsis tacaquirensis, T. taquimbalensis -v- wilkeae),
[Bo-2], [Hi Vol. 2], [Hy],
" -v- glaucus.....

tacnaensis (See Echinopsis tacnaensis),.....
taquimbalensis (See Echinopsis taquimbalensis),
[Bo-2],
" -v- wilkeae (See T. tacaquirensis), [Bo-2],
taratensis (See Echinopsis taratensis),
tarijensis (See Helianthocereus tarijensis, Lobivia formosa ssp. tarijensis, Lobivia tarijensis),
" -v- orurensis (See Helianthocereus antezanae, Helianthocereus orurensis, Lobivia formosa ssp. tarijensis -v- bertramiana).....[Hi Vol. 2],
" -v- poco (See Helianthocereus poco, Lobivia formosa ssp. tarijensis, T. poco), [Hi Vol. 2],
" -v- totorillanus (See Echinopsis totorillana, T. totorillanus)....[Hi Vol. 2],
tarmaensis (See Azureocereus deflexispinus, Echinopsis knuthianus, T. knuthianus),.....[Bo-2],
tenuiarboreus (See Weberbauerocereus cuzcoensis -v- tenuiarboreus),
tenuispinus (See Echinopsis tenuispina),..... [Hi Vol. 2],
" -v- pajonalensis.....[Hi Vol. 2],
tephracanthus (See Eriocereus tephracanthus, Roseocereus tephracanthus, Harrisia tephracantha),.....[Bv-1], [Gr],
terscheckii (See Echinopsis terscheckii), [RR], [Bl], [Bo-2], [Bv-1], [Hi Vol. 2], [Hr], [Hx], [Hy], [Ij], [Af 15-11], [Aj 28-2-1977], [Aj 29-4-1978], [Bi 2-4],
" -v- catamarcensis (See Lobivia andalgalensis, Lobivia huascha -v- andalgalensis),
" -v- montanus.....[Bo-2],
" -f. cristata.....[NN], [Aj 20-12-1969],
terscheckioides (See Echinopsis terscheckioides) [Hi Vol. 2], [Hy],
thelegonoides (See T. thelegonus),
thelegonus (See Echinopsis thelegona).....[C], [RR], [XX], [Bo-II], [Bv-3], [Hb], [Hx], [Hy], [Iu], [Jd],
" -f. cristata.....[N-1980],
torataensis (See Echinopsis torataensis),..... [Hi Vol. 4],
totorensis (See Helianthocereus herzogianus -v- totorensis, T. herzogianus -v- totorensis), [Hi Vol. 2], [Hy],
totorillanus (See T. tarijensis -v- totorillanus),
trichosus (See Echinopsis trichosa)..... [Bo-6], [Iv 12, 1957],

TRICHOCEREUS.....(Cont'd),

tropicus (See Rauhocereus riosaniensis),
tulhuayacensis (See Echinopsis tulhuayacensis), [Bo-2], [Aj 8-7-1957],
tunariensis (See Echinopsis tunariensis), [Hi Vol. 2], [Af 38-2],
uebelmannianus (See Lobivia formosa, Soehrensis formosa, Soehrensia uebelmannianus), [Hi Vol. 3],
uyupampensis (See Echinopsis uyupampensis, T. glaucus -f. pendens).....[Bo-2],
validus (See Echinopsis valida),..... [Bo-2], [Iv 10-1955],
vatteri (See Lobivia huascha, Lobivia callianthus), [Cv-3], [Hi Vol. 2], [Aj 1-10-1949],
" -v- albiflorus.....
" -v- flaviflorus.....
volcanensis (See Echinopsis fabrisii, T. fabrisii).... [Af 51-5],
vollianus (See Echinopsis volliana), [Hi Vol. 2], [Hy],
" -v- rubrispinus.....
weberbauerii (See T. fascicularis, Weberbauerocereus weberbauerii),.....[Bu],
werdermannianus (See T. escayachensis).....[Bo-2], [Hi Vol. 2], [Hy], [Ij], [Aj 7-1937],
" -v- lecoriensis.....
" -v- wilkeae.....
Sp. .....[Bo-6], [Au Vol. 3, Zone 3 Compendium],
Sp. (hybrid).....[Hv],
Sp. (T. huascha (Hybrid).....[Bm-1988],

TURBINICARPUS

dickisoniae (See Neolloydia schmiedickeana -v- dickisoniae, T. schmiedickeanus -v- dickisoniae)

flaviflorus (See Neolloydia schmiedickeana -v- flaviflora, T. schmiedickeanaus -v- flaviflorus), [Ae-1981], [Af 51-3], [Aj 30-1-1979],
gracilis (See Neolloydia schmiedickeana -v- gracilis, T. schmiedickeanus -v- gracilis), [Z], [Ar], [Hx*], [Ae-1981], [Hm 1-1978],
klinkerianus (See Strombocactus klinkerianus, Strombocactus schmiedickeanus -v- klinkerianus, Toumeya klinkeriana, Toumeya schmiedickeana -v- klinkeriana, T. schmiedickeanus -v- klinkerianus), [F], [NN], [Ar], [Ay], [Hu], [Hx], [Hy], [Ae-1982], [Hl 6-1966],

knuthianus (See Echinocactus knuthianus, Gymnocactus knuthianus, Neolloydia knuthiana, Thelocactus knuthianus).....[Af 59-1],
krainzianus (See Neolloydia pseudomacrochele -f. krainziana, Toumeya krainziana, T. pseudomacrochele -v- krainzianus),
[Ar], [Bo-5], [Hx], [Jd], [Af 49-4], [Aw 1-1962], [Hl 2-1967],
laui (See Neolloydia laui, Neolloydia pilispina, Strombocactus laui),
[Z], [Av], [Hx*], [Ae-1981], [Af 49-4], [Af 52-5], [Ah-1985], [Am 6-1980], [Hl 1-1979],
lophophorioides (See Neolloydia lophophorioides, Strombocactus lophophorioides, Toumeya lophophorioides)....[F], [J*], [EE], [NN], [Ar], [Ay], [Bo-5], [Cs], [Hu], [Hx*], [Hy], [Il], [Jd], [Ae-1981], [Af 49-4], [Af 50-4], [Ag-1985], [Am 6-1980], [Am 2-1983], [Aw 1-1989], [Ba 9-1977], [Ba 2-1979], [Dc 1-1980], [Hw-1981],
macrochele (See Neolloydia schmiedickeana -v- macrochele, T. schmiedickeanus -v- macrochele), [F], [Ar], [Ay], [Hx], [Hy], [Jd], [Ae-1982], [Af 49-4],
" -v- schwarzii.....
" -v- " -f. polaskii (See T. polaskii),
papyracanthus (See Pediocactus papyracanthus, Toumeya papyracantha),
peeblesianus (See Navajoa peeblesiana, Pediocactus peeblesianus, Toumeya peeblesiana),
polaskii (See Neolloydia schmiedickeana -v- schwarzii, T. macrochele -v- schwarzii -f. schwarzii, T. schmiedickeanus -v- schwarzii),
[I], [NN], [Ar], [Bo-5], [Hx], [Hy], [Il], [Af 55-6], [Aj 28-4-1977], [Am 2-1983],
" -v- klinkerianus (See Neolloydia schmiedickeana -v- klinkeriana, T. klinkeriana, T. schmiedickeanus -v- klinkerianus).....[Ar],
pseudomacrochele (See Neolloydia pseudomacrochele, Strombocactus pseudomacrochele, Toumeya pseudomacrochele).....[F], [NN], [Ar], [Av], [Bo-5], [Hu], [Hv], [Hx*], [Hy], [Il], [Ae-1981], [Af 52-3], [Ag-1985], [Aw 1-1971], [Aw 3-1975],
" -v- krainzianus (See Neolloydia pseudomacrochele -f. krainziana, Toumeya pseudomacrochele -v- krainziana, T. krainzianus).....[Af 49-4], [Ah-1985], [Ba 3-1979], [Ba 12-1980],

TURBINICARPUS.....(Cont'd),

pseudomacrochele.....(cont'd),
" -v- pseudomacrochele (See Neolloydia pseudo-macrochele),
[Af 49-4], [Am 6-1980],
[Ba 1-1979], [Ba 3-1979],
pseudopectinatus (See Neolloydia pseudopectinata, Normanbokea pseudopectinata, Pelecyphora pseudopectinata),..........[Av], [Hv],
[Ae-1982], [Af 49-4], [Af 58-5], [Ba 4-1979], [Bi 2-3], [Bm -1983], [Hw-1981],
roseiflorus (Invalid name - hybrid ?),
[DD], [Hx*], [Aj 31-6-1980], [Aj 32-5-1981]
[Ba 10-1977], [Hl 5-1980],
schmiedickeanus (See Neolloydia schmiedickeana, Strombocactus schmiedickeanus),
[D], [W], [Z], [NN], [Ar], [Ay], [Bo-5],
[Hx*], [Hy], [Il], [Am-2-1983], [Hl 6-1966], [Hs I-2-1980],
" -v- dickisoniae (See Neolloydia schmiedickeana -v- dickisoniae)....[Af 54-2], [Bi 5-1*],
" -v- flaviflorus (See Neolloydia schmiedickeana -v- flaviflora, T. flaviflorus),
[Af 51-3],
" -v- gracilis (See T. gracilis),
[Ar], [Af 49-4], [Am 6-1980],
" -v- klinkerianus (See Neolloydia schmiedickeana -v- klinkeriana, Strombocactus klinkerianus, Strombocactus schmiedickeanus -v- klinkerianus, Toumeya klinkeriana, Toumeya schmiedickeana -v- klinkeriana, T. klinkerianus),
[Bo-5], [Af 49-4], [Af 51-3], [Af 57-4]
[Am 6-1980], [Ba 11-1978],
" -v- macrochele (See Neolloydia schmiedickeana -v- macrochele, T. macrochele),
[Af 49-4], [Af 58-2], [Am 6-1980],
[Ba 5-1978], [Ba 8-1978],
" -v- " -f. cristata.....[Ar],
" -v- schmiedickeanus (See Neolloydia schmiedickeana -v- schmiedickeana),
[Af 49-4], [Am 6-1980], [Ba 7-1978],
[Flower - Ba 11-1978],
" -v- schwarzii (See Neolloydia schmiedickeana -v- schwarzii, T. schwarzii),
[Af 49-4], [Af 49-5], [Af 59-4],
[Am 2-1983],
" -v- " -f. polaskii (See T. polaskii)....[Hv],
[Ag-1988], [Am 6-1980], [Ba 1-1979],
" -f. cristata.....[Ar], [Hy],

TURBINICARPUS.....(Cont'd),

schwarzii (See Neolloydia schmiedickeana -v- schwarzii, T. schmiedickeanus -v- schwarzii), [F], [H], [Z], [QQ], [Ar], [Ay], [Hu], [Hx*], [Hy], [Aj 31-12-1980], [Am 6-1980], [Flower - [Ba 11-1977], [Ba 1-1979],

valdezianus (See Echinocactus valdezianus, Gymnocactus valdezianus, Neolloydia valdeziana, Normanbokea valdeziana, Pelecyphora valdeziana, Thelocactus valdezianus), [Z], [Ar], [Bt], [Hv], [Ae-1982], [Af 51-6], [Af 58-2], [Ag-1988], [Ah-1985], [Am 2-1983], [Ba 4-1979], [Cu 2-1979],

" -v- albiflorus.....[Hv],

" -f. cristata.....[Af 49-4], [Af 52-2], [Af 56-1] [Ah-1981],

Sp. .....[Hv], [Da 1-1979],

UEBELMANNIA

ammotropha (See U. pectinifera -v- multicostata),

antenaensis (See U. meninensis -v- antenaensis),

brasiliensis (See U. pectinifera),

buiningii.....[DD-524], [Ar], [Ay], [Hv], [Hx], [Jd], [Af 58-1], [Hs II-3-1981],

" -v- nigra.....

centeteria (See U. gummifera),.....

elegans (See U. pectinifera -v- pseudopectinifera),....

flavispina (See U. warasii), [G], [DD-525], [Ay], [Hx*], [Iz], [Hi Vol. 1], [Af 58-1], [Ba 52:9-10, 1973], [Db 5-1979], [Hl 2-1975], [Hs II-3-1981],

flavissima.....

gummifera (See Parodia gummifera), [HH], [Ar], [Hi Vol. 1], [Hx*], [Af 58-1], [Aj 20-7-1969], [Am 8-1981], [Ba 46:161, 1967], [Db 5-1979], [Hl 3-1972],

" -f. cristata.....[Hl 2-1970],

meninensis.....[C], [DD-526], [Gr], [Af 56-3], [Am 8-1981], [Hl 4-1972],

" -v- antenaensis.....

" -v- rubra.....[K], [R], [Z], [DD-527],

pectinifera.....[Ar], [Av], [Ay], [Bt], [Bu], [Hv], [Hx], [Il], [Iz], [Ab-1985], [Ad 3-1983] [Af 56-5], [Af 60-2], [Ah-1985], [Aj 33-2-1982], [Am 8-1981], [Db -26th Anniversary Edition], [Bi Vol. 4, #6],

-v- elegans (See U. elegans, U. pectiniferea -v- pseudopectinifera),

" -v- flavispina (See U. flavispina),

" -v- multicostata (See U. ammotropha),

UEBELMANNIA.....(Cont'd),

pectinifera.....(cont'd),
" -v- pseudopectinifera.....
[G], [Ar], [Hx], [Fig. 4 & 5 are corrected in Af 58-1], [Aj 23-5-1972], [Bi 3-1], [Db 26th Anniv. Issue], [Hl 2-1974], [Hs II-3-1981],
pseudopectinifera (Invalid name - See U. pectinifera -v- pseudopectinifera),
warasii (See U. flavispina),

UTAHIA

sileri (See Pediocactus sileri, Sclerocactus pubispina -v- sileri),
[E], [K], [P], [Q], [DD], [GG], [Ay], [Bo-5], [Bu], [Bv-1], [Dq], [Hx], [Hy], [Af 3-5], [Af 13-6], [Cu 2-1965], [Hl 2-1968],

VATRICANIA

guentheri (See Cephalocereus guentheri, Espostoa guentheri),
[DD], [NN], [Av], [Ay], [Bo-4], [Hi Vol. 2], [Hv], [Hy], [Il], [Jd], [Af 54-4], [Ah-1981],

WEBERBAUEROCEREUS

alboareolatus.....
albus.....[DD], [Hi Vol. 4], [Aj 13-6-1962],
aureispinus.....
cephalomacrostibas (See Haageocereus ferox, Trichocereus cephalomacrostibas),
[Hi Vol. 4],
churinensis.....[Hi Vol. 4], [Hx*], [Aj 13-8-1962],
crassiarboreus (See Trichocereus crassiarboreus, W. cuzcoensis),
cuzcoensis (See W. crassiarboreus),
[Hi Vol. 4],
" -v- tenuiarboreus.....
fascicularis (See Trichocereus fascicularis, W. weberbaueri, W. seyboldianus),
[Bo-2], [Hx*], [Hy], [Hl 1-1972], [Hl 2-1982],
" -v- densispinus.....
" -v- horridispinus (See W. horridispinus, W. weberbaueri -v- horridispinus),
[Aj 23-2-1972],
horridispinus (See W. fascularis -v- horridispinus, W. weberbaueri -v- horridispinus)...[Bo-2],

WEBERBAUEROCEREUS.....(Cont'd),

johnsonii (See Haageocereus johnsonii),
[Hi Vol. 4], [Hy], [Aj 13-5-1962], [Bi 3-2],
longicomus.....[DD], [Hi Vol. 4], [Hx*], [Hy], [N-1980],
marnieranus (See W. rauhii),
rauhii.....[NN], [QQ], [Ay], [Bo-2], [Hy], [Db 9-1980],
" -v- laticornus.....[Bo-2], [Hi Vol. 4],
seyboldianus (See Haageocereus seyboldianus, Trichocereus fascicularis, Trichocereus fascicularis, W. fascicularis, W. weberbaueri),
[Ay], [Bo-2],
torataensis (See Haageocereus torataensis),
[Hi Vol. 4],
weberbaueri (See Trichocereus fascicularis, W. fascicularis),
[Ay], [Bo-2], [Hi Vol. 4], [Hx], [Jd],
[Aw 4-1964], [Hl 3-1982],
" -v- aureifuscus (See W. weberbaueri)....[Bo-2],
" -v- aureispinus.....
" -v- fuscus.....
" -v- horribilis (See W. weberbaueri).....[Bo-2],
" -v- humilis.....(See W. weberbaueri).....[Bo-2],
" -v- horridispinus (See W. fascicularis -v- horridispinus, W. horridispinus),
winterianus.....[DD], [Gr], [Hi Vol. 4], [Hx*], [Hy],
[Il], [N-1980], [Af 50-6],
" -v- australis.....
" -v- flavus.....

WEBEROCEREUS

biolleyi (See Rhipsalis biolleyi),
[I], [NN], [RR], [Ay], [Gr], [Hx*], [Hy],
[Jd], [Af 53-3], [Ag-1983], [Aj 8-8-1957],
[Aj 31-3-1980], [Ba 6-1978], [Hh July 1975],
bradei (See Eccremocactus bradei),
glaber (See Werkleocereus glaber),
mirandae (See Werkleocereus mirandae),
panamensis.....[RR],
tonduzii (See Werkleocereus tonduzii),
trichophorus.....[Jd], [Af 35-6*],
tunilla.....[M*], [DD], [RR], [Bo-2], [Bv-1], [Dq*],
[Hx*], [Hy*], [Aj 32-7-1981],

WEINGARTIA
----
aglaia (See Sulcorebutia tiraquensis -v- bicolorispina)
[Hk 10:54, 1978],
albispina (See Gymnocalycium albispinum),

albaoides (See Sulcorebutia albaoides),
[Hd 4 (8):2, 1983],
" subsp. subfusca (See Sulcorebutia albaoides subsp. subfusca),
[Hd 4 (11): 1, 1983],
albissima (See Sulcorebutia albissima),
[Hg 5 (1):1, 1980],
ambigua (See Neoporteria ambigua),
[Bo-2], [Hy], [Af 23:86, 1951], [Am 8-1981],
arenacea (See Sulcorebutia arenacea),
[Hg 4 (5):69, 1979],
aureiflora (See Sulcorebutia frankiana -v- aureiflora),
aureispina (See Sulcorebutia frankiana -v- aureispina),
backebergiana (See Sulcorebutia steinbachii -v- horrida)
[Hg 5:70, 1977],
brachygraphisa (See W. neocumingii -v- brevispina),
[Bg], [Hj 12 (1):1977],
brevispina (See Sulcorebutia verticillacantha -f. brevispina),
[Hg 5 (2):13, 1980],
buiningiana.....[Bg], [Hi Vol. 2],
caineana (See Sulcorebutia breviflora, Sulcorebutia caineana),
[Hg 5 (1):5, 1980],
callecallensis (See Sulcorebutia callecallensis, Sulcorebutia verticillacantha -v- aureiflora)
[Hd 2 (3):1, 1981],
chilensis (See Neowerdermannia chilensis),
[Hy], [Af 55-3],
chuquichuquiensis (See Rebutia corroana, Sulcorebutia corroana) (Taxon uncertain),
[Hx*], [Hg 8 (4, 5):79, 1983], [Hh Vol. 3, #5, 6, 1979],
cintiensis (See W. fidaiana ssp. cintiensis),
[DD], [Bg], [Hi Vol. 2], [Aj 31-11-1980],
clavata (See Sulcorebutia clavata, Sulcorebutia steinbachii).....[Hg 4 (2):16, 1979],
corroana (See W. neocumingii ssp. pulquinensis -v- corroana, Sulcorebutia corroana, W. chuquichuquiensis) ?
[Gs 19 (82):49, 1964], [Hi 2:650-654, 1980],
croceareolata (See Sulcorebutia verticillacantha -v- cuprea),
[Hg 9 (4):73, 1984],
cumingii (See W. neocumingii),
[I], [Av], [Dq*], [Hx], [Aj 11-11-1960],
[Ea 20, 1937], [He 7 (25):18, 20, 1973],
[He 8 (29):42, 45, 1975],
cylindrica (See Sulcorebutia cylindrica),
[Hg 3 (2):40, 1978],

WEINGARTIA.....(Cont'd),

erinacea (See W. corroana, W. neocumingii ssp. pulquinensis -v- corroana),
[Hi Vol. 2], [Af 54-3], [Cu 23:8, 1961], [Hh 3, (#5 & 6):126, 136, 1979],
" -v- catarirensis (See W. corroana, W. neocumingii -v- pulquinensis, W. pulquinensis -v- corroana),
[Bg], [Cu 23:9, 1961],
[Hl 4-1980],
" -v- erinacea.....[Bg],
fidaiana (See Rebutia fidaiana),
[F], [PP], [Bg], [Bo-3], [Hi Vol. 2], [Hx*],
[Hy], [Jd], [Cu 1, 1968], [Db 9, 1979],
[Hh 3 (#3 & 4), 1978], [Hl 4-1980],
" subsp. cintiensis (See W. cintiensis)...
[Hh 3 (#5 & 6):139, 1979],
flavida (See Sulcorebutia flavissima),
[Hd 2 (6):3, 1981],
flavissima (See Sulcorebutia flavissima),
[Hg 4 (5):69, 1979],
formosa (See Sulcorebutia losenickyana),
[Hg 4 (4):46, 1979],
frankiana (See Sulcorebutia frankiana),
[Hg 95-1978],
glomeriseta (See Sulcorebutia glomeriseta),
[Fy 4:3, 1977], [Hl 4, 1980],
gracilispina.....[Hi Vol. 2],
'hajekyana' (See W. neocumingii subsp. pulquinensis - -v- mairanensis, W. pulquinensis -v- corroana),
[Am 8, 1981],
haseltonii (See Sulcorebutia breviflora),
[Hg 95, 1978],
hediniana (See W. neocumingii ssp. sucrensis -v- hediniana),
[Ar], [Bg], [Bo-3], [Hi Vol. 2], [Hx*], [Hy],
[Jd], [Aj 5-1972], [Aj 8-1980], [Aj 11-1980],
[Am 8-1981], [Ba 5-1976], [He 3 (14):14, 1969],
hoffmanniana (See Sulcorebutia hoffmanniana),
[Hg 5 (1):5, 1980],
inflexiseta (See Sulcorebutia inflexiseta),
[Hg 5 (1):5, 1980],
kargliana.....[Bg], [Jd], [Aj 30:105, 1979], [Aj 32:194-1981],
knizei (See W. neocumingii ssp. pulquinensis -v- corroana),
[Bg], [Fy 4:6, 1977],
" -v- 'bruneispina'.....
krahnii (See Sulcorebutia krahnii),

lanata.....[Ar], [Bg], [Hv], [Aj 29-8-1978], [Am 8-1981]
[Au 16:7, 1961], [Au 37-4-1982], [Hh Vol. 3,-
(5, 6):131, 1979],
" subsp. longigibba (See W. longigibba),
[Hh 3 (5 & 6), 1979],
" " pilcomayoensis (See W. pilcomayoensis),
[Hh Vol. 3, (5&6), 1979],
" " " -f. platygona (See W. platygona),
" " riograndensis (See W. riograndensis),
[Hh Vol. 3, (5&6), 1979],
lecoriensis (See W. westii -v- lecoriensis),
[Bg], [Flower - Aj 31-11-1980], [Am 8-1981]
[Gs 19 (82):47, 1964],
lepida (See Sulcorebutia lepida, Sulcorebutia totorensis -v- lepida),
[Hg 5 (1):5, 1980],
longigibba (See W. lanata ssp. longigibba),
[Bg], [Hi Vol. 2], [Aj 30-9-1979], [Cu 23:8,
1961], [Hl 4, 1980],
'mairanana' (Invalid name - See W. pulquinensis -v-
mairanensis).....[Hg 8 (3):51, 1983],
markusii (See Sulcorebutia markusii)
[Fy 5:2, 1978],
mataralensis.....[Bg], [Hg 9 (5):89, 1984],
menesesii (See Sulcorebutia menesesii),
[Hg 4 (1);6, 1979],
mentosa (See Rebutia mentosa, Sulcorebutia mentosa),
[Hg 4(1):6, 1979],
mizquensis (See Sulcorebutia mizquensis),
[Hg 4(5):69, 1979],
multispina (See W. neocumingii ssp. pulquinensis -v-
multispina, W. pulquinensis).....[Bg], [Hi-
Vol. 2], [Aj 31-11-1980], [Au 16:7, 1961],
muschii (See Sulcorebutia muschii),
[Hg 4(1):6, 1979],
neglecta (See W. neocumingii).....[Hg 8 (3):31, 1983],
neocumingii (See W. cumingii, Gymnocalycium neocumingii)
[C], [T], [DD], [EE], [QQ], [Ar], [Ay], [Bg]
[Bo-3], [Hu], [Hv], [Hx*], [Jd], [Ag-1988],
[Ba 41(9):115, 1962], [He 7 (25):18, 1973],
[Hh Vol.-3, (#5&6), 1979], [Hl 5-1978],
[Hl 4-1980],
" -v- brevispina (See W. brachygraphisa).....
" -v- corroana (See W. corroana, W. erinacea,
W. neocumingii ssp pulquinensis -v-
corroana),
[Hx*], [Hh Vol. 3, #5 & 6, 1979],
" -v- hediniana (See W. neocumingii ssp. sucrensis -v- hediniana, W. hediniana),
[Hh Vol. 3, #5 & 6, 1979],

neocumingii.....(cont'd),
" -v- koehresii.....[Bg], [Aj 32-12-1981], [Hh Vol. 3, #5 & 6, 1979],
" -v- mairanensis (See W. neocumingii ssp. pulquinensis -v- mairanensis), [Hh Vol. 3, #5 & 6, 1979],
" subsp. pulquinensis (See W. pulquinensis), [Hh Vol. 3, #5 & 6, 1979],
" " " -v- corroana (See W. corroana, W. pulquinensis -v- corroana),
" " " -v- mairanensis (See W. mairanana, W. pulquinensis -v- mairanensis)...[Hh Vol. 3, #5 & 6, 1979],
" " " -v- multispina (See W. multispina) [Hh Vol. 3, #5&6, 1979],
" " sucrensis (See W. sucrensis), [Hh Vol. 3, #5 & 6, 1979],
" " " -v- hediniana (See W. hediniana], [Hh 3 (5&6):170, 1979],
" " " -v- trollii (See W. trollii),
neoneumanniana (See W. neumanniana).....[Hy], [Iz],,
neumanniana (See Gymnocalycium neumannianum), [Ay], [Bg], [Bo-3], [Cr], [Hx], [Ig*], [Jd], [Ae-1981], [Ah-1985], [Aj 32-8-1981], [Ba 1-1982], [Cu 2-1979], [Ea 1937:21, 1937], [Hl 4-1980],
" -v- aurantia (Invalid name - See W. neumanniana) [DD],
nigro-fuscata (See Sulcorebutia tiraquensis -v- spinosior).....[Hk 10:113, 1978],
oenantha (See Sulcorebutia oenantha), [Hg 4(5):69, 1979],
oligacantha (See Sulcorebutia tarijensis), [Hg 4(1):2-5, 1979],
pampagrandensis (See Sulcorebutia pampagrandensis), [Hg 4(5):69, 1979],
pasopayana (See Sulcorebutia pulchra -f.-), [Hg 9(3):52, 1984],
perplexiflora (See Sulcorebutia perplexiflora, S. pulchra).....[Hd 3(8):1, 1982],
pilcomayoensis (See Gymnocalycium pilcomayoensis, W. lanata ssp. pilcomayoensis).... [Bg], [Iz], [Ae-1982], [Gs 19(82):44, -1964], [Hh 3, (5-6), 1979],
platygona (See W. lanata ssp. pilcomayensis -f. platygona)........[Bg], [Hw-1981], [Jd], [Gs 19 -(82):50, 1964],

polymorpha (See Sulcorebutia steinbachii -v- polymorpha).....[Hg 1977:70(1977)],
'pruinosa'.....(See W. pulquinensis),
" -v- 'albispina'.....[Bg],
pulchera (See Sulcorebutia pulchra),
[Hg 1978:95(1978)],
pulquinensis (See W. neocumingii ssp. pulquinensis),
[DD], [Bg], [Iz], [Ah-1985], [Aj 27-5-1976], [Aj 31-11-1980], [He 3(14:14, 1969]
[Hl 4-1980],
" -v- corroana (See W. corroana, W. neocumingii -v- corroana)....[Ew 6:29, 1951],
" -v- mairanensis (See W. mairanana, W. neocumingii ssp. pulquinensis -v- mairanensis).....[Bg],
" -v- pulquinensis.....[Bg],
purpurea (See Cinnabarinea purpurea, Sulcorebutia purpurea).....
[Aj 32:270, 1981], [Hh Vol. 1 (5) March, 1974],
pygmaea (See W. kargliana),
[Hi Vol. 2], [Hh Vol. 3, (#5 & 6), 1979],
rauschii (See Sulcorebutia rauschii),
[Hg 95:1978],
riograndensis (See W. lanata ssp. riograndensis),
[DD], [Bg], [Hi Vol. 2], [Hx], [Aj 24-12-1973], [Am 8-1981], [Cu 23:9, 1961],
ritteri (See Sulcorebutia ritteri, Sulcorebutia verticillacantha -v- albispina, Sulcorebutia verticillacantha -v- ritteri),
[Hg 1978:75(1978],
rubro-aurea (See Sulcorebutia rubro-aurea),
saetosa.....[Bg], [Hg 5(4):41, 1980],
saipinensis.....[Bg], [Hg 7(5):71, 1982],
sanguineo-tarijensis (Syn. Sulcorebutia tarijensis),
[Hg 5(3):29, 1980],
saxatilis (See Sulcorebutia verticillacantha -v- albispina).....[Fy 8(1):201, 1981],
sucrensis (See W. neocumingii ssp. sucrensis),
[Bg], [Bt], [Hi Vol. 2], [Hx], [Am 8-1981],
[Au 16:8, 1961],
" -v- trollii (See W. trollii).....
tarabucoensis (See Sulcorebutia tarabucoensis),
[Hg 4(5):69, 1979],
taratensis (See Sulcorebutia taratensis),
[1977:69 (1977)],
tarijensis (See Sulcorebutia tarijensis),
[Hg 3(5):92, 1978],
tiraquensis (See Sulcorebutia tiraquensis),
[Hg 1977:69, (1977)],

WEINGARTIA.....(Cont'd),

torotorensis (See Cinnabarinea torotorensis, Sulcorebutia torotorensis).....[Hx*], [Ae-1981], [Af 43-6-1971], [Hh July, 1973],
totoralensis (See Sulcorebutia hoffmanniana), [Hg 7(3):32, 1982],
totorensis (See Sulcorebutia totorensis), [Hg 4(5):69, 1979],
trollii (See W. neocumingii ssp. sucrensis -v- trollii) [Bg]. [Aj 29-6-1978], [Bm-1989],
tunariensis (See Sulcorebutia tunariensis), [Hg 1978:95(1978)],
vallegrandensis (W. pulquinensis -f.-),
verticillacantha (See Sulcorebutia verticillacantha), [Hg 1977 (5):70, (1977)],
vilcayensis (See W. westii -v- vilcayensis), [Bg], [Gs 19(82):46, 1964],
vizcarrae (See Sulcorebutia vizcarrae), [Hg 9(3):55, 1984],
vorwerkii (See Neowerdermannia vorwerkii),
" -v- erectispina.....
" -v- gielsdorfiana.....
westii (See Gymnocalycium westii), [Bg], [Hi Vol. 2], [Aj 27-11-1976], [Am 8-1981], [Au 13:67, 1958], [Ba 6-1978], [Hh Vol. 3, #5&6, 1979], [Hl 4-1980],
" -v- lecoriensis (See W. lecoriensis), [Hh Vol. 3 (#5 & 6), 1979],
" -v- vilcayensis (See W. vilcayensis),
Sp. Lau 331 A.....[Af 54-2],
Sp. FR-816.....[Hh July, 1973],

WERKLEOCEREUS

glaber.....[RR], [Az], [Af 27-2], [Aj 9-6-1958],
" -v- mirandae.....(See W. mirandae),
imitans (See Eccremocactus imitans), [Ay], [Aj 8-8-1957],
mirandae.....(Chiapas).....
tonduzii.....[G], [M*], [DD], [RR], [Ay], [Bo-2], [Bv-1], [Dq], [Hy*], [Af 30-6], [Aj 8-8-1957], [Aj 31-10-1980],

WIGGINSIA

aciculata (See Notocactus acuatus, W. erinacea),
acuata (See W. erinacea),
acuta (See Notocactus acutus),
arechavaletai (arechavaletae) (See Malacocarpus arechavaletai, Notocactus arechavaletai), [Ar], [Bu], [Hx], [Ba 6-1977],

WIGGINSIA.....(Cont'd),

arechavaletai.....(cont'd),
" -v- aurea.....
" -v- limiticola.....
beltranii (See Malacocarpus beltranii),
bezrucii (See Malacocarpus bezrucii),
calispina (See Malacocarpus calispinus, Malacocarpus arechavaletai, Wigginsia arechavaletai),
catamarcensis (See Malacocarpus catamarcensis),
comantii (See Malacocarpus comantii),
corynodes (See Malacocarpus corynodes, Notocactus corynodes),
[PP], [Ar], [Hy],
courantii (See Malacocarpus courantii, Notocactus courantii, W. tephracantha -v- courantii),
[Hi Vol. 1],
curvispinus (See Horridocactus curvispinus, Malacocarpus curvispinus, Pyrrhocactus curvispinus)
erinacea (See Malacocarpus erinaceus, Malacocarpus tephracanthus, W. tephracantha),
[Hx*], [Hy], [Au 37-4],
" -v- courantii (See W. courantii),
" -v- elatior.....
" -v- tephracantha (See W. tephracantha)
fricii (See Malacocarpus fricii),
[Hl 2-1976],
horstii (See Notocactus horstii, Notocactis neohorstii)
[Ar], [Hi Vol. 1], [Hx*],
" -v- juvenaliformis (See Notocactus horstii -v- juvenaliformis, Notocactus neohorstii -v- juvenaliformis,
[Ar], [Hi Vol. 1],
kovaricii (See Notocactus kovaricii, Notocactus neoarechavaletai -v- kovaricii),
[Hl 6-1975],
langsdorfii (See W. polyacantha).....[Hx*],
leprosorum (See Malacocarpus leprosorum, Notocactus leprosorum),
[Ar], [Hi Vol. 1], [Hx*], [Af 56-5],
leucocarpa (See Malacocarpus leucocarpus),
longispina (See Malacocarpus longispinus),
[Hi Vol. 1],
maassii (maasii) (Echinocactus maassii, Parodia maassii),
macrocantha (See Malacocarpus macrocanthus, W. sellowii -v- macrocanthus).....[Hy], [Hl 4-1972],
macrogona (See Malacocarpus macrogonus),
maldonadensis (See Malacocarpus arechavaletai, Notocactus arechavaletai, Notocactus neoarechavaletai, Wigginsia arechavaletai),
[Hy],

WIGGINSIA.....(Cont'd),

mamillarioides (See Neoporteria mamillarioides),
martinii (See Malacocarpus sessiliflora -v- martinii, Malacocarpus tephracanthus, Wigginsia tephracantha, W. sessiliflora -v- martinii),
neohorstii (See W. horstii, Notocactus neohorstii),
orthacantha (See Echinocactus orthacanthus, Malacocarpus orthacanthus, Notocactus orthocanthus) [Hx*],
pauciareolata (See Malacocarpus pauciareolata, W. sessiliflora -v- pauciareolata),
polyacantha (See Wigginsia langsdorfii), [Hi Vol. 1],
prolifera (See Notocactus prolifera), [Gp*], [Hi Vol. 1],
" -v- longispina (See Malacocarpus prolifera -v- longispina, Notocactus prolifera -v- longispina),
pulcherrimus (See Malacocarpus pulcherrimus, Frailea pulcherrima),
pulvinata (See Notocactus pulvinatus),
reichei (See Echinocactus reichei, Malacocarpus reichei, Neochilenia reichei, Neoporteria reichei, Notocactus reichei, Thelocephala reichei),
rubricostata (See Malacocarpus rubricostatus),
sellowii (See Malacocarpus sellowii, Malacocarpus tephracanthus, Notocactus sellowii, W. tephracantha), [Hi Vol. 1], [Hx], [Hy],
" -v- macrocantha (See Malacocarpus macrocanthus, Wigginsia macrocantha),
sessiliflora (See Malacocarpus sessiliflorus), [Ar], [Hx*], [Hy], [Hw-1989],
" -v- martinii (See Malacocarpus sessiliflorus -v- martinii, Malacocarpus tephracanthus, W. tephracantha),
" -v- pauciareolata (See Malacocarpus pauciareolata).....[Hy],
stegmannii (See Malacocarpus stegmannii, Notocactus sessililorus -v- stegmannii),
strausianus (See Pyrrhocactus strausianus),
tephracantha (See Malacocarpus sellowii, Malacocarpus tephracanthus, W. sellowii), [Ar], [Bu], [Hx],
" -v- courantii (See Malacocarpus courantii, W. courantii),
" -v- depressa (See Malacocarpus tephracanthus -v- depressus),
tuberisulcata (See Horridocactus horridus, Horridocactus tuberisulcatus, Pyrrhocactus horridus, Pyrrhocactus tuberisulcatus),

WIGGINSIA.....(Cont'd),

turbinata (See Notocactus sellowii -v- turbinatus, W. tephracantha),
vorwerkiana (See Malacocarpus vorwerkianus, Notocactus vorwerkianus),
[R], [Ar], [Hx], [Hy],

Sp. .....[Ba 11-1977], [Ba 7-1979],

WILCOXIA

albiflora (See Echinocereus leucanthus),
[E], [I], [DD], [NN], [Av], [Ay], [Az], [Bf], [Bo-4], [Hv], [Hx], [Ih], [Jd], [Af 56-3], [Aj 16-4-1965], [Aw 4-1965],
australis.....[Af 41-5],
kroenleinii.....[Da 2-1980], [Da 4-1980],
leucanthus (See W. albiflora, Echinocereus leucanthus)
mariana (See Peniocereus marianus),
nerispina.....[NN], [Bo-4], [Bt], [Da 4-1978],
papillosa (See Echinocereus papillosus)...[Az], [Cx],
poselgeri (See Cereus tuberosus, Cereus poselgeri, Echinocereus poselgeri, W. tuberosa),
[H], [M*], [S], [Y], [GG], [NN], [PP], [QQ], [RR], [TT], [Av], [Ay], [Bo-4], [Bz], [Cm], [Cx], [Df], [Dq], [Hv], [Hx], [Hy], [Jd], [Af 4-12], [Af 6-2], [Aj 33-11-1982], [Da 2-1980], [Da 4-1980], [Hm 2-1978],
" -f. cristata.....[Ad -],
robusta (See Echinocereus fasciculatus -v- fasciculatus, Echinocereus robustus, Echinocereus fendleri -v- robustus),
schmollii (See Echinocereus schmollii),
[C], [H], [L], [BB], [Az], [Bo-4], [Cx], [Hx*], [Hy], [Il], [Da 4-1978],
" -v- lanata.....[Da 4-1980],
" -v- nerispina (See W. nerispina),
" -v- nigrisseta.....[Da 4-1980],
" -v- nigrispina.....[Da 4-1978],
" -v- serpens.....
senilis.....[CC],
striata (See Neoevansia striata),
[Q], [Ay], [Cx], [Af 20-7], [Da 4-1980],
tamaulipensis.....[Bo-4], [Ba 2-1979],
" -v- brevispina.....[Da 4-1980],
" -f. cristata.....
tomentosa (See W. viperina -v- tomentosa),
[Bo-4], [Hx*],
tuberosa (Invalid name) (See Cereus tuberosus, Echinocereus poselgeri, W. poselgeri).....[Az], [Jd],

WILCOXIA.....(Cont'd),

viperina (See Cullmannia viperina, Peniocereus viperinus),
[CC], [JJ], [RR], [Ay], [Bo-4], [Cx], [Gr],
[Hv], [Hx*], [Aw 3-1970], [Ba 4-1980], [Da 4-1980],
" -v- tomentosa (See W. tomentosa).....[Az],
" -v- viperina.....[Az],
zapilotensis (See Neoevansia zapilotensis

Sp. .....[Bo-6], [Aj 30-5-1979],

WILMATTEA

minutiflora (See Hylocereus microcladus),
[DD - Flower], [RR], [Ay], [Az], [Bo-2],
[Dq*], [Hx*], [Hy], [Jd], [Af 33-1],
[Af 55-2], [Aj 9-9-1958],

Sp. .....[Az],

WINTERIA

aureispina (See Borzicactus aureispinus, Hildewintera aureispina, Winterocereus aureispina),

WINTEROCEREUS.....(Invalid name)

aureispina (See Borzicactus aureispinus, Hildewintera aureispina, Loxanthocereus aureispinus),
[DD], [HH], [NN], [Hy],
" -f. cristata.....[HH],

WITTIA ....(Wittiocactus)

amazonica (See W. panamensis),
[I], [RR*], [Ay], [Hx*], [Hy*], [Jd],
[Aj 27-7-1976], [Aj 29-12-1978], [Aw 2-1986],
costaricensis (See Disocactus himantocladus, Pseudo-rhipsalis himantoclada),
himantoclada (See Disocactus himantocladus, Pseudo-rhipsalis himantoclada),
panamensis (See W. amazonica),
[DD], [RR], [Av], [Bo-2], [Dq*], [Ag-1984],

YUNGASOCEREUS

inquisivensis (See Cleistocactus laniceps, Yungaso-cereus microcarpus).....[Hi Vol. 2],
microcarpus (See Y. inquisivensis),
[Hx*],

ZEHNTNERELLA

chaetacantha.....
" -v- montealtoi.....
matoana.....
polygona.....
squamulosa.....[DD], [RR], [Ay], [Bo-4], [Hi Vol. 1],
[Hx*], [Jd], [Cu 3-1967],

ZYGOCACTUS

candidus (See Epiphyllanthus candidus),
opuntioides (See Epiphyllanthus obovatus, Schlumbergera opuntioides),
[Bu], [Aj 10-5-1959],
truncatus (See Epiphyllum truncatum, Schlumbergera truncata),
[C], [E], [H], [I], [V], [AA], [DD], [LL],
[PP], [TT], [Ar], [Bl], [Bo-2], [Bu], [Bv-1],
[Cj], [Ck], [Cr], [Cz], [Dq - Flower], [Gr],
[Hx*], [Hy], [Ig*], [Af 5-11],
" -v- alba.....[AA],
" -v- altensteinii.....
" -v- crenulatus.....
" -v- delicatus (See Schlumbergera truncata -v- delicata),
[F], [I], [HH], [Ax], [Bo-2],

-Notes-